PRECALCULUS:

A GRAPHING APPROACH

PRECALCULUS:
A GRAPHING APPROACH

DALE VARBERG
Hamline University

THOMAS D. VARBERG
Macalester College

with the help of
Antonio Quesada
University of Akron

Prentice Hall, Englewood Cliffs, New Jersey 07632

Library of Congress Cataloging-in-Publication Data

VARBERG, DALE E.
    Precalculus: a graphing approach/Dale Varberg. Thomas D.
Varberg: with the help of Antonio Quesada.
        p.   cm.
    Includes index.
    ISBN 0-13-010703-4
    1. Functions—Data processing.   2. Graphic calculators.
I. Varberg, Thomas D., (Date)   . II. Title.
QA331.3.V37 1994
512'.13 – dc20

94-9110
CIP

Editor in chief: Jerome Grant
Production editor: Judy Winthrop
Creative director: Paula Maylahn
Art director: Amy Rosen
Interior design: Meryl Poweski
Cover designer: Defranco Design, Inc.
Manufacturing buyer: Trudy Pisciotti
Copy editor: Margo Quinto
Editorial assistant: Joanne Wendelken

© 1995 by Prentice-Hall, Inc.
A Simon & Schuster Company
Englewood Cliffs, New Jersey 07632

Printed in the United States of America
10 9 8 7 6 5 4 3 2 1

ISBN 0-13-010703-4

Prentice-Hall International (UK) Limited, *London*
Prentice-Hall of Australia Pty. Limited, *Sydney*
Prentice-Hall Canada Inc., *Toronto*
Prentice-Hall Hispanoamericana, S.A., *Mexico*
Prentice-Hall of India, Private Limited, *New Delhi*
Prentice-Hall of Japan, Inc., *Tokyo*
Simon & Schuster Asia Pte. Ltd., *Singapore*
Editora Prentice-Hall do Brasil, Ltda., *Rio de Janeiro*

To Jordan Thomas Varberg

CONTENTS

8 SEQUENCES, COUNTING, AND PROBABILITY   *379*

PREFACE

We believe that graphics calculators open a wonderful new door to the learning of elementary mathematics. With this in mind, we have developed each topic in this book so as to maximize use of these devices. They help us to explain the standard material and they allow us to explore ideas that transcend the traditional precalculus course. Most important, their visual and computational excellence make mathematics come alive; they make mathematics fun for students. Our conviction is that every college student who plans to take mathematics should own a graphics calculator. But don't misunderstand. A precalculus course is not just a matter of learning to push the right buttons on a calculator; rather it is about mastering a set of mathematical ideas that are needed in calculus and many fields of science.

Graphics calculators come in many models. This book highlights Texas Instruments calculators but any of the newer models produced by Casio, Hewlett Packard, and Sharp will serve as well. Although computers with appropriate software offer even more mathematical power, we believe graphics calculators are the ideal choice for a precalculus course. They are at exactly the right level—powerful enough to eliminate the drudgery of hand calculation and hand graphing but not so sophisticated as to hide what is going on. Moreover, they are inexpensive and simple to use.

CHOICE OF TOPICS

This book covers most of the topics that have been a part of precalculus courses for decades. The emphasis, however, is different. We emphasize those topics and concepts that are needed in other mathematics courses or in science or in business, especially those that can be explored and illustrated on a graphics calculator. We have deemphasized some manipulative skills (algebraic root finding, simplifying radicals, logarithmic calculations, sophisticated factoring) in favor of visualization, enlarging part of a graph by zooming, data analysis, and modeling of problems from the physical world. Three general principles guided us in the selection of material.

Topics needed in calculus should be included.

Those concepts that can be illustrated clearly on a graphics calculator should be emphasized.

Applications should play a central role.

We chose the title *Precalculus* because many colleges offer a course with this title for students preparing to take calculus. The book is also appropriate for courses with such titles as *Algebra and Trigonometry, Elementary Functions,* or *Functions and Graphs.* The organizing principle for the book is that of a function and its graph. Graphics calculators allow us to study functions in depth: finding their zeros, discovering where they increase and decrease, identifying their maxima and minima, analyzing their asymptotic behavior, and calculating their rates of change. We can view the graphs of several functions at once, find their points of intersection, discover where one is larger than another, and so on. Through the use of parametric representation, we can even watch curves being generated in real time, thereby visualizing motion in a realistic way.

PROBLEM SOLVING

Each section begins with a TEASER, a problem designed to give a flavor of the section and to tease the student into trying to solve it or at least to read the section to see its solution. Every section ends with an extensive set of problems for the student to work. These problem sets are in two parts. Part A (Skills and Techniques) consists of odd-even problem pairs that follow the text and its examples closely. Part B (Applications and Extensions) contains a broad spectrum of problems that make use of the skills learned in Part A in the broad context of mathematical, scientific, and business applications. Part A problems are quite easy; Part B problems are more demanding, growing harder as they approach the last problem, labeled CHALLENGE. Some of the later problems in a set can serve as exploratory projects for students to work, write up, and hand in (see, for example, Problem 62 in Section 7.5).

FLEXIBILITY

The book contains 58 sections, most of which are designed to be covered in one lesson each. Thus, there is plenty of material for a one semester or two-quarter course. Chapter 1 contains relatively simple mathematics, intended to give the student a gentle introduction to the use of a graphics calculator. The real meat of a precalculus course follows in Chapters 2–4, which deal with various classes of functions. After that, an instructor can pick and choose what is to be covered.

PROGRAMMING

Although most graphics calculators are programmable, we delay discussion of this powerful tool until the last chapter of the book, where it fits naturally with our discussion of sequences and series. An instructor may choose to introduce programming earlier if desired.

TEXAS INSTRUMENTS CALCULATORS

As this is being written, three TI graphics calculators are available: TI-81, TI-82, and TI-85. The sophistication and the price of these calculators increase with the model number. The TI-81 is a wonderfully simple but very powerful

calculator; all of the problems in this book can be done on it. The TI-82 is an upgraded version of the TI-81 which handles many more problems automatically but is also somewhat more complicated to operate. Both of these calculators emphasize the kind of mathematics one does in a precalculus course though the TI-82 also handles two of the basic concepts of calculus (derivative and integral). The TI-85 is designed for a calculus course. We will highlight the use of the TI-81 and TI-82 in this book.

ACKNOWLEDGMENTS

We pay tribute to Walter Fleming who gave permission to use material from the book *Precalculus Mathematics* by Walter Fleming and Dale Varberg, Prentice Hall, 1984 and 1989. Antonio Quesada, an expert in the use of graphics calculators, reviewed every page of the manuscript and offered hundreds of suggestions for improvement. A number of mathematics teachers reviewed part or all of the text, among whom are the following: Sharon Abramson, *Nassau Community College;* Maurice Monahan, *South Dakota State University;* William Grimes, *Central Missouri State University;* Philip Montgomery, *University of Kansas;* Robert Hoburg, *Western Connecticut State University;* Arthur Fruhling, *Yuba College;* Jack Porter, *University of Kansas.*

   Priscilla McGeehon first encouraged us to write this book and, as our initial Editor, spurred its development. Later Editor Jerome Grant helped bring the manuscript to its final form. Judy Winthrop steered the book through the production process. Others at Prentice Hall worked on the design and artistic aspects of the book. We thank all of these people for their valuable contributions.

Dale Varberg
Thomas D. Varberg

1 NUMBERS, EQUATIONS, AND INEQUALITIES

Blaise Pascal was sickly as a child and suffered from many ailments throughout his short life of 39 years. Yet he made important contributions in several branches of mathematics, did experimental work in physics, and was a master of French prose. He was deeply religious and believed that faith was ultimately superior to reason. His thinking on spiritual themes appears in the literary classic, Pensées, a book still well worth reading.

Our showcasing of Pascal at the opening of Chapter 1 is for another reason. At age nineteen, he invented the first adding machine to help with his father's work as tax assessor. This primitive device, with its hand-turned wheels, is the precursor of today's electronic calculator. Of course, Pascal would be astounded at the capabilities and the speed of our modern calculators and even more so at those of their big sister, the computer. Computer programmers pay tribute to Pascal's pioneering idea by giving one of the most popular computer languages the name, PASCAL.

Blaise Pascal (1623–1662)

This book highlights the graphics calculator, a device that not only calculates but also displays mathematical results in graphical form. Invented in the 1980s, these pocket-sized computers make the wonders of modern technology available to students in a relatively inexpensive and completely portable form. Our goal is to use graphics calculators to enliven the learning of precalculus mathematics while at the same time enhancing understanding. This first chapter introduces you to these machines in the context of basic mathematics.

TEASER My heart beats 72 times per minute. Measure my lifetime in heartbeats if I live to 85 years and 16 days.

1.1 NUMBERS AND CALCULATIONS

Mathematicians are superb number crunchers. This statement is both profoundly true and definitely false. Most mathematicians detest long division, hate calculating their income taxes, and claim to be unable to balance their checkbooks. On the other hand, they adore number patterns, like to study properties of numbers, and love to build theories based on numbers. And when complicated or tedious calculations are needed, they use calculators and computers to do them. Thus, we begin our study with a quick review of the principal number systems of mathematics. Then we discuss the role of calculators and how we will use them in this course.

NATURAL NUMBERS, INTEGERS, AND RATIONAL NUMBERS

With the **natural numbers** 1, 2, 3, . . . we can count our cousins, our books, and our pocket change. But if we are like most students, we will need 0 or the negatives -1, -2, -3, . . . to indicate our net worth. The natural numbers together with their negatives and 0 constitute the **integers.** The integers can be added, subtracted, and multiplied, but division poses a problem. For while 6/2 and $-12/3$ are perfectly good integers (namely, 3 and -4), 7/2 and $-13/3$ are not. To allow universal division (except by 0), we enlarge our number system to the **rational numbers.** A number is rational if it can be expressed as a ratio (quotient) m/n of two integers (with $n \neq 0$).

Our notation for numbers, called Hindu Arabic decimal notation, allows us to write numbers using only the 10 symbols 0, 1, 2, . . . , 9 and a decimal point. Recall that

$$3425.63 = 3(1000) + 4(100) + 2(10) + 5 + 6\left(\frac{1}{10}\right) + 3\left(\frac{1}{100}\right)$$

The familiar long division process shows that

$$\frac{3}{8} = 0.375 \quad \text{and} \quad \frac{3}{25} = 0.12$$

which are called terminating decimals. On the other hand,

$$\frac{4}{3} = 1.3333\ldots = 1.\overline{3}$$

and

$$\frac{71}{330} = 0.2151515\ldots = 0.2\overline{15}$$

which are called repeating decimals. The bar over a group of digits indicates that the group of digits repeats indefinitely. It can be shown that the rational numbers are precisely those numbers whose decimal expansions either terminate or ultimately repeat.

```
     .8461538
13 √11.00000000
    104
      60
      52
       80
       78
        20
        13
         70
         65
          50
          39
          110
          104
            60
```

Figure 1

■ **Example 1.** (a) Find the decimal expansions of 11/13. (b) Write $0.2\overline{345}$ as a ratio of two integers.

Solution. (a) A simple long division (Figure 1) shows that $11/13 = 0.846153846153\ldots = 0.\overline{846153}$. (b) Here we use a trick. Let $x = 0.2\overline{345}$. Then

$$1000x = 234.5345345\ldots$$

$$x = 0.2345345\ldots$$

Subtraction gives $999x = 234.3$, and so $x = 2343/9990$. If x had repeated in a group of four digits, we would have started with $10000x$ rather than $1000x$. ■

Figure 2

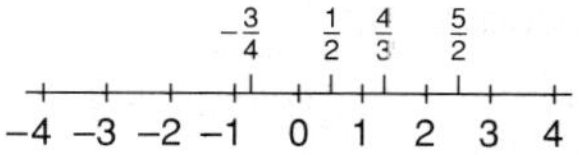

On a horizontal line, label a point (the origin) with the number 0, the point one unit to the right with 1, the point midway between those two points with ½, the point one unit to the left of the origin with −1, and so on (Figure 2).

This process uses up all the rational numbers as labels (that is, as coordinates for points), but leaves many points unlabeled. Consider the point c units to the right of the origin, where c measures the hypotenuse of a right triangle with legs of unit length (Figure 3). As the early Greek mathematicians showed, this point needs an irrational label, for what we call $\sqrt{2}$ is an irrational number. In fact, labeling the line using only the rational numbers leaves out infinitely many points.

ABOUT π

The number π appears throughout mathematics. It is defined as the ratio of the circumference to the diameter of a circle. The rational numbers 22/7 and 3.1416 are often used as approximations to π; however, π is known to be irrational. In August 1987, Gregory and David Chudnovsky used two IBM computers to obtain the decimal expansion of π to more than 1 billion places.

■ **Example 2.** It is known that π is irrational. Show that $n\pi$ is also irrational for any natural number n.

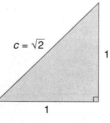

Figure 3

Solution. We use a proof by contradiction. Suppose $n\pi$ is rational; that is, suppose $n\pi = p/q$ where p and q are integers. Then $\pi = p/(nq)$, which contradicts the irrationality of π. ∎

REAL NUMBERS AND THE REAL LINE

We enlarge the number system so that every point on the horizontal line of Figure 2 has a label. The resulting line is called the **real line,** and the associated numbers are called the **real numbers.** Thus, the real numbers are precisely those numbers that can measure lengths, together with their negatives and 0. The real numbers may also be described as those numbers that can be represented in decimal notation. They include the rational numbers (terminating and repeating decimals) and the irrational numbers (nonrepeating decimals). The real numbers are the principal characters in this book, and when we say *number* with no qualifying adjective, we mean real number.

■ **Example 3.** Write a decimal that you know represents an irrational number.

Solution. Technically, this example requires writing down infinitely many digits, which is impossible. We can, however, specify such a decimal by indicating a definite pattern of digits that does not have a repeating group. One example is

$$0.101001000100001000001\ldots$$

Note the introduction of an additional 0 after each successive 1, which eliminates any possibility of a repeating group of digits. Of course, the expansions of $\sqrt{2}$ and π would work as well, but, though their decimal expansions are known to millions of places, no pattern appears, so we cannot write their expansions. ∎

THE COMPLEX NUMBERS

There is one further enlargement of the number system, but it will play a rather minor role in this book. Let i denote a number, called the imaginary unit, which satisfies $i^2 = -1$. Then a number of the form $a + bi$ where a and b are real numbers is called a **complex number.** We agree that $a + 0i = a$, so every real number is automatically a complex number. Figure 4 introduces

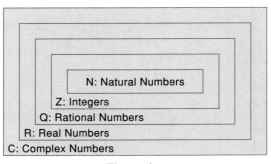

Figure 4

a common notation for these various number systems and illustrates the inclusion relations $\mathbb{N} \subset \mathbb{Z} \subset \mathbb{Q} \subset \mathbb{R} \subset \mathbb{C}$. Here the symbol $\subset$ denotes the phrase *is a subset of*.

CALCULATIONS WITH REAL NUMBERS

We assume that you are familiar with the standard laws obeyed by the real numbers and that you can (if forced) do hand calculations.

1. Commutative laws

$$a + b = b + a \qquad a \cdot b = b \cdot a$$

2. Associative laws

$$a + (b + c) = (a + b) + c \qquad a \cdot (b \cdot c) = (a \cdot b) \cdot c$$

3. Identity elements

$$0 + a = a + 0 = a \qquad 1 \cdot a = a \cdot 1 = a$$

4. Inverses

$$a + (-a) = -a + a = 0 \qquad a(1/a) = (1/a)a = 1 \qquad (a \neq 0)$$

5. Distributive laws

$$a(b + c) = ab + ac \qquad (b + c)a = ba + ca$$

WISE ADVICE

Does a/bc mean $(a/b)c$ or $a/(bc)$? According to our rules, it *should* mean the former. However, this is a good example of where confusion can occur. Therefore, we offer the following Golden Rule:

When in doubt, use parentheses.

CALCULATORS

The calculators of preference in this book are the TI-81 and the TI-82. These calculators have nearly identical keyboards, though the latter is somewhat more versatile. All that we discuss in this book can be done on either of these models. We will use marginal notes to explain differences in operation as they arise. Any other graphics calculator is also appropriate though the reader may need to make occasional references to the user manual.

We use parentheses to indicate which calculations are done first and understand that, in the absence of parentheses, multiplications and divisions are done before additions and subtractions (illustrated in the distributive laws); otherwise operations are done from left to right. Thus,

$$2(13 + 5) - (-4 + 6)/2 = 2 \cdot 18 - 2/2 = 36 - 1 = 35$$

but

$$2 \cdot 13 + 5 - (-4) + 6/2 = 26 + 5 + 4 + 3 = 38$$

Two uses of the minus sign appear in the calculations above. The first minus indicates *subtraction*; the second minus indicates the *negative of*. These two uses of the minus sign are connected by $a - b = a + (-b)$.

■ **Example 4.** Calculate $a = 34 - 3(-40/2 + 6(5 - 2))^3$

Solution. As indicated above, parentheses determine what is done first. Then **unary operations** (operations on one number, such as cubing and taking roots) are performed, followed by multiplications and divisions, and finally additions and subtractions, proceeding from left to right. Thus,

$$a = 34 - 3(-20 + 6 \cdot 3)^3 = 34 - 3(-2)^3 = 34 - 3(-8) = 58 \qquad ■$$

USING A CALCULATOR

In this text, we expect you to do most calculations on a calculator. Calculators vary considerably in the rules they follow, so you will want to study the manual

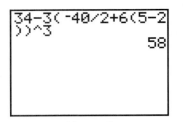

Figure 5

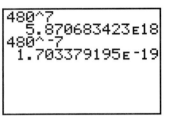

Figure 6

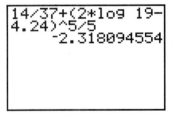

Figure 7

for yours very carefully. We will be using graphics calculators to illustrate most of the ideas in this book. Fortunately, they follow closely the logic and the notation used in hand calculations. Expressions are entered in the order you ordinarily write them and are displayed on the screen (viewing window). When we display a screen, it will be that for the Texas Instruments, Models 81 and 82, but the screen output on other graphics calculators is similar. The screen for the calculation of Example 4 is shown in Figure 5.

All scientific calculators represent very large or very small numbers in scientific notation. A number is in scientific notation if it is written in the form $M \times 10^n$, where $0 \leq |M| < 10$ and n is an integer. Thus,

$$234510000000 = 2.3451 \times 10^{11} = 2.3451\text{E}11$$

and

$$-0.00000000023451 = -2.3451 \times 10^{-10} = -2.3451\text{E}-10$$

Your calculator should display such numbers in the final format.

To raise numbers to powers, you will use the exponential key $\boxed{\wedge}$ (or the $\boxed{x^r}$ key on some calculators). Try calculating $(480)^7$ and $(480)^{-7}$ to see that you get the answers shown in Figure 6.

Incidentally, calculators display decimals up to a specified number of digits (often 10 digits), so they must inevitably round an answer, such as an irrational number, that requires more digits.

■ **Example 5.** Use your calculator to evaluate

$$14/37 + (2 \log 19 - 4.24)^5/5$$

Solution. The proper entry and the answer -2.318094554 appear in the screen of Figure 7. ■

NUMBER SENSE: CHECKING ANSWERS

Your calculator won't make mistakes, but you might give it the wrong instructions. Pressing the wrong keys or pressing them in the wrong order occurs often. Therefore, we should develop number sense. Asked to calculate $31.9 + 4(1002 + 8.4/2)$, Jack and Jill got the two different answers shown in Figure 8. Jill, using number sense, said the answer should be a bit over 4000 and declared that her calculator answer of 4056.7 must be right. Do you recognize Jack's mistake?

Figure 8

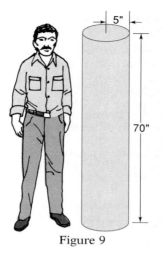

Figure 9

■ **Example 6.** Jack claims that his body's volume is about 1100 cubic inches. Jill says that a little number sense tells her that his figure is wrong. How did she reason?

Solution. She reasoned that Jack's volume is approximately that of a cylinder of radius 5 inches and height 70 inches (Figure 9). Using the formula $V = \pi r^2 h$, she mentally calculated his volume to be about

$$\pi(5)^2(70) \approx 3(25)(70) = 5250 \text{ cubic inches}$$

(From now on, we use the symbol $\approx$ to indicate *is approximately equal to.*) ■

TEASER SOLUTION

To account for leap years, we assume that a year has 365.25 days.

$$85 \text{ years}, 16 \text{ days} \approx ((85)(365.25) + 16)(24)(60)(72)$$
$$= 3{,}220{,}534{,}080 \text{ heartbeats}$$

PROBLEM SET 1.1

A. Skills and Techniques

In Problems 1–10, determine the decimal expansion of the given number. In Problems 11–18, write the given decimal as a ratio of two integers.

1. 7/8
2. 2/9
3. 2/7
4. 9/13
5. 8/9
6. 9/2
7. 17/16
8. 9/14
9. 7/17
10. 5/19
11. 0.6
12. 0.21
13. $0.\overline{6}$
14. $0.2\overline{1}$
15. $0.\overline{123}$
16. $0.1\overline{9}$
17. $1.\overline{23}$
18. $2.\overline{22}$

Indicate whether each number is rational or irrational. Assume π and $\sqrt{2}$ are known to be irrational. You should see simple patterns in Problems 27–30.

19. $\dfrac{3}{4}\pi$

20. $\dfrac{3}{4} + \pi$

21. $\sqrt{2} - 1$
22. $7.\overline{2}$
23. $7.\overline{2} - 3.\overline{41}$
24. $\pi - \dfrac{22}{7}$
25. $1/\sqrt{2}$
26. $\dfrac{3}{4}(1.6\overline{7})$
27. $0.1234567891011\ldots$
28. $0.232332333233332\ldots$
29. $0.23233233322222\ldots$
30. $0.248163264128256\ldots$

Calculate without use of a calculator. Then check that your calculator gives the same answer.

31. $2 - 4(5 - (2 - 3))/2$
32. $((-1 + 4)^2 - 7)^3/8 + 2$
33. $3 - (2(3 - 1)^2 - 2)^2/2$
34. $3/2 + (2/3)(5 - 6/3)^2/4$

Use a calculator to find each of the following. Use number sense to mentally check your answer.

35. $3.124 - (2.345^2 - 1.562^2)/4.512$
36. $(2.45(4.34 - 2.12)^3 - 2.17/3)^2$
37. $\dfrac{7.97 \times 10^{-6}}{2.98} \cdot \dfrac{2.34 \times 10^{-4}}{5.76}$
38. $\dfrac{\sqrt[3]{21.34 \times 10^6}}{\sqrt{41.23}}$
39. $\dfrac{0.0046(1500)}{(1.0046)^{60} - 1}$

40. $\dfrac{0.0078(1950)}{1 - (1.0078)^{-120}}$

41. $\dfrac{\sqrt[3]{41.23}\,\sqrt{39.76} + 8.98}{2.13 + 4.75}$

42. $\dfrac{368 - (4.35 + 20.34/6)}{\sqrt{26.22}}$

As in Example 6, use sensible approximations to estimate the value of each quantity.

43. The volume of your right arm from shoulder to fingertip in cubic inches.

44. The area of an oval-shaped lake that at the extremes is 11 miles long and 3 miles wide.

45. The number of square feet in a figure-eight-shaped sidewalk that is 3 feet wide, assuming the interior circles of the figure eight have diameters of 20 feet.

46. The volume of the trunk of a tree that has a circumference of 10 feet at the base and a circumference of 9 feet at a height of 20 feet if the trunk is 60 feet high.

47. The number of females in a midwestern city that has 50,000 children in its schools (K through 12).

48. The number of minnows in a circular lake of diameter 2 miles and averaging a depth of 50 feet if sampling in many places yielded an average of 21 minnows in every 100,000 cubic feet of water.

B. Applications and Extensions

49. Write $0.\overline{9}$, $0.2\overline{9}$, and $0.34\overline{9}$ as decimals that do not end in all nines. Make a conjecture about decimals that end in all nines.

50. Write $3/4$, $9/16$, $2/25$, $3/40$, and $3/(2^4 5^3)$ as decimals. Make a conjecture about integer fractions that have only twos and fives in their denominators.

51. Suppose that x is rational and y is irrational. Show that $x + y$ is irrational.

52. Suppose that x and y are both nonzero and that x is rational and y is irrational. Show that xy is irrational.

53. Calculate $1 + 2$, $1 + 2 + 4$, $1 + 2 + 4 + 8$, and so on, looking for a pattern. About how much would a person make in 30 working days if the wage started at $1 on the first day and doubled every day thereafter?

54. **Challenge.** The meter is now defined to be the distance traveled by light in 0.000000003335640952 seconds. How long would it take to travel to a star 15.365 light-years away, going at the rate of 50,125 kilometers per hour? Assume that there are exactly 365.25 days in a year.

TEASER A huge bag is filled with pennies, nickels, dimes, and quarters in the proportion 4:3:2:1. How much should you be willing to pay for the privilege of drawing a coin at random, assuming you can keep it?

1.2 ORDER AND AVERAGES

ORDER

On the real line, we introduce an order relation $<$ to stand between two numbers. We say that $a < b$ (read a is less than b) whenever $b - a$ is positive. We interpret $b > a$ (read b is greater than a) to mean exactly the same thing. Thus, $3 < 5$ and $-5 > -8$. We may rephrase this definition geometrically by saying that $a < b$ means that a is to the left of b on the horizontal real line

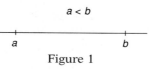

$$a < b$$

Figure 1

(Figure 1). The companion relations $a \leq b$ and $b \geq a$ mean that $b - a$ is either greater than 0 or equal to 0. It is correct to say that $3 \leq 5$; it is also correct to say $3 \leq 3$.

■ **Example 1.** Use the relation $<$ to order, from smallest to greatest, the numbers $\sqrt{2}$, $1.41\overline{4}$, 1.414, $14/10$, $140/99$.

Solution. Use your calculator to discover that

$$\sqrt{2} = 1.41421356 \ldots \text{ and } 140/99 = 1.41414141 \ldots$$

Thus,

$$14/10 < 1.414 < 140/99 < \sqrt{2} < 1.41\overline{4} \qquad ■$$

Manipulations with real numbers using the order relation $<$ depend on the following fundamental properties.

PROPERTIES OF $<$

1. (Trichotomy) Exactly one of $a < b$, $a = b$, and $a > b$ holds.
2. (Transitivity) If $a < b$ and $b < c$, then $a < c$.
3. (Addition) If $a < b$, then $a + c < b + c$.
4. (Multiplication) If $a < b$ and $0 < c$, then $ac < bc$. If $a < b$ and $c < 0$, then $bc < ac$.

Properties 2, 3, and 4 are also valid if $<$ is replaced by $\leq$.

AVERAGES

The (ordinary) **average** of two numbers x_1 and x_2 is defined to be $\bar{x} = (x_1 + x_2)/2$. Geometrically, it is the number midway between x_1 and x_2 on the real line (Figure 2). Similarly, the average of x_1 and $\bar{x}$ is midway between x_1 and $\bar{x}$. This reasoning leads to the remarkable conclusion that, between any two distinct real numbers, there are infinitely many other real numbers, a property referred to as the **denseness** of the real numbers. It also destroys the notion that any one real number can be "just bigger" than another real number. There is no such number.

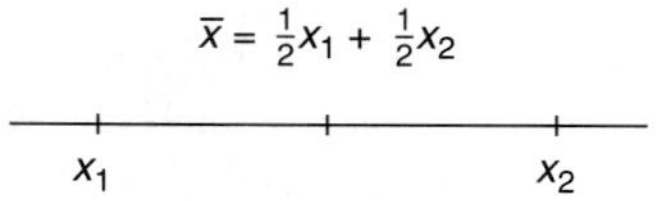

$$\bar{x} = \tfrac{1}{2}x_1 + \tfrac{1}{2}x_2$$

Figure 2

Figure 3

■ **Example 2.** Show that between every pair of distinct rational numbers, there is both a rational and an irrational number (hence, infinitely many of each).

Solution. Let x and y be the two rational numbers with $x < y$. Then $(x + y)/2$ is a rational number between x and y (see Problem 5 in the Problem Set for this section).

Next, choose a natural number n so large that $\sqrt{2}/n < y - x$. It is easy to show that $\sqrt{2}/n$ is irrational (Problem 7) and then that $x + \sqrt{2}/n$ is irrational (Problem 8). Moreover,

$$x < x + \frac{\sqrt{2}}{n} < x + (y - x) = y$$

■

A **weighted average** of two numbers x_1 and x_2 is any number of the form $\hat{x} = wx_1 + (1 - w)x_2$ where $0 < w < 1$. Note that w and $1 - w$ are two positive numbers whose sum is 1.

■ **Example 3.** Show that a weighted average of two numbers is always between the two numbers.

Solution. Assume without loss of generality that x_1 is the smaller of the two numbers (or either of them if the two numbers are equal) and x_2 is the other. We may apply the properties of $\leq$ to obtain

$$x_1 = wx_1 + (1 - w)x_1 \leq wx_1 + (1 - w)x_2 \leq wx_2 + (1 - w)x_2 = x_2$$

The result follows from transitivity. ■

In fact, we can say more than Example 3 asserts. The number $\hat{x} = wx_1 + (1 - w)x_2$ is the $(1 - w)$th fraction of the way from x_1 to x_2. For example, $\hat{x} = (1/3)x_1 + (2/3)x_2$ is two-thirds of the way from x_1 to x_2 (Figure 3). It should be closer to x_2 because we give x_2 the larger weight.

The average of n numbers $x_1, x_2, x_3, \ldots, x_n$ is the number

$$\overline{x} = \frac{x_1 + x_2 + x_3 + \cdots + x_n}{n}$$

and a weighted average is any number of the form

$$\hat{x} = w_1x_1 + w_2x_2 + w_3x_3 + \cdots + w_nx_n$$

where the ws are positive numbers, called weights, with sum 1.

■ **Example 4.** Find the weighted average of -3, -1, 4, and 5 using the corresponding weights $1/10$, $2/5$, $1/5$, and $3/10$.

Solution.

$$\hat{x} = \frac{1}{10}(-3) + \frac{2}{5}(-1) + \frac{1}{5}(4) + \frac{3}{10}(5) = \frac{8}{5} = 1.6$$

■

Note in Example 4 that $\hat{x}$ is between the smallest and the largest of the x-values. This is always the case as you may show (Problems 43 and 44).

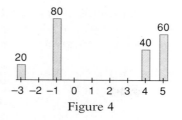

Figure 4

MAKE A GUESS

Your experience with teeter boards should allow you to make an intelligent guess for a center of mass, the balance point. Many people can simply look at Figure 4 and guess that the center of mass is somewhere between 0 and 2. Similarly, one look at Figure 5 suggests that the center of mass is a little larger than 2. Use this kind of intuitive thinking to check on answers you get by calculation.

APPLICATION: CENTER OF MASS

Consider the masses of size 20, 80, 40, and 60 arranged at points with x-coordinates -3, -1, 4, and 5 along a teeter board (Figure 4). According to physics, the balance point for this distribution of masses (that is, the center of mass) is given by

$$\hat{x} = \frac{20(-3) + 80(-1) + (40)(4) + 60(5)}{20 + 80 + 40 + 60}$$

But note that this can be written as

$$\hat{x} = \frac{1}{10}(-3) + \frac{2}{5}(-1) + \frac{1}{5}(4) + \frac{3}{10}(5) = \frac{8}{5} = 1.6$$

which is exactly the expression we got in Example 4. More generally, if m_1, $m_2, \ldots, m_n$ is a distribution of masses located along the x-axis at $x_1, x_2, \ldots, x_n$, and if $M = m_1 + m_2 + \cdots m_n$, then the center of mass is given by

$$\hat{x} = \frac{m_1 x_1 + m_2 x_2 + \cdots + m_n x_n}{M} = \frac{m_1}{M}x_1 + \frac{m_2}{M}x_2 + \cdots + \frac{m_n}{M}x_n$$

Thus, $\hat{x}$ is a weighted average of the xs (note that the weights m_i/M do add to 1).

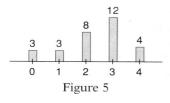

Figure 5

■ **Example 5.** Find the center of mass of the masses 3, 3, 8, 12, and 4 located at points with x-coordinates 0, 1, 2, 3, and 4, respectively (Figure 5).

Solution. The total mass is $M = 3 + 3 + 8 + 12 + 4 = 30$. Thus,

$$\hat{x} = \frac{3}{30}(0) + \frac{3}{30}(1) + \frac{8}{30}(2) + \frac{12}{30}(3) + \frac{4}{30}(4) = \frac{71}{30} \approx 2.367 \qquad ■$$

APPLICATION: GRADE POINT AVERAGE

Jack, a freshman at Hamster U, received 3 credits of F, 3 credits of D, 8 credits of C, 12 credits of B, and 4 credits of A during his first year of college, a total of 30 credits. We ask: What is his grade point average, assuming that $F = 0$, $D = 1, \ldots, A = 4$? This is simply a matter of calculating a weighted average, the weights for the grades 0, 1, 2, 3, and 4 being 3/30, 3/30, 8/30, 12/30, and 4/30, respectively. Thus, Jack's grade point average is the number we calculated in Example 5, namely 2.367.

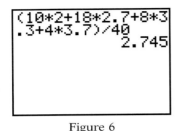

Figure 6

In general, if a student receives m_1 credits at level x_1, m_2 credits at level $x_2, \ldots , m_n$ credits at level x_n, and if $M = m_1 + m_2 + \cdots + m_n$, then her grade point average is given by the boxed formula displayed just before Example 5.

■ **Example 6.** Assume that F corresponds to 0 and A to 4 and that pluses and minuses add or subtract 0.3 (for example, B^- corresponds to 2.7 and C^+ corresponds to 2.3). So far, Jane has received 10 credits of C, 18 credits of B^-, 8 credits of B^+, and 4 credits of A^-. Determine her grade point average.

Solution.

$$\hat{x} = \frac{10}{40}(2) + \frac{18}{40}(2.7) + \frac{8}{40}(3.3) + \frac{4}{40}(3.7) = 2.745$$

The calculator screen for this calculation is shown in Figure 6. ■

APPLICATION: EXPECTED VALUE

Suppose that a rich uncle has offered you a chance to play the following remarkable game over and over as many times as you want. Toss two coins. If both show heads (HH), you get \$1; if both show tails (TT), you get \$2; if you get a combination of a head and a tail, you get \$3. How much would you expect to win on the average on each toss?

Reason this way. If you tossed the two coins 1000 times, you could expect to get HH about 250 times, TT about 250 times, and a combination about 500 times. Thus, your total payoff would be $250(\$1) + 250(\$2) + 500(\$3)$ so that the average per toss would be

$$\frac{250(1) + 250(2) + 500(3)}{1000} = \frac{1}{4}(1) + \frac{1}{4}(2) + \frac{1}{2}(3) = \$2.25$$

In general, if an experiment can result in payoffs $x_1, x_2, \ldots , x_n$ with probabilities $p_1, p_2, \ldots , p_n$ (summing to 1), then the expected payoff (that is, the **expected value**) of the experiment is given by

$$\hat{x} = p_1 x_1 + p_2 x_2 + \cdots + p_n x_n$$

■ **Example 7.** It can be shown that the probabilities for a World Series between two evenly matched baseball teams to end in 4, 5, 6, or 7 games are 1/8, 1/4, 5/16, and 5/16, respectively. Determine the expected number of games in such a series (that is, the average number of games one should expect if thousands of such series were played).

Solution.

$$\hat{x} = \frac{1}{8}(4) + \frac{1}{4}(5) + \frac{5}{16}(6) + \frac{5}{16}(7) = \frac{93}{16} \approx 5.81$$ ■

APPLICATION: MEAN OF A STATISTICAL DISTRIBUTION

Often a set of data is organized as in the chart of Figure 7, which shows the frequency f_i of each score x_i on a 10-point test in a class of 35 students. An important statistic for this set of data is its **mean,** which is the statisticians' name for the ordinary average.

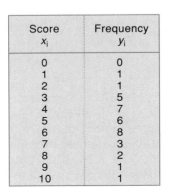

Score x_i	Frequency y_i
0	0
1	1
2	1
3	5
4	7
5	6
6	8
7	3
8	2
9	1
10	1

Figure 7

■ **Example 8.** Find the mean of the distribution of scores in Figure 7.

Solution. We use the boxed formula with the frequency f_i playing the role of the mass m_i. The mean $\hat{x}$ is

$$\frac{0(0) + 1(1) + 1(2) + 5(3) + 7(4) + 6(5) + 8(6) + 3(7) + 2(8) + 1(9) + 1(10)}{35}$$

$$= 180/35 \approx 5.14$$ ■

TEASER SOLUTION

Assuming that you could draw a coin over and over from the bag, you could expect to get an average value of

$$\hat{x} = \frac{4}{10}(1) + \frac{3}{10}(5) + \frac{2}{10}(10) + \frac{1}{10}(25) = 6.4¢$$

on each draw. Therefore, you should be willing to pay anything less than that amount.

PROBLEM SET 1.2

A. Skills and Techniques

In Problems 1–4, use $<$ to order the given numbers from smallest to greatest.

1. $3.\overline{1}416$, π, $22/7$, $3.14159\overline{2}$, 3.141592
2. $\sqrt{3}$, 1.7321, 1.73205, $1.73205\overline{1}$, $\sqrt{631/210.5}$
3. 2222, 222^2, 22^{22}, 2^{222}, $2^{2^{2^2}}$ (*Note:* There is a standard agreement that in a tower of exponents without parentheses, calculation goes from the top down. Thus, $a^{b^c} = a^{(b^c)}$. Be warned, however, that many calculators, including the TI-81 and TI-82, interpret $a\boxed{\wedge}b\boxed{\wedge}c$ as $(a^b)^c$.)
4. 4^5, 5^4, 44^2, $4.5^{4.5}$, $2^{2.2^4}$
5. Show that if x and y are rational numbers, so is their average $(x + y)/2$.
6. Show that if x, y, and w are rational numbers, so is their weighted average $wx + (1 - w)y$.
7. Assuming that $\sqrt{2}$ is irrational and n is a natural number (that is, a positive integer), show that $\sqrt{2}/n$ is irrational.
8. Show that $x + \sqrt{2}/n$ is irrational, assuming that x is rational and n is a natural number.
9. Find, if possible, the largest rational number less than 2.
10. Is the sum of two irrational numbers always irrational? Justify your answer.
11. Calculate the ordinary average of the natural numbers 1 through 10.
12. Calculate the ordinary average of the integers from -5 to 10.
13. Calculate the weighted average of the natural numbers 1 through 10, assuming 1 gets weight 3/20 and 10 gets weight 1/20, and the other numbers get weight 1/10.
14. Calculate the weighted average of the integers -5 through 10 assuming the smallest four get weight x, the middle eight get weight $2x$, and the largest four get weight $3x$. (*Hint:* First determine x.)
15. Suppose $x < y$. Order the following from smallest to largest.

$$\frac{1}{2}x + \frac{1}{2}y, \frac{2}{3}x + \frac{1}{3}y, 0.49x + 0.51y, \frac{8}{15}x + \frac{7}{15}y$$

16. The numbers 1, 2, 3, and 4 are weighted in proportion to their sizes. Determine their weighted average.
17. Consider masses 3, 4, 5, 6, and 7 located at points on a line with coordinates 1, 3, 5, 7, and 9, respectively. Draw a picture representing this situation (like Figure 5), guess at the center of mass, and then determine this center exactly.
18. Repeat Problem 17 but assume the masses are 10, 1, 1, 1, and 1, respectively.
19. Juan, weighing 200 pounds, stands at the left end of a teeter board; his daughter Kaila, weighing 50 pounds, stands 8 feet to the right; and his son Lucas, weighing 75 pounds, stands 3 feet to the right of Kaila. Where should the fulcrum be placed for balance? Assume that the weight of the teeter board is negligible.
20. Each of 10 employees makes $800 a week, but their supervisor makes $3000 a week. What is the average weekly salary of these 11 people?

Determine grade point averages for students having the indicated grades. See Example 6.

21. Amy: 5 credits of A, 6 credits of A^-, 8 credits of B^+, and 5 credits of C.

22. Roberto: 10 credits of A, 6 credits of B, and 10 credits of C^+.

23. Daphne: 5 credits of F, 5 credits of D, and 16 credits of B^+.

24. Arlene: 5 credits of each of A, B, C^+, C, and C^-.

Problems 31–34 are related to Example 7.

25. When a standard die is tossed over and over, what is the expected value of the number that will appear?

26. When two dice are tossed over and over, what total can one expect on the average? (*Hint:* There are 36 equally likely outcomes. For example, a total of 3 has probability 2/36, because a total of 3 corresponds to getting a 2 and 1 or 1 and 2 on the two dice.)

27. Suppose that in tossing two dice, you will receive $4 if you get 7 or 11 but have to pay $1 otherwise. What are your expected winnings per toss? See Problem 26.

28. One hundred thousand people plan to enter a drawing for which the first, second, and third prizes are $10,000, $5000, and $1000. On the basis of expected values, is it worth the cost of the 29¢ postage stamp it takes to enter? Explain.

Find the means of the distributions shown in Problems 29 and 30.

29.

x_i	f_i
0	2
1	3
2	5
3	5
4	2

30.

x_i	f_i
1	7
2	7
3	5
4	3
5	2
6	1
7	1

B. Applications and Extensions

31. Suppose $23.1 < a < 23.2$ and $32.8 < b < 32.9$, where a and b are the altitude and base of a triangle, respectively. What inequality holds for the area A of this triangle?

32. Suppose $10.05 \leq r \leq 10.06$ and $13.55 \leq h \leq 13.66$, where r and h are the radius and height of a cylinder, respectively. What inequality holds for the volume of this cylinder?

33. The **variance** of a set of numbers x_1, x_2, . . . , x_n is the average of the numbers $(x_1 - \bar{x})^2$, $(x_2 - \bar{x})^2$, . . . , $(x_n - \bar{x})^2$, with $\bar{x}$ being the mean. The **standard deviation** (s) is defined as the square root of the variance. Find the mean and the standard deviation of the set of numbers 11, 11.5, 13, 14, 15.

34. Find the standard deviation of the distribution in Problem 29.

35. Most graphics calculators have built-in statistical routines for calculating means and standard deviations. Repeat Problem 33 using your calculator routine.

36. The weights of the 11 members of the first string defensive unit of the football team are 200, 195, 197, 212, 233, 245, 189, 214, 251, 234, and 213 pounds. Determine the mean and standard deviation as in Problem 35.

37. Show that if a, b, c, and d are positive numbers, then $a/b < c/d$ if and only if $ad < bc$.

38. Show that if $0 < a < b$, then $1/b < 1/a$.

39. Show that if $0 \leq a \leq b$, then $a^2 \leq b^2$.

40. Show that if $0 \leq a \leq b$, then $\sqrt{a} \leq \sqrt{b}$.

41. Show that if $a > 0$, then $a + 1/a \geq 2$.

42. Show that if $a > 0$ and $b > 0$, then $\sqrt{ab} \leq (1/2)a + (1/2)b$; that is, the geometric average of two numbers is less than or equal to their arithmetic (that is, their ordinary) average.

43. Show that the weighted average $\hat{x}$ of three numbers $x \leq y \leq z$ satisfies $x \leq \hat{x} \leq z$; that is, the weighted average of three numbers is between the smallest and the largest of the three numbers.

44. **Challenge.** Show that the weighted average of any finite set of numbers is between the smallest and the largest of these numbers.

TEASER Franklin High School reported the following statistics for its senior class. The mean weight of the 320 girls is 130 pounds, the mean weight of the boys is 180 pounds, and the mean weight of all seniors is 160 pounds. The report failed to mention the number of boys, but you can figure it out. Do so.

1.3 EQUATIONS AND THEIR GRAPHS

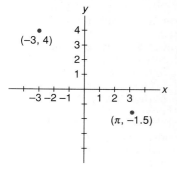

Figure 1

Following Descartes and Fermat, two seventeenth-century French mathematicians, we introduce in the plane two perpendicular real lines (Figure 1). The horizontal axis is called the **x-axis,** the vertical axis is the **y-axis,** and the intersection of the two axes at their 0 points is called the **origin.** Then every point is uniquely specified by giving its two coordinates (x, y), the directed distances from these axes in the x and y directions. Thus, $(-3, 4)$ specifies the point 3 units to the left of the y-axis and 4 units above the x-axis. Given two points (x_1, y_1) and (x_2, y_2) in the plane, we can find the distance d between them by using the familiar **distance formula:**

$$d = \sqrt{(x_2 - x_1)^2 + (y_2 - y_1)^2}$$

■ **Example 1.** Find the distance between the points $(-3, 4)$ and $(\pi, -1.5)$.

Solution. Two correct ways of entering this distance into a calculator are shown in Figure 2. The answer is

$$d = 8.244341109$$

■

```
((π+3)^2+(-1.5-4
)^2)^.5
        8.244341109
√(((π+3)²+(-1.5-4
)²)
        8.244341109
```

Figure 2

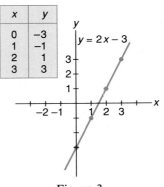

x	y
0	-3
1	-1
2	1
3	3

Figure 3

The introduction of a coordinate system in the plane allows us to transform an equation in x and y (an algebraic object) into a graph (a geometric object). This transformation is the key idea in the subject we call analytic geometry. The **graph** of an equation consists of all points whose coordinates (x, y) satisfy the equation. The traditional way of graphing an equation is a three-step process: find the coordinates of a few points (make a table of values), plot the corresponding points, and connect those points with a smooth curve. We illustrate this process in Figure 3 for the equation $y = 2x - 3$. Although a graphics calculator makes graphing easy (a subject discussed in Section 1.4), we think every student needs to know how to graph in the traditional way.

■ **Example 2.** Sketch the graph of $y = x^2 - x - 6$.

Solution. A table of values, the corresponding points, and the graph are shown in Figure 4. ■

SYMMETRIES OF A GRAPH

The form of an equation can often alert us to properties of the graph. For example, if replacing x by $-x$ gives an equivalent equation (as in $y = x^2 - 5$), then the graph is symmetric with respect to the y-axis (Figure 5). If replacing y by $-y$ gives an equivalent equation (as in $x = y^2$), then the graph is symmetric

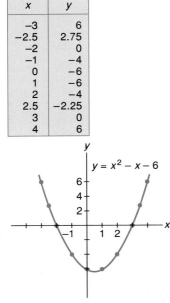

x	y
-3	6
-2.5	2.75
-2	0
-1	-4
0	-6
1	-6
2	-4
2.5	-2.25
3	0
4	6

$y = x^2 - x - 6$

Figure 4

A GRAPH: INFINITELY MANY POINTS

The graph of an equation will normally consist of infinitely many points. A table of values allows us to plot a few of these points. When we connect these points with a smooth curve, we are making a huge assumption, namely, that the graph is well behaved between the plotted points. This assumption is in fact correct for most common functions. Note that the TI-82 has a TABLE key which facilitates making a table of values.

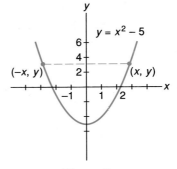

Figure 5

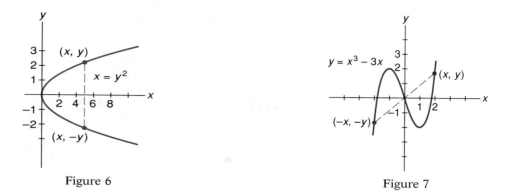

Figure 6

Figure 7

ABSOLUTE VALUE

The symbol ||, called absolute value, occurs in part (d) of Example 3. We will have a lot to say about this later. For now, all you need to know is that

$$|x| = \begin{cases} x \text{ if } x \geq 0 \\ -x \text{ if } x < 0 \end{cases}$$

Thus

$$|3| = 3 \text{ and }$$
$$|-3| = -(-3) = 3$$

Note that

$$|-x| \neq x$$

as you can see by letting $x = -2$. The best we can say is that

$$|-x| = |x|$$

with respect to the x-axis (Figure 6). If replacing x by $-x$ and y by $-y$ gives an equivalent equation (as in $y = x^3 - 3x$), then the graph is symmetric with respect to the origin (Figure 7).

■ **Example 3.** Without actually graphing, determine any symmetries for the graphs of the following equations.

(a) $y^3 = -x^3 + x$ (b) $x^2 + y^2 = 36$ (c) $y = x^2 - 2x + 3$

(d) $y = \dfrac{x^6 - 3x^2}{|x|}$

Solution.

(a) Because $(-y)^3 = -(-x)^3 + (-x)$, that is, $-y^3 = x^3 - x$, is equivalent to the given equation, the graph is symmetric with respect to the origin.
(b) Replacing x by $-x$ or y by $-y$ results in an equivalent equation, so the graph of $x^2 + y^2 = 36$ is symmetric with respect to both axes. It is also symmetric with respect to the origin. Perhaps you recognize that the graph of this equation is a circle of radius 6 centered at the origin.
(c) Replacing x by $-x$ gives $y = x^2 + 2x + 3$; replacing y by $-y$ gives $-y = x^2 - 2x + 3$; doing both gives $-y = x^2 + 2x + 3$. None of these is equivalent to the given equation; the graph fails to have any of the three types of symmetry.
(d) Replacing x by $-x$ does give an equivalent equation; the graph is symmetric with respect to the y-axis. ■

We remark that equation (b) above is a special case of the **standard equation of a circle,** which has the form

$$\boxed{(x - h)^2 + (y - k)^2 = r^2}$$

This is the equation of the circle with center (h, k) and radius r (Figure 8).

SOLVING EQUATIONS ALGEBRAICALLY

Consider the equation $2x - 3 = 0$. To **solve** this equation (or find its solutions) means to find the x-value that makes the equation a true equality. The process

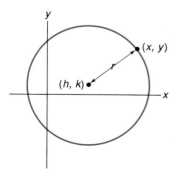

Figure 8

is straightforward. Perform the same operation on both sides (add or subtract, multiply or divide by a nonzero number) until x stands by itself.

$$2x - 3 = 0$$

$$2x = 3 \quad \text{(adding 3 to both sides)}$$

$$x = \frac{3}{2} \quad \text{(dividing both sides by 2)}$$

Note for later reference that $x = 3/2$ is the point where the graph of $y = 2x - 3$ crosses the x-axis (Figure 3).

Next consider the equation $x^2 - x - 6 = 0$. We may rewrite this as $(x - 3)(x + 2) = 0$, which has the two solutions $x = 3$ and $x = -2$ (obtained by setting each factor equal to 0). Note that the points $x = 3$ and $x = -2$ are exactly the points where the graph of $y = x^2 - x - 6$ crosses the x-axis (Figure 4).

Even when we are unable to factor the left side, we can always solve an equation of the form $ax^2 + bx + c = 0$, called a quadratic equation, by use of the **quadratic formula**. This formula says that the two solutions of this equation are given by

$$x = \frac{-b \pm \sqrt{b^2 - 4ac}}{2a}$$

For example, in the equation $x^2 - x - 6 = 0$, $a = 1$, $b = -1$, and $c = -6$, so

$$x = \frac{1 \pm \sqrt{1 + 24}}{2} = \frac{1 \pm 5}{2} = 3, -2$$

■ **Example 4.** Solve the quadratic equation $2x^2 - 2x - 3 = 0$.

Solution. Here $a = 2$, $b = -2$, and $c = -3$. Thus,

$$x = \frac{2 \pm \sqrt{4 + 24}}{4} = \frac{2 \pm \sqrt{28}}{4} = \frac{2 \pm 2\sqrt{7}}{4} = \frac{1 \pm \sqrt{7}}{2} \approx 1.82, -0.82 \quad ■$$

■ **Example 5.** Solve the equation $(2x - 5)/(x + 1) = 4$.

Solution. We use the following steps.

$$2x - 5 = 4x + 4 \quad \text{(multiplying by } x + 1)$$
$$-9 = 2x \quad \text{(subtracting } 2x \text{ and 4)}$$
$$\frac{-9}{2} = x \quad \text{(dividing by 2)} \qquad ■$$

There is danger in multiplying both sides of an equation by a variable expression. For example, multiply $x - 3 = 0$ by $x + 1$ to get $(x + 1)(x - 3) = 0$ and you have introduced the extraneous solution $x = -1$. The problem is that when $x = -1$, you multiplied both sides by 0, which is not legal. We did not make this mistake in the derivation above, because when $x = -9/2$, $x + 1 = -7/2 \neq 0$.

WHY THE GEOMETRIC METHOD WORKS

The equation $2x - 3 = 0$ is the special case of $2x - 3 = y$, where $y = 0$. Thus, to solve $2x - 3 = 0$ is to find out where the graph of $y = 2x - 3$ has $y = 0$. To be precise, it is to find the x-coordinates of those points where the graph meets the x-axis. Similarly, to solve $x^2 - x - 6 = 0$ is to find the x-coordinates of those points where the graph of $y = x^2 - x - 6$ meets the x-axis.

SOLVING EQUATIONS GEOMETRICALLY

If $f(x)$ denotes an expression in x, then the solutions to the equation $f(x) = 0$ are the points (technically, the x-coordinates of the points) where the graph of $y = f(x)$ intersects the x-axis. We pointed this out earlier for the equations $2x - 3 = 0$ and $x^2 - x - 6 = 0$ (see Figures 3 and 4). We will use a graphics calculator to exploit this fact many times later in this book, and even with hand graphing we can use it to solve equations geometrically.

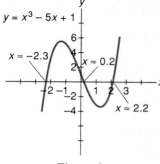

Figure 9

■ **Example 6.** Solve the cubic equation $x^3 - 5x + 1 = 0$.

Solution. Note that there is no simple algebraic method of solving this equation, because we do not know how to factor the left side. However, we can graph the equation $y = x^3 - 5x + 1$ and note where the graph crosses the x-axis. The more carefully we draw the graph, the more accurate will be our answers. Our graph (Figure 9) suggests that there are three answers given approximately by $x = -2.3$, 0.2, and 2.2. ■

SOLVING PAIRS OF EQUATIONS SIMULTANEOUSLY

Next we consider the problem of solving two equations in two unknowns.

■ **Example 7.** Find the common solutions of $y = 2x + 1$ and $y = x^2 - x - 3$.

Solution. We can treat this problem either algebraically or geometrically. To solve algebraically, first set the two expressions for y equal to each other and solve for x.

$$x^2 - x - 3 = 2x + 1$$
$$x^2 - 3x - 4 = 0$$
$$(x - 4)(x + 1) = 0$$
$$x = -1, 4$$

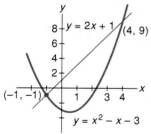

Figure 10

Then substitute in either equation to find y in each case. The two solutions are the ordered pairs $(-1, -1)$ and $(4, 9)$.

To study the problem geometrically, we graph the two equations using the same axes (Figure 10). The intersection points of the two graphs are the solutions we want. You can see that they are the same as the solutions obtained algebraically. ■

AN APPLICATION

A common experience of anyone using mathematics is the need to solve an equation. Here is a simple example.

■ **Example 8.** At noon, Amy left point P walking due north; an hour later Bob left the same point walking due east. Both walked at 4 miles per hour

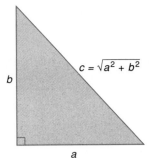

$c = \sqrt{a^2 + b^2}$

b

a

Figure 11

and both carried walkie-talkies with a range of 8 miles. At what time did they lose contact with each other?

Solution. Let t denote the number of hours after noon. Assuming $t \geq 1$, we see that Amy will have walked $4t$ miles and Bob will have walked $4(t - 1)$ miles. Using the **Pythagorean theorem** (Figure 11), we determine the distance between them at time t to be

$$\sqrt{(4t)^2 + (4t - 4)^2}$$

which we set equal to 8. Then we square both sides and simplify to obtain

$$16t^2 + 16t^2 - 32t + 16 = 64$$

$$32t^2 - 32t - 48 = 0$$

$$2t^2 - 2t - 3 = 0$$

This is the equation we solved in Example 4. Its positive solution is 1.82. This means that Amy and Bob will lose contact at 1.82 hours after noon, or at about 1:49. ∎

TEASER SOLUTION

Let x denote the number of boys. Then

$$\frac{320(130) + x(180)}{320 + x} = 160$$

This is equivalent to

$$320(130) + 180x = 160(320 + x)$$

which has the solution $x = 480$.

PROBLEM SET 1.3

A. Skills and Techniques

Find the distance between the given points.

1. $(-1, -1)$ and $(1, 1)$
2. $(5, 3)$ and $(-8, 3)$
3. $(7, \sqrt{2})$ and $(-3, 0)$
4. $(-1, 6)$ and $(4, -6)$
5. $(4.1, 3.7)$ and $(-5.5, 4.6)$
6. $(\pi, \pi + 2)$ and $(\sqrt{3}, -4)$

Determine any symmetries and then graph the given equations.

7. $y = 4$
8. $y = -\pi$
9. $y = 3 - x$
10. $y = 3x + 1$

11. $y = 3x^2 - 5$
12. $y = (x + 3)^2$
13. $y = \dfrac{1}{x^2 + 1} + 2$
14. $y = -\dfrac{1}{3}x^3 + 5x$
15. $y = -2(x + 2)^2 + 4$
16. $y = x(x - 2)^2$

Without actually graphing, determine any symmetries of the graphs of the following equations.

17. $y = 2x$
18. $y = 5x^2$
19. $y = x^3 + 5x$
20. $y^2 = 3x^2$
21. $y = x^4 - 2x^2 - 2$
22. $y = -x^3 + 2x + 2$

23. $|y| = x^4 + 3x^2$
24. $y = (2x + 3)(x - 3)$
25. $y = 3x^2 + 2x$
26. $y = \dfrac{3x}{x^2 + 1}$

Solve each equation, using the algebraic methods of Examples 4 and 5.

27. $3x - 21.2 = -33.5$
28. $0.2x + 1.2 = 0.8x - 3.4$
29. $x^2 - 2x - 15 = 0$
30. $x^2 - 6x - 7 = 0$
31. $\dfrac{x + 5}{x} = 3$
32. $x(x - 2) = 0$
33. $(x - 4)(x + 3) = -6$
34. $\dfrac{3x + 3}{x - 4} = 2$
35. $x^2 + 3x - 5 = 0$
36. $-3x^2 + 7x - 4 = 0$
37. $\dfrac{x^2}{x - 1} = 4x + 1$
38. $x^2 + 4x - 5 = 0$
39. $\dfrac{3}{14}x - \dfrac{7}{3} = \dfrac{1}{7}x + \dfrac{1}{5}$
40. $2 + \dfrac{2}{2 + \dfrac{2}{2 + 2/x}} = \dfrac{1}{2}$

In Problems 41–46, use graphical methods to determine approximate solutions to each equation, as in Example 6.

41. $x^2 - 2x - 7 = 0$
42. $-x^2 + 5x + 1 = 0$
43. $x^3 - 3 = 0$
44. $x^3 - 2x + 3 = 0$
45. $x^4 = -2x + 2$
46. $x^2 = 6x - 10$

Graph the pairs of equations in Problems 47–52. Then find the intersection points of these graphs using algebraic methods. Confirm your answers by looking at the graphs.

47. $y = 2x + 3$ and $y = -3x - 7$
48. $y = 5$ and $y = 3x - 10$
49. $y = x^2 - 5$ and $y = -x^2 + 5$
50. $y = x^2 + 2x + 4$ and $y = 3(x + 2)$
51. $y = x^2 - 3x$ and $y = -x^2 - 2x + 6$
52. $y = -x^3 + x$ and $y = 2x^2 + 2x$

B. Applications and Extensions

53. Determine the two points on the graph of $y = 2x^2 + 5x - 2$ where $y = 1$.
54. Find the intersection points of the circle $x^2 + y^2 = 16$ and the line $y = \sqrt{3}x$.

55. Find the distance between the points of the graph of $y = 2x^2 - 1$ corresponding to $x = -1$ and $x = 2$.
56. Find the distance between the two points of the graph of $y = 2x^2 - 1$ where $y = 7$.
57. Find the distance between the points of intersection of the graphs of $y = x^2 + x - 6$ and $y = 2x + 6$.
58. Find the distance between the points of intersection of the graphs of $x = -y^2 + 2y - 4$ and $x = -3y + 2$.
59. Determine the center and radius of the circle $x^2 - 6x + 9 + y^2 = 36$. (*Hint:* $(x - 3)^2 + y^2 = 6^2$.)
60. Determine the center and radius of the circle $x^2 - 6x + y^2 = 7$. (*Hint:* Add 9 to both sides thus completing the square for the x-terms. In general, to complete the square for $x^2 + ax$, add $(a/2)^2$.)
61. Use the hint in Problem 60 to determine the center and radius of the circle $x^2 - 2x + y^2 - 4y = 6$.
62. Determine the center and radius of the circle $x^2 + 8x + y^2 - 6y = 11$.

For solving word problems, we suggest the following steps. (1) Draw a picture. (2) Assign a letter to the principal unknown. (3) Find an equation involving the unknown. (4) Solve the equation. (5) Check to see that the answer makes sense.

63. When all 300 feet of the string for José's kite are out, the kite is directly above his car 200 feet away. Assume that the string is taut and José holds his end of the string at ground level. How high is the kite?
64. Melinda lives 1 mile east of Ashley's house. Ashley left home, walked 2 miles north and 3 miles east, and arrived at a point 4 miles south of Darlene's house. How far apart do Melinda and Darlene live?
65. Casey leaves Striketown at noon, driving straight north at 65 miles per hour. On his radio he listens to the ball game from Mudville, which is 135 miles straight east of Striketown. If the Mudville radio station transmitter has a range of 200 miles, when does Casey lose his reception?
66. Ed and Trixie are riding on opposite wheels of a double Ferris wheel. Each wheel has a radius of 50 feet, and their centers are separated by 120 feet. If the wheels rotate independently, determine the farthest possible distance between Ed and Trixie. If Ed is at the top of his wheel and Trixie is at the bottom of hers, how far apart are they, assuming the bar connecting the centers is horizontal?
67. Jane can row downstream twice as fast as she can row upstream when the river's current is 2 miles per hour. How fast can she row in still water?
68. A triathalon is a race combining a 1.5-kilometer swim, a 40-kilometer bicycle ride, and a 10-kilometer run. John bicycles four times as fast as he runs and runs six times as fast as he swims. If he finished the entire race in 2 hours, how fast did he bicycle?

69. Determine the length of a belt that runs around two wheels with equations $(x - 3)^2 + (y - 2)^2 = 9$ and $(x - 1)^2 + (y + 5)^2 = 9$.

70. Determine the perimeter of the triangle whose vertices are the centers of the circles $x^2 + y^2 - 4x + 8y = 5$, $x^2 + y^2 + 2x - 2y = 2$, and $x^2 + y^2 - 10y = -21$.

71. Kioko has an average of 88 on the first four exams. What score does she need on the fifth exam to bring her average up to 90?

72. Chan has scores of 84, 89, and 78 on the first three exams. What score does he need on the final exam (which counts twice as much as the earlier exams) to have an average of 90?

73. How many pounds of cashews costing $4.50 per pound should be added to 14 pounds of walnuts costing $3.75 per pound and 16 pounds of peanuts costing $2.20 per pound to obtain a mixture worth $4.00 per pound?

74. **Challenge.** On the surface of the earth, Andrea walked 1 mile south, 1 mile east, 1 mile north, and ended where she started. She did not start at the North Pole. Where did she start? Give all possible solutions.

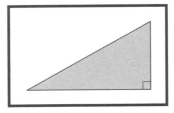

TEASER   The area and perimeter of a right triangle are both 25. Find the lengths of the two legs.

1.4   GRAPHS WITH GRAPHICS CALCULATORS

HOW MUCH IS ENOUGH?

Because we can never show the whole graph of most equations, it is reasonable to ask what we mean by the direction "draw the graph of a certain equation." For example, what does it mean to draw the graph of

$$y = 8x^4 - 4x^2 - x + 12$$

What we draw should show the essential features, such as hills and valleys, and should give an indication of what happens for large $|x|$. Figure 4 is such a graph. Some authors label this the complete graph; others call it a representative graph.

Although computers with graphics capability have been around for many years, it is the recent development of inexpensive hand-held graphics calculators that makes the power of technology easily available to most students. Students can now carry in their pockets and use in their dorm rooms calculators that make graphing easier, more accurate, and certainly more enjoyable than hand graphing. Moreover, these machines are great for experimenting with mathematical ideas and they allow us to ask and answer deep questions.

Several graphics calculators are on the market. Our goal is to write descriptions that apply to most of them and in particular to the TI-81 and TI-82. Keep the owner's manual for your model at hand so you can make any modifications that are necessary.

THE VIEWING WINDOW

Rarely are we able to show the whole graph of an equation, because it will normally stretch across the entire xy-plane. What we should aim to do is to show enough of the graph so that its essential features are visible. In hand graphing, we make this decision when we select the points we will plot and when we choose scales for the two axes. In using a graphics calculator, our first decision is to specify the size of the viewing window, that is, the range of values for x and y that will be displayed.

Every calculator has so-called **default** values for all of its parameters (these values are set by the manufacturer). Typical default range values are $-10 \le x \le 10$, $-10 \le y \le 10$, but they vary with calculator models. You will need to study the owner's manual to discover the default values, to learn how to set new range values, and to see how to return to the default values. On TI

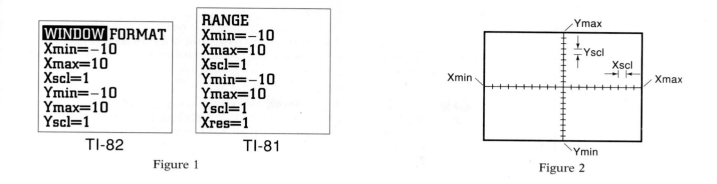

WINDOW FORMAT	RANGE
Xmin=−10	Xmin=−10
Xmax=10	Xmax=10
Xscl=1	Xscl=1
Ymin=−10	Ymin=−10
Ymax=10	Ymax=10
Yscl=1	Yscl=1
	Xres=1

TI-82   TI-81

Figure 1

Figure 2

<table>
</table>

HASH MARKS

As indicated above, **Xscl** and **Yscl** determine the intervals between the hash marks on the axes. You will want to set these values so that the hash marks are nicely spread out, or you may want to leave them out completely (accomplished by setting both scale values to 0). Because the spacing of hash marks is pretty arbitrary, we will say little about it in the rest of the book.

calculators, pressing the (WINDOW) or (RANGE) key will display a menu similar to that in Figure 1. Figure 2 shows the meaning of each item. Typically you can change any of these numbers by moving the blinking cursor (using the arrow keys) to the desired location and then entering new numbers.

We will use the notation [a, b] by [c, d] to describe the size of the viewing window. Thus, [−10, 10] by [−5, 5] indicates a window in which −10 ≤ x ≤ 10 and −5 ≤ y ≤ 5. All calculator-drawn graphs will be displayed in a window with its size indicated beneath the window.

GRAPHING EQUATIONS

Having described how to set the ranges for x and y, we are ready to consider graphing an equation of the form $y = f(x)$. The expression $f(x)$ can be any of the common expressions in x (for example, $x^2 - 3x$ or $3/(x^4 + 2)$) that arise in algebra. To learn how to key in such an expression and then how to instruct your calculator to draw the graph, consult a manual.

■ **Example 1.** Graph $y = x^2 - 6$.

Solution. Set the range values, say to [−10, 10] by [−10, 10]. Enter the expression $x^2 - 6$ and then graph according to the procedure described in the box titled TI Calculators or in your manual. You should obtain a graph very similar to Figure 3. ■

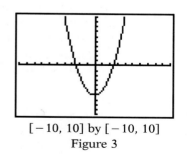

[−10, 10] by [−10, 10]
Figure 3

■ **Example 2.** Graph $y = 8x^4 - 4x^2 - x + 12$.

Solution. Follow the procedure you used in Example 1; that is, set the range values to $[-10, 10]$ by $[-10, 10]$ and graph. Nothing seems to happen. A little thought may suggest that the graph we are asking for is outside the range we have specified. Try changing the range for y to $10 \le y \le 30$ and graph. Now you will get a very thin cup-shaped curve that doesn't show the essential features very well. Change the range for x to $-2 \le x \le 2$ and graph again. The graph (Figure 4) that now appears on the screen shows the true nature of the graph.

■

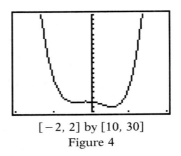

$[-2, 2]$ by $[10, 30]$
Figure 4

Example 2 illustrates the importance of choosing appropriate range values; doing this usually requires some experimentation.

SOLVING EQUATIONS AND ZOOMING

We learned in Section 1.3 that we can solve an equation such as $x^3 - 2x + 6 = 0$ by graphing $y = x^3 - 2x + 6$ and noting where this graph crosses the x-axis. The graph, which you should draw on your calculator (Figure 5), suggests that the given equation has only one solution and that it is very near $x = -2$. But what if we want the answer accurate to three decimal places? We will need to learn how to use the zoom feature of a graphics calculator to meet this demand.

Zooming on a calculator is similar to zooming with a camera; it allows you to zoom in, thereby enlarging a part of the picture that interests you. (You can also zoom out to look at a bigger part of the graph.) To zoom in, first move the cursor to a point near the center of the region of interest. You can move the standard cursor very easily by using the four arrow keys or, what is

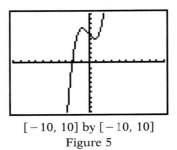

$[-10, 10]$ by $[-10, 10]$
Figure 5

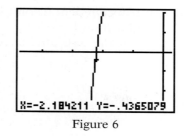

X=-2.184211 Y=-.4365079

Figure 6

better, you can move the trace cursor by pressing the [TRACE] key together with the right and left arrow keys (thereby moving the cursor along the curve). In either case, you will note that values for x and y now appear at the bottom of the window, indicating the coordinates of the cursor. With the cursor in the region of interest, zoom in and the region around the cursor will be enlarged in both the x and y directions by preset factors (the default factors on the TI-81 and TI-82 are 4, and these factors can be changed if desired).

Returning to the situation in Figure 5, move the trace cursor near the intersection of the curve with the x-axis and zoom in using enlargement factors of 4 in each direction. This will give a picture similar to (but perhaps not exactly like) Figure 6. Note the position of the cursor and its coordinates. Again move the trace cursor near the intersection point and zoom in again. Repeat the process until you have the accuracy you want. Note that when the trace cursor is near the x-axis, one step moves it from above the x-axis to below the x-axis (or vice versa) thus trapping the x-value of the solution between the two corresponding x-coordinates of the trace cursor (see the box titled Accuracy). We find that five zooms result in a situation where one step moves the trace cursor from below the x-axis with x-coordinate -2.179996 to above the x-axis with x-coordinate -2.17979. We conclude that the solution to $x^3 - 2x + 6$, accurate to three decimal places, is $x \approx -2.180$.

■ **Example 3.** Find the largest solution of $x^4 - 8x^3 + 22x^2 - 24x + 2 = 0$ accurate to three decimal places.

Solution. Begin by drawing a calculator graph of $y = x^4 - 8x^3 + 22x^2 - 24x + 2$ (Figure 7), which suggests that the solution we want is near $x = 4$. Move the trace cursor near this point on the x-axis, zoom in, and repeat. Once

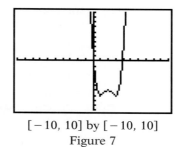

$[-10, 10]$ by $[-10, 10]$
Figure 7

you feel confident using the zooming process, you may wish to increase the magnification factors to 6 or 8, thus moving to greater accuracy at a faster pace. In any case, after a few zooms, you should obtain the solution $x \approx 3.909$.

∎

SOLVING EQUATIONS SIMULTANEOUSLY

We consider next a task already discussed in Section 1.3, that of simultaneously solving two equations in two unknowns.

∎ **Example 4.** Find the common solutions of the following two equations.

$$y = x^3 - x + 2 \qquad y = 0.1x^4 + x - 4$$

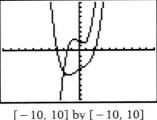

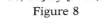

[−10, 10] by [−10, 10]

Figure 8

Solution. A solution consists of an ordered pair (x, y) satisfying both equations. Alternatively, it consists of a point (x, y) on the graphs of both equations. A graphics calculator allows us to overlay graphs, so we can show both graphs in the same viewing window (Figure 8). We note one solution near $x = -4$, and we suspect there may be another off the screen to the right of $x = 3$. Using the zooming process repeatedly, we quickly identify the first solution as $(-2.030, -4.332)$.

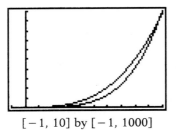

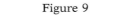

[−1, 10] by [−1, 1000]

Figure 9

To look for another solution, we change the range values to [−1, 10] by [−1, 100] and regraph. This still does not show a second intersection, but it suggests there may be one higher up. After setting the y-range to $-1 \le y \le 1000$ and regraphing, we obtain Figure 9, which shows an intersection in the upper right corner of the screen. Repeated zooming identifies the corresponding solution as $(9.857, 949.795)$. Could there be other solutions still farther off the screen? Knowledge of the general behavior of cubics and quartics (of which we will have more to say in Chapter 2) says no.

∎

AN APPLICATION

Figure 10

Propane gas is often stored in tanks that have the shape of a cylinder with hemispherical ends (Figure 10). The volume of such a tank is the volume of the cylinder ($V = \pi r^2 h$) plus the volume of the two ends, which together make a sphere ($V = (4/3)\pi r^3$).

∎ **Example 5.** Determine the radius x of a propane tank if it is to have a volume of 300 cubic feet and the cylindrical part is to be 10 feet long.

Solution. Our task is to solve the equation

$$\frac{4}{3}\pi x^3 + 10\pi x^2 = 300$$

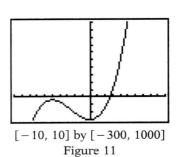

[−10, 10] by [−300, 1000]

Figure 11

To do this, we graph $y = (4/3)\pi x^3 + 10\pi x^2 - 300$ and look for its intersection with the x-axis. Figure 11 suggests that x should be a little less than 3 feet. The zooming process gives $x \approx 2.656$.

∎

TEASER SOLUTION

Denote the legs of the right triangle by x and y (Figure 12). The hypotenuse has length $\sqrt{x^2 + y^2}$. Because the area is 25, $(1/2)xy = 25$, and so $y = 50/x$. Since the perimeter is also 25,

$$x + \frac{50}{x} + \sqrt{x^2 + \frac{2500}{x^2}} = 25$$

or equivalently,

$$x + \frac{50}{x} + \sqrt{x^2 + \frac{2500}{x^2}} - 25 = 0$$

When we solve this equation by graphing using the zooming process, we obtain $x \approx 8.851$. This in turn gives $y = 50/x \approx 5.649$. You should check that these values do satisfy the required conditions.

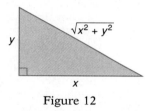

Figure 12

PROBLEM SET 1.4

A. Skills and Techniques

Draw a calculator graph of each equation making sure to display enough of the graph for all essential features to be apparent. Indicate the range values you use.

1. $y = x^3 + x^2 - 3$
2. $y = 0.2x^3 - x^2 + 2$
3. $y = -0.2x^3 - x^2 + 2$
4. $y = x^3 - 3x + 2$
5. $y = x^4 - x^2 - 20$
6. $y = -0.1x^4 + x^2 + 30$
7. $y = 0.1x^4 - x^2 + 2x - 3$
8. $y = x^4 - x^3 - 3x^2 + x - 2$
9. $y = 0.05x^6 - 15x^2 + 5x - 3$
10. $y = 0.2x^5 - 3x^2 + 5x - 2$
11. $y = 40 - 20/(1 + x^2)$
12. $y = -x^4 + (x - 5)^3$
13. $y = \sqrt{x^3} - (x - 3)^2$
14. $y = -\sqrt{40x} + x^2 - 9$

As in Example 3, find the largest solution of each equation (accurate to three decimal places). Note that Problems 15–28 are related to Problems 1–14.

15. $x^3 + x^2 - 3 = 0$
16. $0.2x^3 - x^2 + 2 = 0$
17. $-0.2x^3 - x^2 + 2 = 0$
18. $x^3 - 3x + 2 = 0$
19. $x^4 - x^2 - 20 = 0$
20. $-0.1x^4 + x^2 + 30 = 0$
21. $0.1x^4 - x^2 + 2x - 3 = 0$
22. $x^4 - x^3 - 3x^2 + x - 2 = 0$
23. $0.05x^6 - 15x^2 + 5x - 3 = 0$
24. $0.2x^5 - 3x^2 + 5x - 2 = 0$
25. $20/(1 + x^2) = 40$
26. $(x - 5)^3 = x^4$
27. $(x - 3)^2 = \sqrt{x^3}$
28. $x^2 - 9 = \sqrt{40x}$

In Problems 29–34, find all common solutions of the given pairs of equations, accurate to two decimal places.

29. $y = x^3 - x^2 - 3$ and $y = x^2 - 6$
30. $y = 0.2x^3 - x^2 + 2$ and $y = x^2 + x - 1$
31. $y = -x^3$ and $y = x^4 - x^2 - 5$
32. $y = x^3 + x + 2$ and $y = x^4 - x^2 - 12$
33. $y = \sqrt{x^3} - (x - 3)^2$ and $y = 10 + (0.2x - 0.4)^2$
34. $y = -\sqrt{40x} + x^2 - 9$ and $y = (x - 8)^4$
35. A cylindrical pail of radius x inches and height 10 inches was full of water. Its contents were poured into a cubical tank of side $2x$ units. It took 18 more cubic inches of water to fill this tank. Determine x.
36. A sphere of radius x and a cube of side $x + 6$ have the same volume. Determine x.

B. Applications and Extensions

37. Explain why you know (without graphing) that $x^6 = x^2 - 4$ has no real solutions. Graph $y = x^6$ and $y = x^2 - 4$ to confirm your thinking.
38. Solve the equation of the Teaser Solution algebraically to confirm the book's answer. (*Hint:* Begin by writing in the form $f(x) = \sqrt{g(x)}$. Then square both sides.)
39. The area and perimeter of a rectangle are both 50. Determine the dimensions of the rectangle.
40. Two boats, starting from the same point at the same time and traveling at right angles are 60 miles apart after 2 hours. If one boat travels 6 miles per hour faster than the other, what is the speed of the slower boat?
41. Squares of side x inches are cut from the four corners of a rectangular piece of tin 20 inches by 24 inches (Figure 13). The sides are then turned up to form a box of volume 650 cubic inches. Determine x.

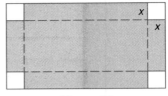

Figure 13

42. The outer dimensions of a closed rectangular box are 6, 8, and 10, all in inches. The box is made of wood x inches thick. If the box holds exactly 400 cubic inches of wild rice, determine x.
43. A spherical shell has thickness 2 centimeters. Determine the outer radius of the shell if the volume of the shell and of the space inside the shell are identical.
44. **Challenge.** One corner of a long strip of paper 16 inches wide is folded over to the opposite edge to determine triangle ABC as shown in Figure 14. Derive a formula for the area of this triangle in terms of x, the length of side AB. Then determine x so that the area is 15 square inches.

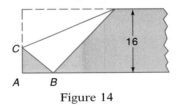

Figure 14

TEASER   Professor Snodgrass has promised a B to anyone who achieves an average score for all exams between 80% and 90%. On the four 1-hour exams, Maria has percentage scores of 88, 82, 81, and 91. Between what two scores must she place on the final exam to achieve a B, assuming the final exam counts as twice an hour exam? Can she still earn an A in the course?

1.5   INEQUALITIES AND ABSOLUTE VALUES

The order relations $<$ and $\leq$ were introduced in Section 1.2. Expressions such as $3x - 2 < 0$ and $-1 \leq 2x \leq 4$ are called **inequalities.** As with equations, our first task is to learn how to solve inequalities. To **solve** an inequality means to determine the values of x that make the inequality true. To do so we usually reduce it to an equivalent inequality of a simple form, typically of a form such

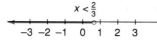

$$x < \frac{2}{3}$$

-3 -2 -1 0 1 2 3

$$-3 \le x < 2$$

-3 -2 -1 0 1 2 3

Figure 1

as $x < 2/3$ or $-3 \le x < 2$. These latter inequalities specify intervals on the real line (Figure 1). An open circle indicates that a point is not included, whereas a filled circle indicates that it is included. In contrast to the situation with an equation in x, where the solution set is usually one number or a small set of numbers, the solution set for an inequality in x is usually an interval of numbers (or perhaps a union of such intervals). But just as there are for solving equations, there are both algebraic and geometric methods for solving inequalities.

SOLVING INEQUALITIES ALGEBRAICALLY

One important tool in solving inequalities is the set of properties of $<$ stated in Section 1.2 and reproduced in the box in the margin. With these properties, we can solve any linear inequality, that is, any inequality of the form $ax + b < 0$.

■ **Example 1.** Solve the inequality $3x - 2 < 0$ and show its solution set on the real line.

Solution. Here are the steps.

$$3x - 2 < 0$$
$$3x < 2 \qquad \text{(adding 2 to both sides)}$$
$$x < \frac{2}{3} \qquad \text{(dividing both sides by 3)}$$

The corresponding interval on the real line is displayed on the first graph in Figure 1. ■

■ **Example 2.** Solve the double inequality $6/7 \le (3x + 15)/7 < 3$ and show its solution set on the real line.

Solution. The given inequality can be rewritten as follows:

$$6 \le 3x + 15 < 21 \qquad \text{(multiplying by 7)}$$
$$-9 \le 3x \qquad < 6 \qquad \text{(subtracting 15)}$$
$$-3 \le x \qquad < 2 \qquad \text{(dividing by 3)}$$

The corresponding set on the real line appears as the second graph in Figure 1. ■

Solving higher degree inequalities algebraically is more difficult because it depends on our ability to factor.

■ **Example 3.** Solve the inequality $x^2 + x - 6 < 0$.

Solution. The given inequality is equivalent to $(x + 3)(x - 2) < 0$. A product is negative if and only if one factor is positive and the other is negative. We organize the possibilities in a sign chart (Figure 2), which shows the sign of each factor.

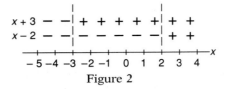

Figure 2

From this sign chart, we see that the two factors have opposite signs precisely on the interval $-3 < x < 2$. ∎

■ **Example 4.** Solve the inequality $(x + 3)(x - 1)^2/(x - 4) \geq 0$ and show the solution set on the real line.

Solution. Again we make a chart (Figure 3) showing the sign of each factor.

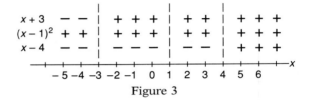

Figure 3

From this chart we conclude that the solution set consists of the union of the two intervals $x \leq -3$ and $x > 4$, together with the point 1. Be sure to note why -3 and 1 are included but 4 is not. The zero points must always be checked carefully. In set notation, we could write the solution set as

$$\{x: x \leq -3\} \cup \{x: x > 4\} \cup \{1\}$$

where the $\cup$ symbol indicates a union. This set is shown geometrically in Figure 4. ∎

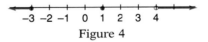

Figure 4

SOLVING INEQUALITIES GEOMETRICALLY

The algebraic method described above works fine for factored inequalities but fails miserably in other situations. The geometric method we describe now works beautifully in most cases. We begin with an example much like Example 3.

■ **Example 5.** Solve the inequality $x^2 + x - 8 < 0$.

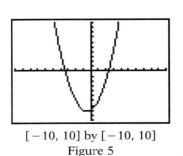

$[-10, 10]$ by $[-10, 10]$

Figure 5

Solution. We use our calculator to graph $y = x^2 + x - 8$ (Figure 5). The solution set is the x-interval where $y < 0$, that is, the interval where the graph drops below the x-axis. We can see from Figure 5 that the interval $-3.5 < x < 2.5$ is approximately what we want. To achieve better accuracy, we use the zooming process described in Section 1.4. A more accurate solution is the interval $-3.372 < x < 2.372$. ■

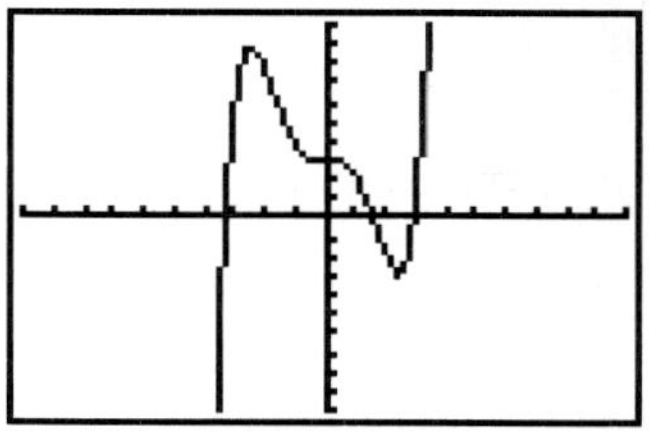

Of course, we do not need to use the geometric method in Example 5, because the quadratic formula gives us the exact endpoints of the solution interval. They are $(-1 \pm \sqrt{33})/2$. However, no such simple algebraic formula exists for our next example.

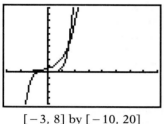

■ **Example 6.** Solve the inequality $0.1x^5 - x^3 + 3 \geq 0$.

Solution. Using a calculator, we graph

$$y = 0.1x^5 - x^3 + 3$$

Of course, what we show in Figure 6 is only part of the graph but a little experimenting convinces us that the left and right branches really do keep going downward and upward, respectively. This graph leads to our first estimate of the solution; it consists of the union of the two intervals $-3.3 \leq x \leq 1.5$ and $x \geq 3$. The zooming procedure produces the more accurate intervals $-3.292 \leq x \leq 1.589$ and $x \geq 2.977$. ■

■ **Example 7.** Solve the inequality $x^5 > 4^x$.

Solution. Begin by graphing $y = x^5$ and $y = 4^x$ using the range values $[-3, 8]$ by $[-10, 20]$ (Figure 7). This graph could easily mislead you, for it suggests that the inequality holds for all x greater than about 1.5. Experimenting with larger range values, however, may convince you that these two curves eventually intersect to the right of $x = 1.5$. In fact, if you set Ymax = 50,000, you will discover an intersection of the two curves at about $x = 7$. To pin down the solution, you may use zooming to approximate the x-coordinates of the intersection points. Alternatively, you may use zooming to approximate the solutions of $x^5 - 4^x = 0$ (see the box titled Alternatives). You should obtain as the solution to the inequality the interval $1.527 < x < 7.038$. ■

ABSOLUTE VALUES AND INEQUALITIES

We introduced the absolute value symbol in Section 1.3. Recall that $|x| = x$ if $x \geq 0$ and $|x| = -x$ if $x < 0$. Thus, $|5.2| = 5.2$, but $|-5.2| = -(-5.2) = 5.2$. Clearly $|x|$ is always a nonnegative number. Here are the most important properties of absolute value.

Figure 6: $[-10, 10]$ by $[-10, 10]$

Figure 7: $[-3, 8]$ by $[-10, 20]$

ALTERNATIVES

To solve $f(x) = g(x)$ geometrically, two methods suggest themselves.

1. Graph $y = f(x)$ and $y = g(x)$ in the same plane. Determine the x-coordinates of their intersection points, using zooming.

2. Graph $y = f(x) - g(x)$ and determine the x-coordinates of its intersections with the x-axis, using zooming.

It is important to understand both methods. Sometimes one is easier to execute than the other.

PROPERTIES OF $|\ldots|$
1. $\quad |-x| = |x|$
2. $\quad |xy| = |x||y|$
3. $\quad |x/y| = |x|/|y|$
4. $\quad |x + y| \leq |x| + |y|$ (Triangle inequality)

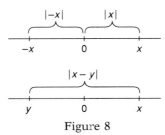

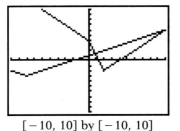

[−10, 10] by [−10, 10]

Figure 9

Geometrically, $|x|$ is the distance between x and 0 on the real line. More generally, $|x - y|$ is the distance between x and y on the real line (Figure 8). Thinking geometrically leads to the following facts.

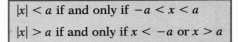

$|x| < a$ if and only if $-a < x < a$

$|x| > a$ if and only if $x < -a$ or $x > a$

■ **Example 8.** Rewrite each inequality without using the absolute value symbol.

(a) $|x| \leq 4$ (b) $|x - 3| < 5$ (c) $|x + 2| > 3$ (d) $\left|\dfrac{x}{3} - 2\right| \leq 2.4$

Solution.
(a) $-4 \leq x \leq 4$
(b) Think geometrically. The distance between x and 3 is less than 5. Thus, $-2 < x < 8$. Alternatively, rewrite $|x - 3| < 5$ as $-5 < x - 3 < 5$ and then add 3 to all three members.
(c) Note that $|x + 2| = |x - (-2)|$. The distance between x and -2 is greater than 3. Thus, $x < -5$ or $x > 1$.
(d) Rewrite the given inequality as $-2.4 \leq x/3 - 2 \leq 2.4$. Then add 2 to get $-0.4 \leq x/3 \leq 4.4$. Finally, multiplying by 3 gives $-1.2 \leq x \leq 13.2$. ■

■ **Example 9.** Solve the inequality $|2x - 4| - |x| < |0.5x + 4| - 3$.

Solution. It is possible but tedious to solve this inequality algebraically by dividing it into four problems. To do this, split the real line into four intervals using the points where each absolute value is 0 (namely 2, 0, and -8); then solve the inequality on each interval.

We prefer to graph $y = |2x - 4| - |x|$ and $y = |0.5x + 4| - 3$ in the same plane (Figure 9) and observe that the first graph is below the second on the approximate x-interval $0.8 < x < 9.6$. (Note that TI calculators denote $|\ldots|$ with ABS; some calculators do not have a built-in absolute value.) With zooming, we can improve the solution to $0.857 < x < 10.000$. ■

An Application

In the real world, perfectly constructed objects do not exist. A manufacturer can only hope to meet requirements within a specified error.

■ **Example 10.** A manufacturer is required to make metal balls having a volume of 16 cubic centimeters with an error of at most 0.10 cubic centimeters. Determine the possible values for the radius x.

Solution. Recalling that the formula for the volume of a ball of radius x is $V = (4/3)\pi x^3$, we state the condition that must be satisfied as

$$\left|\frac{4}{3}\pi x^3 - 16\right| \leq 0.1$$

This condition is equivalent to

$$-0.1 \leq \frac{4}{3}\pi x^3 - 16 \leq 0.1$$

We transform this to a solution using the following steps.

$$-0.3 \leq 4\pi x^3 - 48 \leq 0.3$$

$$47.7 \leq \quad 4\pi x^3 \leq 48.3$$

$$\frac{47.7}{4\pi} \leq \quad x^3 \leq \frac{48.3}{4\pi}$$

$$\sqrt[3]{\frac{47.7}{4\pi}} \leq \quad x \leq \sqrt[3]{\frac{48.3}{4\pi}}$$

$$1.560 \leq \quad x \leq 1.566 \qquad \blacksquare$$

TEASER SOLUTION

The final exam is to be weighted twice as heavily as the hour exams, so we will give it weight 1/3 and the four hour exams weight 1/6. Thus, for Maria to get a B, we want the following inequality to hold.

$$80 < \frac{1}{6}(88) + \frac{1}{6}(82) + \frac{1}{6}(81) + \frac{1}{6}(91) + \frac{2}{6}x < 90$$

This inequality can be successively transformed as indicated.

$$80 < \frac{342 + 2x}{6} < 90$$

$$480 < 342 + 2x < 540$$

$$138 < \quad 2x < 198$$

$$76 < \quad x < 99$$

A score between 76 and 99 will give Maria a B, and a 99 or better will give her an A.

PROBLEM SET 1.5

A. Skills and Techniques

Solve the given inequality algebraically and show its solution set on the real line.

1. $5x - 2 > 6$
2. $3x - 3 < 9$
3. $4x - 6 \geq 2x + 1$
4. $2x + 3 \geq \frac{7}{3} - x$

5. $3x + \frac{2}{3} \leq \frac{5}{2}x - 1$
6. $0.4x + 1.9 < 1.2x - 3.5$

Solve the given double inequality algebraically and show its solution on the real line.

7. $-6 \leq \frac{2}{3}x + 2 < 5$
8. $1 \leq \frac{1}{2}x - 2 \leq 2$

9. $-3 < \dfrac{4x + 11}{3} < 0$

10. $\dfrac{3}{5} < \dfrac{2x - 5}{5} < 9$

11. $x - 5 < 5x + 3 < x$

12. $x \leq \dfrac{4x - 2}{3} < x + 1$

Use a sign chart to solve each inequality. See Examples 3 and 4.

13. $(x - 3)(x - 5) > 0$
14. $x^2 - x \leq 0$
15. $x^2 + 5x - 6 \leq 0$
16. $x^2 - 8x - 9 > 0$
17. $x^3(x + 3) > 0$
18. $(x^3 + 2x)(x + 4) \leq 0$
19. $\dfrac{x(x + 1)}{x - 2} \geq 0$
20. $(x^2 - 3x + 2)(x^2 - 4) < 0$
21. $(1 - x)(x - 4)(x + 3) < 0$
22. $\dfrac{x}{x^2(x - 1)} \geq 0$

Solve each inequality graphically as in Examples 5 and 6. Give your answer accurate to three decimal places.

23. $x^2 + 2x - 2 < 0$
24. $2x^2 + 2x - 11 > 0$
25. $x^3 - 5x + 5 > 0$
26. $x^4 - 2x^2 - 1 \leq 0$
27. $\dfrac{x^4}{7} - \dfrac{x^3}{5} + x^2 + x \leq 0$
28. $x^3 + 0.1x^2 - 0.5x + 0.15 > 0$
29. $x^5 - 2x^4 + 30x^3 - 44x^2 + 5x \geq 0$
30. $x^4 - 2x^2 + (0.5)^x > 0$

Solve each inequality accurate to two decimal places.

31. $-x^3 - x^2 > -5x + 4$
32. $(x - 2)^4 < x^2 + 2x$
33. $x^3 - x < \dfrac{3}{2x}$
34. $11x^3 - 5x < x^4 + 2x$
35. $x^3 < 2^x$
36. $x^3 < 1.5^x$
37. $x^3 > 3^x$
38. $\pi^x < |x|^\pi$

Write each inequality without use of absolute values.

39. $|3x| \leq 9$
40. $|x| \geq 4$
41. $|x + 1.5| < 2.5$
42. $|x - 4| < 1.3$
43. $|2x - 9| > 4$
44. $\left| \dfrac{2}{3}x - 6 \right| \leq 2$

45. $|0.9x - 1.2| \geq 2.1$

46. $\left| 3x - \dfrac{4}{3} \right| > \dfrac{2}{3}$

Solve graphically.

47. $|2x - 1| \leq |3x + 5|$

48. $\left| \dfrac{2}{3}x - 1 \right| \geq |3x + 2| - 5$

49. $|2x| + |x - 2| + |x + 3| < x^2$

50. $|x^2 + 2x - 5| \leq 8 - \left(\dfrac{x}{2} \right)^2$

Problems 51–54 are related to Example 10.

51. The volume of a sphere has been measured as 72 cubic centimeters with an error of at most 3%. Specify the range for the radius x.

52. You are instructed to make a cylindrical barrel with height twice its diameter. The barrel must be able to hold between 15 and 20 cubic feet of water. What range of barrel heights can you use?

53. If you measure the length of a side of a cube to be 3.45 feet with a possible error of 0.02 feet, what can you say about the volume V?

54. If you measure the radius of a sphere to be 5.25 centimeters with an error of at most 0.03 centimeters, what are the limits on its surface area S? Recall that $S = 4\pi r^2$.

B. Applications and Extensions

55. Second State Bank now employs eight tellers with an average annual salary of $32,000, but it plans to add an additional teller. What range of salary can be offered if the president says the average teller salary must be between $30,000 and $33,000?

56. Longview State College's football stadium, which seats 4500 people, always sells out for home games. Ticket prices are $5 in advance or $8 at the gate. How many tickets should be reserved for gate sales if the school needs to gross at least $24,000 per game?

57. A baseball thrown straight up with an initial velocity of 110 feet per second from an initial height of 50 feet will have height $s = -16t^2 + 110t + 50$ after t seconds. During what time interval will the ball's height be above 150 feet? Between heights 150 and 200 feet?

58. Bill cut a 30-meter rope into two pieces by making the cut x meters from one end. With the piece of length x he formed a square and with the other a circle. What values of x will ensure that the combined areas of the square and the circle will exceed 50 square meters?

59. Company A will loan out a car for $35 per day plus 10¢ per mile, whereas Company B charges $30 per day plus 12¢ per mile. I need a car for 5 days. For what range of mileage will I be ahead financially by renting from Company B?

60. Maureen's bicycle has 27 inch diameter tires. Her tenth gear has a gear ratio of 3.2 to 1, which means her tires make 3.2 revolutions for each revolution of the pedals. What is her range of speeds in tenth gear if she is able to keep her pedal pace between 50 and 70 revolutions per minute? Give your answer in miles per hour (1 mile is 5280 feet).

61. Three circles A, B, and C have their centers at (4, 6). Circle A is tangent to the y-axis, circle B is tangent to the x-axis, and circle C goes through the origin.
(a) For what values of x does $(x, 3)$ lie inside A?
(b) For what values of y does $(6, y)$ lie inside B?
(c) For what values of x does $(x, 2x)$ lie outside C?

62. For what positive numbers is it true that the sum of a number and its reciprocal is greater than 2?

63. Andy Smart knows that one arm of his two-pan balance is slightly longer than the other, so it does not weigh accurately. When a customer orders 2 pounds of coffee, he puts a 1-pound weight in the left pan, balances it with coffee in the right pan, and then pours the coffee into a sack. Next he puts the 1-pound weight in the right pan, balances it with coffee in the left pan, and pours that coffee into the sack. That is 2 pounds of coffee, he says. Show that he actually gives the customer more than 2 pounds. (*Hint:* Let the length of one arm of the balance be 1 unit and the other $1 + a$ units. Determine how much coffee is actually weighed out each time.)

64. **Challenge.** Consider the familiar right triangle with sides 3, 4, and 5. Of course, $3^2 + 4^2 = 5^2$. Calculate $3^3 + 4^3$, $3^4 + 4^4$, . . . and compare it with 5^3, 5^4, ... Make a conjecture about an inequality between $3^n + 4^n$ and 5^n for an integer $n > 2$. Make a similar conjecture about a right triangle with sides a, b, and c. Prove your conjecture.

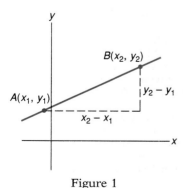

TEASER A wheel whose rim has the equation $x^2 + y^2 = 25$ is spinning rapidly counterclockwise. A speck of dirt leaves the rim at (3, 4) and flies off along the tangent line. How high up does it hit on the wall $x = -9$? Distances are in feet.

1.6 LINES

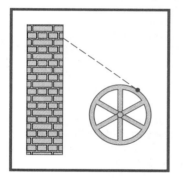

Figure 1

We have been using the word *line* without actually defining this term. We do not attempt a definition here either except to say that for us the term **line** will always mean what Euclid called a straight line, and we always assume that a line stretches infinitely far in both directions. The part of a line between two points is called a **line segment;** a line segment is the shortest curve connecting those two points. What we want to do in this section is to study some of the properties of lines and their segments; in particular, we will develop various forms for the equation of a line.

THE SLOPE OF A LINE

Place a line in the coordinate system and pick two distinct points $A(x_1, y_1)$ and $B(x_2, y_2)$ on it (Figure 1). From point A to point B, there is a **rise** (vertical

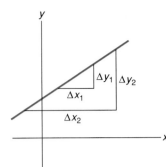

Figure 2

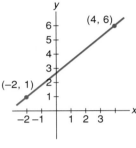

Figure 3

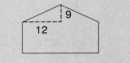
change) of $\Delta y = y_2 - y_1$ units and a **run** (horizontal change) of $\Delta x = x_2 - x_1$ units. We define the **slope** of the line to be

$$m = \frac{\text{rise}}{\text{run}} = \frac{\Delta y}{\Delta x} = \frac{y_2 - y_1}{x_2 - x_1}$$

Several questions immediately arise. Does the number m depend on which points A and B we pick? The answer is no; for as we know from elementary geometry, the ratios of corresponding sides in similar triangles are equal (see Figure 2). Note that it doesn't even matter if A is to the left or the right of B since

$$\frac{y_2 - y_1}{x_2 - x_1} = \frac{y_1 - y_2}{x_1 - x_2}$$

What is important is that we take the coordinates in the same order in both numerator and denominator.

Another obvious question to ask is what we do if the line is vertical, in which case A and B will have the same x-coordinates, making the denominator 0 in the expression for m. The answer is: We leave the notion of slope undefined for vertical lines.

■ **Example 1.** Determine the slope of the line (Figure 3) that passes through the points $(-2, 1)$ and $(4, 6)$.

Solution. A simple calculation gives

$$m = \frac{6 - 1}{4 - (-2)} = \frac{5}{6}$$

This means that this line always rises 5 units in a run of 6 units, it rises 2.5 units in a run of 3 units, and so on. ■

Just a little thought convinces us that a line rising to the right has positive slope and that the steeper the line the larger the slope. A horizontal line has 0 slope. A line falling to the right has negative slope. Figure 4 shows the slopes for several lines all going through the point $(2, 1)$.

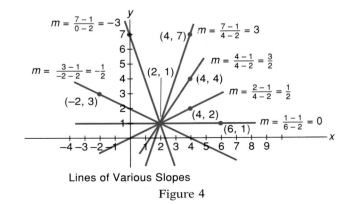

Lines of Various Slopes

Figure 4

THE POINT-SLOPE FORM

Our principal goal is to obtain equations for lines. Consider again the line passing through $(-2, 1)$ and $(4, 6)$, which is shown in Figure 3. Consider any other point (x, y) on this line. Calculating the slope using the points $(-2, 1)$ and (x, y) must give the slope 5/6 that we obtained in Example 1, that is,

$$\frac{y - 1}{x + 2} = \frac{5}{6}$$

or equivalently,

$$y - 1 = \frac{5}{6}(x + 2)$$

We call this the point-slope form for the equation of this line. Some will quibble with our use of the word *the* since we would obtain the different-looking equation $y - 6 = (5/6)(x - 4)$ if we use the point $(4, 6)$ in our equation in place of $(-2, 1)$. These equations are equivalent, however, because both simplify to $5x - 6y = -16$.

What we have done in this special case can be done in general. The **point-slope form** for the equation of a line passing through (x_1, y_1) with slope m is

$$\boxed{y - y_1 = m(x - x_1)}$$

■ **Example 2.** Find the point-slope form for the equation of the line passing through $(-4, 3)$ and $(1, -4)$. Then determine the y-coordinate of the point on this line whose x-coordinate is 13.

Solution. First we calculate the slope to be

$$m = \frac{-4 - 3}{1 - (-4)} = \frac{-7}{5} = -\frac{7}{5}$$

Thus the equation of the line is $y - 3 = -(7/5)(x + 4)$ and the y-coordinate of the point with x-coordinate 13 is

$$y = 3 - \frac{7}{5}(13 + 4) = -20.8$$

■

THE SLOPE-INTERCEPT FORM

Suppose a line crosses the y-axis at q (technically at $(0, q)$) and has slope m (Figure 5). Applying the point-slope form to this situation yields $y - q = m(x - 0)$, that is,

$$\boxed{y = mx + q}$$

This is called the **slope-intercept form** for the equation of a line. Whenever we see an equation written in this form, we know it is the equation of a line with slope m and y-intercept q.

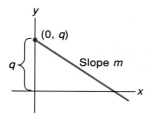

Figure 5

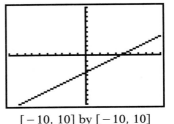

$[-10, 10]$ by $[-10, 10]$

Figure 6

■ **Example 3.** Find the slope and y-intercept of the line with equation $3x - 4y = 14$. Then graph this line.

Solution. We solve this equation for y, obtaining $y = (3/4)x - 14/4$. The slope is 3/4 and the y-intercept is $-7/2$. Our calculator gives Figure 6 as the graph.

■

THE GENERAL LINEAR FORM

Unfortunately, vertical lines do not have equations of either the point-slope or slope-intercept form because they do not have slopes. But vertical lines do have equations. For example, the vertical line with x-intercept a has the equation $x = a$. It would be nice if there were a form that covered the equations of all lines. There is; it is the **general linear form**

$$Ax + By + C = 0$$

The equation of a nonvertical line can be written in this form ($mx - y + q = 0$) and so may the equation of a vertical line ($x + 0y - a = 0$).

A useful fact about lines with equations in general linear form is the formula for the distance d from the point (x_1, y_1) to the line $Ax + By + C = 0$, namely,

$$d = \frac{|Ax_1 + By_1 + C|}{\sqrt{A^2 - B^2}}$$

We won't prove this **point-line formula** now (it's a little tricky); rather we choose to illustrate it.

■ **Example 4.** Find the distance from the point $(-2, 3)$ to the line with equation $y = 4x - 7$.

Solution. First we write the equation in the general linear form $4x - y - 7 = 0$. Then

$$d = \frac{|4(-2) - 3 - 7|}{\sqrt{4^2 + (-1)^2}} = \frac{18}{\sqrt{17}} \approx 4.37$$

■

PARALLEL AND PERPENDICULAR LINES

You should not be surprised by the following fact.

Two nonvertical lines are parallel if and only if their slopes are equal.

But you may be surprised at the next statement.

Two nonvertical lines are perpendicular if and only if their slopes m_1 and m_2 are negative reciprocals; that is, if and only if $m_2 = -1/m_1$.

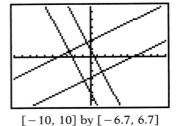

$[-10, 10]$ by $[-6.7, 6.7]$

Figure 7

To provide evidence for these facts, we have drawn in Figure 7 calculator graphs of $y = 0.5x - 3$, $y = 0.5x + 2$, each with slope 0.5, and $y = -2x + 1$,

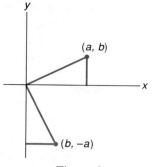

Figure 8

CALCULATOR HINT

Windows with the same size units in the x and y directions can be achieved on the TI-81 and TI-82 by selecting **ZSquare** from the ZOOM menu.

$y = -2x - 5$, each with slope -2. Our viewing window has units of approximately equal length on the x- and y-axes (otherwise, perpendicularity doesn't make sense). Note that lines in the first pair are parallel as are those in the second pair and that lines in the second pair appear to be perpendicular to those in the first pair. Figure 8 suggests a way of proving the perpendicularity criterion.

■ **Example 5.** (a) Find the distance between the parallel lines $3x - 4y = 7$ and $-6x + 8y = 16$. (b) Then find the equation of the line perpendicular to both of them that passes through $(-2, 3)$.

Solution.

(a) That the two given lines are parallel is clear when you solve them for y, because this shows that both have slope 3/4. To use the point-line formula given above, we first find a point (any point) on the second line. Substitute a value for x, say $x = 0$, and solve for y, thus obtaining the point $(0, 2)$ on the second line. Now apply the formula to obtain

$$d = \frac{|3(0) - 4(2) - 7|}{\sqrt{3^2 + (-4)^2}} = \frac{15}{5} = 3$$

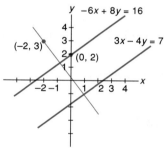

Figure 9

(b) The given lines have slope 3/4, so a line perpendicular to them will have slope $-4/3$. The line through $(-2, 3)$ with this slope has equation (point-slope form) $y - 3 = -(4/3)(x + 2)$. In general linear form, we have $4x + 3y - 1 = 0$.

Figure 9 shows all three lines ■

THE MIDPOINT FORMULA AND ITS GENERALIZATIONS

Our next goal is to obtain the formula for the midpoint M of the line segment joining the points $A(x_1, y_1)$ and $B(x_2, y_2)$. Look at Figure 10. It should be apparent to you that M has coordinates that are the averages of the coordinates for the two points; that is,

$$M = \left(\frac{1}{2}(x_1 + x_2), \frac{1}{2}(y_1 + y_2) \right)$$

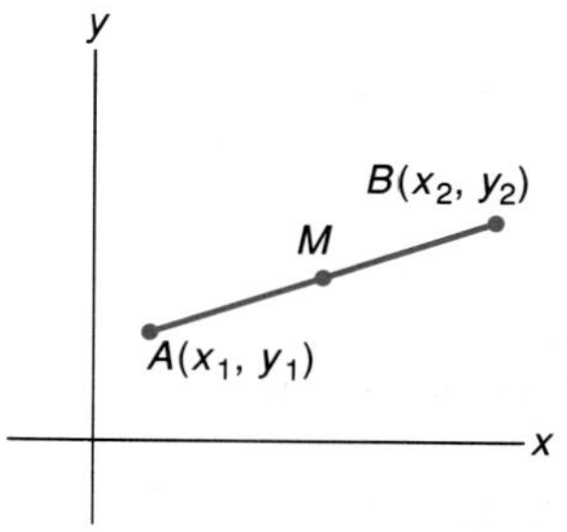

Figure 10

This is called the **midpoint formula.**

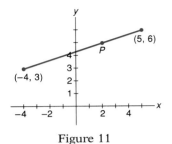

Figure 11

More generally, let w be a number between 0 and 1. We may be interested in the point P on this line segment that is the $(1 - w)$th fraction of the way from $A(x_1, y_1)$ to $B(x_2, y_2)$. We want to give weight w to the first point and weight $1 - w$ to the second point. Thus we have the formula

$$P = (wx_1 + (1 - w)\,x_2,\ wy_1 + (1 - w)\,y_2)$$

■ **Example 6.** Find the point P two-thirds of the way from $(-4, 3)$ to $(5, 6)$. See Figure 11.

Solution. Using the formula immediately above, we see that

$$P = \left(\frac{1}{3}(-4) + \frac{2}{3}(5), \frac{1}{3}(3) + \frac{2}{3}(6)\right) = (2, 5)$$

■

AN APPLICATION

Here is a business application that most of us are likely to face.

■ **Example 7.** Acme Car Rental charges a daily rate of \$27 plus 30¢ for each mile exceeding 100 miles. Beta Car Rental charges a daily rate of \$40 plus 31¢ for each mile exceeding 200. For both rental firms, write a formula for the cost C of renting a car for 5 days and driving x miles. Then evaluate C for a 5-day trip of length 2100 miles to determine which firm will give you the best price and by how much.

Solution. Here are the cost formulas.

$$\text{Acme: } C = \begin{cases} 5(27), 0 \le x \le 500 \\ 5(27) + 0.30(x - 500) = 0.30x - 15, x > 500 \end{cases}$$

$$\text{Beta: } C = \begin{cases} 5(40), 0 \le x \le 1000 \\ 5(40) + 0.31(x - 1000) = 0.31x - 110, x > 1000 \end{cases}$$

The costs for a 5-day trip of 2100 miles are as follows.

$$\text{Acme: } C = 0.30(2100) - 15 = \$615$$

$$\text{Beta: } C = 0.31(2100) - 110 = \$541$$

Thus, you will be \$74 better off by renting from Beta. ■

TEASER SOLUTION

Refer to Figure 12 and note that the tangent line to the circle at $(3, 4)$ is perpendicular to the radius through $(3, 4)$. The latter has slope $4/3$; thus the tangent line has slope $-3/4$. The equation of the tangent line is $y - 4 = -(3/4)(x - 3)$. Setting $x = -9$ in this equation gives $y = 4 - (3/4)(-9 - 3) = 13$. If we ignore the effect of gravity, the speck of dirt will hit the wall at a point 13 feet above the horizontal line through the axis of the wheel.

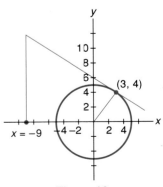

Figure 12

PROBLEM SET 1.6

A. Skills and Techniques

Determine the slope of the line that passes through each pair of points.

1. $(-5, 4)$ and $(4, 3)$
2. $(4, 0)$ and $(0, -3)$
3. $(\pi, 4)$ and $(\sqrt{2}, 4)$
4. $(-5, 4)$ and $(-5, \pi)$
5. The points on the graph of $y = x^2$ where $x = -2$ and $x = 3$.
6. The points on the graph of $y = x^3 - 4$ where $x = 1$ and $x = 3$.
7. The points where $y = x^3$ and $y = 2x^2$ intersect.
8. The points where $y = x^2$ and $x^2 + y^2 = 6$ intersect.

Determine the equation of the line that satisfies the following conditions. Write your answer in the slope-intercept form $y = mx + b$.

9. It goes through $(4, -5)$ and $(-3, 2)$.
10. It goes through $(\pi, 5)$ and $(\pi + 2, -12)$.
11. It goes through $(4, -5)$ and is parallel to $2x - 3y = 7$.
12. It goes through $(-2, -6)$ and is perpendicular to $5x + 2y = 7$.
13. It has the same y-intercept as $y = 3x - 2$ and is perpendicular to this line.
14. It is parallel to the line $4x - 3y = 1$ and has x-intercept 5.
15. It has $x = -6$ and $y = -4$ as its two intercepts.
16. It goes through $(-2, 6)$ and the intersection of $3x - 4y = -5$ with $2x + y = 4$.
17. It goes through $(1, -1)$ and the intersection of $y = 2x$ with $x^2 + y^2 = 10$ in the third quadrant.
18. It goes through $(6, -6)$ and is parallel to the x-axis.
19. Its y-intercept is 5 and it cuts off a region of area 4 in the first quadrant.
20. It is tangent to the circle $x^2 + y^2 = 25$ at $(3, -4)$.
21. It goes through the centers of the circles $x^2 + 2x + y^2 = 0$ and $x^2 - 4x + y^2 + 6y = 0$.
22. It goes through the intersections of $y = x^2 - 10x + 3$ and $y = -x^2 - 8x + 7$.

Determine each of the following distances.

23. Between $(-1, 2)$ and the line $3x - 4y = 7$.
24. Between $(3, -2)$ and the line $5x + 12y = 3$.
25. Between the parallel lines $5x + 12y = 3$ and $5x + 12y = 8$.
26. Between the parallel lines $y = 2x - 1$ and $y = 2x + 9$.
27. Between the line $y = -3x + 7$ and the origin.
28. Between the line $y = 2x - 1$ and the center of the circle $x^2 + y^2 - 4y = 0$.

Problems 29–32 are related to Example 6.

29. Find the point one-third of the way from $(1, 0)$ to $(7, 10)$.
30. Find the point three-fourths of the way from $(-2, 2)$ to $(6, -10)$.
31. Find the equation of the circle with the line segment connecting $(-2, 3)$ and $(6, 9)$ as diameter.
32. Find the equation of the line through the midpoint of the segment connecting $(-1, -3)$ and $(-5, 7)$ and perpendicular to this segment.
33. Jessica works for base salary of $1200 per month plus $40 for every computer she sells. She has been offered a similar job in another city paying $1000 per month plus $50 for each computer sold. Write formulas for her total monthly salary in each case and determine how many computers she must sell each month to make a move worthwhile.
34. ABC company is selling 400 dingbats a week at a price of $5 each. A research firm estimates that it can increase its weekly sales by 20 units for each 10¢ drop in the unit price. Write a formula for the number of units N it can sell per week in terms of the unit price x if $x \geq 400$ and estimate its sales when its unit price is set at $4.20.
35. The value of a house is increasing linearly. It was worth $70,000 in 1970 and $80,000 in 1990. What will it be worth in 2020?
36. A car is depreciating in value linearly. It cost $20,000 and was worth $12,000 3 years later. What will it be worth after an additional 4 years?

B. Applications and Extensions

37. Find the value of k for which the line $4x + ky = 6$
    (a) passes through $(-2, 1)$
    (b) is parallel to the y-axis
    (c) is parallel to the x-axis
    (d) is parallel to $6x - 9y = 11$
    (e) has equal x- and y-intercepts
    (f) is perpendicular to the line $2x - 6y = 7$
38. Find the value of k so that the line $kx - 3y = 20$
    (a) is parallel to the line $y = 2x - 9$
    (b) is parallel to the x-axis
    (c) is perpendicular to the line $y = 3x - 5$
    (d) is 5 units from the origin
39. Find the equation of the perpendicular bisector of the line segment connecting $(2, 3)$ and $(5, -7)$.
40. Find the center of the circle that circumscribes the triangle with vertices $(0, 8)$, $(6, 2)$, and $(12, 14)$. (*Hint:* The center of this circle is the intersection of the perpendicular bisectors of the sides.)
41. Show that the diagonals of a parallelogram bisect each other. (*Hint:* Place the parallelogram in the

coordinate system so that three of its vertices are $(0, 0)$, $(a, 0)$, and (b, c).)

42. Show that the line segment joining the midpoints of two sides of a triangle is parallel to the third side and half as long.

43. Find x and y so that $(-1, 2)$ is three-fourths of the way along the segment joining (x, y) to $(3, -7)$.

44. The tangent line to the circle $x^2 + y^2 = 25$ at $(3, 4)$ meets the x-axis at $(a, 0)$. Determine a.

45. A line through $(13, 0)$ is tangent to the circle $x^2 + y^2 = 26$ at (a, b) in the first quadrant. Determine a and b.

46. The larger roof section of the building in Figure 13 has a 5:12 pitch. Determine the total area of the roof.

47. The horizontal distance from A to B is 6000 meters. The road between them is straight and flat except for one hill shaped like an inverted vee, which has a grade of 15%. If this hill is 100 meters high, how far did Patrick run in going from A to B?

48. **Challenge.** Matilda spilled 1000 salt crystals (points) on the floor. Prove that there is a line with exactly 500 crystals on each side of it.

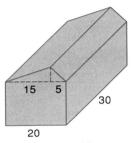

15   5

30

20

Figure 13

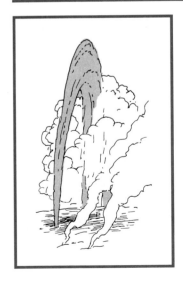

One of the chief goals of science is to discover relationships between variables. Relationships like that between the distance s in feet a body falls and the elapsed time t in seconds ($s = 16t^2$) or between the period T in seconds of a pendulum and its length d in centimeters ($T = 2\pi\sqrt{d/980}$) are well established. Other relationships like that between the score on a math test and the time spent studying for it are less precise but still worth thinking about. A standard problem in business is to relate the number of items sold to the price charged per item and then to relate the total profit to this price.

Many relationships that occur in the world of experience are nonlinear, but fortunately some of them are approximately linear. In this section, we will learn how to analyze experimental data that appear to have an approximately linear character. In particular, we will determine the line that "best" fits a set of data. This process, called **linear regression,** is easily carried out on a graphics calculator.

PLOTTING DATA

Suppose we are interested in knowing how a variable y (the **dependent variable**) depends on another variable x (the **independent variable**). For example, we might be interested in how the weight y of an adult man depends on his height x.

Our first task is to collect some data. We take a sample of size n and record the corresponding x- and y-values in a table, thus generating a set

$$\{(x_1, y_1),\ (x_2, y_2),\ (x_3, y_3),\ \ldots,\ (x_n, y_n)\}$$

of ordered pairs. Next, we plot this set of ordered pairs in the xy-plane, either by hand or preferably using a graphics calculator (most graphics calculators make this an easy task). A visual examination of the plotted data (called a **scatter plot**) may suggest a possible linear relationship, a possible nonlinear relationship, or no relationship at all (Figure 1).

Student number	Height x (inches)	Weight y (pounds)
1	66	170
2	73	179
3	67	164
4	63	129
5	68	142
6	60	128
7	69	184
8	66	152
9	70	176
10	76	223
11	63	144
12	70	159
13	71	195
14	70	162
15	68	162
16	72	188

Figure 2

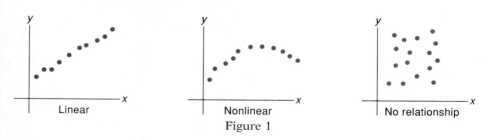

Figure 1

■ **Example 1.**   Wayne Shorter has measured the heights and weights of the 16 men on his dorm floor with the results shown in Figure 2. Make a scatter plot of these data with x as height and y as weight.

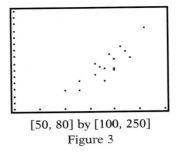

[50, 80] by [100, 250]

Figure 3

Solution. First enter the 16 pairs of *xy*-data (refer to your calculator's manual). Then select appropriate range values and plot the data. The resulting scatter plot is shown in Figure 3. Note that there is a tendency toward a linear relationship. ∎

LINEAR REGRESSION (LEAST SQUARES)

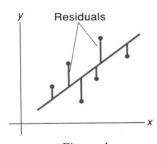

Figure 4

As we look at Figure 3, our inclination is to superimpose a straight line of positive slope with about half of the points on each side of it. Unfortunately, infinitely many different such lines can be drawn. Which of them is the line that best fits the data? The answer depends on the criterion we choose to determine "best." One choice would be to minimize the sum of the vertical distances (called the **residuals**) shown in Figure 4, but this leads to complicated formulas. The more popular choice is the criterion of **least squares.** According to this criterion, the best line $y = a + bx$ is the line that minimizes the sum of the squares of the residuals, that is, the line that minimizes

$$S = \Sigma(y_i - a - bx_i)^2$$

Here the symbol Σ is the standard mathematical symbol for summation, and in this case it stands for the sum of all n terms of the indicated form.

∎ **Example 2.** Calculate the sums S_1 and S_2 for line 1 ($y = x$) and line 2 ($y = 1 + x/2$) using the four data points shown in Figure 5, and determine which line fits the data better.

Solution. Line 1 has $a = 0$ and $b = 1$, so

$$S_1 = (1-0-1)^2 + (2-0-3)^2 + (2-0-4)^2 + (6-0-5)^2 = 6$$

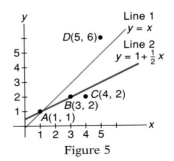

Figure 5

In Section 1.6, we wrote
the equation of a line in
the form $y = mx + q$
with m being the slope
and q the y-intercept.
Here we write the
corresponding equation
in the form

$$y = a + bx$$

We do this because the
TI-81 uses b and a rather
than m and q in
connection with linear
regression. The TI-82
uses this form but also
allows the user to choose
the form

$$y = ax + b$$

CALCULATOR HINT

To graph the data points
and the regression line
together, you should
enter the regression
equation (**RegEQ** from
the VARS menu) after
Y1 = in the Y= menu.
Then on the TI-82, press
STAT PLOT, turn on **Plot 1**,
and press GRAPH. On the
TI-81, select **Scatter**
from the STAT menu.

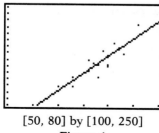

[50, 80] by [100, 250]

Figure 6

Line 2 has $a = 1$ and $b = 1/2$, so

$$S_2 = \left(1 - 1 - \frac{1}{2}\right)^2 + \left(2 - 1 - \frac{3}{2}\right)^2 + (2 - 1 - 2)^2 + \left(6 - 1 - \frac{5}{2}\right)^2 = 7.75$$

Since $S_1 < S_2$, we conclude that line 1 fits the data better than line 2. ∎

Example 2 still leaves unanswered the question of which of all lines fits the given data best. This question is most easily answered using the methods of calculus. Here, we simply state the results. The line that best fits the data using the criterion of least squares is the line $y = a + bx$, where b and a are determined by the formulas

$$b = \frac{n\Sigma x_i y_i - \Sigma x_i \Sigma y_i}{n\Sigma x_i^2 - (\Sigma x_i)^2} \qquad a = \frac{\Sigma y_i - b\Sigma x_i}{n}$$

■ **Example 3.** Determine the line that best fits the four data points of Figure 5, first by a hand calculation and then using a calculator.

Solution.

$$b = \frac{4(1 \cdot 1 + 3 \cdot 2 + 4 \cdot 2 + 5 \cdot 6) - (1 + 3 + 4 + 5)(1 + 2 + 2 + 6)}{4(1^2 + 3^2 + 4^2 + 5^2) - (1 + 3 + 4 + 5)^2}$$

$$= \frac{37}{35}$$

$$a = \frac{11 - (37/35)(13)}{4} = -\frac{24}{35}$$

Thus, the line that fits the data best is the line

$$y = -\frac{24}{35} + \frac{37}{35}x \approx -0.69 + 1.06x$$

A calculator will provide this information, once the data are entered, if you simply press the correct keys. Some calculators also automatically show $r \approx 0.81$. We will discuss that parameter later in the section. ■

■ **Example 4.** Use a calculator to determine the regression line (the least squares line) for the height/weight data of Example 1. Then plot these data and the least squares line in the same plane.

Solution. This is simply a matter of entering the data and pressing appropriate keys (check your manual). The required line is $y = -199.24 + 5.35x$. Your calculator should be able to show the data points and the regression line together as in Figure 6. For later reference, we note that $r = 0.87$. ■

PREDICTION

A principal reason for fitting a line to a set of data is to enable us to predict values of the dependent variable y for new values of x.

■ **Example 5.** Wayne Shorter has learned that two new men, Tim and Sam, will be moving to his dorm floor. Help Wayne predict their weights, given that Tim is 73 inches tall and Sam is 64 inches tall.

Solution. We simply substitute these two values for x in the regression equation $y = -199.24 + 5.35x$ of Example 4.

$$\text{Tim:} \quad y = -199.24 + 5.35(73) \approx 191$$

$$\text{Sam:} \quad y = -199.24 + 5.35(64) \approx 143 \qquad \blacksquare$$

People make two kinds of predictions on the basis of regression lines: **interpolation** and **extrapolation.** Interpolation involves predicting y for values of x between x-values in the data set; extrapolation involves predicting y for values of x beyond those in the data set. Interpolation tends to be reliable; extrapolation can be quite unreliable as we now illustrate.

■ **Example 6.** Over the past 10 years, Farmer Brown has kept careful records of the amount of irrigation water used and the final corn yield y (Figure 7). Determine the regression equation and use it to predict the yield next year when he expects to be allotted 8 inches of irrigation water. Would it be reasonable to use this model to predict the yield with 50 inches of irrigation water?

Year	Irrigation water x (inches)	Corn yield y (bushels/acre)
1984	6	102
1985	9	133
1986	9	128
1987	12	154
1988	11	176
1989	7	122
1990	4	98
1991	14	155
1992	14	189
1993	10	150

Figure 7

[0, 18] by [0, 200]

Figure 8

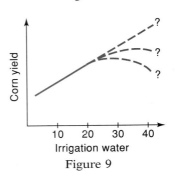

Figure 9

Solution. A calculator gives $y = 62.7 + 8.13x$ as the regression line. This line and the data are plotted in Figure 8. Substituting $x = 8$ gives $y = 128$, suggesting that Farmer Brown can expect 128 bushels per acre next year, a reasonable interpolated result. An x-value of 50 would lead to a predicted yield of 469 bushels per acre, which any corn farmer knows is nonsensical. A little thought and Figure 9 show the danger of extrapolating far beyond the range of the data. The parameter r, which we discuss next, has the value 0.90 in this example. ■

THE CORRELATION COEFFICIENT

More than 100 years ago, Karl Pearson introduced the **correlation coefficient** r as a measure of the tendency for two variables x and y to be related in a linear way. It is defined by the complicated formula shown in the box titled Correlation and is disussed at length in any statistics book. Calculating r by

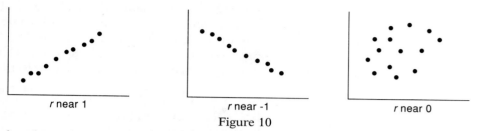

r near 1 r near -1 r near 0

Figure 10

hand is tedious, but fortunately most graphics calculators do this for us automatically once the data are entered. The value of r is always between -1 and 1. A value of r near 1 indicates that the data lie close to a rising line; a value of r near -1 indicates that the data lie close to a falling line; and a value of r near 0 suggests that no tendency toward a linear relationship exists (Figure 10).

You may confirm these interpretations of r by referring to Figure 5 ($r = 0.81$), Figure 6 ($r = 0.87$), Figure 8 ($r = 0.90$), and by studying our next example.

■ **Example 7.** Plot the data in Figure 11, guess at the value of the correlation coefficient, and then use your calculator to determine r.

x	y
1	5
2	1
3	4
3	3
4	4
5	2
6	3
6	1

Figure 11

Solution. The plotted data are shown in Figure 12 and suggest a slight tendency toward a falling line. We would guess r to be negative and quite small. A calculator gives $r = -0.44$. ■

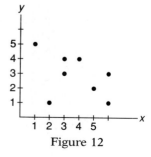

Figure 12

TEASER SOLUTION

A table of the Old Faithful data is shown in Figure 13. Let x denote the duration of an eruption and y the time interval to the next eruption. After we entered these data, a calculator produced the regression line $y = 31.5 + 11.2x$. The plotted data and the regression line are shown in Figure 14. The predicted value (an interpolation) for the time interval to the next eruption after 6:33 is

$$y = 31.5 + 11.2(2.7) \approx 61.7 \text{ minutes}$$

We predict that the next eruption will occur at about 7:35.

Time of eruption	Duration x (minutes)	Interval to next eruption y (minutes)
12:04	1.8	54
12:58	4.5	69
2:07	3.5	77
3:24	1.7	48
4:12	4.7	92
5:44	1.7	49
6:33	2.7	?

Figure 13

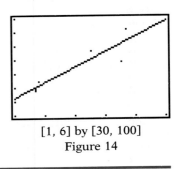

[1, 6] by [30, 100]
Figure 14

PROBLEM SET 1.7

A. Skills and Techniques

Make a scatter plot of each of the following sets of data using the information in Figure 15. (Note: Plotting y vs. x means to use y as the dependent variable and x as the independent variable.) Then tell whether the plotted data suggest a linear relationship, a nonlinear relationship, or no relationship at all between the two variables. Also calculate the correlation coefficient in each case.

x	w	y	z	a	f	g
1	2	10	86	−1	21	17
2	5	11	81	−4	30	3
3	6	11	74	−3	36	9
5	8	15	71	−2	45	14
8	12	22	59	0	49	18
9	22	26	52	2	54	13
10	20	28	50	3	63	9
12	25	36	40	1	70	17
15	28	50	31	2	69	14
20	41	99	12	−3	75	10

Figure 15

1. w vs. x
2. z vs. w
3. y vs. x
4. f vs. z
5. z vs. x
6. x vs. y
7. a vs. x
8. a vs. f
9. g vs. a
10. f vs. x

Plot the given data and the two lines together in the same plane. Conjecture which of the two lines best fits the data. Then test your conjecture by calculating the sum S of the squares of the residuals for each line.

11. $\{(1, 2), (2, 2.5), (2, 4), (4, 6), (5, 4.5)\}$; line 1: $y = x$; line 2: $y = 1.4x$
12. $\{(1, 1), (2, 2), (3, 5), (4, 5), (5, 4)\}$; line 1: $y = x + 1/2$; line 2: $y = (9/5)x - 2$
13. $\{(2, 9), (4, 9), (6, 5), (8, 6), (10, 3)\}$; line 1: $y = 12 - x$; line 2: $y = 9 - (1/2)x$
14. $\{(5, 20), (10, 19), (15, 12), (20, 11), (25, 9)\}$; line 1: $y = 25 - (4/5)x$; line 2: $y = 16 - (1/5)x$

Determine the equation of the line that best fits the data, first by a hand calculation and then by linear regression with a calculator. Check to see that the two results agree.

15. Data of Problem 11
16. Data of Problem 12
17. Data of Problem 13
18. Data of Problem 14

Age (years)	Price (dollars)					
	Alpha	Beamer	Cheapster	Deluxe	Everlast	Flop
2	12000	47200	6000	14800	13500	9900
4	9600	39400	5100	13300	12700	6300
6	7800	35200	4000	9900	10900	3700
8	5700	29900	2900	7400	10000	1900
10	4500	24700	1700	6200	9100	800

Figure 16

Problems 19–24 are related to Figure 16, which gives price vs. age data for six models of used cars. For each model, (a) determine the linear regression equation for the dependence of price on age and (b) predict the prices of cars that are 5 and 12 years old.

19. Alpha model
20. Beamer model
21. Cheapster model
22. Deluxe model
23. Everlast model
24. Flop model

For each relationship, conjecture which of the following numbers would be closest to the correlation coefficient that could be expected between the indicated variables: 0.9 or 0 or −0.9.

25. The height of an oak tree and its base circumference.
26. The age of an oak tree and its height.
27. The length of a student's shadow at 4:00 P.M. and the student's height.
28. The monthly cost of a student's phone bills and the student's height.
29. The weight of a pencil and its length.
30. The price of a seat at the opera and the number of rows from the stage.
31. The length of time required to drive from New York to Chicago and the average driving speed.
32. The amount of tread left on an automobile tire and the number of miles driven.

Find the regression line and the correlation coefficient. Then plot the data and the regression line together.

33.

x:	1	2	3	4	5	6
y:	2	5	8	12	10	14

34.

x:	2	4	6	8	10	12
y:	19	18	12	9	9	2

35.

x:	2.4	3.3	7.9	8.9	11.0	16.5
y:	−5.4	−0.1	11.5	21.5	25.8	41.3

36.

x:	−110	−100	−90	−80	−70	−60
y:	56	45	45	31	17	−2

37.

x:	16	14	22	9	8	1
y:	−34	−32	−57	−21	−3	2

38.

x:	5.16	5.35	5.40	5.71	5.82	5.97
y:	98.6	94.4	34.2	43.3	44.4	17.3

B. Applications and Extensions

39. According to Ohm's law, the relationship between the electric current I in amps and the voltage drop V in volts across a resistor of resistance R in ohms is linear: $V = IR$. Given below are voltage and current measurements for a resistor of unknown resistance. Find the equation of the line that best fits the data and determine the resistance of the resistor. Use voltage as the dependent variable.

Current (I):	0.066	0.149	0.269	0.373	0.725
Voltage (V):	9.0	20.1	36.3	50.5	98.4

40. Repeat Problem 39 for the following set of measurements.

Current (I):	0.45	0.68	0.99	1.23	1.46
Voltage (V):	45	67	100	121	149

41. The position x of a ball moving along a line at a constant velocity v varies linearly with time t: $x = vt + x_0$, where x_0 is the ball's initial position. The data given in Figure 17 are the measured positions in centimeters of three different balls at various times in seconds. (a) Use linear regression to find the velocities and initial positions of the three balls. (b) Predict the position of each ball at time $t = 30$ seconds.

Time t (s)	Position x (cm)		
	Ball 1	Ball 2	Ball 3
5.0	23	67	101
10.0	58	78	200
15.0	91	91	298
20.0	129	101	399
25.0	151	113	501

Figure 17

42. Troy has made measurements of the average mass of a U.S. nickel (in grams) as a function of its date of minting. Determine the equation of the straight line that best fits Troy's data and graph it together with the data on your calculator. Use the regression equation to predict the masses of a new nickel and of one that has been in circulation for 50 years.

Years in circulation:	5	10	15	25	40
Average mass (g):	5.59	5.56	5.53	5.48	5.45

43. Figure 18 shows the total U.S. farm employment in millions of people for the period 1935–1975 (Source: U.S. Department of Agriculture). (a) Make a scatter plot of the data to see if a linear relationship appears to exist. (b) Guess at the correlation coefficient and then calculate it. (c) Determine the regression equation and plot its graph together with the data. (d) Predict the farm employment for the year 1980 from the regression equation and compare the result with the actual employment of 3.7 million.

Year	U.S. farm employment (millions)
1935	12.7
1940	11.0
1945	10.0
1950	9.9
1955	8.4
1960	7.1
1965	5.6
1970	4.5
1975	4.3

Figure 18

44. Challenge. Michelle, a strong swimmer, has given Wanda a 4-minute head start in a 1000-meter swimming race. Various distances and split times for the two swimmers are given below. (a) Find the linear regression line for each swimmer's distance d as a function of time t. (Note that Michelle started swimming at time $t = 4.0$ min.) (b) Will Michelle overtake Wanda, and, if so, at what distance and time?

Distance (m):	100	200	300	400	500	600
Michelle (min):	6.4	8.9	11.3	13.6	16.1	18.8
Wanda (min):	2.6	5.5	7.8	10.4	14.1	17.7

CHAPTER 1 REVIEW PROBLEM SET

In Problems 1–10, write True or False in the blank. If false, provide a counterexample (a specific example that shows that the statement is sometimes false).

_____ **1.** $\sqrt{x^2} = x$
_____ **2.** $N.9\overline{9} = N + 1$ where N is a nonnegative integer
_____ **3.** If $a < b$, then $|a| < |b|$.
_____ **4.** If $a > b > 0$, then $a^2 > b^2$.
_____ **5.** If $a < b < 0$, then $1/a > 1/b$.
_____ **6.** $|-x| = x$
_____ **7.** $a/b/c = a/(b/c)$ provided b and c are not 0.
_____ **8.** If in a set of 10 numbers, all but one are greater than 6, then their average is greater than 6.

_____ **9.** If Mary averaged 86 on her first four 100-point tests, she will never be able to average 90 or better by doing well on the fifth 100-point test.
_____ **10.** The lines $y = 5x + a$ and $y = -0.2x - b$ are perpendicular.

11. Define the term *rational number*.

12. Assuming that $\sqrt{3}$ is irrational, show that $4 - \sqrt{3}$ is irrational.

13. Calculate $3/8 - (1/2)(4(6 - 4)(4 - 2)^2)/8$ by hand and then check your result with a calculator.

14. Use a calculator to evaluate each of the following:
(a) $4.32(3.29 + 3.89)^4 - 7.11$
(b) $(5.43 \times 10^{-7})(6.89 \times 10^6)\sqrt{(3.14 \times 10^{-1})}$

(c) $\dfrac{\sqrt[3]{5.54}\sqrt{8.93} + \sqrt{5.42}}{(2.13 \times 10^{-3})^2}$

(d) $\dfrac{(1.065)^{23} - (1.023)^{65}}{1 - (1.25)^{-17}}$

15. Explain how you would estimate the number of seeds in a large watermelon without counting all of them.
16. Estimate the volume of air (in cubic feet) inside a 60 foot long school bus.
17. Express 5/11 and 5/13 as repeating decimals.
18. Express $0.\overline{468}$ and $3.2\overline{45}$ as ratios of two integers.
19. Which is larger: 16/5 or 13/4?
20. Order the following numbers from least to greatest.

$$\dfrac{29}{20}, 1.44, 1.\overline{4}, \sqrt{2}, \dfrac{130}{89}$$

21. Show that if $a < b$, then $a < (2a + b)/3$.
22. If $(a + b)^2 = a^2 + b^2$, what conclusion do you draw about a and/or b?
23. Find the center of mass of the masses depicted in Figure 1.

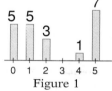

Figure 1

24. Assuming that each course has the same weight, guess who has the higher grade point average: Fred, with 6 As, 5 Bs, 6 Cs, and 1 D, or Wilma, with 9 As, 3 Bs, 4 Cs, and 2 Fs. Then, using $A = 4$, $B = 3$, and so on, determine each student's average.
25. In chess, each side has eight pawns of value 1, two knights and two bishops of value 3, two castles of value 5, a queen of value 10, and an invaluable king. (a) What is the average value of the non-king pieces? (b) What is the average value of the non-king pieces, excluding the pawns?
26. Find the distance between each pair of points.
    (a) (6, 2) and (1, −10)
    (b) (2.3, 5.2) and (−4.4, 8.9)
27. Determine any symmetries and then sketch the graphs of each equation by hand.
    (a) $y = 2x$
    (b) $y = -3x + 3$
    (c) $y = 3x^2 + 2$
    (d) $y = (x - 4)^2 + 1$
    (e) $y = x^3 - x$
    (f) $y = x(x + 2)^2$
28. Without actually graphing, determine any symmetries in the graph of each equation.
    (a) $y = -5x$
    (b) $y = 3x^4 + 2x^2$
    (c) $y = x(x^3 - 5x)$
    (d) $|y| = 3x^2 + 1$
    (e) $y = \dfrac{x}{x^3 + 2}$
    (f) $y^2 = 2x^3 - 2x$

Solve each equation using algebraic methods.

29. $1.2x + 4.3 = -3.2x - 8.6$
30. $2x^2 + 11x = 21$
31. $(x - 4)(x + 2) = 5$
32. $\dfrac{3x + 5}{2x - 7} = 3$
33. Find the exact intersection points of the graphs of the equations $y = x^3 - x$ and $y = 2(x^2 + x)$.
34. Determine algebraically the points on the graph of $y = x^2 - 3x - 2$ where $y = 2$.

Solve each inequality using algebraic methods.

35. $5x - 3 < 0$
36. $x^2 - 2x - 24 < 0$
37. $x^2 + 4x + 4 > 0$
38. $\dfrac{2x + 1}{x - 3} \geq 0$
39. $1 - \dfrac{4}{x^2} < 0$
40. $\dfrac{x(x - 2)}{x^2 + 1} \leq 0$
41. $6 < 2x - 8 < 10$
42. $0.5 > 0.1x > 0.25$

Use a graphics calculator in Problems 43–50, giving any answers accurate to two decimal places.

43. Graph $y = x^3 - 2x^2 - 2x$ and determine x so that $y = 7$.
44. Graph $y = x^4 - 8x^3 - 51x^2 + 158x + 440$, making sure you show all essential features.
45. Find the two solutions of $-x^4 + 5x^2 - x - 6 = 0$.
46. Find the largest solution of $2.5x^3 - 8.3x + 5.4 = 0$.
47. Find all common solutions of $y = 2x^2 - 5$ and $y = x + 3$.
48. Find the common solution of $y = -(x - 5)^3$ and $y = x^3 - 4.5x^2$.
49. Solve the inequality $1.2x^3 - 6.3x^2 + 4.4 > 0$.
50. Solve the inequality $x^2 + 1 < -2x^2 + 5x + 5$.
51. Write the inequalities $|x - 9| < 2.5$ and $|2x + 3| < 5$ without using the absolute value symbol.
52. Write $3 \leq x \leq 12$ as a single inequality, using the absolute value symbol.
53. A rectangle has a perimeter of 20 feet. Assuming that one side is x feet long, write an expression for the area of the rectangle in terms of x.
54. An open rectangular box is to be made from a square piece of cardboard 12 centimeters on a side

by cutting identical squares of side length x from each of the four corners. Express the volume of the box in terms of x.

55. A cylindrical gasoline storage tank will be built inside a square plot, 80 feet on a side, which is to be enclosed by a dike high enough to hold the gasoline in case of a leak. How high must the dike be if the cylinder has radius 20 feet and height 50 feet?

56. In a 4-mile walk, Jenny covered the first 2 miles in 40 minutes. How fast must she walk the rest of the way to average 3.5 miles per hour?

Write the equation of the line satisfying the given conditions in the form $Ax + By + C = 0$.

57. It is vertical and passes through $(4, -3)$.
58. It goes through $(2, 3)$ and $(5, -1)$.
59. It goes through the origin and is parallel to the line $5x + 7y = 2$.
60. It has y-intercept 6 and is perpendicular to the line $2x + 3y = 7$.
61. It has x-intercept 2 and y-intercept -1.
62. It is tangent to the circle $x^2 + y^2 = 25$ at the point $(3, -4)$.

Problems 63–66 refer to the points $A(3, -1)$, $B(5, 3)$, and $C(-1, 9)$.

63. Sketch the triangle ABC and find its perimeter.
64. Write the equations of the three sides of this triangle in the form $Ax + By + C = 0$.
65. Find the midpoints D and E of AB and AC and verify that DE is parallel to BC and half as long.
66. Find the equation of the circle that has AB as a diameter.
67. Find the point one-third of the way from $(2, 0)$ to $(5, 10)$.

68. Find the point three-fourths of the way from $(8, -2)$ to $(6, 6)$.

Problems 69–73 refer to Figure 2, which presents data for the heights of three trees measured on January 1 in certain years over the period 1980–1992.

69. For each tree, obtain the linear regression equation for height h in terms of time t (let t be the number of years after 1980). Also give the correlation coefficient in each case.
70. Which tree had the highest average growth rate over the 12-year period and what was this growth rate?
71. Predict when the apple tree will reach 9 meters in height.
72. Predict the height of the elm tree on January 1, 2000.
73. In what year and at what height is the elm tree predicted to be as tall as the apple tree (Figure 2)?

Year	Tree Height (meters)		
	Apple	Elm	Oak
1980	6.9	4.2	14.0
1982	7.1	4.6	14.0
1984	7.2	5.0	14.1
1987	7.9	5.5	14.2
1990	8.1	6.4	14.2
1992	8.1	6.9	14.4

Figure 2

74. A machine costing $50,000 will depreciate to a value of $1500 by the end of 10 years. Write an expression for its value V at the time t years after purchase.

2

FUNCTIONS AND THEIR GRAPHS

Two Frenchmen, Pierre de Fermat and René Descartes, are given credit for introducing the idea of coordinate geometry. It has been called one of the most fruitful veins of thought ever struck in mathematics. Because this invention allows us to transform a curve into an equation, we can bring the power of algebra to bear on geometric problems. Conversely, we can take an algebraic equation and by the magic of plotting its graph in the coordinate plane make the equation accessible to our visual intuition. This interplay between algebra and geometry is at the heart of all that we do in this book.

de Fermat was trained as a lawyer but made mathematics his hobby. He worked on problems of maxima and minima, initiated with Pascal the first study of probability, and is famous for his results in number theory. His so-called last theorem stood as an unproved conjecture until a proof was finally announced in 1993. In a

René Descartes (1596–1650)   Pierre de Fermat (1601–1665)

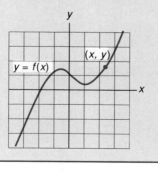

book written in 1629, but not published until 1679, de Fermat makes explicit use of coordinates and applies algebra to the study of geometric curves.

Descartes is best known as the first great modern philosopher but he also was a biologist, a physicist, and a mathematician. He thought that mathematics could unlock the secrets of the world. One of the appendices to his book, Discours de la méthode . . . has the title La Géométrie. It was published in 1637. In it, he proposes the basic ideas of what we now call coordinate geometry. In his honor, the familiar pair of numbers (x, y) that we use to label a point are called Cartesian coordinates. A set of such coordinate pairs determines what we call a Cartesian graph. Our first task in Chapter 2 is to introduce the concept of a function and its graph. Then we use a graphics calculator to display and interpret graphs for polynomial and other simple functions.

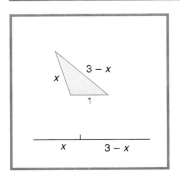

TEASER Cut a 3-foot stick into two pieces by making the cut at distance x from one end. Use these two pieces together with a 1-foot ruler to make a triangle of area $A(x)$. Write a formula for $A(x)$ and determine the domain and range of this function. (*Hint:* Heron's formula says that the area of a triangle of sides a, b, and c and semiperimeter $s = (a + b + c)/2$ is $\sqrt{s(s - a)(s - b)(s - c)}$.)

2.1   THE FUNCTION CONCEPT

All scientists are interested in the concept of relating one variable to another. It may be the simple notion of how the area of a circle is related to its radius. It may be as complicated as the relationship between cigarette smoking and life span for an individual. But whenever this relationship can be precisely specified, we have what mathematicians call a function.

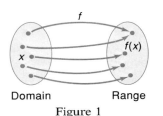

Domain   Range

Figure 1

> **Definition.** A **function** f is a rule that associates with each element x in one set, called the **domain** of the function, a single value $f(x)$ from another set. The set of values so obtained is called the **range** of the function. Figure 1 illustrates these concepts.

This definition is very general. The function might be the rule that assigns to each person in your mathematics class a grade. In this case, the domain consists of the members of your class, the rule is the method of determining grades, and the range is the set of grades given (A, A⁻, B⁺, B, and so on). Most functions considered in this book will have sets of numbers for their domain and range.

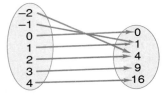

The squaring function

Figure 2

■ **Example 1.** Let the domain for a function be the set $\{-2, -1, 0, 1, 2, 3, 4\}$ and let the rule be "square." Draw a schematic picture of this function and specify its range.

Solution. A picture of this function is shown in Figure 2. The range is the set $\{0, 1, 4, 9, 16\}$. ■

Note two important facts. First, the domain and the rule together determine the range. Second, two or more elements of the domain may be paired with the same element in the range but not vice versa (Figure 3).

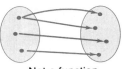

Not a function

Figure 3

Letters, such as f, g, and h, are used for functions. The symbol $f(x)$ is read "f of x" or "f at x" and stands for the value f assigns to x. Thus, if $f(x) = x^3$, then $f(2) = 2^3 = 8$ and $f(-1/2) = (-1/2)^3 = -1/8$.

■ **Example 2.** Let $f(x) = (x^2 - 1)/x$ with domain the set of all nonzero real numbers. Calculate: (a) $f(2)$, (b) $f(1)$, (c) $f(1/2)$, (d) $f(-5/4)$, (e) $f(\sqrt{2})$, (f) $f(f(\sqrt{2}))$, (g) $f(0.1)$, (h) $f(x^2)$, (i) $f(a - 1)$.

Solution.

(a) $f(2) = \dfrac{2^2 - 1}{2} = \dfrac{3}{2}$

(b) $f(1) = \dfrac{1^2 - 1}{1} = 0$

(c) $f(1/2) = \dfrac{1/4 - 1}{1/2} = -\dfrac{3/4}{1/2} = -\dfrac{3}{2}$

(d) $f(-5/4) = \dfrac{25/16 - 1}{-5/4} = \left(\dfrac{9}{16}\right)\left(-\dfrac{4}{5}\right) = -\dfrac{9}{20}$

(e) $f(\sqrt{2}) = \dfrac{2 - 1}{\sqrt{2}} = \dfrac{1}{\sqrt{2}}$

(f) $f(f(\sqrt{2})) = f(1/\sqrt{2}) = \dfrac{1/2 - 1}{1/\sqrt{2}} = -\sqrt{2}/2$

(g) $f(0.1) = \dfrac{0.01 - 1}{0.1} = \dfrac{-0.99}{0.1} = -9.9$

(h) $f(x^2) = \dfrac{(x^2)^2 - 1}{x^2} = \dfrac{x^4 - 1}{x^2}$

(i) $f(a - 1) = \dfrac{(a - 1)^2 - 1}{a - 1} = \dfrac{a^2 - 2a + 1 - 1}{a - 1} = \dfrac{a^2 - 2a}{a - 1}$ ■

MORE ON DOMAIN AND RANGE

The domain for a function should always be specified but in fact usually is not. It is understood that, if no domain is mentioned, it is assumed to be the natural domain. The **natural domain** for a function is the largest set of real numbers for which the function rule makes sense and gives real number values. Thus, if someone refers to the squaring function f and does not specify a domain, it is understood to be the set $\mathbb{R}$ of all real numbers.

■ **Example 3.** Determine the natural domain for the function f with the rule $f(x) = \sqrt{(3x + 1)}/x^2$.

Solution. First, we must insist that $x \neq 0$ to avoid division by 0. Second, we require that $3x + 1 \geq 0$, which is equivalent to $x \geq -1/3$. This makes $f(x)$ a real number. Thus, the domain is $\{x: x \geq -1/3, x \neq 0\}$, read the set of xs such that x is greater than or equal to $-1/3$ and x not equal to 0. ■

The problem of determining the range of a function is often quite difficult;

we will have more to say about it later after we have discussed graphing. But sometimes the range is quite obvious. Take $f(x) = x^2$ with its natural domain of all real numbers. It should be clear that we can get any nonnegative number as a value for this function (and that we cannot get any negative number). The range is the set of all nonnegative real numbers. Similarly, the range for the function $g(x) = x^2 + 6$ is the set of all real numbers greater than or equal to 6; that is, it is the set $\{y: y \geq 6\}$.

■ **Example 4.** Let $h(x)$ be the *n*th digit in the decimal expansion of 5/13. (a) Calculate $h(3)$, $h(8)$, and $h(33)$. (b) Determine the domain and range of h.

Solution.
(a) $5/13 = 0.384615384615 \ldots = 0.\overline{384615}$. Thus, $h(3) = 4$ and $h(8) = 8$. Since $33 = 5(6) + 3$, $h(33) = h(3) = 4$.
(b) The domain is the set $\mathbb{N}$ of natural numbers and the range is the set $\{3, 8, 4, 6, 1, 5\}$. ■

GRAPHS OF FUNCTIONS

We discussed the concept of the graph of an equation in Section 1.3. By the **graph of a function** f, we mean the graph of the equation $y = f(x)$.

■ **Example 5.** Draw the graph of the function $f(x) = x^3 + x - 1$ and determine its domain and range.

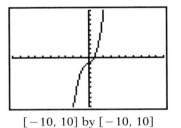

[−10, 10] by [−10, 10]

Figure 4

Solution. Clearly, we may substitute any real number for x; so the domain is the set $\mathbb{R}$ of all real numbers. The graph is the graph of $y = x^3 + x - 1$; our calculator produced the graph shown in Figure 4. Experimenting with the (WINDOW) or (RANGE) menu and a little thought should convince you that the left arm keeps going down and that the right arm keeps going up, suggesting that the range of f is also the set $\mathbb{R}$ of all real numbers. ■

■ **Example 6.** (a) Draw the graph of $f(x) = 3x/(x - 2)$. (b) Determine the domain and range of f. (c) Indicate where f is negative, that is, where f has negative values. (d) Describe the behavior of f near $x = 2$.

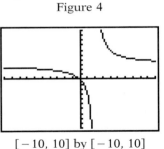

[−10, 10] by [−10, 10]

Figure 5

Solution. (a) Our calculator gave the graph shown in Figure 5. (b) The domain is the set $\{x: x \neq 2\}$, and the range is the set $\{y: y \neq 3\}$. The latter is suggested by the graph but can be made more convincing by solving the equation $y = 3x/(x - 2)$ for x.

$$y = \frac{3x}{x - 2}$$
$$xy - 2y = 3x$$
$$x(y - 3) = 2y$$
$$x = \frac{2y}{y - 3}$$

This shows that for each y except 3 there is an x that will give this value for the function. (c) Our graph suggests that f is negative on the interval $0 < x <$

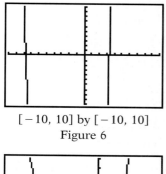

$[-10, 10]$ by $[-10, 10]$

Figure 6

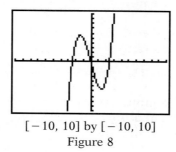

$[-15, 10]$ by $[-500, 500]$

Figure 7

2. This can be confirmed algebraically. (d) Note that $f(2.01) = 6.03/0.01 = 603$ and $f(2.001) = 6.003/0.001 = 6003$. The closer x is to 2 on the right of 2, the larger is $f(x)$. Conversely, the closer x is to 2 on the left of 2, the smaller (that is, the more negative) is $f(x)$. ∎

■ **Example 7.** Use your calculator to draw the graph of $f(x) = 0.2x^4 + 3x^3 + 12x^2 - 20x - 200$ and determine where f is negative.

Solution. Use the range values $[-10, 10]$ by $[-10, 10]$ and your calculator will give the unenlightening Figure 6. After trying various values in the (WINDOW) or (RANGE) menu, you may be led to something like Xmin = -15, Xmax = 10, Ymin = -500, and Ymax = 500 leading to the graph shown in Figure 7.

Figure 7 indicates that f is negative on the approximate interval $-8 < x < 3.5$. When we apply the zooming procedure, we obtain the more accurate interval $-7.729 < x < 3.320$. ∎

EVEN AND ODD FUNCTIONS

Figures 8 and 9 show the graphs of two functions, namely, $g(x) = x^3 - 6x$ and $h(x) = 0.1x^4 - 2x^2 + 5$. Note their symmetries.

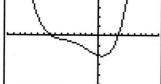

$[-10, 10]$ by $[-10, 10]$

Figure 8

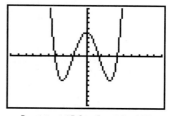

$[-10, 10]$ by $[-10, 10]$

Figure 9

We say that f is an **odd function** if $f(-x) = -f(x)$; we say that f is an **even function** if $f(-x) = f(x)$. You may check that g is odd whereas h is even. But recalling the discussion of symmetry in Section 1.3, we see that the first condition is exactly that for the graph to be symmetric with respect to the origin and the second condition is that for symmetry with respect to the y-axis.

■ **Example 8.** Determine which of the following functions are odd, which are even, and which are neither.

$$f(x) = x^2/(x^4 + 1), \quad g(x) = x^3 |x|, \quad h(x) = x^5 + 1, \quad k(x) = |x^5|$$

Solution.

$$f(-x) = \frac{(-x)^2}{(-x)^4 + 1} = \frac{x^2}{x^4 + 1} = f(x)$$
$$g(-x) = (-x)^3|-x| = -x^3|x| = -g(x)$$
$$h(-x) = (-x)^5 + 1 = -x^5 + 1$$
$$k(-x) = |(-x)^5| = |-x^5| = |x^5| = k(x)$$

Thus, g is odd, f and k are even, and h is neither. ∎

Figure 10

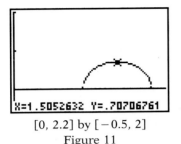

X=1.5052632 Y=.70706761

[0, 2.2] by [−0.5, 2]

Figure 11

TEASER SOLUTION

This problem would be pretty hard (though still solvable) if it were not for the hint about Heron's formula. Refer to Figure 10 and note that the 3 sides of our triangle are 1, x, and $3 - x$, making $s = 2$. Thus,

$$A(x) = \sqrt{2(1)(2 - x)(-1 + x)}$$

This formula suggests a domain of $1 \le x \le 2$ (to make the values real numbers), and this is exactly what we need to make a triangle out of our three line segments (a degenerate one when $x = 1$ or $x = 2$). The range for A is the interval $0 \le y \le \sqrt{2}/2 \approx 0.707$. The maximum value of $\sqrt{2}/2$ corresponds to the case of an isosceles triangle with sides 1, 1.5, and 1.5. You can confirm that this is so by graphing the function A. Figure 11 shows such a graph with the trace cursor at the maximum point of the graph.

PROBLEM SET 2.1

A. Skills and Techniques

Each of the following functions is to have as domain the set $\{-3, -1, 0, 1, 3, 5\}$. Draw a schematic picture of each function and determine its range. See Example 1.

1. The function described by the rule "triple and add 3."
2. The function described by the rule "square and subtract -2."
3. $f(x) = x^2 - 2x$
4. $f(x) = |x| - 1$
5. $g(x) = \dfrac{2}{x^2 + 1}$
6. $h(x) = \dfrac{x^4}{2x - 5}$
7. $d(t) = t^2 - |t|$
8. $f(t) = -(t^2 - 1)(t^2 - 9)$

For the following functions, calculate: (a) $f(0)$, (b) $f(2)$, (c) $g(\sqrt{3})$, (d) $f(f(2))$, (e) $g(f(2))$.

9. $f(x) = \sqrt{x}$ and $g(x) = x^4$
10. $f(x) = |x - 4|$ and $g(x) = \dfrac{3}{x}$
11. $f(t) = \dfrac{t^2 - 1}{t^2 + 1}$ and $g(x) = \sqrt{3x}$
12. $f(x) = \sqrt{24/(x^2 + 2)}$ and $g(t) = \dfrac{24}{t^2 + 2}$
13. $f(w) = |-w^3 + w^2|$ and $g(w) = 3$
14. $f(x) = x^3 - 5x$ and $g(x) = f(x^2)$

In Problems 15–18, determine the natural domain of the given function.

15. $f(x) = \dfrac{x^2 - 7}{x}$
16. $g(x) = \dfrac{-x^3 - x^2 + 1}{|x| - 2}$
17. $h(x) = \sqrt{(x^2 + 1)/(x^2 - 1)}$
18. $k(t) = \dfrac{\sqrt{t}}{t^4 - 81}$

Determine both the (natural) domain and the range of the functions in Problems 19–22.

19. $f(n)$ is the nth digit in the decimal expansion of $3/111$.
20. $g(n)$ is the nth digit in the decimal expansion of $13/600$.
21. $f(x) = 3|x| - 2$
22. $g(x) = 0.1x^2 + 2.5$

Answer as best you can each question for the function f whose graph is shown. (a) What is the value of $f(2.5)$? (b) For what values of x is $f(x) = 0$? (c) For what values of x is $f(x) < 0$?

23.

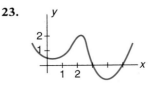

24.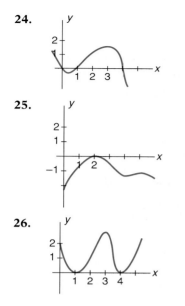

25.

26.

For the functions f and g whose graphs are shown, determine the values of x where each statement holds. (a) $f(x) = 0$, (b) $f(x) > 0$, (c) $g(x) = 0$, (d) $g(x) \le 0$, (e) $f(x) = g(x)$, (f) $f(x) > g(x)$.

27.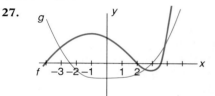

28.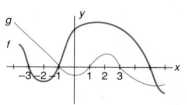

Use a calculator to draw the graph of each function, determine its domain and range, and specify where the function is negative.

29. $f(x) = x^3 - 2x$

30. $f(x) = |x| - 5$

31. $g(x) = \dfrac{5x - 2}{x}$

32. $h(x) = \dfrac{4x}{x - 2}$

33. $f(x) = 0.15x^3 - 3x^2 - 2x - 12$

34. $g(x) = -x^4 + 12x^2 - 7x + 3$

Use algebra, as in Example 8, to determine whether the given function is even, odd, or neither.

35. $f(x) = \dfrac{|x|}{x}$

36. $g(x) = x^3 + x$

37. $h(x) = 3x^5 + x - 5$

38. $k(x) = \dfrac{3x^4 - 2x^2}{7x}$

39. $f(t) = |t^3| + 2t^2$

40. $s(t) = |t| - t$

B. Applications and Extensions

In each of Problems 41–46, sketch a graph for a function satisfying the given information.

41. $f(0) = 1.5$, $f(0.5) = 1.9$, $f(2) = 3.2$, $f(3) = 3.5$, $f(4) = 2.8$

42. $g(0) = 1.5$, $g(1) = 0.5$, $g(1.5) = -0.5$, $g(2) = -1$, $g(2.5) = 0$

43. $f(x) = -1$ for $x < 0$, $f(0) = 3$, $f(x) = 2$ for $x > 0$

44. $f(x) < 0$ for $x \ne 1$, $f(1) = 0$

45. $g(x) > 0$ for $0 < x < 2$ and for $x > 4$, $g(x) \le 0$ otherwise

46. $f(x) = 0$ if x is an integer, $f(x) > 0$ otherwise

In Problems 47–56, write a formula of the form $f(x) = \dots$ for a function that satisfies the stated conditions.

47. It squares a number, divides the result by 3, and finally adds 6.

48. It adds 3 to a number, squares the result, and finally divides by 4.

49. It gives the volume of a box of width x, length twice its width, and height 7.

50. It gives the volume of a sphere of diameter x.

51. It gives the diameter of a sphere whose volume is x.

52. It gives the radius of a cylinder whose height is 6 and whose volume is x.

53. It gives the area of an equilateral triangle of perimeter x.

54. It gives the area of a regular hexagon inscribed in a circle of radius x.

55. $f(1) = 2$ and $f(2) = f(3) = 0$

56. $f(0) = 1$ and $f(-2) = f(-1) = f(1) = f(2) = 0$

57. A ship leaves port at 10:00 A.M. sailing west at 18 miles per hour. At noon, a second ship leaves the same port sailing south at 24 miles per hour. Express the distance $d(t)$ between the ships in terms of t, the number of hours after noon. Then calculate $d(6)$.

58. A 2-mile race track has the shape of a rectangle with semicircular ends of radius x (Figure 12). Write a formula for the area $A(x)$ of the region inside the track and determine the domain and range of A.

Figure 12

59. Recall that **prime numbers** are the positive integers with exactly two positive integer divisors; that is, they are the integers 2, 3, 5, 7, 11, 13, ... Let $f(x)$ be the number of primes less than x. Calculate $f(19.5)$ and $f(28)$. Determine the domain and range of this function.

60. Let $g(x)$ be the distance to the nearest prime. Calculate $g(15)$ and $g(34.2)$. Determine the domain and range of this function.

61. Let $f(x) = 2x/(x - 2)$. Find and simplify: (a) $f(1.9)$ (b) $f(2a)$ (c) $f(1/a)$ (d) $(f(x + h) - f(x))/h$.

62. Let $g(x) = (x^2 - 2)/(2x)$. Find and simplify: (a) $g(1/3)$ (b) $g(1/x)$ (c) $g(2x - 2)$ (d) $(g(x + h) - g(x))/h$.

63. Let $f(x) = (3x - 4)/(x - 3)$. Find and simplify $f(f(x))$ and then calculate $f(f(3.4567))$ the easy way.

64. **Challenge.** Find the formula for the perimeter $P(A)$ of an equilateral triangle with area A. Then calculate $P(16\sqrt{3})$.

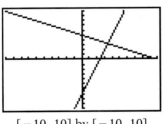

TEASER Rita sells major appliances. Her monthly salary is $2500 plus $30 for each appliance she sells. The payroll department deducts 32% of her salary for taxes and pension and then $15 that she has pledged to the United Way. Write a simple formula for her take-home pay $P(x)$, assuming she sells x appliances, and calculate $P(20)$.

2.2 LINEAR FUNCTIONS

The simplest of all functions, but very important, are the linear functions. A function f is **linear** if it has the form $f(x) = mx + q$ for some constants m and q. The graph of the linear function f is a line with slope m and y-intercept q, a subject we studied in detail in Section 1.6. Note that if $m = 0$, a linear function becomes a constant function, and its graph is a horizontal line.

■ **Example 1.** Draw the graphs of $f(x) = 3x - 7$ and $g(x) = -(1/2)x + 5$ in the same plane and determine the point of intersection of these graphs.

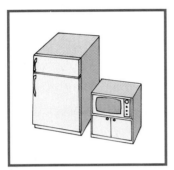

$[-10, 10]$ by $[-10, 10]$
Figure 1

Solution. The required graphs are the graphs of the lines $y = 3x - 7$ and $y = -(1/2)x + 5$, graphs that are easy to draw by hand or with a calculator. Our calculator gave the graphs shown in Figure 1.

We can read the approximate coordinates of the intersection point from the graphs, and we could certainly get them to great accuracy by zooming. The algebraic method is so simple, however, that it should be used. Set the two expressions for y equal to each other, obtaining in successive steps

$$3x - 7 = -\frac{1}{2}x + 5$$
$$6x - 14 = -x + 10$$
$$7x = 24$$
$$x = \frac{24}{7}$$

Substituting this x in either equation gives $y = 23/7$. The intersection point is $(24/7, 23/7) \approx (3.43, 3.29)$. ■

PROPERTIES OF LINEAR FUNCTIONS

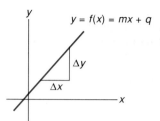

Figure 2

A line is unique among curves in having the same steepness (slope) at each of its points. This fact gives rise to two intimately related properties of linear functions.

1. **(Proportionate change property)** *For a linear function, a change in the domain variable x always produces a proportionate change in the value of the function.* To be precise, if x changes by an amount Δx, then the value of the linear function $y = f(x) = mx + q$ changes by an amount $\Delta y = m\,\Delta x$. This is because the ratio $\Delta y/\Delta x$ is the constant m (Figure 2).

    For example, if $y = f(x) = 3x - 5$, then a change in x of amount 2 produces a change in y of amount 3(2), a change in x of amount 5 produces a change in y of amount 3(5), and a change in x of amount Δx produces a change in y of amount $\Delta y = 3(\Delta x)$.

2. **(Constant rate of change)** *For a linear function, the rate at which the function changes with respect to the domain variable x is constant.* More precisely, let $y = f(x) = mx + q$. Then the rate of change of y with respect to x is the constant m.

    For example, if Carole is driving so that her distance $s(t)$ from home in miles at time t hours is given by $s(t) = 50t + 30$, then the rate of change of distance with respect to time (her speed) is 50 miles per hour.

■ **Example 2.** The selling price $p(t)$ of a hupmobile has grown linearly according to the formula $p(t) = 1200 + 60t$ dollars, where t stands for the number of years after 1960. (a) Assuming this pattern continues, how much will the price grow during the 3-year period 1996–1999? (b) How fast is price changing with respect to time?

Solution.
(a) The price in any 3-year period will grow by 60(3) = $180.
(b) The rate of change of price with respect to time is $60 per year.   ■

APPLICATION: STRAIGHT-LINE DEPRECIATION

Most machines depreciate (lose value) as they age. There are several methods of figuring depreciation on the basis of curves, but the most common is the straight-line method, which we will use in our next example.

■ **Example 3.** A car costing $18,000 will have a value of $600 at the end of 12 years. (a) Write a formula for its value $V(t)$ when it is t years old ($0 \le t \le 12$). (b) Draw the graph of this function. (c) Determine the car's value at the end of 3.5 years. (d) When is the car's value between $8000 and $10,000? (e) How much value does it lose every 2.5 years? (f) What is the rate of change of value with respect to time?

Solution.
(a) The slope m is given by

$$m = \frac{18{,}000 - 600}{0 - 12} = -1450$$

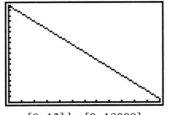

[0, 12] by [0, 18000]

Figure 3

Thus, $V(t) = 18,000 - 1450t$.

(b) A calculator-drawn graph appears as Figure 3.

(c) $V(3.5) = 18,000 - 1450(3.5) = \$12,925$.

(d) We could use the graph to approximate the required time interval but we can do better algebraically. Check that the following algebraic steps are correct.

$$8000 < 18000 - 1450t < 10000$$
$$-10000 < -1450t < -8000$$
$$\frac{-10000}{-1450} > t > \frac{-8000}{-1450}$$
$$5.517 < t < 6.897$$

The value of the car will be between \$8000 and \$10,000 between year 5.52 and year 6.90.

(e) $\Delta V = -1450(2.5) = -\$3625$. The car loses \$3625 every 2.5 years.

(f) The rate of change of value with respect to time is $-\$1450$ per year. ■

APPLICATION: WORKING ON COMMISSION

Salespeople are commonly paid a basic salary plus a commission based on the amount they sell.

■ **Example 4.** Harvey sells men's suits. He is paid a basic monthly salary of \$1200 plus \$25 for every suit he sells. (a) Write the formula for his monthly wage $W(x)$ in a month in which he sells x suits. (b) How many suits must he sell in order to earn a wage of at least \$3000? (c) Find his wage in a month when he sold 50 suits. (d) How much would his wage have increased if he had sold 55 rather than 50 suits? (e) What is the rate of change of his wages with respect to the number of suits sold?

Solution.

(a) $W(x) = 1200 + 25x$

(b)
$$1200 + 25x \geq 3000$$
$$25x \geq 1800$$
$$x \geq 72$$

Harvey will need to sell at least 72 suits.

(c) $W(50) = 1200 + 25(50) = \$2450$

(d) $\Delta W = 25(5) = \$125$

(e) The rate of change of W with respect to x is \$25 per suit. ■

APPLICATION: COST OF PRODUCTION

A manufacturer classifies its costs into two categories: fixed costs and variable costs (indirect costs and direct costs are other terms). The fixed costs consist of such things as real estate taxes, insurance, and building maintenance; the variable costs are those that depend directly on the number of items produced, including the cost of materials and the cost of labor involved in producing and selling these items.

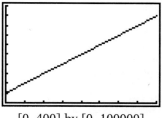

[0, 400] by [0, 100000]

Figure 4

■ **Example 5.** A plant making one type of TV set has fixed weekly costs of $10,000 and variable costs of $200 per unit. (a) Write a formula for the total weekly cost $C(x)$ in a week when x TV sets are produced. (b) Draw the graph of C. (c) Determine the total cost in a week when 150 sets were produced. (d) How much will it cost to increase production by 25 sets?

Solution.

(a) $C(x) = 10,000 + 200x$

(b) The graph of C is shown in Figure 4 for $0 \le x \le 400$.

(c) $C(150) = 10,000 + 200(150) = \$40,000$

(d) $\Delta C = 200(25) = \$5000$ ■

Economists use the term **marginal cost** to indicate the cost of increasing production by one unit (alternatively as the rate of change of cost with respect to the number produced). Thus, in Example 5, the marginal cost is $200 per unit.

APPLICATION: BREAK-EVEN ANALYSIS

The total revenue $R(x)$ that a manufacturer obtains is just the number x of items sold times the price p per item, that is, $R(x) = px$. Assuming that all items manufactured can be sold at the asking price p, the profit $P(x)$ in manufacturing x items is given by $P(x) = R(x) - C(x)$. Here $C(x)$ is the aforementioned total cost of producing x items. A concept important to all manufacturers is the break-even point, that is, the number x_0 of items that must be sold to just break even. Above this point, there will be a profit; below it, there will be a loss (negative profit).

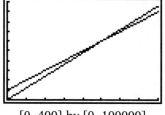

[0, 400] by [0, 100000]

Figure 5

■ **Example 6.** Assume that the manufacturer of Example 5 sells TV sets to retail stores for $240 per unit. (a) Determine $R(x)$, the weekly revenue. (b) Draw graphs of the cost function C and the revenue function R in the same plane for $0 \le x \le 400$. (c) Determine the break-even point. (d) Evaluate the profit in a week when 290 sets were manufactured and sold.

Solution.

(a) $R(x) = 240x$

(b) Figure 5 shows calculator-generated graphs of C and R and Figure 6 shows the same graphs properly labeled.

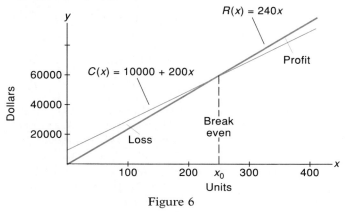

Figure 6

(c) The break-even point x_0 is determined by solving the equation $R(x) = C(x)$, which is easily done algebraically as follows.

$$240x = 10{,}000 + 200x$$
$$40x = 10{,}000$$
$$x = 250$$

Thus, $x_0 = 250$.

(d)
$$P(290) = R(290) - C(290)$$
$$= 240(290) - 10{,}000 - 200(290) = \$1600 \qquad \blacksquare$$

TEASER SOLUTION

$$P(x) = 2500 + 30x - 0.32(2500 + 30x) - 15$$
$$= 0.68(2500 + 30x) - 15$$
$$= 1700 + 20.4x - 15 = 1685 + 20.4x$$
$$P(20) = 1685 + 20.4(20) = \$2093$$

PROBLEM SET 2.2

A. Skills and Techniques

In Problems 1–6, graph each pair of linear functions and guess at the coordinates of the intersection point. Then find this point exactly by algebraic methods.

1. $f(x) = 2x + 8$, $g(x) = -3x - 2$
2. $f(x) = -5x + 7$, $g(x) = 2$
3. $f(x) = \dfrac{3}{2}x - \dfrac{5}{2}$, $g(x) = 2x + \dfrac{1}{2}$
4. $f(x) = \dfrac{2}{7}x + \dfrac{5}{4}$, $g(x) = -\dfrac{3}{7}x - \dfrac{7}{4}$
5. $f(x) = 1.2x - 2.6$, $g(x) = -2.4x + 0.8$
6. $f(x) = 2\pi x$, $g(x) = -\pi x + 6\pi$
7. Purple Cab charges $3 plus an additional 40¢ for each mile of the trip. (a) Write a formula for the cost $C(x)$ of a taxi ride of x miles. (b) Compute the cost of a 12-mile ride. (c) For what length ride is the cost less than $10?
8. ABC leases cars for $150 per month plus 18¢ per mile. (a) Write a formula for the cost $C(x)$ of leasing a car for 3 months and driving it x miles. (b) Compute the cost to a person who leased for 3 months and drove 12,550 miles. (c) How many miles can a person drive during 3 months and still keep the cost under $800?
9. The Kowalskis bought their house in 1980 for $85,000, and it has increased in value by $6000 a year since then. (a) Write a formula for the value

$V(t)$ of their house t years after 1980. (b) State the rate of change of value with respect to time.
10. Sara is 120 miles due east of Denver and is driving straight east at 63 miles per hour. (a) Write a formula for her distance $d(t)$ from Denver t hours from now. (b) State the rate of change of distance with respect to time.
11. For tax purposes, the Wheeler Bicycle Company is depreciating its frame-welding robot costing $98,000 to zero value over a 7-year period, using the straight-line method. (a) Write a formula for the robot's value $V(t)$ when it is t years old. (b) Calculate the robot's value after 4 years. (c) During what time interval is its value between $30,000 and $60,000? (d) State the rate of change of value with respect to time in dollars per year.
12. A machine that cost $44,000 new is expected to have a junk value of $2000 after 8 years. Assume straight-line depreciation. (a) Write a formula for the value $V(t)$ of the machine t years after purchase. (b) Compute its value after 6.5 years. (c) During what time interval is its value between $20,000 and $30,000? (d) Determine the rate of change of value with respect to time in dollars per year.
13. The R-rating of fiberglass insulation varies linearly with its thickness. A 5-inch batt of insulation has an R-rating of 16. Dwight Danbury's house has an uninsulated ceiling with an R-rating of 3. (a) Express the R-rating $R(x)$ of his ceiling in terms of x, the number of inches of fiberglass Dwight puts in.

(b) To obtain an R-rating of 36, how many inches of insulation must he put in? (c) State the rate of change of R-rating with respect to the thickness of the insulation.

14. The percentage $P(T)$ of tomato seeds that germinate depends in a linear way on the temperature T (within certain bounds). At 12°C, 40% germinate, whereas at 17°C, 70% germinate. (a) Write a formula for $P(T)$. (b) At what temperature does the percentage reach 100%? (c) State the rate of change of P with respect to T.

15. Insurance salesperson Shirley Adams is paid an annual salary of $22,000 plus 0.9% of the face value of all policies she sells. (a) Write a formula for her income $I(x)$ in a year in which she sells x dollars of insurance policies. (b) Calculate her income for a year in which she sells $600,000 of insurance. (c) How much would she have to sell to make her commission equal her base salary?

16. Carlos works for the Wacky Widget Company earning $350 per week plus 50¢ for each widget he sells. (a) Express his weekly income $I(x)$ in terms of the number x of widgets he sells each week. (b) Calculate his income in a week when he sold 170 widgets. (c) State the rate of change of income with respect to the number of widgets sold.

17. Ben is selling popcorn at the state fair. His fixed cost is $50 per day for booth rental, and his variable cost is 30¢ per bag of popcorn. (a) Write a formula for his daily cost $C(x)$ if he sells x bags of popcorn per day. (b) Graph $C(x)$ for $0 \le x \le 500$. (c) How much does his cost rise for every 100 bags he sells?

18. A small toy company has weekly fixed costs of $2400 plus $1.50 for each unit produced. (a) Express the company's weekly costs $C(x)$ in terms of x, the number of units produced each week. (b) Graph this cost function for $0 \le x \le 5000$. (c) What is the marginal cost?

19. Perform a break-even analysis for Problem 17, assuming Ben sells popcorn for 80¢ a bag. (a) Determine Ben's daily revenue $R(x)$. (b) Superimpose the graph of R on the graph of C. (c) Determine the break-even point. (d) Calculate Ben's profit on a day in which he made and sold 300 bags of popcorn.

20. The toy company of Problem 18 sells its toys at a price of $2.25 per unit. (a) Determine $R(x)$, the weekly revenue. (b) Determine the weekly profit $P(x)$. (c) Superimpose the graph of R on the graph of C. (d) Determine the break-even point. (e) If the company lost money by making and selling only 2000 units one week, how many units must it make and sell the next week to recoup its loss?

B. Applications and Extensions

In Problems 21–24, determine the formula for $f(x)$, given its graph.

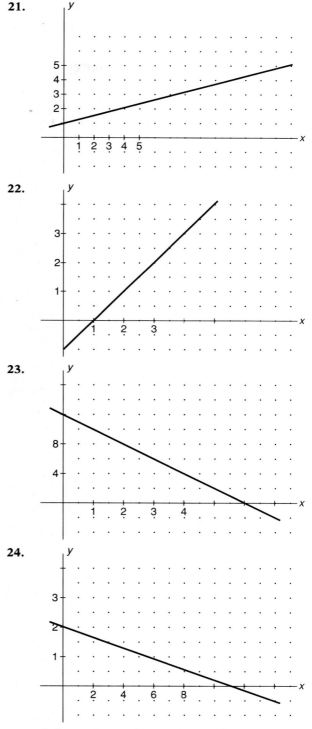

21.

22.

23.

24.

25. If f is a linear function with $f(-3) = 11$ and $f(2) = 5$, write the formula for $f(x)$.

26. Show that there is no linear function whose graph goes through (3, 10), (21, 22), and (60, 49).

27. Let $y = g(x)$ where g is a linear function with $g(2) = 5$, and suppose the rate of change of y with respect to x is -3. Determine $g(x)$.

28. Let $M = F(T)$, where F is a linear function with $F(-2) = 2$, and suppose the rate of change of M with respect to T is 3/5. Determine $F(T)$.

29. Water freezes at 32°F and at 0°C; it boils at 212°F and at 100°C. (a) Express the Celsius temperature C in terms of the Fahrenheit temperature F. (b) Calculate the Celsius temperature when it is 72°F. (c) At what Fahrenheit temperature do the two scales give the same reading? (d) What is the rate of change of C with respect to F?

30. Let $f(x) = x^2 - 3x + 11$ and let g be a linear function that increases by the same amount as f on the interval $2 \le x \le 10$ and $g(2) = f(2)$. Give the formula for $g(x)$.

31. Suppose it costs \$65,000 to produce 500 units per week and the (constant) marginal cost is \$90. (a) Determine the fixed weekly cost. (b) What does it cost to produce 501 units per week? (c) What does it cost to produce x units per week?

32. A bicycle company has fixed monthly costs of \$9000. Its total costs were \$42,990 in a month it produced 132 bicycles. What are the company's variable costs per bicycle?

33. A computer manufacturer has annual fixed costs of \$8 million and variable costs of \$400 per unit. If it broke even last year on revenues of \$24 million, how many units did it sell and at what price?

34. The Universal Umbrella Company plans to close one of its two factories. The factory in Fairville has fixed weekly costs of \$7800 and variable costs of \$1.75 per unit, whereas the Plainville factory has fixed costs of \$10,000 and variable costs of \$1.25 per unit. (a) Write a formula for the total weekly cost at each factory and graph the formulas in the same plane. (b) Which factory should be closed if the company expects to sell 4300 units per week?

35. At noon, two motorists are 500 miles apart driving toward each other at 62 and 57 miles per hour,

respectively. Write a formula for $s(t)$, the distance between them at t hours after noon, and determine the rate of change of s with respect to time.

36. Tom was 500 feet down the road running at 15 feet per second when Joel started chasing him at 20 feet per second. At what rate is Joel closing the distance between them, and how long will it take Joel to catch up?

37. The graph of a linear function f goes through (5, 4) and this line together with the positive x- and y-axes enclose a region of area 245/6 (Figure 7). Determine $f(x)$.

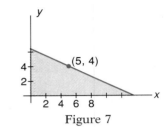

Figure 7

38. The graph of a linear function f is tangent to the circle $x^2 - 4x + y^2 - 10y = -4$ at (5, 9). Determine $f(x)$.

39. Suppose that f and g are linear functions. Show that h is also linear, given that $h(x) = f(g(x))$.

40. Suppose that f is a linear function such that $f(f(x)) = 4x + 12$. Determine $f(x)$.

41. Show that an odd linear function f must have the form $f(x) = mx$.

42. Show that a linear function f satisfying $f(u + v) = f(u) + f(v)$ must have the form $f(x) = mx$.

43. Let f be a linear function. Show that for any numbers x_1 and x_2,

$$f\left(\frac{1}{3}x_1 - \frac{2}{3}x_2\right) = \frac{1}{3}f(x_1) + \frac{2}{3}f(x_2)$$

44. **Challenge.** Conjecture an appropriate generalization of Problem 43 that involves n points $x_1, x_2, \ldots, x_n$ and n positive weights $w_1, w_2, \ldots, w_n$ that sum to 1. Then prove this generalization.

2.3 QUADRATIC FUNCTIONS

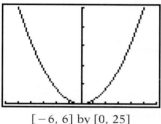

[−6, 6] by [0, 25]

Figure 1

A **quadratic function,** also called a second-degree function, is a function of the form $f(x) = ax^2 + bx + c$ where a, b, and c are constants and $a \neq 0$. The graph of such a function is always a parabola, a special cup-shaped curve about which we will have much to say later. The simplest of all quadratic functions is $f(x) = x^2$, whose graph is the subject of our first example.

■ **Example 1.** Draw the graph of $y = f(x) = x^2$. Then determine the distance between the points on this graph corresponding to $x = -1$ and $x = 4$.

Solution. We use a calculator to draw the graph after setting the WINDOW or RANGE values to $[-6, 6]$ by $[0, 25]$. We get the typical parabolic shape shown in Figure 1. The points mentioned have coordinates $(-1, 1)$ and $(4, 16)$; the distance between them is

$$d = \sqrt{(4 + 1)^2 + (16 - 1)^2} = \sqrt{250} \approx 15.81.$$ ■

Note that the graph of $f(x) = x^2$ is symmetric with respect to the y-axis. A parabola always has an **axis of symmetry.** The point where a parabola crosses its axis of symmetry (the origin in Example 1) is called the **vertex** of the parabola.

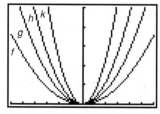
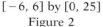

[−6, 6] by [0, 25]

Figure 2

THE GRAPH OF $F(X) = AX^2$

Increasing the magnitude of a narrows the graph; of $f(x) = ax^2$; changing its sign flips the graph across the x-axis.

■ **Example 2.** Using the same axes, draw the graphs of $f(x) = 0.5x^2$, $g(x) = 0.9x^2$, $h(x) = 1.5x^2$, and $k(x) = 3x^2$. Then use another set of axes to draw the graphs of $g(x) = 0.9x^2$, $G(x) = -0.9x^2$, and $K(x) = -3x^2$.

Solution. The first four graphs are shown in Figure 2. The last three appear in Figure 3. ■

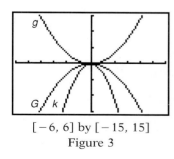

[−6, 6] by [−15, 15]

Figure 3

The Graphs of $F(X) = X^2 + K$ and $G(X) = (X - H)^2$

Adding k to a function shifts the graph k units vertically; replacing x by $x - h$ shifts the graph h units horizontally. In both cases, the shape and orientation of the graph are preserved.

■ **Example 3.** Using the same axes, draw the graphs of $f(x) = x^2$ and $F(x) = x^2 + 3$. Then use another set of axes to draw the graph of $g(x) = x^2$ and $G(x) = (x - 4)^2$.

Solution. The first two graphs are shown in Figure 4; the second pair in Figure 5.

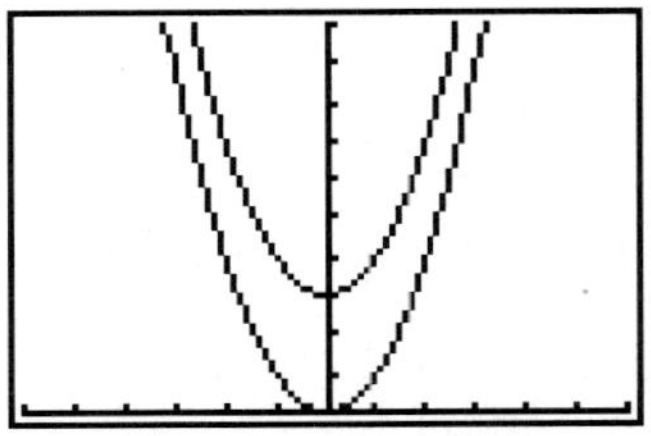

$[-6, 6]$ by $[0, 10]$
Figure 4

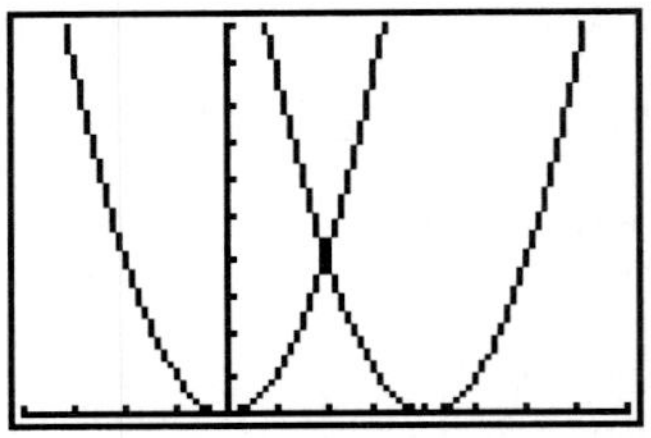

$[-4, 8]$ by $[0, 10]$
Figure 5

Note that adding 3 to a function shifts the graph upward 3 units; replacing x by $x - 4$ shifts the graph 4 units to the right. ■

■ **Example 4.** Explain how the graph of $F(x) = (x + 2)^2 - 5$ is related to the graph of $f(x) = x^2$. Then draw the graph of both functions using the same axes.

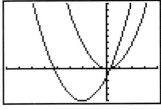

$[-8, 4]$ by $[-5, 10]$
Figure 6

Solution. The graph of F has the same shape and orientation as that of f but is shifted 2 units left and 5 units down. Figure 6 shows both graphs. Note that the vertex of the graph of F is at $(-2, -5)$. ■

The Graph of $F(X) = AX^2 + BX + C$

First we remind you that the expression $x^2 + px$ can be completed to a perfect square by the addition of $(p/2)^2$. Thus,

$$ax^2 + bx + c = a\left(x^2 + \frac{b}{a}x\right) + c$$

$$= a\left(x^2 + \frac{b}{a}x + \frac{b^2}{4a^2}\right) + c - \frac{b^2}{4a}$$

$$= a\left(x + \frac{b}{2a}\right)^2 + c - \frac{b^2}{4a}$$

From this we conclude that the graph of $F(x) = ax^2 + bx + c$ has the same shape as the graph of $f(x) = ax^2$. It will be a parabola that opens up or down depending on whether a is positive or negative. Its vertex has x-coordinate $x = -b/(2a)$, a fact worth highlighting.

The vertex of the parabola $y = ax^2 + bx + c$ is at $x = -b/(2a)$.

■ **Example 5.** Before graphing the parabola with equation $y = -2x^2 + 8x + 8$, tell whether it turns up or down and find both coordinates of the vertex.

Solution. Since $a < 0$, the parabola turns down. The x-coordinate of the vertex is $x = -8/(2(-2)) = 2$. The corresponding y-coordinate is best found by substitution in the equation: $y = -2(2)^2 + 8(2) + 8 = 16$. ■

APPLICATION: MAXIMUM HEIGHT OF A PROJECTILE

Physicists tell us that if an object is shot straight up from an initial height of s_0 feet and with an initial speed of v_0 feet per second, it will be at height $s = -16t^2 + v_0 t + s_0$ feet after t seconds (assuming that air resistance plays a negligible role). Note that this is the equation of a parabola.

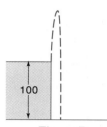

100

Figure 7

■ **Example 6.** A steel ball is shot straight up from the top edge of a building 100 feet high with an initial speed of 400 feet per second (Figure 7). (a) How high will this ball get? (b) When will it hit the ground? (c) During what time interval will it be at least 1200 feet high?

Solution. The height s after t seconds is given by $s = -16t^2 + 400t + 100$, the equation of a downward opening parabola. The t-coordinate of the vertex is $-400/(2(-16)) = 12.5$.
(a) The maximum height is the s-coordinate of the vertex, that is,

$$s_{max} = -16(12.5)^2 + 400(12.5) + 100 = 2600 \text{ feet}$$

(b) The ball will hit the ground when $s = 0$, that is, when

$$-16t^2 + 400t + 100 = 0$$

The positive solution of this equation, according to the quadratic formula, is

$$t = \frac{-400 - \sqrt{160000 - 4(-16)(100)}}{-32} = 25.248 \text{ seconds}$$

(c) The ball will have height more than 1200 when

$$-16t^2 + 400t + 100 > 1200$$

and this will occur between the two solutions of the equation $-16t^2 + 400t - 1100 = 0$. Again, we apply the quadratic formula, obtaining $t = 3.146$ and $t = 21.854$. ■

APPLICATION: BRIDGE CABLES

The support cables for a suspension bridge take the shape of a parabola.

■ **Example 7.** The cables for a suspension bridge drop from towers 100

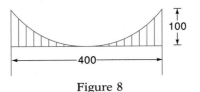

Figure 8

meters apart. Write the xy-equation of a cable, assuming the vertex is at the origin, and use it to determine the length of a supporting strut 50 meters from a tower.

Solution. The equation has the form $y = ax^2$. The point $(200, 100)$ is on the parabolic cable, which means that

$$100 = a(200)^2$$

Thus, $a = 1/400$, and the equation of the parabola is $y = (1/400)x^2$. At $x = 150$, $y = (1/400)(150)^2 = 56.25$. The strut is 56.25 meters long. ∎

APPLICATION: SLOPE AND RATE OF CHANGE

Recall a fact from Section 2.2, which was observed there for linear functions but is true in general.

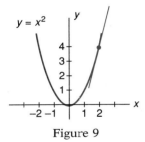

Figure 9

slope = rate of change

For a linear function $f(x) = mx + q$, the slope and rate of change have the constant value m for all x. Our goal is to make sense of one and hence both of these concepts for nonlinear functions.

Consider once again the nonlinear function $f(x) = x^2$ whose graph is displayed in Figure 9. At each point on this graph, there is a **tangent line** (the line that most closely approximates the graph in a neighborhood of the point). The **slope of the curve** at this point is defined to be the slope of the corresponding tangent line. Of course, this slope changes as the point moves along the curve. For example, at $x = -1$, the slope is negative, whereas at $x = 2$, the slope is positive and quite large. Is there a way of calculating these slopes and the corresponding rates of change?

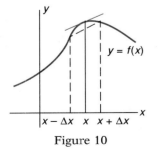

Figure 10

Consider the graph $y = f(x)$ of an arbitrary function with a smooth graph (no breaks or corners). Let Δx be a small number and consider the line (Figure 10) through the points with x-coordinates $x - \Delta x$ and $x + \Delta x$. This line has slope

$$m_x = \frac{f(x + \Delta x) - f(x - \Delta x)}{2\,\Delta x}$$

and this number should be a very good approximation to the actual slope at x. In fact, the smaller we take Δx the better the approximation (though the limitations of a calculator can undermine this assertion when actual calculations are made).

■ **Example 8.** Find the slope of the curve $y = f(x) = x^2$ at $x = 2$ (see Figure 9).

Solution. Choosing an appropriate Δx (not too big, not too small) is somewhat tricky, but let us agree to use $\Delta x = 0.01$ in examples and problems. We are thus led to make the following calculation.

$$\frac{f(2.01) - f(1.99)}{0.02} = \frac{4.0401 - 3.9601}{0.02} = 4$$

We conclude that the slope at $x = 2$ is (approximately) 4. ■

■ **Example 9.** Calculate the slope of the curve $y = f(x) = 0.4x^2 - 3x$ at $x = -2$ and $x = 6$ (see Figure 11).

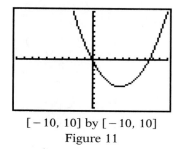

$[-10, 10]$ by $[-10, 10]$
Figure 11

Solution. The values to be calculated are

$$\frac{0.4(-1.99)^2 - 3(-1.99) - (0.4(-2.01)^2 - 3(-2.01))}{0.02}$$

and

$$\frac{0.4(6.01)^2 - 3(6.01) - (0.4(5.99)^2 - 3(5.99))}{0.02}$$

Our calculator gives -4.6 and 1.8 as the two answers. ■

TEASER SOLUTION

(a) The t-coordinate of the vertex is $t = -b/2a = -128/-32 = 4$ and the corresponding s-coordinate is $s = -16(4)^2 + 128(4) + 48 = 304$. The ball reached a height of 304 feet.

(b) The ball hit the ground when $s = 0$, that is, when $-16t^2 + 128t + 48 = 0$ or equivalently when $t^2 - 8t - 3 = 0$. The quadratic formula gives a positive solution to this equation of $4 + \sqrt{19} \approx 8.36$. The ball hit the ground after 8.36 seconds.

(c) We use our calculator to approximate the slope at $x = 8.36$ to be -139.52. Thus, the ball hit the ground at a rate of -139.52 feet per second (negative because the ball was falling), that is, with a speed of 139.52 feet per second. Note that this is a greater speed than the speed with which the ball was thrown upward, as it should be. Why?

PROBLEM SET 2.3

A. Skills and Techniques

Draw the graph of each function and guess at the distance between the points corresponding to $x = -2$ and $x = 1$. Then calculate this distance accurate to two decimal places.

1. $f(x) = 3x^2$
2. $g(x) = 0.2x^2$
3. $s(x) = -\dfrac{1}{8}x^2$
4. $h(x) = -15x^2$

Sketch (by hand) each set of graphs using the same axes. Confirm your results by using a graphics calculator.

5. $f(x) = 3x^2$, $g(x) = 0.3x^2$, $h(x) = -0.3x^2$, and $k(x) = -4x^2$
6. $f(x) = -2x^2$, $g(x) = -\dfrac{1}{2}x^2$, $h(x) = \dfrac{1}{2}x^2$, and $k(x) = 5x^2$
7. $f(x) = x^2 + 2$ and $g(x) = x^2 - 3$
8. $f(x) = 3x^2 - 5$ and $g(x) = -3x^2 + 5$
9. $f(x) = x^2$ and $g(x) = (x + 2)^2$
10. $f(x) = x^2$ and $g(x) = \left(x - \dfrac{1}{2}\right)^2$

Explain how the graph of the second function is related to the graph of the first. Then use your calculator to draw their graphs as a check on your answer.

11. $f(x) = x^2$ and $F(x) = (x - 3)^2 - 3$
12. $g(x) = -2x^2$ and $G(x) = -2(x + 1)^2 + 5$
13. $h(x) = 4x^2$ and $H(x) = -4(x - 3)^2 + 2$
14. $k(x) = 0.2x^2$ and $K(x) = 0.4(x - 1)^2 - 3$

Mentally decide whether you think the following pairs of graphs will intersect in two, one, or zero points. Then draw their graphs to check your answer.

15. $f(x) = 2x^2$ and $g(x) = 0.5x^2$
16. $f(x) = x^2 - 2$ and $g(x) = -x^2 + 2$
17. $f(x) = 2x^2$ and $g(x) = 0.5x^2 + 3$
18. $f(x) = x^2 + 2$ and $g(x) = -x^2 - 2$
19. $f(x) = x^2$ and $g(x) = (x - \pi)^2$
20. $f(x) = 0.7x^2$ and $g(x) = 0.6x^2 + 1$
21. $f(x) = -7x^2$ and $g(x) = -9x^2 - 2$
22. $f(x) = -x^2$ and $g(x) = (x + 1)^2$

The graph of each of the following is a parabola. Determine its vertex and orientation without actually graphing.

23. $f(x) = 5x^2 + 20x - 3$
24. $g(x) = \dfrac{7}{2}x^2 - 14x$
25. $f(x) = -0.75x^2 + 3.75x - 0.25$
26. $g(x) = -6x^2 + 4x - 1$
27. $f(t) = 4t^2 - t$
28. $g(u) = -0.3u^2 - 1.2u - 1.2$

Problems 29–36 are related to Examples 6 and 7. Assume that a falling body obeys the equation $s = -16t^2 + v_0 t + s_0$ where s is the height in feet at time t seconds.

29. A ball is shot straight up from the edge of a bridge with an initial velocity of 176 feet per second. (a) When does the ball reach maximum height? (b) What is this height (above the bridge)? (c) When will the ball hit the river 120 feet below the bridge? (d) During what interval is the ball more than 480 feet above the bridge?

30. A projectile is shot straight up from the ground with a velocity of 960 feet per second. (a) How high will it go? (b) When will it hit the ground? (c) During what time interval will it be more than 3200 feet high?

31. Elena plans to throw a baseball from the sidewalk to her brother Ricardo who is leaning out a window 140 feet above her. How fast must she throw the ball for it to just reach him?

32. If you toss a ball 4 feet in the air, what initial velocity are you giving it?

33. The parabolic cable for a suspension bridge is attached to two towers at points 400 feet apart and 90 feet above the horizontal bridge deck. The cable drops to a point 10 feet above the deck. Find the xy-equation of the cable assuming its vertex is at $(0, 10)$.

34. The curve shown in Figure 12 is part of a parabola. Determine the distance PQ.

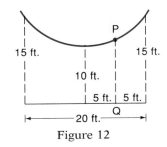

Figure 12

35. The parabolic cable for a suspension bridge drops to the bridge deck from points on towers 160 feet above the deck. If the towers are 600 feet apart, how far from a tower would we find an 80-foot supporting strut?

36. Write the xy-equation of a bridge cable that has its vertex at the center of the bridge deck and has two 70-foot supporting struts placed 280 feet apart. Then determine the distance between the towers if the cable is attached at points 110 feet above the deck.

Using $\Delta x = 0.01$ as in Examples 8 and 9, find the slope of the curve $y = f(x)$ at the indicated point. (You may wish to use the nDeriv feature on the TI-81 or TI-82 to do this.)

37. $f(x) = 3x^2 - 3$ at $x = 1$
38. $f(x) = -x^2 + x + 1$ at $x = 4$

39. $f(x) = \dfrac{5}{2}x^2 + \dfrac{3}{2}x$ at $x = -0.3$

40. $f(x) = 0.7x^2 - 0.3x + 0.2$ at $x = 2$

41. $f(x) = -2.4x^2 - 1.8$ at $x = -5$

42. $f(x) = (x + 3)^2$ at $x = 4$

B. Applications and Extensions

In Problems 43–50, we show the graph of a quadratic function f. Determine the formula for f(x). Note in Problems 49 and 50 that $f(x) = ax^2 + bx + c$ where c is the y-intercept.

43.

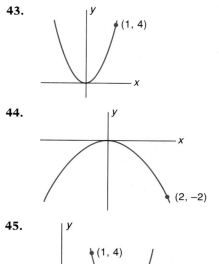

44.

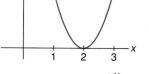

45.

46.

47.

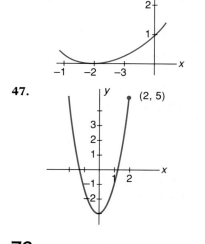

48.

49.

50.

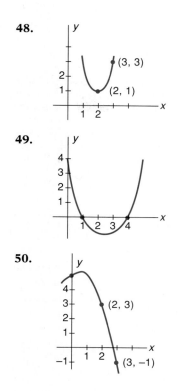

51. The parabola $y = a(x - 2)(x - 8)$ passes through the point $(10, 40)$. Find its vertex.

52. We call $d = b^2 - 4ac$ the discriminant of the parabola $y = ax^2 + bx + c$. If $d > 0$, what can we say about the number of x-intercepts of the parabola? (*Hint:* Think about the quadratic formula.)

53. For what value of c will the minimum value of $f(x) = x^2 + 8x + c$ be 3?

54. For what values of k does the parabola $y = x^2 - kx + 9$ have two x-intercepts?

55. A projectile is shot straight up from the ground with an initial velocity v_0. How do the projectile's maximum height and the time to reach this height change if the initial velocity is doubled?

56. Find a formula in terms of a and b for the area of the triangle shown in Figure 13, assuming $c = (a + b)/2$.

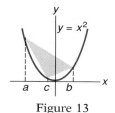

Figure 13

57. Let $s = t^2 - 3t + 7$ represent the x-coordinate (in meters) of an object traveling along a horizontal line after t seconds. Determine its velocity at time $t = 2$.

58. If $m = 3u^2 + 4u$, how fast is m changing with respect to u when $u = 2$?

59. A retailer has learned from experience that if she charges x dollars apiece for toy trucks, she can sell $300 - 10x$ of them each month. The trucks cost her $12 each. (a) Write a formula for her total monthly profit P on trucks in terms of x. (b) Determine what she should charge for each truck to maximize her profit. (c) Calculate the rate of change of P with respect to x when $x = 20$.

60. **Challenge.** Let $f(x) = ax^2 + bx + c$ be an arbitrary quadratic function. Show that the slope formula

$$m_x = \frac{f(x + \Delta x) - f(x - \Delta x)}{2\Delta x}$$

gives the same value for every choice of Δx. (*Note:* This is definitely not true for higher-degree polynomial functions.)

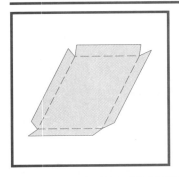

TEASER An open box is to be made from a piece of sheet metal 12 inches by 16 inches by cutting squares of side x inches from each of the four corners and turning up the sides. Determine the maximum volume and the value of x that gives this volume.

2.4 MORE ON GRAPHICS CALCULATORS

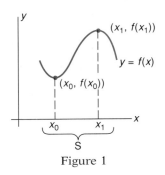

Figure 1

We are all engaged in optimization. Students want to maximize their grade point averages; professors want to minimize their work loads; workers want to maximize their salaries. If the quantity to be optimized can be expressed in terms of a variable x, we may be able to use mathematics to solve these problems. We begin by clarifying our terminology (Figure 1).

> **Definition.** Let f have domain S. Then $f(x_0)$ is the **minimum value** for f on S if $f(x) \geq f(x_0)$ for all x in S. Similarly, $f(x_1)$ is the **maximum value** for f on S if $f(x) \leq f(x_1)$ for all x in S. In this case, $(x_0, f(x_0))$ and $(x_1, f(x_1))$ are called a **minimum point** and a **maximum point,** respectively.

As a simple example, consider $f(x) = (x - 2)^2 - 4$ on the interval $0 \leq x \leq 5$ (see Figure 2). Then $f(2) = -4$ is the minimum value and $f(5) = 5$ is the maximum value. Also, $(2, -4)$ is the minimum point; $(5, 5)$ is the maximum point. Ordinarily, we cannot determine optimum values and optimum points by such simple reasoning; here is where graphics calculators can help us.

OPTIMIZATION WITH GRAPHICS CALCULATORS

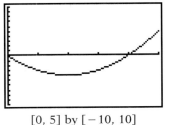

[0, 5] by [−10, 10]

Figure 2

We begin by graphing the function we want to optimize. Then we scan the graph to determine minimum and maximum points approximately. Finally, we zoom to determine these points accurately.

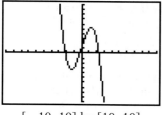

$[-10, 10]$ by $[10, 10]$

Figure 3

■ **Example 1.** Determine the minimum and maximum points for $f(x) = -x^3 + 5x + 1$ for the interval $0 \leq x \leq 4$.

Solution. A calculator-generated graph of this function is shown in Figure 3. We see immediately that the minimum point for the interval $0 \leq x \leq 4$ occurs at the right end. Since $f(4) = -43$, this point has coordinates $(4, -43)$.

The maximum point for the given interval appears to occur near $x = 1$. To identify it more precisely, we move the trace cursor as near as we can to the maximum point and then zoom in over and over. Some screens in the process are shown in Figure 4. A study of these screens suggests that the maximum point is approximately $(1.29, 5.30)$. ■

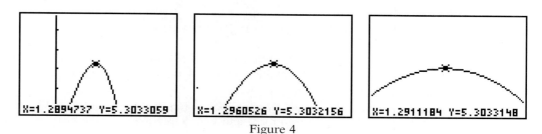

Figure 4

Study Figure 4 again. It shows clearly what happens when we zoom in on a smooth curve over and over. The curve begins to look more and more like a straight line; this tendency makes accurate graphical determination of optimum points difficult. Is there a way to overcome this problem? Yes, by stretching the curve more and more in the vertical direction. We illustrate in our next example.

■ **Example 2.** Find the minimum point of $f(x) = -x^3 + 5x + 1$ for the interval $-4 \leq x \leq 0$.

Solution. This is the same function studied in Example 1 but now with a different interval. Reproduce its graph, move the trace cursor near the required minimum point (Figure 5), and get ready to zoom in. But this time, set the zoom factors so that the enlargement in the y-direction is much larger than in the x-direction (for example, set XFact = 4, YFact = 12). Now when we zoom in, the curve will appear to approach a straight line at a much slower rate than in Example 1 thus allowing easier and more accurate identification

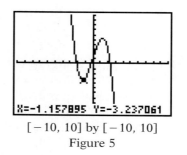

$[-10, 10]$ by $[-10, 10]$

Figure 5

of the optimum point. After several steps, we conclude that the minimum point is approximately $(-1.291, -3.303)$. ■

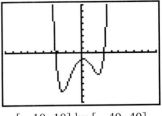

$[-10, 10]$ by $[-40, 40]$
Figure 6

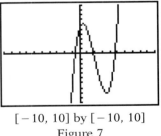

■ **Example 3.** Find the minimum point of $f(x) = 0.05x^6 - 5x^2 + 3x - 5$.

Solution. Since no interval is specified, we take it to be the whole real line. Experimenting with range values suggests setting Xmin $= -10$, Xmax $= 10$, Ymin $= -40$, and Ymax $= 40$. The resulting graph is shown in Figure 6 and leads us to believe that there is a minimum point near $(-2.5, -30)$.

To identify this point more accurately, we use the magnification factors (XFact $= 4$, YFact $= 12$) and zoom in repeatedly. After several steps, we obtain $(-2.473, -31.561)$. ■

TI-82 HINT

The zooming procedure described in the text for finding maxima and minima works fine on both the TI-81 and TI-82. Owners of the TI-82 will discover automatic routines for finding maxima and minima included in the CALC menu. Once they understand the zooming method, they should feel free to use these routines.

INCREASING AND DECREASING FUNCTIONS

It should be pretty obvious what we mean by the words *increasing* and *decreasing* but to make sure we give a definition.

> **Definition.** We say that f is **increasing** on an interval S if $x_1 < x_2$ implies that $f(x_1) < f(x_2)$ for x_1 and x_2 in S. Similarly, we say that f is **decreasing** on S if $x_1 < x_2$ implies that $f(x_1) > f(x_2)$ for x_1 and x_2 in S.

An important question is to determine where a function is increasing and where it is decreasing. While the complete analysis of this question requires the methods of calculus, we can usually approximate the correct answer by simply studying the graph of the function. A graphics calculator with the zoom feature will be an invaluable aid.

$[-10, 10]$ by $[-10, 10]$
Figure 7

■ **Example 4.** Determine where the function $f(x) = x^3 - 6x^2 + 5x + 5$ is decreasing.

Solution. The calculator-generated graph shown in Figure 7 suggests that the interval $0.5 \le x \le 3.5$ is approximately what we want. To get better accuracy, we use the zooming process to determine the x-coordinates of the maximum point and minimum point at the ends of this interval (see Examples 2 and 3). We obtain as our best three-decimal-place answer the interval $0.472 \le x \le 3.528$. ■

That was straightforward enough. Our next example shows that this subject is more subtle than Example 4 suggests.

$[-10, 10]$ by $[-10, 10]$
Figure 8

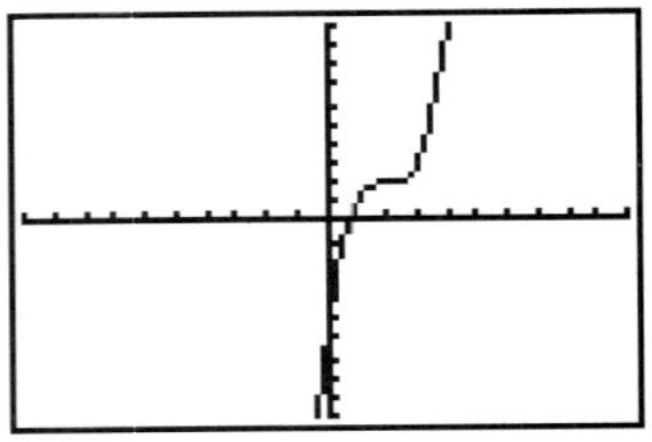

■ **Example 5.** Determine if the function $f(x) = x^3 - 6x^2 + 11.97x - 6$ ever decreases and if so where.

Solution. The graph of this function is shown in Figure 8. Our first impression is that f increases everywhere, but we should at least check out what appears to be a level spot near $x = 2$. When we do, using the zooming process (with XFact $= 4$ and YFact $= 12$), we discover that f decreases on the interval $1.90 \le x \le 2.10$. (The techniques of calculus are needed to demonstrate conclusively that this is the correct answer.) ■

CONCAVE UP AND CONCAVE DOWN

Where does a curve open up (hold water) and where does it open down (spill water)? We need to be more precise about our terminology.

> **Definition.** We say that f is **concave up** on an interval S if the inequality
> $$f(ux_1 + vx_2) \leq uf(x_1) + vf(x_2)$$
> holds for all x_1 and x_2 in S and all positive weights u and v summing to 1. It is **concave down** if the reverse inequality holds.

Study Figure 9 to absorb the meaning of this definition. In essence, it says that a function is concave up on an interval if the line segment joining two points of the graph of f is always on or above the graph of f. It is concave down when this line segment is always on or below the graph. Determining where a function is concave up or concave down is another important question best treated with the tools of calculus. However, careful study of the graph of a function will give an approximate answer to this question.

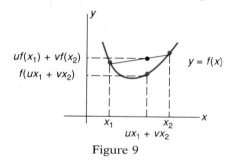

Figure 9

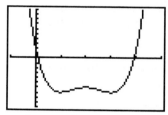

$[-1, 5]$ by $[-10, 10]$

Figure 10

■ **Example 6.** Determine where the graph of $g(x) = x^4 - 8x^3 + 22x^2 - 24x + 2$ is concave down.

Solution. The calculator-generated graph shown in Figure 10 suggests that f is concave down on the interval $1.5 \leq x \leq 2.5$. The zooming process does not help much in improving accuracy. ■

AN APPLICATION

This application is based on the well-known fact that gas mileage for a vehicle decreases when its speed increases.

■ **Example 7.** Suppose that the cost of operating a certain truck (gas, oil, maintenance, and so on) when driven at x miles per hour is $30 + 0.35x$ cents. At what speed should this truck be driven to minimize cost if the driver gets $11 per hour?

Solution. Let's determine the total cost $C(x)$ in cents of driving the truck 1 mile at speed x. The time to travel 1 mile is $1/x$, and therefore the cost of the driver for 1 mile is $1100/x$ cents. Thus,

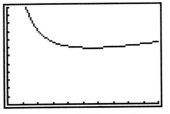

Figure 11

$$C(x) = 30 + 0.35x + \frac{1100}{x}$$

The graph of this function on the interval $0 < x \leq 100$ is shown in Figure 11. When we minimize, using the zooming process (see Examples 2 and 3), we find the minimum point to be (56.06, 69.24). The truck should be driven at about 56 miles per hour, which gives a cost of about 69¢ per mile. ∎

TEASER SOLUTION

The volume $V(x)$ of the box (see Figure 12) is given by

$$V(x) = x(16 - 2x)(12 - 2x)$$

with domain $0 \leq x \leq 6$. The graph of V appears in Figure 13. The zooming process shows that there is a maximum volume of 194.07 cubic inches when $x = 2.26$ inches.

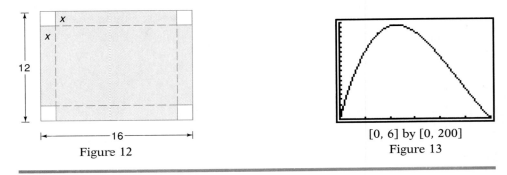

Figure 12

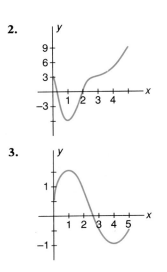

[0, 6] by [0, 200]
Figure 13

PROBLEM SET 2.4

A. Skills and Techniques

Problems 1–4 show the graph of a function f on the interval $0 \leq x \leq 5$. Determine as best you can (a) the maximum value of f, (b) the minimum value of f, (c) the x-interval(s) on which f is increasing.

1.

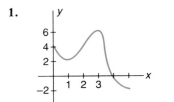

2.

3.

4.

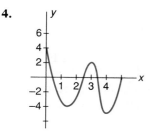

Without graphing, determine the coordinates of the maximum point for each function on the given interval. Then confirm your answer by graphing.

5. $f(x) = 2x + 5$ on $-2 \le x \le 3$
6. $g(x) = -4x - 14$ on $-20 \le x \le -10$
7. $h(x) = -\frac{3}{2}x^2 + 4$ on $-3 \le x \le 3$
8. $k(x) = 7 - (x - 4.3)^2$ on $0 \le x \le 10$
9. $f(x) = -1 - (x - 2.1)^2$ on $0 \le x \le 10$
10. $g(x) = -x^2 + 6x + 19$ on $0 \le x \le 10$

Use graphing and zooming to find the maximum and minimum points of each function on the given interval, accurate to two decimal places.

11. $f(x) = \frac{1}{3}x^3 - \frac{1}{2}x^2 - x$ on $-1.5 \le x \le 2$
12. $g(x) = -0.1x^3 + 2x + 1$ on $-5 \le x \le 5$
13. $h(x) = \frac{1}{2}x^4 - 20x^3 + 201x^2 + 50$ on $-5 \le x \le 25$
14. $k(x) = 0.01x^5 - 0.1x^4 - 5x^2 + 100$ on $-6 \le x \le 12$
15. $f(x) = \frac{0.2x^3 - 5x + 6}{x^2 + 7}$ on $-5 \le x \le 5$
16. $g(x) = \frac{x^2 + 1}{x^4 + 10}$ on $0 \le x \le 5$

Find the indicated point accurate to two decimal places.

17. The minimum point of $f(x) = x^4 - 100x$
18. The minimum point of $f(x) = 0.03x^4 + 0.05x^3 - 5x^2 + 2$
19. The maximum point of $g(x) = -0.01x^6 - 0.01x^5 + x^2$
20. The maximum point of $h(x) = -\frac{1}{60}x^6 + \frac{1}{50}x^5 + 10$

Determine, as in Examples 4 and 5, the intervals on which the given function is decreasing.

21. $f(x) = x^3 + x^2 - 5x$
22. $f(x) = 0.4x^3 - 5x^2 + 12x + 6$
23. $g(x) = -0.015x^5 + 6x^2 + 8$
24. $h(x) = \frac{1}{5}x^5 - \frac{5}{2}x^3 + x^2 + 4x$
25. $f(x) = \frac{1}{5}x^5 - \frac{1}{2}x^4 - \frac{5}{3}x^3 + 3x^2$
26. $k(x) = x^3 + 6x^2 + 13x$

27. $f(x) = x^3 - 2x^2 + 1.35x$
28. $f(x) = \dfrac{x^3 + x}{x^2 + 0.1}$

Use a graph, as in Example 6, to discover where the graph of each function is concave up.

29. $f(x) = -x^4 + 9x^2 + x + 20$
30. $g(x) = -4x^2 + 11x - 2$
31. $g(x) = 0.2x^3 + 1.2x^2 - 5x$
32. $f(x) = -x^4 + 10x^3 - 18x^2 - 22x$
33. $h(x) = x^4 - x^3 + x^2 - 6$
34. $f(x) = -0.2x^4 + 0.5x^3 + 2x^2 - 5x + 5$

Turn each problem into the mathematical problem of maximizing or minimizing a function f on a certain domain. Give this function and the appropriate domain. Then solve the problem.

35. Harry plans to fence a rectangular pasture along a straight river (Figure 14). Assuming the riverside does not require fence and that he has 1800 feet of fence available, what dimensions will maximize the area of the pasture?

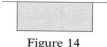

Figure 14

36. If Harry of Problem 35 chooses to use the 1800 feet of fence to make three identical pens along the river as shown in Figure 15, what dimensions would maximize the area of each pen?

Figure 15

37. If your body burns $150 + 2x^2$ calories per hour when bicycling at x miles per hour, at what speed should you ride to minimize the number of calories consumed on a 50-mile trip?
38. Redo Problem 37 assuming this time that your body uses $200 + 4x^{3/2}$ calories per hour.

B. Applications and Extensions

39. Determine the minimum point for $f(x) = (0.5x^3 + 1)/x^2$ for $x > 0$.
40. Determine the x-interval on which $g(x) = (3 + 8x)/(1 + x^2)$ is increasing.
41. If the sum of two numbers is 30, what is the smallest possible value for the sum of their squares?
42. If the difference of two numbers is 5, what is the smallest possible value for the sum of their squares?

43. The rectangle shown in Figure 16 has two sides along the coordinate axes and a vertex (x, y) on the curve $y = 9 - x^2$. Determine x and y so that the area of the rectangle is maximized.

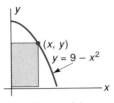

Figure 16

44. Repeat Problem 43 but with the curve $y = 9 + x - x^3$.

45. Consider Problem 43 for the curve $(9 + x^2)/x^2$. Find (if possible) the maximum and minimum area for the rectangle.

46. A 30-foot rope is cut into two pieces of length x and $30 - x$ ($0 \leq x \leq 30$). The piece of length x is formed into a circle and the other piece into a square. What value of x minimizes their combined area? Maximizes this area?

47. A closed rectangular box is to have square ends and is to contain 16 cubic feet. What dimensions will minimize its surface area?

48. Two sides of a triangle are to be 10 inches long. How long should the third side be to make the triangle have maximum area?

49. An apartment complex has 90 units. When the monthly rent is $500 per unit, all units are rented. It is estimated that for each $10 increase in rent, one apartment unit will become vacant. What rent should be charged to maximize the total revenue?

50. The people planning a fundraising event feel confident that they can sell all 1100 tickets if they charge $20 per ticket but believe they will fail to sell 25 tickets for every $1 increase in price. How should they set the ticket price to maximize their revenue?

51. Let f be a function that is concave up on an interval set S and let α_1, α_2, and α_3 be positive weights with sum 1. Show that for any numbers x_1, x_2, and x_3 in S, $f(\alpha_1 x_1 + \alpha_2 x_2 + \alpha_3 x_3) \leq \alpha_1 f(x_1) + \alpha_2 f(x_2) + \alpha_3 f(x_3)$. (*Hint:* $\alpha_1 x_1 + \alpha_2 x_2 + \alpha_3 x_3 = (\alpha_1 + \alpha_2)((\alpha_1 x_1 + \alpha_2 x_2)/(\alpha_1 + \alpha_2)) + \alpha_3 x_3$.)

52. **Challenge.** State an appropriate generalization of Problem 51. Then try to prove it.

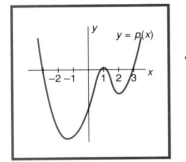

TEASER Determine the fourth-degree polynomial $p(x)$ whose graph crosses the x-axis at $x = -3$ and $x = 3$ and just touches this axis at $x = 1$, given that $p(0) = -18$.

2.5 POLYNOMIAL FUNCTIONS

We have studied linear functions $f(x) = ax + b$ and quadratic functions $f(x) = ax^2 + bx + c$. We could now go on to cubic functions, quartic functions, and so on. Rather we choose to jump to the general case of a polynomial function. A (real) **polynomial** $p(x)$ is an expression of the form

$$p(x) = a_n x^n + a_{n-1} x^{n-1} + \cdots + a_1 x + a_0$$

where n is a nonnegative integer and the as (called coefficients) are real numbers. If $a_n \neq 0$, then n is the **degree** of the polynomial. Thus, $p(x) = 2x^4 + x^3 + 3x - 1$ is a fourth-degree, or quartic, polynomial, and $p(x) = x^7 + \pi$ is a seventh-degree polynomial. On the other hand, $p(x) = x^{1.5} - 2x$ is not a polynomial and neither is $p(x) = x^2 + 3/x$. We have previously used polynomials in many of our examples, but now we want to study them as a class, looking

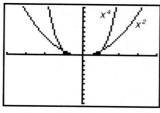

$[-4, 4]$ by $[-10, 10]$
Figure 1

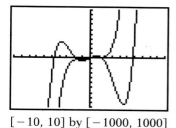

$[-4, 4]$ by $[-10, 10]$
Figure 2

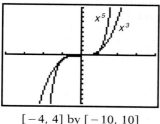

$[-10, 10]$ by $[-1000, 1000]$
Figure 3

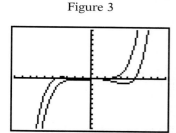

$[-10, 10]$ by $[-10000, 10000]$
Figure 4

for general properties and patterns. Note that the natural domain for a real polynomial is the set $\mathbb{R}$ of all real numbers.

THE GRAPH OF A POLYNOMIAL

To learn something about the graphs of polynomials, we look first at the graph of $p(x) = x^n$ for several values of n.

■ **Example 1.**   Draw the graph of $p(x) = x^n$ for $n = 2, 3, 4,$ and 5.

Solution. Figure 1 shows the graph for $n = 2$ and 4 with the range values $[-4, 4]$ by $[-10, 10]$; Figure 2 does the same for $n = 3$ and 5. It would be a good idea to try other range values to get a more complete picture of these graphs.   ■

Example 1 leads us to make two observations about the graph of $p(x) = x^n$ that are both true and very important.

1.   *If n is even, both arms of the graph of $f(x) = x^n$ point up; if n is odd, the left arm points down and the right arm up.*

2.   *The larger n is the more rapidly the graph rises or falls as the distance from the origin increases.*

Next we illustrate how the behavior of the leading term (the highest-degree term) determines the behavior of the graph of a polynomial at distances far from the origin.

■ **Example 2.**   Draw the graphs of $p(x) = x^5 - x^4 - 30x^3 + 80x + 3$ and $q(x) = x^5$ in the same plane.

Solution. Figure 3 shows these graphs first for the range values $[-10, 10]$ by $[-1000, 1000]$. Figure 4 shows the same thing but with Ymin $= -10,000$ and Ymax $= 10,000$.   ■

On the basis of this example and thinking about the relative sizes of x^n for various values of n, we make a third observation.

3.   *The behavior of the graph of a polynomial far from the origin is similar to that of the graph of the leading term. We say that the leading term of a polynomial $p(x)$ dominates the other terms for large $|x|$.*

Observations 1, 2, and 3 are true in general but we do not claim to have proved them. You should try many other examples and work the problem set to add evidence. Rigorous proofs would take us too far afield.

ZEROS OF POLYNOMIAL FUNCTIONS

We call c a **zero** of the function f if $f(c) = 0$, that is, if c is a solution to the equation $f(x) = 0$. A great deal is known about the zeros of a polynomial function, a topic we consider now.

The number c is a zero of the polynomial $p(x)$ if and only if $x - c$ is a factor of $p(x)$.

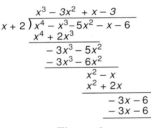

Figure 5

Proof. Suppose $x - c$ is a factor of $p(x)$, meaning that $p(x) = (x - c)q(x)$ where $q(x)$ is another polynomial. Then

$$p(c) = (c - c)q(c) = 0$$

Conversely suppose that $p(c) = 0$. The long division process shows that if we divide $p(x)$ by $x - c$, obtaining a quotient $q(x)$ and a constant remainder R as illustrated in Figure 5, then

$$p(x) = (x - c)q(x) + R$$

This in turn implies that $p(c) = R$, and since $p(c) = 0$, R must be 0. Thus $p(x) = (x - c)q(x)$.

■ **Example 3.** Determine a fourth-degree polynomial whose only zeros are ± 2, 0, and 7.

Solution. One such polynomial is

$$x(x - 7)(x - 2)(x + 2) = (x^2 - 7x)(x^2 - 4) = x^4 - 7x^3 - 4x^2 + 28x$$

Any constant multiple of this polynomial would work as well. ■

■ **Example 4.** Factor the polynomial $p(x) = x^4 - x^3 - 5x^2 - x - 6$ given that -2 and 3 are zeros of $p(x)$.

Figure 6

Solution. Since -2 is a zero, $x + 2$ is a factor. Either long division (Figure 6) or synthetic division (assuming you know this process) can be used to show that

$$p(x) = (x + 2)(x^3 - 3x^2 + x - 3)$$

Similarly, $x - 3$ must be a factor of $x^3 - 3x^2 + x - 3$. The other factor, obtained by long division or synthetic division, is $x^2 + 1$. Thus

$$p(x) = (x + 2)(x - 3)(x^2 + 1)$$

(Using only real polynomials, we cannot factor further. If we allow the use of complex numbers, however, we can write $p(x) = (x + 2)(x - 3)(x - i)(x + i)$.) ■

The situation just illustrated is typical as is clear from one form of a very famous theorem.

Any real polynomial can be factored into a product of real linear and real quadratic factors (the latter having no real zeros).

This theorem, which had been conjectured for a long time, was finally proved rigorously by Carl Gauss in 1799. Many different proofs are known, but all of them are above the level of this book. For a good discussion, see William Dunham, "Euler and the fundamental theorem of algebra," *College Mathematics Journal* 22, 1991, pp. 282–293.

It follows from the fundamental theorem of algebra that a polynomial of degree n can have at most n zeros. It may have considerably fewer. For example the polynomial $(x + 4)^3(x + 1)^2(x - 1)$, which is of degree 6, has only three zeros, namely -4, -1, and 1. It is said to have -4 as a zero of *multiplicity* 3, -1 as a zero of multiplicity 2, and 1 as a simple zero.

RELATING ZEROS TO THE GRAPH

We already know from Section 1.3 that the real zeros of $f(x)$ are the x-coordinates of the points where the graph of $y = f(x)$ intersects the x-axis. But for polynomials we can say more.

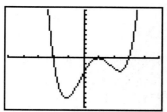

$[-6, 6]$ by $[-10, 10]$

Figure 7

■ **Example 5.** Draw the graph of $p(x) = (1/10)(x + 4)^3(x + 1)^2(x - 1)$.

Solution. After setting the range values $[-6, 6]$ by $[-10, 10]$, we obtain the graph shown in Figure 7. ■

Note another fact that is illustrated in Example 5.

4. *The graph of a polynomial is tangent to the x-axis at a real zero of multiplicity greater than 1 and crosses the x-axis at a real zero whose multiplicity is odd.*

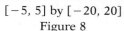

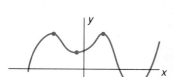

$[-5, 5]$ by $[-20, 20]$

Figure 8

■ **Example 6.** Draw the graph of $p(x) = x^4 - 3x^3 - 3x^2 + 11x - 6$. Use this graph to determine its real zeros with their multiplicities.

Solution. The graph (Figure 8) suggests that there are real zeros of odd multiplicity at (approximately) $x = -2$ and $x = 3$ together with a zero of even multiplicity at $x = 1$. Since the number of zeros with their multiplicities must total 4, we conclude that 1 is a double zero whereas -2 and 3 are simple zeros. You may check algebraically that what the graph suggests is actually the case. ■

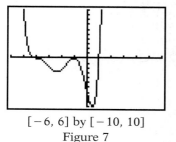

Turning points

Figure 9

TURNING POINTS

We call the maximum points and minimum points of a polynomial **turning points** (Figure 9). Note that we include what should technically be called *local* maximum and minimum points, that is points that give a maximum value or minimum value with respect to points nearby. Now we ask the question: How many turning points can an nth-degree polynomial have? Figures 3, 4, and 7 may suggest an answer.

5. *A polynomial of degree n has at most n − 1 turning points. In fact, the number of turning points is either n − 1 or is less than this by an even number.*

Note that if $c_1, c_2, \ldots, c_n$ are n distinct real numbers, then the graph of $p(x) = (x - c_1)(x - c_2) \ldots (x - c_n)$ will cross the x-axis at each of these numbers and must therefore have a turning point in between each adjoining pair of them. In this case, the nth-degree polynomial will have $n - 1$ turning points.

TEASER SOLUTION

A fourth-degree polynomial whose graph crosses the x-axis at ± 3 and just touches at 1 has the form $p(x) = a(x - 3)(x + 3)(x - 1)^2$. But we are given that $p(0) = -18$ so $-18 = a(-3)(3)(-1)^2$. Thus $a = 2$ and

$$p(x) = 2(x - 3)(x + 3)(x - 1)^2 = 2(x^2 - 9)(x^2 - 2x + 1)$$
$$= 2x^4 - 4x^3 - 16x^2 + 36x - 18$$

PROBLEM SET 2.5

A. Skills and Techniques

Determine before graphing each function the direction the arms of each graph should point. Then confirm your answer by graphing.

1. $f(x) = x^7$
2. $f(x) = x^6$
3. $f(x) = -x^5$
4. $f(x) = 4x^3$
5. $f(x) = -6x^4$
6. $f(x) = -0.01x^7$

Using the same axes, graph the given polynomial and its leading term. Experiment with various range values in order to make these graphs illustrate that the leading term dominates a polynomial for large $|x|$. For example, a choice of $[-10, 10]$ by $[-10, 10]$ in Problem 7 is poor. A choice of $[-25, 25]$ by $[-10,000, 10,000]$ is better.

7. $f(x) = x^3 - 3x^2 + 5$
8. $f(x) = -5x^3 - 22x$
9. $f(x) = 2x^4 - 12x^3 + 2x$
10. $f(x) = 0.14x^5 - 0.2x^4 - 3x^2 - 10x + 2$
11. $f(x) = -\dfrac{1}{8}x^5 - \dfrac{1}{2}x^3 + 89x^2 + 76$
12. $f(x) = -x^6 + 2x^5$

Write a polynomial $P(x)$ that satisfies the given conditions. Then graph this polynomial to confirm that it meets the requirements. Assume that the leading coefficient is 1 unless otherwise specified.

13. $P(x)$ is of third degree with zeros ± 1 and 3.
14. $P(x)$ is of third degree with zeros $-7, -5$, and 0.
15. $P(x)$ is of fourth degree with zeros $-3, -1, 0$, and 5.
16. $P(x)$ is of fourth degree with zeros ± 2 and ± 1.

17. $P(x)$ has leading term $2x^5$ and zeros 0, 1, 2, 3, and 5.
18. $P(x)$ has leading term $-x^3$ and zeros ± 2 and 7.
19. $P(x)$ is of third degree, has zeros $-1/2, 1$, and 3, and $P(0) = 9$.
20. $P(x)$ is of fourth degree, has zeros ± 3 and $\pm \sqrt{3}$, and $P(0) = -81$.
21. $P(x)$ is of fourth degree and has ± 1 as its only zeros with 1 being a simple zero.
22. $P(x)$ is of third degree and has 2 as its only zero.
23. $P(x)$ has as its only zeros -2 (multiplicity 3) and 3 (multiplicity 2).
24. $P(x)$ is of fourth degree and has simple zeros at ± 3 and no other real zeros.

Factor each polynomial into linear and quadratic factors as guaranteed by the fundamental theorem of algebra.

25. $f(x) = x^3 - 8x^2 + 17x - 10$, given that 1 is a zero.
26. $f(x) = x^3 + (5/2)x^2 + (1/2)x - 1$, given that 1/2 is a zero.
27. $g(x) = x^4 - 5x^3 - 25x^2 + 65x + 84$, given that 3 and 7 are zeros.
28. $g(x) = x^4 + (10/3)x^3 + 4x^2 + 10x + 3$, given that -3 and $-1/3$ are zeros.

By studying the form of the polynomial, identify all the x-values where the graph of the polynomial will cross the x-axis, become tangent to the x-axis as it crosses it, or just touch but not cross the x-axis. Then graph the polynomial to confirm your answer.

29. $f(x) = (x + 3)(x - 2)(x - 4)$
30. $f(x) = x(x^2 - 9)$
31. $f(x) = (x + 2)(x^2 + 9)$
32. $f(x) = (x + 2)^2(x^2 + 9)$
33. $f(x) = (x + 2)^2\left(x - \dfrac{1}{2}\right)x$

34. $f(x) = (x + 2)^3(x - 1)^2$

35. $f(x) = (2x + 7)^3(3x - 1)\left(\dfrac{1}{3}x\right)$

36. $f(x) = (x^2 - 2)^2(x^2 - x - 12)$
37. $f(x) = x^6 - x^5 - 6x^4$
38. $f(x) = x^3 - 3x^2 + 3x - 1$

B. Applications and Extensions

Problems 39–44 show graphs of a fifth-degree polynomial $p(x)$. Write its formula in factored form, assuming the leading coefficient is 1. Check by graphing your answer.

39.

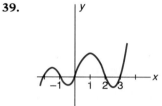

40.

41.

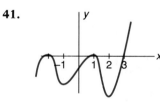

42.

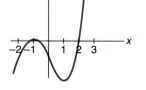

43.

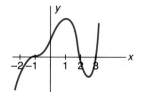

44.

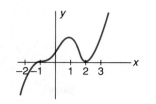

45. Determine the third-degree polynomial $p(x)$ with simple zeros of -1, 2, and 4 satisfying $p(5) = 36$.
46. A polynomial $p(x)$ has leading term $-2x^3$ with simple zeros -1, 1, and c. If $p(2) = 12$, find $p(x)$.
47. Determine k given that 1 is a zero of the polynomial $3x^3 - 2x^2 + kx - 3$.
48. Factor $9x^3 + 18x^2 - x + k$, given that -2 is a zero.

In Problems 49–52, determine the number of turning points of $p(x)$, assuming that $a < b < c$.

49. $p(x) = (x - a)(x - b)(x - c)$
50. $p(x) = (x - a)^3(x - b)^2(x - c)$
51. $p(x) = (x - a)^2(x - b)^2(x - c)^2$
52. $p(x) = (x - a)^4(x - b)(x - c)$

In Problems 53–58, explain why each figure cannot be the graph of a polynomial with leading term $3x^4$.

53.

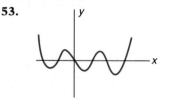

54.

55.

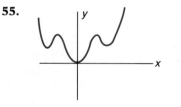

56.

57.

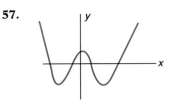

58.

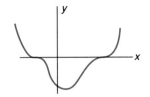

59. If $y = 2x^3 - 4x^2 + 5x - 6$, how fast is y changing with respect to x when $x = 2$? See Section 2.3.

60. At what rate does the volume of a sphere change with respect to its radius x when this radius is 20 inches?

61. An object is moving along a line so that its distance s from the origin after t seconds is $t^3 - 3t^2 + 4$ meters. Determine its velocity at time $t = 3$.

62. In Problem 61, determine the velocity at time $t = 1$. What is the significance of the negative sign?

63. Show that if c is a zero of $p(x) = x^4 + 3x^3 - 5x^2 + 3x + 1$ then $1/c$ is also a zero. What property of the coefficients makes this true? Can you state a general theorem?

64. **Challenge.** Take an ordinary sheet of 8½ by 11-inch paper and fold a corner over to the opposite side, thus determining triangle *DEF* (Figure 10) with area $A(x)$. Experiment with several folds to see how A varies with x.

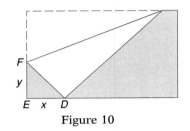

Figure 10

(a) $A(x)$ is a third-degree polynomial. Find it.
(b) Check that 0 and 8.5 are zeros of $A(x)$. Why must this be so?
(c) Determine x in the interval $0 \le x \le 8.5$ so that $A(x)$ is maximized.

TEASER A canner of fruit juices wishes to make cylindrical cans that hold exactly one liter (1000 cubic centimeters). What dimensions for such a can require the least amount of material; that is, what dimensions give a can of least surface area?

2.6 RATIONAL FUNCTIONS

A rational function is a ratio of two polynomial functions. More precisely, f is a **rational function** if $f(x) = p(x)/q(x)$ where $p(x)$ and $q(x)$ are polynomials. Examples of rational functions are

$$f(x) = \frac{2x - 3}{x + 1} \qquad g(x) = \frac{x^3 + 3x - 1}{x^2 + 1} \qquad h(x) = \frac{x^2 - x - 6}{x^2 - 4}$$

The third of these functions introduces a complication we wish to avoid. Note that the numerator can be factored as $(x - 3)(x + 2)$ and the denominator as $(x - 2)(x + 2)$; thus we can cancel the common factor of $x + 2$ in numerator and denominator—or can we? It is true that for almost all purposes, we can replace the function h by the function $\hat{h}$ with the formula

$$\hat{h}(x) = \frac{x - 3}{x - 2}$$

But h and $\hat{h}$ are not identical since h is undefined at $x = -2$ but $\hat{h}(-2) = 5/4$. Because we wish to avoid further discussion of this technical point, all rational functions discussed in this section are in reduced form; that is, they have no nontrivial common polynomial factors in numerator and denominator. Our goal is to study the graphs of reduced rational functions.

HORIZONTAL ASYMPTOTES

To begin, consider the following rational function.

$$f(x) = \frac{3x^2 - 4}{x^2 + 1}$$

Figures 1 and 2 show two views of its graph.

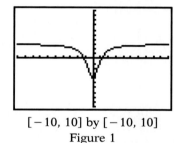
$[-10, 10]$ by $[-10, 10]$
Figure 1

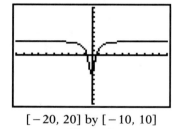
$[-20, 20]$ by $[-10, 10]$
Figure 2

These figures lead us to conjecture that the graph of $y = f(x)$ snuggles more and more closely to the line $y = 3$ as the distance from the origin increases. More evidence is provided by tracing the graph far out in either horizontal direction, all the time watching the y-value of the trace cursor. (*Note:* On the TI-81 and TI-82, the viewing rectangle automatically pans to the right or left as the cursor crosses the edge, so you can continue to view the graph.)

We need a notation and a language for describing the behavior we observe. Let the horizontal arrow → be an abbreviation for the word *approaches;* and let $x \to \infty$, read "x approaches infinity," be an abbreviation for the phrase "x grows larger and larger without bound." Then we can describe the right-end behavior of the function f by saying $f(x) \to 3$ as $x \to \infty$. Similarly, we describe the left-end behavior of f by saying $f(x) \to 3$ as $x \to -\infty$. In addition, we call the line $y = 3$ a horizontal asymptote for the graph of f.

In general, when the graph of a function gets closer and closer to a line as its distance from the origin increases, that line is called an **asymptote** for the graph. In hand graphing, we often draw asymptotes for a graph as dotted lines; they are not part of the graph of the function, but they are guidelines that help us draw the graph and explain its behavior.

■ **Example 1.** Draw the graph of $f(x) = (4x^2 - 8x)/(x^2 + 1)$ both by hand and using a calculator on the interval $-10 \le x \le 10$.

Solution. To get a handle on the end behavior of $f(x)$, divide numerator and denominator by x^2 (the highest power in the expression for $f(x)$) to obtain

$$f(x) = \frac{4 - 8/x}{1 + 1/x^2}$$

In this form, it is pretty clear that $f(x) \to 4$ as $x \to \infty$ or as $x \to -\infty$. Thus, the line $y = 4$ is a horizontal asymptote for the graph (in both directions). Our first step in drawing a hand graph is to draw in $y = 4$ as a dotted line. Then after getting a small table of values, we sketch the graph shown in Figure 3. Of course, it is much easier to press a few keys and obtain the calculator-drawn graph shown in Figure 4. ■

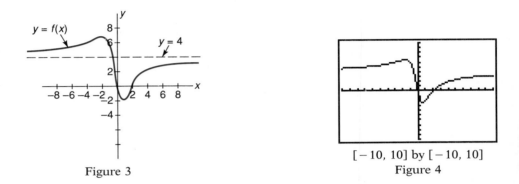

$[-10, 10]$ by $[-10, 10]$

Figure 3   Figure 4

Example 1 illustrates an important fact. The end behavior of the graph of a rational function is determined by the leading terms in numerator and denominator; the graph for large $|x|$ will behave like the quotient of the leading terms.

■ **Example 2.** Determine the equation of the horizontal asymptote for the graph of

$$f(x) = \frac{3x^6 + 2x^5 + 5}{2x^6 + x^4 + 9}$$

Solution. The graph of this function behaves like the graph of $y = 3x^6/2x^6$ when $|x|$ is large. The line $y = 3/2$ is a horizontal asymptote. ■

VERTICAL ASYMPTOTES

Consider next the function $f(x) = 1/(x - 2)$. The number 2 is not in the domain of this function, but numbers arbitrarily close to 2 are. Note the following values of f.

$$
\begin{array}{ll}
f(1.9) = -10 & f(2.1) = 10 \\
f(1.99) = -100 & f(2.01) = 100 \\
f(1.999) = -1000 & f(2.001) = 1000
\end{array}
$$

They show that $f(x)$ becomes negatively unbounded as x approaches 2 from the left and that $f(x)$ becomes positively unbounded as x approaches 2 from the right. To abbreviate these statements we write

$$f(x) \to -\infty \text{ as } x \to 2^-$$

and

$$f(x) \to \infty \text{ as } x \to 2^+$$

The superscripts $-$ and $+$ on a number serve to indicate the direction of approach to that number.

With the aid of a few more function values, we are able to make the hand-drawn graph in Figure 5 and confirm it with the calculator-drawn graph of Figure 6.

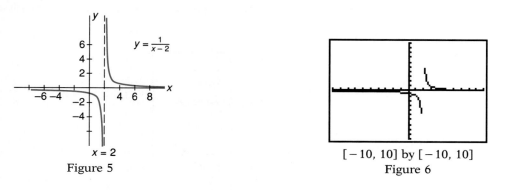

$y = \frac{1}{x-2}$

$x = 2$

Figure 5

$[-10, 10]$ by $[-10, 10]$

Figure 6

Referring to Figures 5 and 6, we arrive at the conclusion that the line $x = 2$ is a vertical asymptote and that the line $y = 0$ is a horizontal asymptote.

■ **Example 3.** Use your calculator to draw the graph of

$$f(x) = \frac{3x^2 + 5}{x^2 - 4x + 4} = \frac{3x^2 + 5}{(x - 2)^2}$$

Then identify its asymptotes.

Solution. The graph is shown in Figure 7. There is a vertical asymptote at $x = 2$ (where the denominator is 0) and a horizontal asymptote at $y = 3$ (obtained by looking at the quotient of the leading terms). ■

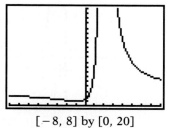

$[-8, 8]$ by $[0, 20]$

Figure 7

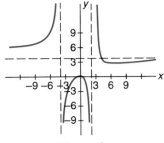

Figure 8

Example 4. Draw the graph of

$$f(x) = \frac{4x^2}{x^2 + 2x - 8} = \frac{4x^2}{(x - 2)(x + 4)}$$

and determine its asymptotes.

Solution. Figure 8 shows a hand-drawn graph with its asymptotes $x = -4$, $x = 2$, and $y = 4$ serving as guidelines. Figures 9 and 10 show the corresponding calculator-drawn graph in two different modes: the dot mode and the connected mode. In dot mode (available on some graphics calculators including the TI-81 and TI-82), the calculator simply plots points. In connected mode (which we have consistently used till now), the calculator attempts to connect these points and may introduce a connecting segment that should not be there, as in the left part of Figure 10. From now on, we will often choose dot mode in graphing functions whose graphs have jumps (discontinuities). ■

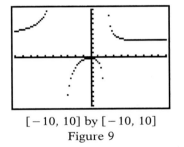

[−10, 10] by [−10, 10]
Figure 9

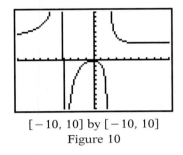

[−10, 10] by [−10, 10]
Figure 10

SLANT ASYMPTOTES

Consider the following function.

$$f(x) = x - 1 + \frac{4}{x - 2} = \frac{x^2 - 3x + 6}{x - 2}$$

Since $4/(x - 2) \to 0$ as $|x| \to \infty$, the first formula for f suggests that the graph of $y = f(x)$ should be very similar to the graph of $y = x - 1$ for large $|x|$. This is confirmed in Figures 11 and 12, which show the graph of $y = f(x)$ by itself

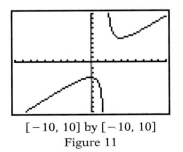

[−10, 10] by [−10, 10]
Figure 11

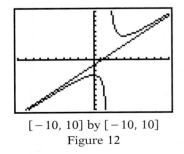

[−10, 10] by [−10, 10]
Figure 12

90 • Functions and Their Graphs

and then with the graph of $y = x - 1$ superimposed. The line $y = x - 1$ is called a **slant asymptote** for the graph.

When can we expect slant asymptotes? They will occur in the situation where the degree of the numerator is exactly one more than that of the denominator.

$$\begin{array}{r}
2x + 1 \\
x^2 + x - 6 \,\overline{)\, 2x^3 + 3x^2} \\
\underline{2x^3 + 2x^2 - 12x} \\
x^2 + 12x \\
\underline{x^2 + x - 6} \\
11x + 6
\end{array}$$

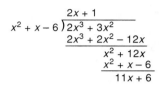

Figure 13

■ **Example 5.** Determine all the asymptotes for the graph of

$$f(x) = \frac{2x^3 + 3x^2}{x^2 + x - 6} = \frac{x^2(2x + 3)}{(x - 2)(x + 3)}$$

Solution. From the factored form of the denominator, we conclude that there are vertical asymptotes at $x = -3$ and at $x = 2$. There are no horizontal asymptotes. The long division in Figure 13 implies that

$$f(x) = 2x + 1 + \frac{11x + 6}{x^2 + x - 6}$$

from which we conclude that the line $y = 2x + 1$ is a slant asymptote. After setting the range values at $[-10, 10]$ by $[-20, 20]$, we draw the graph of $y = f(x)$ and then this graph with $y = 2x + 1$ superimposed (Figures 14 and 15), both in dot mode. ■

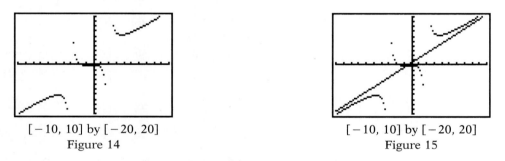

$[-10, 10]$ by $[-20, 20]$
Figure 14

$[-10, 10]$ by $[-20, 20]$
Figure 15

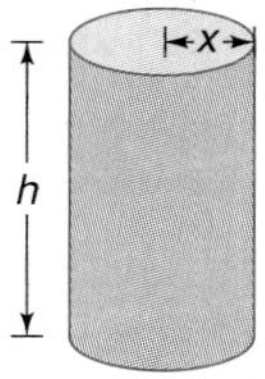

Figure 16

Teaser Solution

Recall the formulas $V = \pi r^2 h$ for the volume of a cylinder, $A = \pi r^2$ for the area of a circle, and $C = 2\pi r$ for the circumference of a circle. Then refer to Figure 16 and note that since the volume is to be 1000 cubic centimeters, $\pi x^2 h = 1000$, which implies that $h = 1000/(\pi x^2)$. The surface area $S(x)$ of the cylinder consists of the area of the circular top and bottom and the curved side (actually a rectangle when opened up), which leads to the formula

$$\begin{aligned} S(x) &= 2\pi x^2 + 2\pi x h \\ &= 2\pi x^2 + \frac{2000}{x} \end{aligned}$$

The graph of this function for the range values [0, 10] by [0, 2500] is shown in Figure 17, suggesting a minimum when $x \approx 5$. The zooming process leads to the more accurate $x = 5.42$ centimeters. The corresponding value for h is $h = 1000/(5.42^2\pi) \approx 10.84$ centimeters.

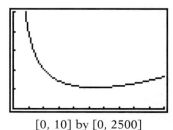

[0, 10] by [0, 2500]
Figure 17

PROBLEM SET 2.6

A. Skills and Techniques

In each problem, determine the equation of any horizontal asymptote. Then confirm your answer by drawing the graph of the given function with appropriate range values to suggest its end behavior.

1. $f(x) = \dfrac{-8x^2 - 5}{x^2 + 4}$

2. $g(x) = \dfrac{9x^2 - 2x - 3}{3x^2 + 4x + 2}$

3. $g(x) = \dfrac{7x^2 + 7x - 2}{0.5x^2 - 0.25x + 1}$

4. $f(x) = \dfrac{x^2 - 3x + 5}{5x^2 + 4}$

5. $f(x) = \dfrac{x^4 - 5x^3 - 2x^2 + 4}{3x^4 + x^2 + 1}$

6. $g(x) = \dfrac{-2x^4 - 7x^2 - 4x - 7}{0.1x^4 + 3}$

7. $h(x) = \dfrac{3x + 1}{x^2 + 2}$

8. $h(x) = \dfrac{2x^3 - 9}{x^4 + x^2 + 3}$

9. $g(x) = \dfrac{(-3x + 2)^4}{(2x^2 + 1)^2}$

10. $f(x) = \dfrac{(x^2 + 2)^2}{3x^4 + 2}$

11. $f(x) = \dfrac{x^6 - 2x^3 + 1}{(x^2 + 1)^2}$

12. $h(x) = \dfrac{(x^2 + 1)^2}{3x^2 + 2}$

In each problem determine the equations of all horizontal and vertical asymptotes. Confirm your answers by drawing the graph.

13. $f(x) = \dfrac{x}{x + 3}$

14. $f(x) = \dfrac{3x}{7 - x}$

15. $f(x) = \dfrac{x}{(x + 3)^2}$

16. $f(x) = \dfrac{3x^2}{(7 - x)^2}$

17. $f(x) = \dfrac{2x^2}{(x - 3)(x + 1)}$

18. $f(x) = \dfrac{5x}{(x - 2)(x + 4)}$

19. $f(x) = \dfrac{-3x^2}{(x + 2)(x - 6)}$

20. $f(x) = \dfrac{(x - 3)^2}{x^2 + 2x + 1}$

21. $f(x) = \dfrac{18x^4}{(x + 5)^2(2x - 8)^2}$

22. $f(x) = \dfrac{5x(x + 1)^3}{(x^2 - 4)^2}$

23. $f(x) = \dfrac{x^3}{(x + 0.1)(x - 0.1)^2}$

24. $f(x) = \dfrac{(x + 1)^2}{x^3 - x^2 - 6x}$

In each problem, determine the equations of all asymptotes (horizontal, vertical, and slant). Then draw the graph to confirm your answer.

25. $f(x) = 2x + \dfrac{1}{x^2 + 1}$

26. $f(x) = -\dfrac{1}{2}x + \dfrac{x}{x^2 + 4}$

27. $f(x) = -x + 3 + \dfrac{3}{x - 2}$

28. $f(x) = \dfrac{3}{2}x - 2 - \dfrac{4}{x + 3}$

29. $f(x) = \dfrac{3x^2 - 7}{2x}$

30. $f(x) = \dfrac{-2x^2 + x + 1}{x + 1}$

31. $f(x) = \dfrac{2x^2 - 4x - 3}{x + 2}$

32. $f(x) = \dfrac{-5x^2 + 30x + 3}{x - 5}$

33. $f(x) = \dfrac{3x^3 - 2x^2 - 2x + 5}{x^2 + 2}$

34. $f(x) = \dfrac{x^3 - 6x^2 + 2x - 11}{(x - 3)^2}$

B. Applications and Extensions

In Problems 35–40, the equations of the asymptotes are given, together with some other information. Sketch a graph of a function that satisfies these data.

35. $y = 1, x = 2; f(-1) = 1/2, f(3) = 2; f$ is concave down for $x < 2, f$ is concave up for $x > 2$.

36. $y = -1, x = -2; f(-3) = 0, f(0) = -1/2; f$ is concave up for $x < -2$ and for $x > -2$.

37. $y = 1.5, x = -3, x = 2; f(-5) = 2, f(0) = 0, f(4) = 2.5; f$ is concave up on the intervals where it is defined.

38. $y = 2, x = 1, x = 4; f(-3) = 3, f(3) = 0, f(6) = 3; f$ is concave down on $1 < x < 4$ and concave up elsewhere where defined.

39. $y = 2x - 1, x = 1; f(0) = -3, f(2) = 4; f$ is concave down for $x < 1$ and concave up for $x > 1$.

40. $y = x + 1, x = 2; f(1) = 3.5, f(3) = 6; f$ is concave up on the intervals where it is defined.

In Problems 41–44, tell what $f(x)$ approaches in each of the following cases. Graph to confirm your answer.
(a) $x \to \infty$ (b) $x \to -\infty$ (c) $x \to 4^+$ (d) $x \to -5^+$

41. $f(x) = \dfrac{x^2}{(x - 4)(x + 5)}$

42. $f(x) = \dfrac{2x}{(x - 4)(x + 5)}$

43. $f(x) = \dfrac{-x + 1}{(x - 4)(x + 5)}$

44. $f(x) = \dfrac{x^2 - 20}{(x - 4)(x + 5)}$

45. Determine the equations of the asymptotes of $f(x) = x^n/(x^2 + 1)$ for $n = 0, 1, 2, 3,$ and 4.

46. Determine the equations of all asymptotes of $f(x) = x^4/(x^n - 1)$ for $n = 1, 2, 3,$ and 4.

47. State the condition on the degrees of numerator and denominator of a rational function that holds if its graph has a horizontal asymptote.

48. State the condition on the degrees of numerator and denominator of a rational function that holds if its graph has no asymptotes.

49. A manufacturer of gizmos has yearly fixed costs of $30,000 and direct costs of $500 per gizmo. Write a formula for the average cost $A(x)$ of a gizmo when the manufacturer makes x gizmos in a year. Graph $A(x)$ and indicate two conclusions you can draw from the graph.

50. A salt solution with a concentration of 50 grams of salt per liter of solution flows into a large tank at the rate of 20 liters per minute (Figure 18). The tank contained 400 liters of pure water at $t = 0$.

Figure 18

(a) Write a formula for $S(t)$, the amount of salt in the tank after t minutes.

(b) Write a formula for the concentration $c(t)$ of salt in the tank after t minutes and draw its graph.

(c) What happens to $c(t)$ as $t \to \infty$?

51. An open rectangular box (Figure 19) with a square base of length x is to have a volume of 265 cubic centimeters. (a) Write a formula for the outer surface area $S(x)$ of this box. (b) What is the appropriate domain for this function? (c) Determine the dimensions of the box that require the least material to make.

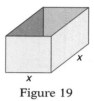

Figure 19

52. Determine the value of x in $1 < x < 5$, where

$$F(x) = \frac{6k}{x - 1} - \frac{12k}{x - 5}$$

achieves its minimum value, k being a positive constant.

53. Determine the domain and range of the function $f(x) = (x - 1)(x - 4)/(x - 5)$.

54. Determine the domain and range of the function $f(x) = (3x^2 + 8)/((9x - 3)(x + 5))$.

55. Find a formula for $f(x)$, given that f is a rational function whose graph has exactly two asymptotes, namely, $y = 2x + 3$ and $x = 3$, and $f(2) = 5$. More than one answer is possible.

56. Repeat Problem 55 with asymptotes $y = x - 5$ and $x = 4$ and with $f(8) = 5$.

57. Repeat Problem 55 with asymptotes $y = -x + 2$, $x = -2$, and $x = 1$ and with $f(3) = 0$.

58. **Challenge.** A large pool has three drains A, B, and C of different sizes. It takes drain A 3 hours longer than drain C to empty the pool, and it takes drain B 2 hours longer than drain C to empty the pool. When all three drains are open, the pool empties in 6 hours. How long does it take drain C to empty the pool by itself?

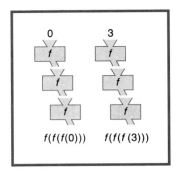

TEASER   Let $f(x) = (2x - 3)/(x - 1)$. Calculate $f(f(f(0)))$, $f(f(f(3)))$, and $f(f(f(-1)))$. From this calculation, conjecture a result. Prove it.

2.7   COMBINATIONS OF FUNCTIONS

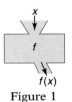

Figure 1

Think of a function as a machine (f) that accepts an input x and produces an output $f(x)$ (Figure 1). If we have two such machines, we may combine them in various ways (Figure 2).

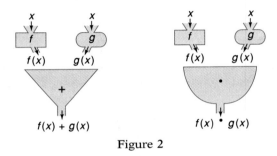

Figure 2

Thus, associated with two functions f and g, we have the sum function $f + g$, the difference function $f - g$, the product function $f \cdot g$, and the quotient function f/g defined by

$$(f + g)(x) = f(x) + g(x)$$

$$(f - g)(x) = f(x) - g(x)$$

$$(f \cdot g)(x) = f(x) \cdot g(x)$$

$$\left(\frac{f}{g}\right)(x) = \frac{f(x)}{g(x)}$$

■ **Example 1.**   Let $f(x) = |x - 1|$ and $g(x) = x^2$. Calculate each of the following: (a) $(f + g)(3)$ (b) $(f - g)(4)$ (c) $(f \cdot g)(5)$ (d) $(f/g)(-2)$ (e) $(f \cdot (f - g))(4)$.

Solution.
(a) $(f + g)(3) = f(3) + g(3) = 2 + 9 = 11$
(b) $(f - g)(4) = f(4) - g(4) = 3 - 16 = -13$
(c) $(f \cdot g)(5) = f(5) \cdot g(5) = 4 \cdot 25 = 100$

(d) $\left(\dfrac{f}{g}\right)(-2) = \dfrac{f(-2)}{g(-2)} = \dfrac{3}{4}$

(e) $(f \cdot (f - g))(4) = f(4) \cdot (f - g)(4) = 3(-13) = -39$ ■

Use of the four arithmetic operations in combining functions is quite straightforward; we move on to composition of functions, which is considerably more difficult.

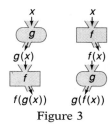

$f(g(x)) \qquad g(f(x))$

Figure 3

COMPOSITION OF FUNCTIONS

Machines can often be hooked together in tandem to create more-complicated machines (Figure 3); this is the idea behind composition. Given two functions f and g, we may define a new function $f \circ g$, called the **composite of f with g,** by

$$(f \circ g)(x) = f(g(x))$$

The domain of $f \circ g$ consists of those inputs x in the domain of g for which $g(x)$ is in the domain of f.

■ **Example 2.** Let $f(x) = x^2 - 1$ and $g(x) = 1/x$.
(a) Calculate $(f \circ g)(2)$ and $(g \circ f)(2)$.
(b) Write formulas for $(f \circ g)(x)$ and for $(g \circ f)(x)$.
(c) Determine the domains for $f \circ g$ and $g \circ f$.

Solution.

(a) $(f \circ g)(2) = f(g(2)) = f\left(\dfrac{1}{2}\right) = \dfrac{1}{4} - 1 = -\dfrac{3}{4}$

$\quad (g \circ f)(2) = g(f(2)) = g(3) = \dfrac{1}{3}$

(b) $(f \circ g)(x) = f(g(x)) = f\left(\dfrac{1}{x}\right) = \left(\dfrac{1}{x}\right)^2 - 1 = \dfrac{1 - x^2}{x^2}$

$\quad (g \circ f)(x) = g(f(x)) = g(x^2 - 1) = \dfrac{1}{x^2 - 1}$

(c) The domain of $f \circ g$ is $\{x: x \neq 0\}$; the domain of $g \circ f$ is $\{x: x \neq \pm 1\}$.

Note that $f \circ g \neq g \circ f$; the commutative law does not hold for composition. ■

■ **Example 3.** In Figure 4, we show the graphs of f and g on $0 \leq x \leq 6$. Sketch in the graph of $f + g$. Use these graphs to calculate each of the following:

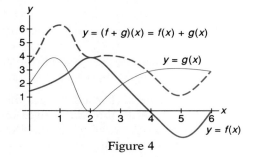

Figure 4

(a) $(f + g)(3)$ (b) $(f + g)(5)$ (c) $(f \circ g)(1)$ (d) $(f \cdot g)(0)$
(e) $((f/g) - (g/f))(1)$ (f) $(g \circ g)(2) - (g \cdot g)(2)$ (g) $(g \circ f)(2) + (f \circ g)(2)$.

Solution. The graph of $f + g$, obtained by adding y-coordinates, is shown as a dotted curve in Figure 4.

(a) $(f + g)(3) = 2 + 2 = 4$
(b) $(f + g)(5) = -1.5 + 3 = 1.5$
(c) $(f \circ g)(1) = f(4) = 0$
(d) $(f \cdot g)(0) = \left(\dfrac{3}{2}\right)(2) = 3$
(e) $\left(\left(\dfrac{f}{g}\right) - \left(\dfrac{g}{f}\right)\right)(1) = \dfrac{2}{4} - \dfrac{4}{2} = -\dfrac{3}{2}$
(f) $(g \circ g)(2) - (g \cdot g)(2) = g(0) - 0 \cdot 0 = 2 - 0 = 2$
(g) $(g \circ f)(2) + (f \circ g)(2) = g(4) + f(0) = 3 + \dfrac{3}{2} = \dfrac{9}{2}$ ∎

The reverse of composing functions is decomposing them, an extremely important operation in calculus. For example, suppose $F(x) = \sqrt{x^2 + 3}$. One way to decompose this function is to let $g(x) = x^2 + 3$ and $f(x) = \sqrt{x}$. Then $F = f \circ g$; since $F(x) = f(g(x)) = f(x^2 + 3) = \sqrt{x^2 + 3}$.

■ **Example 4.** Let $F(x) = (x^4 - 2)^3$. Decompose this function (a) as a composite of two functions and (b) as a composite of three functions.

Solution. Here is one set of answers to this problem.
(a) Let $g(x) = x^4 - 2$ and $f(x) = x^3$. Then $F = f \circ g$.
(b) Let $h(x) = x^4$, $g(x) = x - 2$, and $f(x) = x^3$. Then $F = f \circ g \circ h$.

Note that the decomposition in (b) mimics the way you would go about calculating $F(x)$ for any particular x: First raise x to the fourth power, then subtract 2, and finally cube the result. ■

HORIZONTAL AND VERTICAL SHIFTS

We have already discussed the topic of horizontal and vertical shifts in connection with quadratic functions (Section 2.3). What was true there is true in general.

1.   *The graph of $y = f(x - h)$ is the graph of $y = f(x)$ shifted h units horizontally.*
2.   *The graph of $y = f(x) + k$ is the graph of $y = f(x)$ shifted k units vertically.*

■ **Example 5.** Sketch the graphs of $y = x^4$, $y = (x + 6)^4$, $y = x^4 + 3$, and $y = (x - 5)^4 - 1$ in the same plane.

Solution. These four graphs, all having the same shape, are shown in Figure 5. Note that the fourth one involves a horizontal shift of 5 units and a vertical shift of -1 unit. It would be a good idea to check these graphs on your calculator. ■

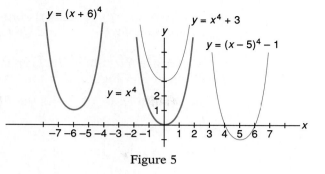

Figure 5

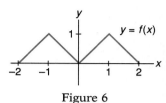

Figure 6

Here is another example illustrating the same phenomenon.

■ **Example 6.** Figure 6 shows the graph of $y = f(x)$ on the interval $-2 \le x \le 2$. Sketch the graphs of $y = f(x + 0.5)$ and $y = f(x) + 1.5$.

Solution. The first graph is shifted 0.5 units left. The second graph is shifted 1.5 units up. The results are shown in Figure 7. ■

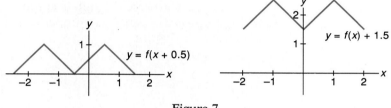

Figure 7

STRETCHES AND SHRINKS

Again what happened with quadratic functions in Section 2.3 provides a good model for the general case.

3. *The graph of $y = af(x)$ is similar to the graph of $y = f(x)$ but stretched vertically if $a > 1$ and shrunk vertically if $0 < a < 1$. The graph of $y = -af(x)$ is the graph of $y = af(x)$ reflected across the x-axis.*

■ **Example 7.** Consider again the graph of $y = f(x)$ shown in Figure 6. Sketch the graphs of $y = f(x)$, $y = 3f(x)$, $y = (1/2) f(x)$, and $y = -(1/2) f(x)$.

Solution. The four graphs are shown in Figures 8 and 9. ■

Figure 8   Figure 9

4. *The graph of y = f(cx) is similar to the graph of y = f(x) but compressed horizontally if c > 1 and stretched horizontally if 0 < c < 1.*

■ **Example 8.** Show the graphs of $y = f(x)$, $y = f(3x)$, and $y = f(x/2)$ for the function of Figure 6 on the interval $-2 \leq x \leq 2$.

Solution. Figure 10 shows the three graphs. ■

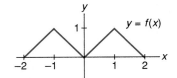

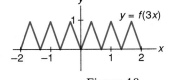

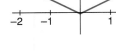

Figure 10

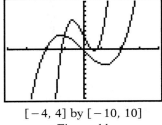

$[-4, 4]$ by $[-10, 10]$
Figure 11

■ **Example 9.** Let $f(x) = x^3 - 4x$. Explain how the graph of $y = f(2x) + 3$ is related to the graph of $y = f(x)$. Then use your calculator to confirm your statement.

Solution. The second graph is compressed horizontally by a factor of 2 and shifted upward 3 units as compared with the first graph. These results are confirmed in Figure 11. ■

COMBINATIONS OF FUNCTIONS ON A CALCULATOR

So far, we have discussed putting functions together mainly in terms of algebra and hand-drawn graphs. Now we illustrate some of these ideas using a graphics calculator.

■ **Example 10.** (a) Draw the graphs of $y_1 = f(x) = 1.3x - 2$, $y_2 = g(x) = 0.1x(x^2 - 3x - 11)$, and $y_3 = y_1 + y_2 = f(x) + g(x)$ on the same screen. (b) Do the same with y_3 replaced by $y_4 = y_1 \cdot y_2$.

Solution. The results are shown in Figures 12 and 13. Refer to the box titled Calculator Hints to see how this is accomplished on the TI-81 and TI-82. ■

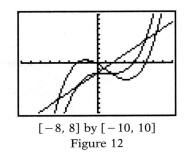

$[-8, 8]$ by $[-10, 10]$
Figure 12

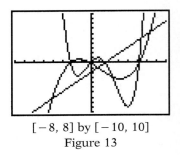

$[-8, 8]$ by $[-10, 10]$
Figure 13

■ **Example 11.** Let $y_1 = f(x) = 1.3x - 2$ and $y_2 = g(x) = 0.2x^4 + x^3 - 2$. Draw the graph of $y_4 = g(f(x)) = 0.2\,y_1^4 + y_1^3 - 2$.

Solution. This is done in Figure 14. ■

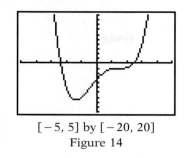

$[-5, 5]$ by $[-20, 20]$
Figure 14

TEASER SOLUTION

Since $f(x) = (2x - 3)/(x - 1)$, it follows that:

$$f(f(f(0))) = f(f(3)) = f\left(\frac{3}{2}\right) = 0$$

$$f(f(f(3))) = f\left(f\left(\frac{3}{2}\right)\right) = f(0) = 3$$

$$f(f(f(-1))) = f\left(f\left(\frac{5}{2}\right)\right) = f\left(\frac{4}{3}\right) = -1$$

On this basis, we conjecture that $f(f(f(x))) = x$. We leave the actual algebraic demonstration to you, noting only that you should discover that the result is valid provided $x \neq 1$ and $x \neq 2$.

PROBLEM SET 2.7

A. Skills and Techniques

Let $f(x) = x + 3$, $g(x) = |x|/2$, $h(x) = (x + 1)/x$, and $k(x) = -x^2 + 2$. Evaluate each of the following.

1. (a) $(f + g)(2)$
   (b) $(f - g)(-4)$
   (c) $(f \cdot g)(0)$
   (d) $\left(\dfrac{f}{g}\right)(3)$

2. (a) $(f + g)(-1)$
   (b) $(g - f)(8)$
   (c) $(f \cdot g)\left(-\dfrac{1}{2}\right)$

   (d) $\left(\dfrac{g}{f}\right)(-12)$

3. (a) $(f + h)(-3)$
   (b) $(g \cdot k)(-2)$
   (c) $\left(\dfrac{k}{f}\right)(4)$
   (d) $\left(\dfrac{k}{g}\right)(\sqrt{2})$

4. (a) $(g - f)(-2\pi)$
   (b) $\left(\dfrac{k}{f}\right)(-1)$
   (c) $(k \cdot h)(2)$
   (d) $(k - g)(12)$

5. (a) $\left(\dfrac{k}{g}\right)(4)$

(b) $\left(\dfrac{k}{g} + f\right)(4)$

(c) $\left(\dfrac{k}{k/g}\right)(4)$

(d) $(h \cdot (f + k))(3)$

6. (a) $(h + f)(1)$

(b) $((h + f) \cdot g)(1)$

(c) $\left(\dfrac{h + f}{k}\right)(1)$

(d) $\left(\dfrac{h + f}{k}\right)(2)$

In Problems 7–10, evaluate $f \circ g$ and $g \circ f$ at the indicated value of x. Then write formulas for $(f \circ g)(x)$ and $(g \circ f)(x)$. Finally, determine the domains of $f \circ g$ and $g \circ f$.

7. $f(x) = 2x, g(x) = 1 - x; x = -5$

8. $f(x) = \sqrt{x}, g(x) = x^2 + 5; x = 2$

9. $f(x) = \dfrac{x}{1 - x}, g(x) = 3x - 2; x = 0$

10. $f(x) = \dfrac{2}{x^2}, g(x) = |4x|; x = \sqrt{2}$

11. Figure 15 shows the graph of the functions f and g on the interval $0 \le x \le 4$. Sketch the graph of $f \cdot g$ and evaluate each of the following.

(a) $(f \cdot g)(3)$

(b) $(f - g)(1)$

(c) $(f \circ g)(2)$

(d) $(g \circ f)(2)$

(e) $\left(f \circ \left(\dfrac{f}{g}\right)\right)(4)$

(f) $((f \circ g) - (g \cdot f))(0)$

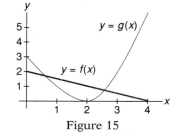

Figure 15

12. Figure 16 displays the graphs of the functions F and G on the interval $-3 \le x \le 3$. Sketch the graph of $F - G$ and evaluate each of the following.

(a) $(F - G)(-2)$

(b) $(F + G)(2)$

(c) $(G \circ G)(-3)$

(d) $\left(G + \left(\dfrac{F}{G}\right)\right)(3)$

(e) $(F \circ (F - G))(-2)$

(f) $(F \circ (F \circ F))(3)$

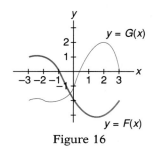

Figure 16

Decompose each function as the composite: (a) of two functions (b) of three functions. See Example 4 and note that answers may differ.

13. $F(x) = \sqrt{x^2 + 3}$

14. $F(x) = \dfrac{1}{x^3 - 2}$

15. $F(x) = (x - 4)^2 + 3$

16. $F(x) = (x^3 - 7)^2$

17. $F(x) = \dfrac{1}{(x + 1)^2}$

18. $F(x) = \sqrt{(3 - x)^5 + 2}$

Sketch the graphs of the three functions in the same plane. Then confirm your graphs using a calculator.

19. $f(x) = x^3, g(x) = x^3 - 4, h(x) = (x + 7)^3$

20. $f(x) = -x^4, g(x) = -x^4 + 5, h(x) = -(x - 5)^4$

21. $f(x) = \sqrt{x}, g(x) = \sqrt{x} + 5, h(x) = \sqrt{x + 5}$

22. $f(x) = \sqrt{|x|}, g(x) = \sqrt{|x|} - 6, h(x) = \sqrt{|x - 6|}$

Explain in words how the graphs of g and h are related to the graph of f. Then confirm your answers by using a calculator to draw their graphs.

23. $f(x) = x^2, g(x) = 3x^2, h(x) = \dfrac{1}{2}x^2$

24. $f(x) = x^4, g(x) = -8x^4, h(x) = \dfrac{1}{4}x^4$

25. $f(x) = \sqrt{25 - x^2}, g(x) = -2\sqrt{25 - x^2},$

$h(x) = \sqrt{25 - \left(\dfrac{x}{2}\right)^2}$

26. $f(x) = |x^2 - 16|, g(x) = \dfrac{1}{2}|x^2 - 16|,$

$h(x) = \left|\left(\dfrac{1}{2}x\right)^2 - 16\right|$

27. $f(x) = 0.4x^3 - x^2 - 2, g(x) = 0.4(x - 3)^3 - (x - 3)^2 - 2, h(x) = 0.4(2x)^3 - (2x)^2 - 2$

28. $f(x) = \dfrac{x^3 - 30x}{x^2 + 2}, g(x) = \dfrac{(x + 4)^3 - 30(x + 4)}{(x + 4)^2 + 2},$

$h(x) = \dfrac{(3x)^3 - 30(3x)}{(3x)^2 + 2}$

In Problems 29–32, the graph of f is shown. Draw the graphs of: (a) 3 f(x − 2)   (b) f(2x) + 2

29.

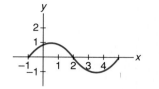

30.

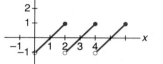

31.

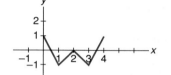

32.

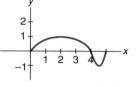

Use a graphics calculator in Problems 33–36 as in Examples 10 and 11.

33. Given that $f(x) = 1.5x + 1$ and $g(x) = 0.2x^3 - 5x + 2$, draw the graphs of $f - g$ and $f \cdot g$.

34. Given that $f(x) = 0.4x^2 + 0.5$ and $g(x) = -x^2 + 2x - 2$, draw the graphs of $f + g$ and g/f.

35. Given that $f(x) = 0.6x^2 + 5x + 0.5$ and $g(x) = -x^2 + 3x - 2$, draw the graph of $f \circ g$.

36. Given that $f(x) = 1.5x + 1$ and $g(x) = 0.2x^3 - 5x + 2$, draw the graph of $g \circ f$.

B. Applications and Extensions

37. Let f be a given function. Write in terms of f a formula $g(x) = \ldots$ for the function g whose graph is shifted 2 units to the left and is compressed vertically by a factor of 8 when compared with the graph of f.

38. Follow the instructions of Problem 37 if the graph of g is shifted upward 3 units and compressed horizontally by a factor of 2.

39. Let $f(x) = 1/(x - 1)$ and $g(x) = \sqrt{x + 1}$. Specify the domains of $f, g, f \circ g$, and $g \circ f$.

40. Let $f(x) = \sqrt{9 - x^2}$ and $g(x) = 1/x$. Specify the domains of $f, g, f \circ g$, and $g \circ f$.

41. Let $g(x) = 3x - 2$ and $(f \circ g)(x) = -9x + 4$. Determine $f(x)$, given that f is a linear function.

42. If f is a linear function and $f(x^2 - 1) = 3x^2 - 1$, determine $f(x)$.

43. The **difference quotient** $\dfrac{f(x + h) - f(x)}{h}$ plays an important role in calculus. Write and then simplify this quotient for (a) $f(x) = x^2 + 2x$ and (b) $f(x) = 1/x$.

44. Follow the instructions of Problem 43 for (a) $f(x) = x^3$ and (b) $f(x) = (x + 1)/x$.

45. Let f be an even function (meaning $f(-x) = f(x)$) and g be an odd function (meaning $g(-x) = -g(x)$) with both functions having $\mathbb{R}$ as their domains. Determine which of the following are even, which are odd, and which are neither.

(a) $f \cdot g$ (b) $\dfrac{f}{g}$ (c) $g \cdot g$ (d) $g \cdot g \cdot g$

(e) $f + g$ (f) $g \circ g$ (g) $f \circ g$ (h) $g \circ f$

46. Show that any function f having $\mathbb{R}$ as its domain can be written as the sum of an even function and an odd function. (*Hint:* Let $f(x) = g(x) + h(x)$ where $g(x) = (f(x) + f(-x))/2$. Determine h, and then show that g is even and h is odd.)

47. At noon, Soon Lee left point A walking east at 3.5 miles per hour. Exactly 1 hour later, Marcia left point A walking north at 4 miles per hour. (a) Refer to Figure 17 and write a formula for the distance $d = d(t)$ between them t hours after noon ($t \geq 1$). (b) At what time are they 20 miles apart (to the nearest minute)?

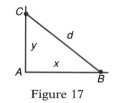

Figure 17

48. Figure 18 shows a box in which the length x is twice the width y and three times the height z. (a) Write a formula for the perimeter $P = P(x)$ of triangle ABC in terms of x. (b) Determine x to two decimal places if $P(x) = 20$.

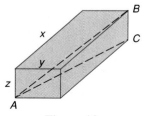

Figure 18

49. Repeated composition of functions—$f(x)$, $f(f(x))$, $f(f(f(x)))$, and so on—plays a fundamental role in certain parts of mathematics and science (for example, fractal theory). The TI-82 allows us to make such calculations easily as we now show. Enter $Y_1 = f(x) = (x - 3)/(x + 1)$ in the ⃞ menu. Then perform the following operations.

$$\boxed{2} \;\boxed{\text{STO→}}\; \text{X} \;\boxed{\text{ENTER}}$$
$$\boxed{\text{Y-VARS}}\;\boxed{1}\;\boxed{1}\;\boxed{\text{STO→}}\; \text{X} \;\boxed{\text{ENTER}}$$
$$\boxed{\text{ENTER}}$$
$$\boxed{\text{ENTER}}$$
$$\boxed{\text{ENTER}}$$
$$\vdots$$

Repeat with 3.2 replacing 2, then with 0.7 replacing 2, and so on. Make a conjecture about $f(f(f(x)))$. Prove your conjecture algebraically. Note: Use the same procedure on the TI-81 but delete the second ⃞ in the second row.

50. **Challenge.** Follow the procedure in Problem 49 for the function $f(x) = 2.5(x - x^2)$ using as starting values (replacing 2) the numbers 0.3, 0.6, 0.9, 1.1, and then lots of other values. From your experimenting, make a conjecture about what happens to this function when composed repeatedly.

$$x \longrightarrow x^3 \longrightarrow x^3 + 1 \longrightarrow \frac{x^3 + 1}{3} = y$$

$$y \longrightarrow ? \longrightarrow ? \longrightarrow ? = x$$

TEASER   Take a number x, cube it, add 1, divide the result by 3, and call the answer y. Write a formula for the function $y = f(x)$ that this description determines. Then write a formula for $f^{-1}(y)$, that is, for the function that will take y as input and return x as output.

2.8   INVERSE FUNCTIONS

If I put on a shoe, I can take it off again. The second operation undoes the first and restores the situation to its original condition. If I put on a shoe and then tie its laces, I can undo the result by first untying the laces and then taking off the shoe. Note that to undo a process, I must undo each step but in reverse order.

Take a number x, add 2 to it, and divide the result by 6, giving an answer $y = f(x)$. In symbols,

$$y = f(x) = \frac{x + 2}{6}$$

To undo this calculation, take the answer y, multiply it by 6, and subtract 2. If we call the undoing function g, then g acts on y according to the formula

$$g(y) = 6y - 2$$

As a check, note that:

$$g(f(4)) = g(1) = 4$$
$$g(f(-14)) = g(-2) = -14$$

and in general

$$g(f(x)) = g\left(\frac{x + 2}{6}\right) = 6\left(\frac{x + 2}{6}\right) - 2 = x$$

■ **Example 1.**   Let $y = f(x) = \sqrt[3]{x} + 6$. Write a formula for the undoing function $g(y)$, the function that takes y as input and gives x as output. Check your answer by showing that $g(f(x)) = x$.

Solution. We must undo the operations of cube rooting and adding 6 but in reverse order. In other words, we must first subtract 6 and then cube the result. Thus,

$$g(y) = (y - 6)^3$$

To check, note that

$$g(f(x)) = g(\sqrt[3]{x} + 6)$$
$$= (\sqrt[3]{x} + 6 - 6)^3$$
$$= (\sqrt[3]{x})^3 = x$$

■

Our goal in this section is to study undoing functions; our special name for them is **inverse functions.** Not every function has an inverse function; some operations cannot be undone in an unambiguous way, as we shall see.

ONE-TO-ONE FUNCTIONS HAVE INVERSE FUNCTIONS

One-to-one

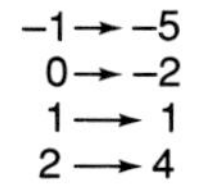

$$-1 \rightarrow -5$$
$$0 \rightarrow -2$$
$$1 \rightarrow 1$$
$$2 \rightarrow 4$$

Figure 1

Not one-to-one

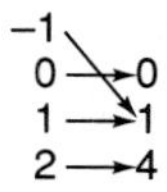

$$-1$$
$$0 \rightarrow 0$$
$$1 \rightarrow 1$$
$$2 \rightarrow 4$$

Figure 2

Consider the two functions diagrammed in Figures 1 and 2. The first one can be undone in an unambiguous way; just retrace the arrows backward. If we try to undo the second function, however, we are caught in a dilemma. What number will the undoing function assign to 1? There are two possible answers (1 and -1), and there is no way to choose between them. Remember that a function must give *one* output for each input.

Thinking about this simple example leads us to make a definition. A function f is **one-to-one** if different inputs produce different outputs, that is, if $x_1 \neq x_2$ implies $f(x_1) \neq f(x_2)$. A function can be undone if and only if it is one-to-one.

There is a simple graphical criterion for determining whether a function is one-to-one. Assume that $y = f(x)$ is graphed in a plane with the x-axis horizontal and the y-axis vertical. *Then f is one-to-one if and only if every horizontal line meets the graph in at most one point.*

■ **Example 2.** Draw the graphs of $f(x) = x^3 - 2x + 2$ and $g(x) = 5/(x - 2)$. Which of these functions has an inverse?

Solution. Calculator-drawn graphs of these two functions are shown in Figures 3 and 4. Clearly the first function fails to satisfy our criterion. The second function does appear to be one-to-one, but since we can examine only a part of the graph we should perhaps state that our answer is tentative. We will have more to say about this function later (Example 3). ■

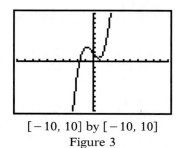

[-10, 10] by [-10, 10]
Figure 3

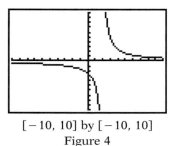

[-10, 10] by [-10, 10]
Figure 4

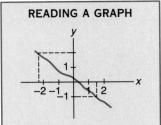
We are ready for a formal definition.

> **Definition.** Let f be a one-to-one function with domain X and range Y. Then the function g with domain Y and range X that satisfies
> $$g(f(x)) = x$$
> for all x in X is called the **inverse function** for f.

Mathematicians and scientists use a special symbol for an inverse function, namely, f^{-1}. Note a possible source of confusion. The symbol f^{-1}, used in connection with a function, does not mean $1/f$. It will always mean the function that undoes f, that is, the function satisfying $f^{-1}(f(x)) = x$. Note also that if f is one-to-one, so that f^{-1} exists, then f^{-1} is also one-to-one and can be undone by f. Thus, two relations hold for every one-to-one function f.

$$f^{-1}(f(x)) = x \text{ for all } x \text{ in } X$$
$$f(f^{-1}(y)) = y \text{ for all } y \text{ in } Y$$

Figure 5   Figure 6

The facts are illustrated schematically in Figures 5 and 6.

FINDING A FORMULA FOR F^{-1}

Suppose that f is one-to-one with formula $y = f(x)$. To find the formula for f^{-1} we must know the way to go from y back to x. To put it in algebraic language, we must solve the equation $y = f(x)$ for x in terms of y. Let us illustrate for the function F of Example 2.

■ **Example 3.** Let $y = F(x) = 5/(x - 2)$, $x \neq 2$. Find the formula for $F^{-1}(y)$.

Solution. Study the following algebraic manipulations.

$$y = \frac{5}{x - 2}$$
$$(x - 2)y = 5$$
$$xy - 2y = 5$$
$$xy = 5 + 2y$$

Thus,

$$F^{-1}(y) = \frac{5 + 2y}{y}$$

We can check our answer by calculating $F^{-1}(F(x))$ and $F(F^{-1}(y))$.

$$F^{-1}(F(x)) = F^{-1}\left(\frac{5}{x-2}\right) = \frac{5 + 2(5/(x-2))}{5/(x-2)} = \frac{5x - 10 + 10}{5} = x$$

$$F(F^{-1}(y)) = F\left(\frac{5+2y}{y}\right) = \frac{5}{((5+2y)/y) - 2} = \frac{5y}{5 + 2y - 2y} = y$$

Note that the domain of F is $\{x : x \neq 2\}$, and its range is $\{y : y \neq 0\}$. These sets are the corresponding range and domain of F^{-1}, respectively. ■

It is time to make an important point about the formulas for functions. The formulas $f(x) = 2x$, $f(t) = 2t$, and $f(y) = 2y$ say exactly the same thing. They tell us that f takes an input and doubles it. Similarly the formulas

$$F^{-1}(y) = \frac{5 + 2y}{y}$$

$$F^{-1}(t) = \frac{5 + 2t}{t}$$

$$F^{-1}(x) = \frac{5 + 2x}{x}$$

all describe the same function. It happens that we often prefer to describe a function using x as the domain variable. Thus, we offer the following three-step procedure for finding the formula for $f^{-1}(x)$.

PROCEDURE FOR FINDING $F^{-1}(X)$

1. Solve the equation $y = f(x)$ for x in terms of y.
2. Use $f^{-1}(y)$ to name the resulting expression in y.
3. Replace y by x to get the formula for $f^{-1}(x)$.

■ **Example 4.** Use the three-step procedure to find $f^{-1}(x)$ for the function $f(x) = \sqrt[3]{x} + 6$ of Example 1.

Solution.

Step 1: $y = \sqrt[3]{x} + 6$

                $\sqrt[3]{x} = y - 6$

                $x = (y - 6)^3$

Step 2: $f^{-1}(y) = (y - 6)^3$

Step 3: $f^{-1}(x) = (x - 6)^3$ ■

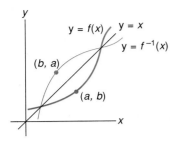

Figure 7

THE GRAPHS OF F AND F^{-1}

Since for one-to-one functions, $y = f(x)$ and $x = f^{-1}(y)$ are equivalent, that is, they determine the same ordered pairs, the graphs of these two equations are exactly the same. Suppose however, we want to compare the graphs of $y = f(x)$ and $y = f^{-1}(x)$, where you will note that we have used x as the domain variable in both cases. To get $y = f^{-1}(x)$ from $x = f^{-1}(y)$, we need to interchange the roles of x and y. To perform this interchange graphically is to reflect across the line $y = x$, as shown in Figure 7.

 Example 5. Let $f(x) = 3x - 4$ and $g(x) = (x^3 - 2)/5$. Determine formulas for f^{-1} and g^{-1}. Show the graphs of $y = f(x)$ and $y = f^{-1}(x)$ in the same plane. Do the same for $y = g(x)$ and $y = g^{-1}(x)$.

Solution. The formulas for f^{-1} and g^{-1} can be obtained directly or by the three-step process. They are

$$f^{-1}(x) = \frac{x + 4}{3}$$

$$g^{-1}(x) = \sqrt[3]{5x + 2}$$

The required graphs are shown in Figures 8 and 9. ∎

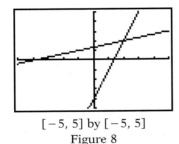

[−5, 5] by [−5, 5]

Figure 8

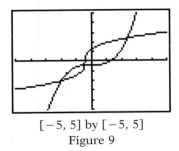

[−5, 5] by [−5, 5]

Figure 9

RESTRICTING THE DOMAIN TO OBTAIN AN INVERSE

We have emphasized that not all functions have inverse functions; only one-to-one functions have this property. But many important functions that are not one-to-one on their natural domains become one-to-one when we restrict their domains in an appropriate way. Consider $y = f(x) = x^2$, which is not one-to-one and does not have an inverse function (Figure 10). If we restrict

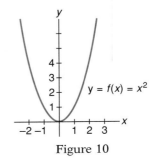

Figure 10

its domain to $\{x: x \geq 0\}$, however, the resulting function f is one-to-one and its inverse f^{-1} is given by $f^{-1}(x) = \sqrt{x}$, as is shown in Figure 11.

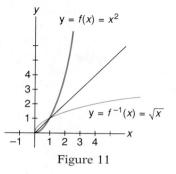

Figure 11

■ Example 6. Restrict the domain of $f(x) = x^2 - 4x$ so that f is one-to-one. Then obtain a formula for $f^{-1}(x)$.

Solution. The graph of $y = f(x)$ is shown in Figure 12. We could restrict the domain in a number of different ways, but the most natural is to pick the right branch of the graph, that is, to make the domain $\{x: x \geq 2\}$. Then we use the three-step procedure.

Step 1:
$$y = x^2 - 4x$$
$$y + 4 = x^2 - 4x + 4 = (x - 2)^2$$
$$x - 2 = \pm\sqrt{y + 4}$$
$$x = 2 \pm \sqrt{y + 4}$$
$$x = 2 + \sqrt{y + 4} \quad \text{(because } x \geq 2\text{)}$$

Step 2: $\quad f^{-1}(y) = 2 + \sqrt{y + 4}$

Step 3: $\quad f^{-1}(x) = 2 + \sqrt{x + 4}$

Note that in the first step we initially found two values of x for each y (as we expect without restriction on x). The fact that we had restricted x to $x \geq 2$ allowed us to excise the minus sign. ■

GRAPHS OF INVERSE FUNCTIONS ON A CALCULATOR

As we have seen, the graph of the inverse function, assuming it exists, is the graph of the original function reflected across the line $y = x$. Can we make this reflection without knowing the equation of the reflected graph? Yes. The TI-82 has an automatic routine for reflecting a graph across the line $y = x$ (see the box titled Graphs of Inverses). This reflection can also be accomplished on either the TI-81 or TI-82 by making use of the concept of **parametric form,** a topic we discuss briefly now.

Imagine that the coordinates of a moving point (x, y) are given by the equations $x = g(t)$ and $y = h(t)$. The variable t may be thought of as representing time and is called a **parameter.** As t increases, the point (x, y) moves around the plane and determines a curve (this is discussed in detail in Section 7.5).

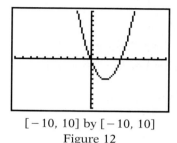

$[-10, 10]$ by $[-10, 10]$
Figure 12

GRAPHS OF INVERSES ON THE TI-82

To obtain the graph of $\mathbf{Y_1} = f(x)$ and its reflection across the line $y = x$, enter $\mathbf{Y_1} = f(x)$ in the usual way and return to the home screen. Then press the following sequence of keys.

(DRAW) (8) (Y-VARS) (1) (1)
(ENTER)

Whether or not the reflected graph is the graph of a function depends on whether or not the function f is one-to-one.

Most graphics calculators have a parametric mode that permits parametric graphing (see the manual for your calculator).

To graph a function f, we may replace the equation $y = f(x)$ by the two equations $x = t$ and $y = f(t)$ and graph in parametric mode. The resulting curve can then be reflected about the line $y = x$ by interchanging the roles of y and x, that is, by graphing $y = t$ and $x = f(t)$. The latter curve may or may not be the graph of a function (depending, of course, on whether f was one-to-one or not); if f was one-to-one, then the curve is the graph of f^{-1}.

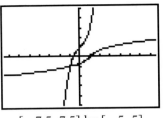

$[-7.5, 7.5]$ by $[-5, 5]$

Figure 13

■ **Example 7.** Using a graphics calculator, in parametric mode, draw the graph $y = f(x) = x^3 + x + 1$ and its inverse $y = f^{-1}(x)$ in the same plane.

Solution. First we write $x = t$ and $y = t^3 + t + 1$ and graph. Then we write $y = t$ and $x = t^3 + t + 1$ and graph again. The latter is the graph of f^{-1}. The results appear as Figure 13. ■

TEASER SOLUTION

$$x \rightarrow x^3 \rightarrow x^3 + 1 \rightarrow \frac{x^3 + 1}{3} = y$$

$$y \rightarrow 3y \rightarrow 3y - 1 \rightarrow \sqrt[3]{3y - 1} = x$$

Figure 14

The description—take a number x, cube it, add 1, and divide the result by 3—determines the function with formula $y = f(x) = (x^3 + 1)/3$. The inverse function f^{-1} has the formula $f^{-1}(y) = \sqrt[3]{3y - 1}$. See Figure 14.

PROBLEM SET 2.8

A. Skills and Techniques

The equation $y = f(x)$ determines a function that can be undone by another function g. Write a formula for g(y) and check that your answer is correct by showing that g(f(x)) = x.

1. $f(x) = 2x + 7$
2. $f(x) = 3x - 4$
3. $f(x) = \dfrac{x - 4}{3}$
4. $f(x) = 3(x + 2)$
5. $f(x) = \sqrt{x} + 1$
6. $f(x) = 5\sqrt{x}$
7. $f(x) = \sqrt{x + 1}$
8. $f(x) = \sqrt{5x}$
9. $f(x) = x^3 - 5$
10. $f(x) = \dfrac{1}{2}\sqrt[3]{x} - 11$
11. $f(x) = \dfrac{(x - 3)^3}{9}$
12. $f(x) = (2\pi(x + 4))^3$

Draw the graph of each of the following and use the horizontal line test to decide whether it is one-to-one. Assume that each function has its natural domain.

13. $f(x) = x^2 + 2x + 2$
14. $f(x) = -x^3 + x^2 - x$
15. $f(x) = x^5 + x^3 + x$
16. $f(x) = x^3 - 2x^2 + x$
17. $f(x) = \dfrac{x}{x - 5}$
18. $f(x) = \dfrac{x + 1}{x}$
19. $f(x) = \dfrac{5x^2 - 2x - 5}{x^2 + 2}$
20. $f(x) = \sqrt{x - 4}$
21. $f(x) = \sqrt{x} - 4$
22. $f(x) = x^2 + \sqrt{x} - 4$
23. $f(x) = x^2 - \sqrt{x} - 4$
24. $f(x) = x^3 - 0.1\sqrt{x} - 4$
25. $f(x) = x^3 - 0.5x - 4$
26. $f(x) = x^3 + 0.5x - 4$

Use the three-step rule (see Examples 3 and 4) to determine whether the function f has an inverse function, and, if it does, give the formula for $f^{-1}(x)$. Assume that each function has its natural domain.

27. $f(x) = -2x + 2$

28. $f(x) = \frac{1}{2}x + \frac{1}{3}$

29. $f(x) = \frac{x - 3}{x}$

30. $f(x) = \frac{x}{x - 3}$

31. $f(x) = \frac{x + 2}{x - 3}$

32. $f(x) = \frac{x - 5}{x + 1}$

33. $f(x) = (x + 2)(x - 3)$

34. $f(x) = \frac{5}{x^2 + 1}$

35. $f(x) = 3 + \sqrt{x - 3}$

36. $f(x) = \frac{\sqrt{x - 3}}{5}$

Each function f has an inverse function. Find the formula for f^{-1}. Then use your calculator to graph f, f^{-1}, and $y = x$ in the same plane. Observe the reflection property noted in the text.

37. $f(x) = 4x + \frac{7}{4}$

38. $f(x) = x^3 + 3$

Each of the following functions also has an inverse function, but you won't be able to find the formula for it. Nevertheless, use your calculator to draw the graph of f and f^{-1} in the same plane.

39. $f(x) = x^3 - x^2 + 2x$
40. $f(x) = x^5 - x^4 + x - 2$
41. $f(x) = \sqrt{x + 8} + 0.2x$
42. $f(x) = 0.1x^3 - 2 + \sqrt{x}$

Graph the given function f. Then restrict the domain of f so that it has an inverse function. Write a formula for f^{-1}. Since this can be done in more than one way, we suggest that you always pick the right-hand branch.

43. $f(x) = 5 - x^2$

44. $f(x) = \frac{1}{5}x^2 + 11$

45. $f(x) = (x - 3)^2 + 5$
46. $f(x) = x^2 - 2x - 4$

47. $f(x) = \frac{10}{x^2 - 2x + 4}$

48. $f(x) = \frac{-8}{x^2 + 4x + 5}$

B. Applications and Extensions

In Problems 49–52, we show the graph of a one-to-one function f. Read $f^{-1}(0)$, $f^{-1}(-2)$, and $f^{-1}(3)$ from this graph. Then graph $y = f^{-1}(x)$.

49.

50.

51.

52.

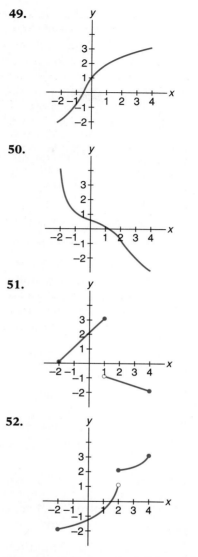

53. Draw the graph of $f(x) = (x^3 + 2)/(x^3 + 3)$, which may lead you to think that f does not have an inverse function. Yet show algebraically that it does and give the formula for $f^{-1}(x)$.

54. Let $f(x) = (x - 3)/(x + 1)$. Find $f^{-1}(x)$ and $f(f(x))$. What conclusion do you draw?

55. The function $f(x) = a(x^3 + x^2 + x + 1)$ is one-to-one provided $a \neq 0$. Determine a, given that $f^{-1}(30) = 2$.

56. Show that if f and g have inverses, then $f \circ g$ has an inverse and $(f \circ g)^{-1} = g^{-1} \circ f^{-1}$.

57. What must be true about the graph of f if f is self-inverse (meaning that $f^{-1} = f$)? Give formulas for two simple functions that are self-inverse.

58. Show algebraically that $f(x) = x/(x - 1)$ is self-inverse.

59. Let $f(x) = (ax + b)/(cx + d)$ with $bc - ad \neq 0$. What condition connecting a and d will make f self-inverse?

60. Challenge. Let $f_1(x) = x$, $f_2(x) = 1/x$, $f_3(x) = 1 - x$, $f_4(x) = 1/(1 - x)$, $f_5(x) = (x - 1)/x$, and $f_6(x) = x/(x - 1)$. Note that

$$f_4(f_3(x)) = \frac{1}{1 - f_3(x)} = \frac{1}{1 - (1 - x)} = \frac{1}{x} = f_2(x)$$

that is, $f_4 \circ f_3 = f_2$. In fact, if we compose any two of these six functions, we will get one of the six functions. Complete the composition table (Figure 15) and then use it to find each of the following (which will also be one of the six functions).

(a) $f_3 \circ f_3 \circ f_3 \circ f_3 \circ f_3$
(b) $f_1 \circ f_2 \circ f_3 \circ f_4 \circ f_5 \circ f_6$
(c) f_6^{-1}
(d) $(f_3 \circ f_6)^{-1}$
(e) F if $f_2 \circ f_5 \circ F = f_5$

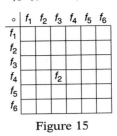

Figure 15

TEASER Using the absolute value function as the building block, create a calculator graph of the sawtooth function shown in the figure.

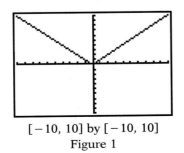

2.9 SPECIAL FUNCTIONS

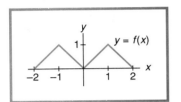

A MISSING ABSOLUTE VALUE KEY

Fortunately, the TI-81 and TI-82 have an absolute value key, denoted by [ABS]. If your calculator fails to have such a key, don't panic; think of a substitute. Here is one.

$$|x| = \sqrt{x^2}$$

Thus, in Example 1,

$$|x - 3| - 4$$
$$= \sqrt{(x - 3)^2} - 4$$

We have met the **absolute value function** several times in earlier discussions. Recall that it is defined by a two-part formula

$$ABS(x) = |x| = \begin{cases} x & \text{if } x \geq 0 \\ -x & \text{if } x < 0 \end{cases}$$

■ **Example 1.** Use your calculator to draw the graphs of $y = |x|$, $y = |x - 3| - 4$, and $y = -|x + 2| + 3$.

Solution. The three graphs are shown in Figures 1, 2, and 3. As we expect from the discussion in Section 2.7, the second graph is shaped like the first but is shifted 3 units right and 4 units down. The third graph is shifted 2 units left, reflected about the x-axis, and shifted up 3 units. ■

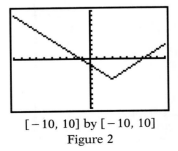

[−10, 10] by [−10, 10]
Figure 1

[−10, 10] by [−10, 10]
Figure 2

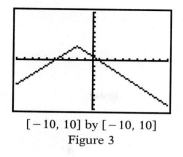

[−10, 10] by [−10, 10]
Figure 3

PIECEWISE DEFINED FUNCTIONS

Figure 4

THE TI-81 AND TI-82

On these calculators, piecewise defined functions are entered using the inequality operators <, ≤, >, and ≥ from the TEST menu. For example, the function of Example 2 is entered as

$(-x + 2)(x \le 2)$
$+ (2x - 8)(x > 2)$

The multiplier $(x \le 2)$ has the value 1 if $x \le 2$ and the value 0 otherwise. Similarly, the multiplier $(x > 2)$ has the value 1 if $x > 2$ and the value 0 otherwise.

The absolute value function is the simplest of what are called **piecewise defined functions.** These are functions that are defined by different formulas on different parts of their domains. Such functions arise often in calculus and in real life, as we shall see. Here is an example.

■ **Example 2.** Sketch the graph of the following piecewise defined function.

$$f(x) = \begin{cases} -x + 2 & \text{if } x \le 2 \\ 2x - 8 & \text{if } x > 2 \end{cases}$$

Solution. The graph is shown in Figure 4. Note the use of an open circle and a filled circle at break points in hand-drawn graphs. The y-value at such points is the y-coordinate of the filled circle. ■

Can the graph of a piecewise defined function be drawn using a calculator? Yes (on many calculators), but you may have to study a manual to learn how to do it.

■ **Example 3.** Use a calculator to graph the function f of Example 2.

Solution. The box titled The TI-81 and TI-82 gives a brief explanation of the procedure for entering a piecewise defined function on TI calculators. Using it led to Figure 5, which we consider to be unsatisfactory. Because our calculator was set in connected mode, it tried to eliminate the discontinuity at $x = 2$. A much better graph, obtained using dot mode, is shown in Figure 6. You should note, however, that you cannot tell from a calculator-drawn graph the value of the function at a discontinuity point. ■

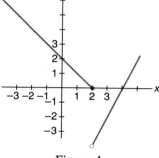

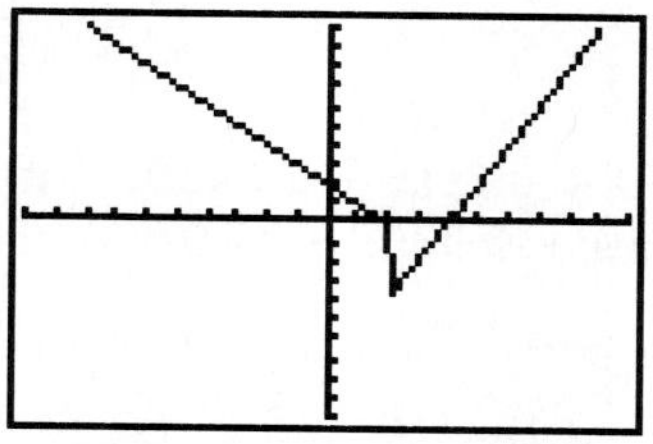

[−10, 10] by [−10, 10]
Figure 5

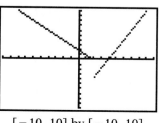

[−10, 10] by [−10, 10]
Figure 6

DOUBLE INEQUALITIES

On the TI-81 and TI-82, note that $(-3 \le x)(x \le 3)$ will have the value 1 when both inequalities hold, 0 otherwise. Thus it is the appropriate multiplier for the domain $-3 \le x \le 3$.

For the rest of this section, we suggest that you keep your calculator in dot mode.

■ **Example 4.** Use your calculator to draw the graph of the function.

$$f(x) = \begin{cases} -x - 6 & \text{if } x < -3 \\ 9 - x^2 & \text{if } -3 \le x \le 3 \\ x - 6 & \text{if } x > 3 \end{cases}$$

Solution. Figure 7 gives a reasonable calculator-drawn graph of this function. ■

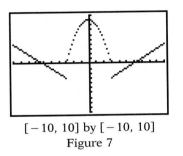

$[-10, 10]$ by $[-10, 10]$
Figure 7

■ **Example 5.** For the year 1990, the IRS tax rate for a single taxpayer was listed as shown in Figure 8. If $T(x)$ denotes the tax on an income of x dollars, write a three-part formula for T for the domain $0 < x \le 97{,}620$ and draw its graph. Determine the tax on a taxable income of \$54,000.

Schedule X—Use if your filing status is **Single**

If the amount on Form 1040, line 37, is: Over—	But not over—	Enter on Form 1040, line 38	of the amount over—
\$0	\$19,450	.......15%	\$0
19,450	47,050	\$2,917.50 + 28%	19,450
47,050	97,620	10,645.50 + 33%	47,050

Figure 8

Solution.

$$T(x) = \begin{cases} 0.15x & \text{if } 0 < x \le 19{,}450 \\ 2917.50 + 0.28(x - 19{,}450) & \text{if } 19{,}450 < x \le 47{,}050 \\ 10{,}645.50 + 0.33(x - 47{,}050) & \text{if } 47{,}050 < x \le 97{,}620 \end{cases}$$

The graph is shown in Figure 9 and $T(54{,}000) = \$12{,}939$. ■

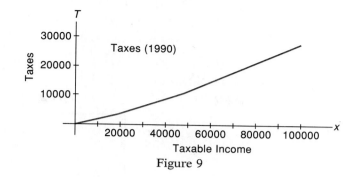

Taxes (1990)

Taxable Income

Figure 9

THE GREATEST INTEGER FUNCTION

A function often met in calculus courses is the **greatest integer function,** symbolized by [[ ]]. By [[x]] we mean the greatest integer less than or equal to x. Thus [[6.4]] = 6, [[6]] = 6, and [[−6.4]] = −7. Some calculators have built in this function but may denote it by a different symbol (the TI-81 and TI-82 denote it by *int*).

■ **Example 6.** Draw the graphs of $y = [[x]]$, $y = 2[[x/3]]$, and $y = [[2x − 1]]$.

Solution. The three graphs, all drawn using a graphing calculator in dot mode, are shown in Figures 10, 11, and 12. ■

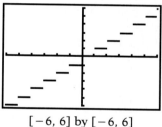

[−6, 6] by [−6, 6]
Figure 10

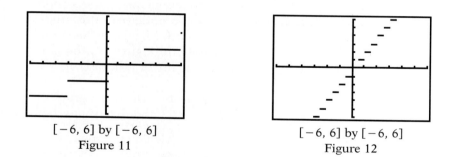

[−6, 6] by [−6, 6]
Figure 11

[−6, 6] by [−6, 6]
Figure 12

A typical real-world example of a function whose behavior is similar to the greatest integer function is the postage stamp function, which in 1994 used the following rule: 29¢ for the first ounce and an additional 23¢ for each additional ounce or fraction thereof. Here is another example.

■ **Example 7.** A parking lot charges $3 for the first hour and 50¢ for each additional half hour or fraction thereof, with a maximum of $7 for a whole day. Write a formula for $C(x)$, the cost of parking x hours, using the greatest integer function. Then draw the graph of C for the interval $0 < x \le 10$.

Solution. A possible formula for $C(x)$ is as follows.

$$C(x) = \begin{cases} 3.00 & \text{if } 0 < x \le 1 \\ 3.00 + 0.5[[2x - 1]] & \text{if } 1 < x < 4.5 \\ 7.00 & \text{if } 4.5 \le x \le 24 \end{cases}$$

Its graph is shown in Figure 13. ∎

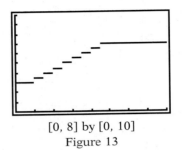

[0, 8] by [0, 10]
Figure 13

PERIODIC FUNCTIONS

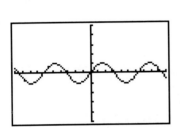

[−10, 10] by [−5, 5]
Figure 14

A periodic function is a function whose graph repeats itself in regular cycles. More precisely, f is a **periodic function** if there is a positive number p such that $f(x + p) = f(x)$ for all x in the domain of f. The smallest such positive number is called the **period** of a periodic function. The premier examples of periodic functions are the trigonometric functions to be studied in detail in Chapter 4. To anticipate this study, try graphing $y = \sin x$ in radian mode (every graphics calculator has a built-in sin function). You should obtain the graph shown in Figure 14, the graph of a periodic function with period 2π. There are plenty of other examples.

∎ **Example 8.** Draw the graphs of $f(x) = x - [[x]]$ and $g(x) = (x - [[x]])^2$ for $-3 \le x \le 3$ and determine their periods.

Solution. Calculator-drawn versions of these graphs are shown in Figures 15 and 16. Both functions are periodic with period 1. ∎

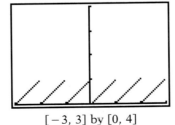

[−3, 3] by [0, 4]
Figure 15

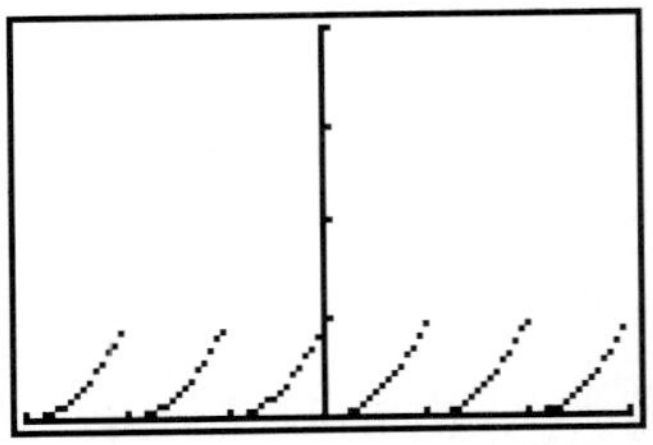

[−3, 3] by [0, 4]
Figure 16

■ **Example 9.** Refer to Example 8 and let

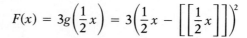

$$F(x) = 3g\left(\frac{1}{2}x\right) = 3\left(\frac{1}{2}x - \left[\left[\frac{1}{2}x\right]\right]\right)^2$$

Indicate how the graph of F is related to the graph of g. Then draw the graph of F and compare it with Figure 16 to confirm your answer.

Solution. The multiplier 1/2 on x spreads out the graph horizontally by a factor of 2, giving this graph a period of 2. The multiplier of 3 in front of g stretches the graph vertically by a factor of 3. A calculator-drawn graph of F is shown in Figure 17. ■

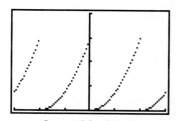

$[-3, 3]$ by $[0, 4]$
Figure 17

TEASER SOLUTION

To obtain the left half of the desired graph reflect the graph of $|x|$ about the x-axis, then shift it 1 unit left and 1 unit up. To get the right half, do the same thing but with a right shift. The corresponding function is given by the two-part formula

$$f(x) = \begin{cases} -|x + 1| + 1 & \text{if } -2 \le x \le 0 \\ -|x - 1| + 1 & \text{if } 0 < x \le 2 \end{cases}$$

You will need to consult a manual to learn how to enter this function into your calculator. On the TI-81 or TI-82, you would enter

$$(-\text{ABS}(x+1) + 1)(-2 \le x)(x \le 0) + (-\text{ABS}(x-1) + 1)(0 < x)(x \le 2)$$

Figure 18 displays the resulting graph.

$[-2, 2]$ by $[-1, 2]$
Figure 18

PROBLEM SET 2.9

A. Skills and Techniques

Try to predict the nature of the graph of each function. Then use your calculator to draw the graph.

1. $f(x) = 3|x|$
2. $f(x) = |0.5x|$
3. $f(x) = 2|x - 3| - 4$
4. $f(x) = 0.5|x + 2| + 1$
5. $f(x) = |x^2 - x - 6|$
6. $f(x) = |0.3x(x - 2)(x + 4)|$
7. $f(x) = |x|^3$
8. $f(x) = |x|^3 - 6|x| + 2$
9. $f(x) = x - |x| + 2$
10. $f(x) = 3x/|x|$

Graph each function in Problems 11–22, either by hand or using a calculator (dot mode).

11. $f(x) = \begin{cases} -x - 2 & \text{if } x < 1 \\ -3 & \text{if } x \ge 1 \end{cases}$

12. $f(x) = \begin{cases} x & \text{if } x < 3 \\ 3 & \text{if } x \ge 3 \end{cases}$

13. $f(x) = \begin{cases} 0.5x + 3 & \text{if } x \le -2 \\ 3x - 2 & \text{if } x > -2 \end{cases}$

14. $f(x) = \begin{cases} x + 3 & \text{if } x < 2 \\ -x + 7 & \text{if } x \geq 2 \end{cases}$

15. $f(x) = \begin{cases} x - 1 & \text{if } x \leq -4 \\ -5 & \text{if } -4 < x < 6 \\ 1 - x & \text{if } x \geq 6 \end{cases}$

16. $f(x) = \begin{cases} x + 2 & \text{if } x < -2 \\ 2x & \text{if } -2 \leq x < 2 \\ x - 2 & \text{if } x \geq 2 \end{cases}$

17. $f(x) = \begin{cases} x + 4 & \text{if } x < 0 \\ x^2 & \text{if } 0 \leq x \leq 2 \\ 4 & \text{if } x > 2 \end{cases}$

18. $f(x) = \begin{cases} -x & \text{if } x < -2 \\ \frac{1}{2}x^3 + 2x & \text{if } -2 \leq x < 0 \\ \frac{1}{4}x^3 - 2x^2 + 9 & \text{if } x \geq 0 \end{cases}$

19. $f(x) = \frac{3}{2}[[x - 1]]$

20. $f(x) = 2\left[\left[\frac{1}{3}x\right]\right] + 1$

21. $f(x) = 3\left[\left[-\frac{1}{2}x\right]\right]$

22. $f(x) = |2 - [[2x]]|$

23. Acme Car Rental fees for each day are $40 plus 30¢ per mile for each mile beyond 100 miles. Write the two-part formula for the cost $C(x)$ of renting a car for 4 days and driving a total of x miles. Then graph this function.

24. Geraldo's salary each month is $1000 plus $25 for each suit he sells plus an additional $5 bonus for each suit that he sells beyond 50 suits in a month. Write a formula for his salary $S(x)$ in a month in which he sells x suits and graph the result.

25. Janson Bus Company offers one-day excursion trips for the total cost of $150 plus $25 for each passenger numbered 1–30 plus $20 for each passenger numbered 31–85. Write a formula for the cost $C(x)$ of taking a group of x passengers on an excursion, $0 < x \leq 85$.

26. Giselle left Hampton at noon heading due north at 50 miles per hour. Two hours later, Jane left the same place heading due west at 60 miles per hour. Write a formula for $D(t)$, the distance between them t hours after noon.

In Problems 27–38, draw the graph of each function (dot mode) and decide whether it is periodic. If it is, determine its period. Note that your calculator has keys for sin, cos, and tan.

27. $f(x) = \sin 3x$
28. $f(x) = \sin \pi x$
29. $f(x) = \cos(0.5x)$
30. $f(x) = \tan x$
31. $f(x) = [[2x]] - 2x$
32. $f(x) = (x - [[x]])^3$
33. $f(x) = [[x^2]] - x^2$
34. $f(x) = [[2x]] - x$
35. $f(x) = [[5x - 1]] + 1 - 5x$
36. $f(x) = [[1 - 2x]] + 2x$
37. $f(x) = |x - [[x + 0.25]]|$
38. $f(x) = |2x - [[2x + 1]]|$

39. Parking meters in Longview impose a charge of 25¢ for anything less than an hour, then 25¢ for each additional hour or fraction thereof, and allow entering up to 6 quarters. Write a formula in terms of [[ ]] representing the charge for x hours and graph the result.

40. Parking lots in Chesterville charge 10¢ for anything less than 12 minutes plus 5¢ for each additional 12-minute segment (or part thereof) up to a maximum of 2 hours. Write a formula for the cost $C(x)$ of parking x minutes using [[ ]]. Graph the result and find the cost of parking for 1 hour and 31 minutes.

B. Applications and Extensions

41. Using | |, write a formula for each function graphed below.

(a)

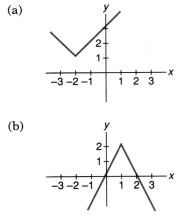

(b)

42. Using [[ ]], write a formula for each function graphed below.

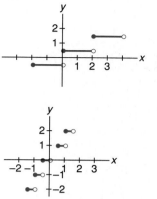

43. For the graph shown below, determine (a) $f(4)$, (b) $f(0.9)$, (c) $(f \circ f)(-1)$, (d) $f(x)$ in terms of [[ ]].

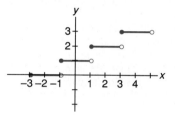

44. For the graph shown below, determine (a) $g(-1)$, (b) $g(-1.1)$, (c) $(g \circ g \circ g)(4)$, (d) $g(x)$.

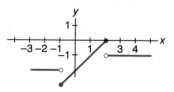

45. Michelle's medical policy pays each year the following percentage of her total medical costs: 0% of the first $1000, 80% of the next $5000, and 100% of costs above that. Write a formula for the amount $A(x)$ that Michelle will pay in a year where she has $x of medical expenses and determine the maximum amount she will have to pay in a year.

46. Standard group ticket rates to a baseball game are $14 each for the first 20, $9 each for the next 20, and $4 each beyond 40. When the manager negotiates with what he expects to be a large group, he sometimes offers a special rate of $6.25 per ticket. Write formulas $C(x)$ and $C^*(x)$ for the standard rate and the special rate for a group of size x and determine which is the cheaper rate for a group of size 150 and by how much.

47. In a certain state, the income tax rates are as follows. The first $10,000 is exempt, the next $10,000 is taxed at 6%, and all above that is taxed at 9%. Write a multipart formula for the total tax $T(x)$ on an income of $x and graph the result.

48. Consider the U.S. 1994 postage stamp function: 29¢ for less than one ounce and 23¢ for each additional ounce or fraction thereof. Write a formula, using [[ ]] for the cost of a package weighing x ounces. Graph the result and find the cost of sending a package weighing 5.3 ounces.

49. If $f(x) = x$ on $-1 < x \le 2$ and f is periodic, determine (a) $f(3)$ (b) $f(3.5)$ (c) $f(11)$.

50. Let $f(n)$ be the nth digit in the decimal expansion of 5/13. (a) State the natural domain for this function. (b) Determine its range. (c) Is f periodic? (d) If so, give its period.

51. Let $f(x) = x + [[x]]$ with domain $-1 \le x < 2$. (a) Determine the range of f. (b) Is f one-to-one? (c) If so, evaluate $f^{-1}(-1.5)$.

52. Is $f(x) = 5$ with domain $\mathbb{R}$ periodic? If so, what is its period?

53. For what values of a is $f(x) = x + a[[x]]$ with domain $\mathbb{R}$ one-to-one?

54. Let $\lceil x \rceil$, called the **ceiling function,** be defined as the smallest integer greater than or equal to x. (a) Evaluate $\lceil 3.1 \rceil$, $\lceil 3.9 \rceil$, and $\lceil 4 \rceil$. (b) For what values of x is $\lceil x \rceil = [[x + 1]]$?

55. Let $\langle\!\langle x \rangle\!\rangle$ denote the **distance to the nearest integer function** so that $\langle\!\langle 1.3 \rangle\!\rangle = 0.3$ and $\langle\!\langle 1.8 \rangle\!\rangle = 0.2$. (a) Sketch its graph by hand and then convince yourself that $\langle\!\langle x \rangle\!\rangle = |x - [[x + 0.5]]|$. (b) Draw the graph of $y = \langle\!\langle 0.2x \rangle\!\rangle$ using your calculator (connected mode) and determine its period.

56. **Challenge.** Use your calculator to draw a reasonable graph of $y = 2 \langle\!\langle x \rangle\!\rangle / 2^{[[x]]}$ for the domain $0 \le x < \infty$. Then calculate what you believe to be the total area of the region between this graph and the x-axis.

CHAPTER 2 REVIEW PROBLEM SET

In Problems 1–10, write True or False in the blank. If false, tell why.

_____ **1.** If $f(x) = |x + 3|$, then $f(-3.5) = f(0.5)$.
_____ **2.** The graphs of $f(x) = (x^2 - 4)/(x - 2)$ and $g(x) = x + 2$ are the same.
_____ **3.** The natural domain of $\sqrt{x^2 - x - 12}$ is the set $[x: -3 < x < 4]$.
_____ **4.** The range of the greatest integer function $g(x) = [[x]]$ is the set of all integers.
_____ **5.** The function $f(x) = (x + 3)^2$, $x \le -3$, is one-to-one.
_____ **6.** If $f(x) = x^3$, then $f(f(x)) = x^6$.
_____ **7.** The graph of $f(x) = (x - 2)/(x^2 + 4)$ has no horizontal asymptotes.
_____ **8.** The graph of $f(x) = (x - 2)(x^2 + 4)/(x + 2)$ crosses the x-axis only once.
_____ **9.** If $g(4)$ is in the domain of f, then 4 is in the domain of $f \circ g$.
_____ **10.** The functions $f(x) = \sqrt[3]{x} - 2$ and $g(x) = x^3 + 2$ are inverse functions.

11. Let $f(x) = x^2 - 4$ and $g(x) = 1/\sqrt{x}$. Calculate, if possible.
(a) $g(0.81)$
(b) $f(\sqrt{7})$
(c) $f(g(0.01))$
(d) $f(2) \cdot g(0)$
(e) $g(g(16))$
(f) $f(0)/g(4)$

12. Determine the natural domain of f if $f(x) = \sqrt{x}/(x - 1)$.

Figure 1 shows the graph of $y = f(x)$ with domain $-3 \le x \le 4$. In Problems 13–20, determine as best you can the required information.

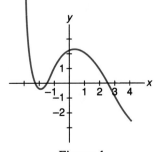

Figure 1

13. $f(-0.8)$ and $f(f(3.5))$
14. The value of x where $f(x) = 3.3$

15. Where $f(x) \ge 0$
16. Where f is increasing
17. Where f is concave down
18. The minimum value of f on the interval $-3 \le x \le 4$
19. The slope at $x = 2.5$
20. All solutions to $f(x) = 0$
21. Which of the following functions are even, which are odd, and which are neither?
(a) $f(x) = 3x^3 - 2x$
(b) $f(x) = -0.25x^4 - 5$
(c) $f(x) = x(x^2 - 5)$
(d) $f(x) = (x - 2)^3$
(e) $f(x) = |x^3|(x^2 - 5)$
(f) $f(x) = \dfrac{x^3}{x^2 + 1}$

22. Explain how the graph of $f(x) = (x + 2)^2 + 4$ is related to the graph of $g(x) = x^2 - 1$.
23. For each parabola, find the coordinates of the vertex without graphing and tell whether it turns up or down.
(a) $f(x) = -3x^2 - 6x + 1$
(b) $g(x) = x^2 + 4x$
(c) $h(x) = -0.5x^2 + 2x - 9$
(d) $k(x) = (x - 2)^2 + 6$
24. Find the minimum value of $f(x) = x^3 - 2x^2 - 5x + 7$ on the interval $0 \le x \le 5$, accurate to two decimal places.
25. Find the minimum and maximum points of $g(x) = x^6 - 20x^5 + 11x^3 + 3$ on the interval $-1 \le x \le 1$.
26. Find the maximum point on the graph of $f(x) = -2x^4 + 8x^2 - 3x$ with domain $x \le 0$ and determine where this function is decreasing.
27. Find the minimum point on the graph of $g(x) = x^4 - 7x^3 + 11x^2 + 2x$ and determine approximately where this function is concave down.
28. Solve the equation $f(x) = g(x)$ given that $f(x) = x^3 - 3x^2 + 2$ and $g(x) = x^2 + 5x + 4$.
29. Determine the slope of the graph of $f(x) = x^3 - 3x^2$ at $x = -2$.
30. Factor $p(x) = x^3 - 2x^2 - 29x - 42$ into linear factors given that $c = -2$ is a zero of $p(x)$.
31. Determine a fifth-degree polynomial whose only zeros are 1, 2, and 3 with 1 and 2 being simple zeros.
32. In each case determine all real zeros and give their multiplicities.
(a) $f(x) = (x - 1)^2(x + 2)(x - 3)$
(b) $g(x) = 3x^3(x^2 - 9)^2$
(c) $h(x) = x^3 + 8x^2 + 16x$
(d) $k(x) = (x^2 + 1)(x^2 - 1)^3$
33. Sketch the best graph you can of each function.
(a) $f(x) = a(x - 2)(x + 3)^2$, $a < 0$
(b) $f(x) = a(x + 3)^3(x - 2)(x - 4)$, $a > 0$

In Problems 34–37, draw the graph of each function and determine any horizontal, vertical, and slant asymptotes.

34. $f(x) = \dfrac{2x}{x + 5}$

35. $g(x) = \dfrac{3x^2}{x^2 + 5}$

36. $h(x) = \dfrac{x^3 - 9x^2 + 5x + 7}{x^2 - 2x + 5}$

37. $k(x) = \dfrac{-2x^3 + 5x^2 - 4}{x^2 - 9}$

38. Write a formula for a rational function that satisfies all of the following conditions.
(a) It has zeros at -2 and 3.
(b) Its only asymptotes are $x = 0$ and $y = 1$.

39. Write a formula for a function with a slant asymptote of $y = 2x - 1$ that also has a vertical asymptote at $x = 3$.

40. Let $f(x) = 2x + 3$ and $g(x) = x^3$. Write formulas for each of the following.
(a) $(f + g)(x)$
(b) $(g - f)(x)$
(c) $(f \cdot g)(x)$
(d) $(f/g)(x)$
(e) $(f \circ g)(x)$
(f) $(g \circ f)(x)$
(g) $(f \circ f)(x)$
(h) $(g \circ g \circ g)(x)$

41. If $f(x) = \sqrt[3]{x - 2} + 3$ and $g(x) = (x - 3)^3$, write a formula for each of the following.
(a) $f(x + 2)$
(b) $g(f(x))$
(c) $g^{-1}(x)$
(d) $f^{-1}(x)$

42. If $f(x)$ is obtained by adding 3 to x and then dividing the result by 2, find a formula for $f^{-1}(x)$.

43. Decide whether $g(x) = x/(x + 2)$ has an inverse function and if so find the formula for $g^{-1}(x)$.

44. Graph $y = f(x) = (1/2)x^3 - 4$ and $y = f^{-1}(x)$ in the same plane.

45. Determine $f(x)$ and $g(x)$ if $h(x) = \sqrt{x^3 - 7}$ is decomposed as $h(x) = (f \circ g)(x)$.

46. Show that 2 is not in the range of $f(x) = (2x + 1)/(x + 2)$.

47. Determine the natural domain of $f(x) = \sqrt{9 - x^2}/(x + 2)$.

48. Indicate a good way to restrict the domain of $f(x) = (2x - 5)^2$ so that f has an inverse. Then determine a formula for the inverse of this f.

49. Assume that f is an even function, that g is an odd function, and that both are defined for all real numbers. Show that $f \cdot g$ is an odd function and that $f \circ g$ is an even function.

Use your calculator to draw graphs of each of the functions in Problems 50–56.

50. $f(x) = \dfrac{1}{2}|x^2 - 20| - 5$

51. $f(x) = \begin{cases} x^2 - 4 & \text{if } x \le -1 \\ x^3 + 4 & \text{if } x > -1 \end{cases}$

52. $f(x) = \begin{cases} \dfrac{10}{x} & \text{if } x \le -2 \\ -x^2 - 1 & \text{if } -2 < x < 2 \\ -\dfrac{10}{x} & \text{if } x \ge 2 \end{cases}$

53. $f(x) = [[2x]]$
54. $f(x) = [[5 - 2x]]$
55. $f(x) = ([[x]] - x)^3$

56. $f(x) = \dfrac{1}{2x - [[2x]]}$

57. Draw the graph of a periodic function of period 2 that satisfies $f(x) = x^2$ on $-1 \le x \le 1$.

58. A silo has the shape of a cylinder topped by a hemisphere (Figure 2). The cylindrical part has height 10 meters and radius x meters. (a) Find formulas for the volumes $V_c(x)$ and $V_h(x)$ of the cylindrical and hemispherical parts. (b) Solve $V_c(x) = V_h(x)$ for $x > 0$. (c) What value of x maximizes $V_c(x) - V_h(x)$?

$h = 10\text{m}$

x

Figure 2

59. Pedro is on an island at point A, 3 miles from the nearest point B on the straight shoreline of a large lake (Figure 3). He will row toward point C, x miles down the shore from B, at 2.5 miles per hour and then walk to his home D, 12 miles along the shore from B, at 3.5 miles per hour. (a) Write formulas in terms of x for S, the total distance traveled, and T, the time required, and graph them over the domain $0 \le x \le 12$. (b) Find the minimum and maximum travel times over this domain.

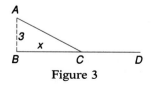

A

3

x

B C D

Figure 3

60. Find the point on the parabola $y = x^2 - x$ that is closest to the point $(4, -2)$ and determine the distance between these two points. Give answers accurate to two decimal places.

3

EXPONENTIAL AND LOGARITHMIC FUNCTIONS

Can you imagine anyone writing the equivalent of over 75 books on mathematics? That is the contribution of Leonhard Euler, the most prolific mathematical writer of all time. Born near Basel, Switzerland, Euler held positions at the University of Basel, St. Petersburg Academy of Sciences, and Berlin Academy of Sciences. When he died, it was said that all the mathematicians of Europe were his students.

Blindness during the last 17 years of his life seems not to have hampered his output. He had a prodigious memory and he is said to have done a calculation of 50 decimal places in his head. Yet he found time for his 13 grandchildren and often spent his evenings playing games with them or reading the Bible to them.

We introduce Euler in connection with Chapter 3 because he first recognized the importance of the number e and the exponential function e^x. He showed that e and e^2 are irrational numbers and discovered the remarkable relation $e^{i\pi} = -1$. Much of what is taught today in a standard calculus course follows the ideas of Euler.

The graph of

$$cosh\ x = (e^x + e^{-x})/2$$

is called a catenary; it is the form of a hanging chain. The inverted catenary is the form of the strongest possible arch. The Finnish-American architect, Eero Saarinen, used this latter fact in his design of the Gateway Arch in St. Louis. It is 630 feet high and 630 feet wide at its base. This arch, which celebrates the westward expansion of the United States, was completed in 1965.

Eero Saarinen (1910–1961)   Leonhard Euler (1707–1783)

The Gateway Arch

$y \approx 699 - 69\ cosh\ (0.0095x)$

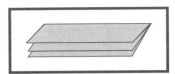

TEASER A very long sheet of paper 0.01 inch thick is folded in half over and over, creating a higher and higher stack of paper. If the sheet is folded 40 times, how high will the stack be? Guess between 10 feet, 5000 feet, and thousands of miles. Then work out the answer.

3.1 EXPONENTS AND THEIR PROPERTIES

We have assumed that you are familiar with integral exponents, and we have used them freely in earlier chapters. Our goal is to extend to rational exponents and to real exponents while preserving the familiar laws of exponents. We begin with a review.

Let b be any number and let n be a positive integer. Then we define b^n by

$$b^n = \underbrace{b \cdot b \cdot b \cdots b}_{n \text{ factors}}$$

Thus, $b^3 = b \cdot b \cdot b$ and $b^5 = b \cdot b \cdot b \cdot b \cdot b$. On the basis of this definition, we obtain the laws of exponents. For the moment, we assume that m and n are positive integers and in law 3 that $m > n$.

LAWS OF EXPONENTS

1. $b^m \, b^n = b^{m+n}$
2. $(b^m)^n = b^{mn}$
3. $\dfrac{b^m}{b^n} = b^{m-n}$
4. $(ab)^n = a^n \, b^n$
5. $\left(\dfrac{a}{b}\right)^n = \dfrac{a^n}{b^n}$

ZERO AND NEGATIVE INTEGRAL EXPONENTS

In extending the notion of exponent beyond the positive integers, we are motivated by our desire to preserve these laws. Thus b^0 must satisfy

$$b^0 \, b^2 = b^{0+2} = b^2$$

which implies that

$$\boxed{b^0 = 1 \quad (b \neq 0)}$$

Moreover

$$b^{-n} \, b^n = b^{-n+n} = b^0 = 1$$

which means that

$$b^{-n} = \frac{1}{b^n} \quad (b \neq 0)$$

With these understandings, the laws of exponents continue to hold, and the restriction in law 3 is removed.

■ **Example 1.** Simplify each of the following, writing the answer without negative exponents. (a) -5^{-2} (b) $(-5)^{-2}$ (c) $(b^{-5} b^2)^{-2}$ (d) $(a^{-1} b^{-2})^3$

Solution.

(a) $-5^{-2} = -\dfrac{1}{5^2} = -\dfrac{1}{25}$

(b) $(-5)^{-2} = \dfrac{1}{(-5)^2} = \dfrac{1}{25}$

(c) $(b^{-5} b^2)^{-2} = (b^{-5+2})^{-2} = (b^{-3})^{-2} = b^{(-3)(-2)} = b^6$

(d) $(a^{-1} b^{-2})^3 = a^{(-1)(3)}b^{(-2)(3)} = a^{-3}b^{-6} = \left(\dfrac{1}{a^3}\right)\left(\dfrac{1}{b^6}\right) = \dfrac{1}{a^3b^6}$ ■

■ **Example 2.** Simplify and write the answer without negative exponents.

(a) $\dfrac{8ab^{-2}c^3}{(2a)^2 b^{-4} c^2}$ (b) $\left(\dfrac{(2xy^{-2})^3(x^2y^{-1})^2}{2xy^3}\right)^2$ (c) $\dfrac{(a^{-1} + b^{-1})^{-1}}{ab}$

Solution. The laws of exponents imply that we may move a factor from numerator to denominator or vice versa by changing the sign of its exponent, a fact we often find useful.

(a) $\dfrac{8ab^{-2}c^3}{(2a)^2b^{-4}c^2} = \dfrac{8ab^{-2}c^3}{4a^2b^{-4}c^2} = \dfrac{8b^{-2}b^4c^3c^{-2}}{4a^{-1}a^2} = \dfrac{2b^2c}{a}$

(b) $\left(\dfrac{(2xy^{-2})^3(x^2y^{-1})^2}{2xy^3}\right)^2 = \dfrac{(2xy^{-2})^6(x^2y^{-1})^4}{4x^2y^6} = \dfrac{64x^6y^{-12}x^8y^{-4}}{4x^2y^6} = \dfrac{16x^6x^8x^{-2}}{y^6y^{12}y^4} = \dfrac{16x^{12}}{y^{22}}$

(c) $\dfrac{(a^{-1} + b^{-1})^{-1}}{ab} = \dfrac{(1/a + 1/b)^{-1}}{ab} = \dfrac{((b + a)/ab)^{-1}}{ab} = \dfrac{ab/(a + b)}{ab} = \dfrac{1}{a + b}$ ■

ROOTS AND RATIONAL EXPONENTS

In earlier chapters we used square roots and cube roots without explanation, assuming that you were familiar with these concepts from previous courses. Here we make explicit definitions of these roots and their generalizations.

There are two problems with square roots, problems that carry over to all even roots. First, every positive number has two square roots. For example, the two square roots of 4 are -2 and 2. The symbol $\sqrt{}$ is used to denote the positive square root; thus $\sqrt{4} = 2$, and the two square roots of 5 are $-\sqrt{5}$ and $\sqrt{5}$. Second, a negative number does not have a real square root. For example, the two square roots of -4 are $-2i$ and $2i$, imaginary numbers. Neither of these problems occurs with cube roots or with any odd roots. Every real number has exactly one real cube root. For example, the cube root of -8

WARNING

Because the square root symbol $\sqrt{}$ is often misused, we emphasize that when $a > 0$,

$\sqrt{a}$ *denotes the positive square root of a.*

Thus, $\sqrt{9} = 3$, $\sqrt{16} = 4$, and $\sqrt{21} \approx 4.5826$. Note also that $\sqrt{a^2} = |a|$.

is -2 and the cube root of 27 is 3. These considerations lead us to the following general definition in which n is a positive integer.

> **Definition.** If n is odd, $\sqrt[n]{b}$ is the unique real number satisfying $(\sqrt[n]{b})^n = b$. If n is even and $b \geq 0$, $\sqrt[n]{b}$ is the unique nonnegative real number satisfying $(\sqrt[n]{b})^n = b$.

Thus, $\sqrt[4]{81} = 3$ and $\sqrt[5]{-32} = -2$.

With roots well defined, we are ready to continue our development of exponents. Again we are motivated by our desire to preserve the laws of exponents stated at the beginning of this section. If n is a positive integer, law 2 requires that $(b^{1/n})^n = b^{(1/n)n} = b^1 = b$, which means that

$$b^{1/n} = \sqrt[n]{b}$$

For example, $9^{1/2} = \sqrt{9} = 3$ and $8^{1/3} = \sqrt[3]{8} = 2$. Next if m and n are positive integers with no common divisors other than ± 1, we define

$$b^{m/n} = (\sqrt[n]{b})^m = \sqrt[n]{b^m}$$

provided this definition gives a real number, that is, provided $b > 0$ when n is even. Finally, we define

$$b^{-m/n} = \frac{1}{b^{m/n}}$$

Thus, b^q is well defined for all rational numbers q (at least when $b \geq 0$) and, moreover, the laws of exponents continue to hold.

■ **Example 3.** Simplify and write the answer without negative exponents.

(a) $9^{-2}9^{-2/3}9^{7/6}$ (b) $\left(\dfrac{a^{-2}b^{2/3}}{\sqrt{2}\,b^{-1/2}}\right)^4$ (c) $(a^2b^{-1/4})^2(2^{2/3}a^{-1/3}b^{-1/2})^3$

Solution.

(a) $9^{-2}9^{-2/3}9^{7/6} = 9^{(-12-4+7)/6} = 9^{-9/6} = 9^{-3/2} = \dfrac{1}{9^{3/2}} = \dfrac{1}{(\sqrt{9})^3} = \dfrac{1}{27}$

(b) $\left(\dfrac{a^{-2}b^{2/3}}{\sqrt{2}\,b^{-1/2}}\right)^4 = \dfrac{a^{-8}b^{8/3}}{4b^{-2}} = \dfrac{b^{8/3}b^2}{4a^8} = \dfrac{b^{14/3}}{4a^8}$

(c) $(a^2b^{-1/4})^2\,(2^{2/3}a^{-1/3}b^{-1/2})^3 = a^4b^{-1/2}2^2a^{-1}b^{-3/2} = 4a^3b^{-2} = \dfrac{4a^3}{b^2}$ ■

■ **Example 4.** Perform the indicated operations and simplify.

(a) $(a^{1/2} - b^{1/2})(a^{1/2} + b^{1/2})$
(b) $(a^{2/3} + b^{2/3})^3$
(c) $\dfrac{(x + 2)^{2/3}}{x} - \dfrac{1}{(x + 2)^{1/3}}$

Solution.

(a) Recall that $(x - y)(x + y) = x^2 - y^2$. Thus,

$$(a^{1/2} - b^{1/2})(a^{1/2} + b^{1/2}) = (a^{1/2})^2 - (b^{1/2})^2 = a - b$$

(b) Recall that $(x + y)^3 = x^3 + 3x^2y + 3xy^2 + y^3$. Thus,

$$(a^{2/3} + b^{2/3})^3 = a^2 + 3a^{4/3}b^{2/3} + 3a^{2/3}b^{4/3} + b^2$$

(c) $\dfrac{(x + 2)^{2/3}}{x} - \dfrac{1}{(x + 2)^{1/3}} = \dfrac{(x + 2)^{2/3}(x + 2)^{1/3} - x}{x(x + 2)^{1/3}} = \dfrac{x + 2 - x}{x(x + 2)^{1/3}} = \dfrac{2}{x(x + 2)^{1/3}}$ ■

REAL EXPONENTS

To set the stage for the next extension, consider the problem of defining 2^π. Recall that π is an irrational number whose decimal expansion begins as follows.

$$\pi = 3.1415926535\ldots$$

The numbers 3, 3.1, 3.14, 3.141, ... are all rational and so 2^3, $2^{3.1}$, $2^{3.14}$, $2^{3.141}$, ... are well defined real numbers. The chart in Figure 1 shows (calculator) values for the first few of these numbers and suggests that they are converging to a definite number, the number that we hope is 2^π. We don't claim to have proved anything, only to have made an observation. In fact, only the concepts of calculus allow us to make a rigorous definition of 2^π or of 2 raised to any irrational exponent. In calculus, you will see a proper definition of a^x for $a > 0$ and any real number x and will learn that the laws of exponents continue to hold.

LAWS OF EXPONENTS

1. $a^x a^y = a^{x+y}$
2. $(a^x)^y = a^{xy}$
3. $\dfrac{a^x}{a^y} = a^{x-y}$
4. $(ab)^x = a^x b^x$
5. $\left(\dfrac{a}{b}\right)^x = \dfrac{a^x}{b^x}$

CALCULATING a^x

There are no problems in calculating a^x on the TI-81 or TI-82 as long as a is positive. Simply press $a \wedge x$. But watch out if $a < 0$. Pressing $(-1) \wedge (1/2)$ or $(-1) \wedge (2/3)$ gives an error message. We expect this for the first since it is an imaginary number. But the second has the value 1. To avoid this problem when $a < 0$, write

$$a^{m/n} = (a^{1/n})^m$$

and press the keys

$$(a \wedge (1/n)) \wedge m$$

---

q	2^q
3	8
3.1	8.5741877
3.14	8.815240927
3.141	8.821353305
3.1415	8.824411082
3.14159	8.824961595
3.141592	8.824973829
3.1415926	8.824977499
$\downarrow$	$\downarrow$
π	2^π

Figure 1

Fortunately, calculators are programmed to give good approximations to a^x for $a > 0$ and x any real number, but be careful in case a is negative (see the box titled Calculating a^x).

■ **Example 5.** Use a calculator to approximate each of the following as decimals.

(a) $(3.12)^{3/4}$ (b) 3^π (c) $\pi^{1-\pi}$ (d) $(\pi + 1)^{\sqrt{2}}$ (e) $(-4.32)^{4/3}$

Solution. (a) $(3.12)^{3/4} \approx 2.34756$ (b) $3^\pi \approx 31.54428$ (c) $\pi^{1-\pi} \approx 0.08616$ (d) $(\pi + 1)^{\sqrt{2}} \approx 7.46116$ (e) $(-4.32)^{4/3} = ((-4.32)^{1/3})^4 \approx 7.03577$ ■

Folding a piece of paper 0.01 inch thick once doubles the thickness to $2(0.01)$ inches; folding it again doubles its thickness to $2 \cdot 2(0.01) = 2^2(0.01)$ inches. After 40 folds, the stack of paper will have a height of $2^{40}(0.01)$ inches. Of course, we can use a calculator to evaluate this number and convert it to feet or miles. But let's use number sense to get a ball park answer. We use the approximations $2^{10} \approx 1000$, 1 foot ≈ 10 inches, and 1 mile ≈ 5000 feet.

$$
\begin{aligned}
(0.01)2^{40} &= (0.01) \cdot 2^{10} \cdot 2^{10} \cdot 2^{10} \cdot 2^{10} \text{ inches} \\
&\approx 10^{-2} \cdot 10^3 \cdot 10^3 \cdot 10^3 \cdot 10^3 \text{ inches} \\
&= 10^{10} \text{ inches} \\
&\approx 10^9 \text{ feet} \\
&\approx \frac{10 \cdot 10^8}{5 \cdot 10^3} \text{ miles} \\
&= 2 \cdot 10^5 \text{ miles} \\
&= 200{,}000 \text{ miles}
\end{aligned}
$$

If you use a calculator, you will get 173,534 miles.

PROBLEM SET 3.1

A. Skills and Techniques

Simplify the following expressions, writing your answer without negative exponents.

1. -4^{-2}
2. $(-3)^{-4}$
3. $(-4)^{-2}$
4. -3^{-4}
5. $4^{-5}4^34^2$
6. $6^56^{-3}6^{-4}$
7. $\dfrac{4^0 + 0^4}{4^{-1}}$
8. $\dfrac{2^{-2} - 4^{-3}}{(-2)^2 + (-4)^0}$
9. $-a^{-4} + (-a)^4$
10. $(a^3b^2)^{-1}$
11. $(b^2b^7)^{-3}$
12. $(x^2x^{-4})^{-2}$
13. $(a^3b^2a^{-3}b^4)^2$
14. $(2a^{-4}b^3a^2)^3$
15. $\left(\dfrac{3a^2bc}{b^3}\right)^2$
16. $\left(\dfrac{a^4b}{(ab)^2}\right)^2$

17. $\dfrac{a^2b^{-3}}{(a^3b)^{-3}}$
18. $\left(\dfrac{a^{-3}}{a^{-2}b^2}\right)^{-2}$
19. $\left(\dfrac{3x^2y^{-2}}{2x^{-1}y^4}\right)^{-3}$
20. $\left(\dfrac{(2x^{-1}y^2)^2}{2xy}\right)\left(\dfrac{x^{-3}}{y^3}\right)$
21. $\left(\dfrac{x^3y^4}{x^{-2}}\right)^{-1}(xy^4)$
22. $\left(\left(\dfrac{x^{-2}}{2}\right)^3(4xy^{-1})^2\right)^2$
23. $(x + x^{-1})^2$
24. $(x^{-1} + y^{-1})^{-1}(x + y)$

Rewrite using exponents instead of radicals and simplify.

25. $\sqrt[3]{x^2}$
26. $\sqrt[5]{x^{-3}}$
27. $(\sqrt[3]{x})^2$
28. $\sqrt{x^2}$
29. $\dfrac{\sqrt[4]{y}}{\sqrt[3]{y}}$

30. $y^3\sqrt[3]{y}$
31. $-\sqrt{(x+y)^3}$
32. $\sqrt[3]{(x-y)^2}$
33. $(\sqrt[4]{x^3}\,\sqrt[3]{y^2})^2$
34. $\sqrt[3]{x^2}\,\sqrt{x^{12}}$
35. $\sqrt{(a-b)^4}$
36. $\sqrt{(a-b)^2}$
37. $(\sqrt{a}-\sqrt{b})^2$
38. $\sqrt[3]{\sqrt{x^{12}}}$

Write without radicals or negative exponents and simplify.

39. $(25^{1/6})^3$
40. $(-8)^{2/3}$
41. $\dfrac{\pi^2\pi^{-3/4}}{\sqrt[4]{\pi}}$
42. $\dfrac{(2^{1/5})^{10}2^{-3}}{4}$
43. $\left(\dfrac{a^{2/3}b^{4/3}}{\sqrt[6]{b}\,b^3}\right)^6$
44. $(a^3b^2c^{-1})^{1/3}(a^2b^{-3/4})^8$
45. $\dfrac{x^{1/4}(8x)^{-2/3}}{x^{-3/4}}$
46. $\dfrac{(\sqrt[3]{x}\,\sqrt[5]{y})^{-3}}{(xy)^{1/5}}$
47. $\left(\dfrac{2x^{-1/2}}{y}\right)^4\left(\dfrac{y}{x}\right)^{-2}$
48. $(xy^{-2/3})^3(xy^{1/2})^2$
49. $y^{2/3}(3y^{4/3}-y^{-5/3})$
50. $x^{-3/4}(x^{7/4}+2x^{-1/4})$
51. $(a^{3/2}+b^{3/2})^2$
52. $(x^{1/2}+y^{-2})^2$
53. $(x^\pi x^{1/2}x^{1-\pi})^3$
54. $(x^{\sqrt{8}}y^{\sqrt{2}})^{\sqrt{2}}$

A problem that arises in calculus is to write expressions like the following as a single fraction. Do so and simplify.

55. $(x-3)^{1/2}-\dfrac{x}{(x-3)^{1/2}}$
56. $(x+5)^{1/3}-\dfrac{x}{(x+5)^{2/3}}$
57. $(x^2+2)^{2/3}-\dfrac{x^2}{(x^2+2)^{1/3}}$
58. $2x(x^2+8)^{1/2}-\dfrac{x^3}{(x^2+8)^{1/2}}$

With the help of your calculator, write each number as a decimal to five-decimal-place accuracy. You may want to review the box titled Calculating a^x. Also, remember that a^{b^c} means $a^{(b^c)}$.

59. $(2\pi)^{\sqrt{2}}$
60. $(\sqrt{2})^{\pi-1}$
61. $(34+\pi)^{-0.23}$
62. $(\sqrt{2}-1)^{1.45}$
63. $(2.34)^{2/3}$
64. $(4.5)^{2/3}$
65. $(-2.34)^{2/3}$
66. $(-4.5)^{4/5}$
67. $(4.5)^{1.2\pi}$
68. $(2.3)^{2^{-2.5}}$

B. Applications and Extensions

In Problems 69–72, use a calculator to arrange each set of numbers from least to greatest.

69. $\pi^2-2^\pi,\ 2^{0.1},\ (\sqrt{2})^{1/\pi},\ \dfrac{3^\pi}{\pi^3}$
70. $\pi^{1/2},\ \sqrt{2}^{\sqrt{2}},\ \pi^{1/\pi},\ \pi-\pi^{1/\pi}$
71. $(\sqrt{2})^{\sqrt{7.5}},\ (\sqrt{3})^{\sqrt{3}},\ 2^{\sqrt[3]{2\pi}},\ 0.5\pi^{\sqrt{2}}$
72. $(\sqrt{5})^{\sqrt{5}},\ \sqrt{\pi^\pi},\ 2(\sqrt{2})^\pi,\ (90\pi)^{1/\pi}$

Solve for x in Problems 73–76.

73. $4^{x+1}=\left(\dfrac{1}{2}\right)^{2x}$
74. $6^{x^2-x}=36$
75. $9^{2x/3}=27^{(4x+1)/6}$
76. $2^{4x}4^{2x+3}=64^{(x-1)/2}$
77. With a calculator, evaluate $\sqrt{2+\sqrt{3}}$ and $(\sqrt{2}+\sqrt{6})/2$. Make a conjecture. Prove it.
78. Let a and b be positive numbers with $a\neq b$. Experiment to conjecture which is larger: $\sqrt{2a+2b}$ or $\sqrt{a}+\sqrt{b}$. Prove your conjecture.
79. Experiment to conjecture what happens to $n^{1/n}$ as the integer $n\to\infty$.
80. Experiment to conjecture what happens to x^x as the positive real number $x\to 0$.
81. Consider $z=x^y$. Give examples of each of the following.
    (a) x, y, and z are rational.
    (b) x and y are rational but z is irrational.
    (c) x is irrational but y and z are rational.
82. **Challenge.** Prove that there exist irrational numbers x and y such that x^y is rational. (*Hint:* Consider $\sqrt{2}^{\sqrt{2}}$ and $(\sqrt{2}^{\sqrt{2}})^{\sqrt{2}}$.)

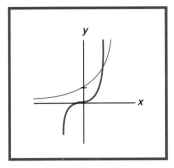

3.2   EXPONENTIAL FUNCTIONS

We can understand the development in the previous section a little better by drawing the graph of $y = 2^x$ for x in three different domains (Figure 1). First we restrict x to the set of integers, then to the set of rational numbers, and finally to the set of real numbers. Note the corresponding transformation from a totally disconnected curve to a curve with holes to a nice continuous curve.

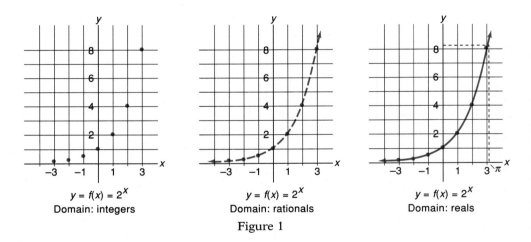

| $y = f(x) = 2^x$ | $y = f(x) = 2^x$ | $y = f(x) = 2^x$ |
| Domain: integers | Domain: rationals | Domain: reals |

Figure 1

What we have done for $y = 2^x$ can be done for $y = b^x$ for any $b > 0$. The corresponding function $f(x) = b^x$ with $b > 0$ and $b \neq 1$ is called an **exponential function** with base b. From Section 3.1, we know that an exponential function satisfies these three properties.

$$b^x b^y = b^{x+y} \quad \frac{b^x}{b^y} = b^{x-y} \quad (b^x)^y = b^{xy}$$

GRAPHS OF EXPONENTIAL FUNCTIONS

In Figures 2 and 3, we show the graphs of several exponential functions.

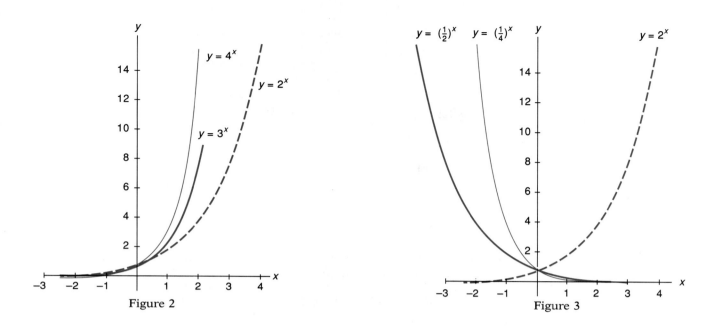

Figure 2

Figure 3

On the basis of these graphs and some thinking, we make several assertions about exponential functions.

1. The function $f(x) = b^x$ has the set $\mathbb{R}$ of real numbers as its domain and the set of positive real numbers as its range.
2. The graph of f is increasing and concave up if $b > 1$; it is decreasing and concave up if $0 < b < 1$. In either case, f is one-to-one and has an inverse.
3. If $1 < a < b$ and $x > 0$, then b^x grows more rapidly than a^x.
4. The graph of f goes through $(0, 1)$ and has $y = 0$ as a horizontal asymptote.

■ **Example 1.** Draw the graphs of $f(x) = 1.5^x$ and $g(x) = 1.5^{x-4} - 5$ in the same coordinate plane. Do these graphs ever intersect?

Solution. The second graph has the same shape as the first but is shifted 4 units right and 5 units down. Thus, they will never intersect. These graphs are shown in Figure 4. ■

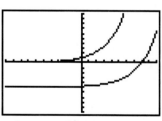

$[-10, 10]$ by $[-10, 10]$

Figure 4

THE NATURAL EXPONENTIAL FUNCTION

You are familiar with the special number $\pi = 3.14159\ldots$ that plays such a significant role in mathematics. Another special number is the number $e = 2.71828\ldots$ named after the great eighteenth-century mathematician Leonhard Euler, who first recognized its importance. Like π, the number e is an irrational number and so has a nonrepeating infinite decimal expansion. You will see its real significance when you study calculus, but we can give some hints of its importance in a precalculus course.

To specify this new number, consider the table in Figure 5, which shows values of $(1 + 1/n)^n$ for increasing values of n. It can be shown (and is shown

n	$\left(1+\frac{1}{n}\right)^n$
1	2
2	2.25
3	2.37037037
10	2.59374246
100	2.70481383
1000	2.71692393
10000	2.71814593
100000	2.71826824
$\downarrow$	$\downarrow$
∞	$e = 2.71828..$

Figure 5

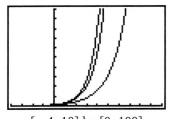

AN EXPERIMENT

Graph $y = (1 + 1/x)^x$ for the range values [0, 2000] by [2.65, 2.75]. Then trace this curve by moving the trace cursor farther and farther to the right, all the time watching the y-value at the bottom of the screen. What do you observe?

$[-4, 10]$ by $[0, 100]$

Figure 6

in calculus) that this sequence of numbers converges toward a number that we call e, that is,

$$\left(1 + \frac{1}{n}\right)^n \to e \text{ as } n \to \infty$$

This we take as the definition of e, and using it we can obtain the decimal expansion of e to any desired degree of accuracy. In fact, this expansion is known to hundreds of thousands of places, but for us the approximation $e \approx 2.71828$ will be good enough.

The function $f(x) = e^x$ is called the **natural exponential function**. The adjective *natural* suggests the fundamental role this function plays; some have called it the most important of all functions. Because of its importance, all graphics calculators have a special key for this function.

■ **Example 2.** Use a calculator to draw the graphs of $f(x) = 2^x$, $g(x) = e^x$, and $h(x) = 3^x$ in the same plane.

Solution. The graphs are shown in Figure 6. Note that the graph of g is between the other two graphs. ■

To suggest one of the special characteristics of the natural exponential function, we are going to calculate the slope of the curve $y = e^x$ at several points. Then we will make a critical observation.

■ **Example 3.** Calculate the slope of the curve $y = f(x) = e^x$ at $x = -1, 0, 2,$ and 4 and make a conjecture.

Solution. To approximate these slopes, we need to calculate

$$\frac{f(x + \Delta x) - f(x - \Delta x)}{2 \Delta x} = \frac{e^{x + \Delta x} - e^{x - \Delta x}}{2 \Delta x}$$

at $x = 0, 1, 2,$ and 4 for a very small Δx, say $\Delta x = 0.001$. Thus, at $x = -1$, we must calculate

$$\frac{e^{-1+0.001} - e^{-1-0.001}}{0.002} = \frac{e^{-0.999} - e^{-1.001}}{0.002}$$

(On TI calculators, this can be done automatically using the nDeriv feature.) When we make the four required calculations, we obtain the numbers in the second column of the table in Figure 7. Note how close these numbers are to those in the third column. They certainly suggest a conjecture (see also Problem 51).

x	Slope of $y=e^x$ at x	e^x
-1	0.3678795	0.3678794
0	1.0000002	1.0000000
2	7.3890573	7.3890561
4	54.5981592	54.5981500

Figure 7

The slope of the curve $y = e^x$ at x is e^x. Equivalently, the rate of change of e^x with respect to x is e^x.

This conjecture is true. A major task in calculus is to prove it and to explore its many consequences. ▪

All exponential functions are related. In fact, all can be expressed in terms of one of them, and this one is usually taken to be the natural exponential function. We illustrate.

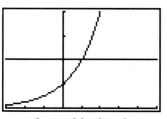

[−3, 5] by [0, 4]

Figure 8

■ **Example 4.** Determine k so that $2^x = e^{kx}$.

Solution. By the laws of exponents $e^{kx} = (e^k)^x$, so our problem is to determine k so that $(e^k)^x = 2^x$, that is, so that $e^k = 2$. This will become a trivial problem once we learn about logarithms, but for now we solve the problem by graphing $y = e^k$ and $y = 2$ to determine their intersection point (Figure 8). The zooming process leads to the solution $k \approx 0.693$. We conclude that $2^x \approx e^{0.693x}$. ▪

What we have just demonstrated for 2^x can be done for any exponential function b^x. Thus, the study of exponential functions is the study of the function $f(x) = e^{kx}$ for various values of k.

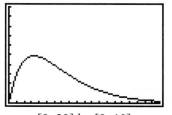

[0, 20] by [0, 10]

Figure 9

■ **Example 5.** Find the maximum point on the graph of $y = 4xe^{-0.3x}$ for $0 \le x \le 20$.

Solution. Figure 9 shows the graph and suggests that there is a maximum near $x = 3$. The maximization process (Section 2.4) identifies the maximum point as (3.33, 4.91). Note that $e^{-0.3x}$ approaches 0 fast enough as x becomes large to overwhelm the factor $4x$, which by itself would increase in size. ▪

EXPONENTIAL FUNCTIONS
OR POLYNOMIAL FUNCTIONS

The behavior of processes for large x is of great interest to scientists. For example, a computer scientist might want to know how the time T required to alphabetize a list of x items grows with x. A question often asked about such a process is: Does it grow exponentially or polynomially (or neither)? The answer is important because the behavior of exponential functions is quite different from that of polynomial functions for large x even though both functions grow rapidly. In fact, exponential functions grow overwhelmingly faster than polynomial functions. It is shown in calculus that:

If $b > 1$, b^x ultimately grows faster than x^n for any n.

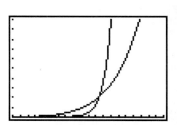

[10, 30] by [0, 10^{10}]

Figure 10

■ **Example 6.** Choose a large enough domain and range to show that the graph of $y = e^x$ ultimately surpasses the graph of $y = x^7$.

Solution. We choose the domain $10 \le x \le 30$ and the range $0 \le y \le 10^{10}$. The graphs in Figure 10 suggest what we are asked to show. ▪

TEASER SOLUTION

Figure 11 shows the graphs of $y = 2^x$ and $y = x^5$ for the range values $[-10, 10]$ by $[-10, 10]$. Zooming shows that one intersection point is at about $(1.177, 2.261)$. The preceding discussion convinces us that there should be another intersection point farther to the right. After some experimentation, we select the range values $[0, 30]$ by $[0, 10^7]$ and obtain the graph in Figure 12. The intersection point can be found by zooming in on this graph. Alternatively, we may graph $y = 2^x - x^5$ as shown in Figure 13 and zoom in on the point where the graph crosses the x-axis. This gives the x-coordinate of the point we are after; the y-coordinate is easily obtained by substitution. Either way, we find the intersection point to be approximately $(22.440, 5,690,000)$.

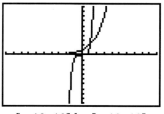

$[-10, 10]$ by $[-10, 10]$
Figure 11

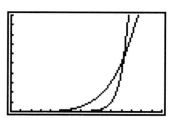

$[0, 30]$ by $[0, 10^7]$
Figure 12

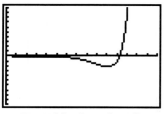

$[0, 30]$ by $[-10^7, 10^7]$
Figure 13

PROBLEM SET 3.2

A. Skills and Techniques

Calculate each expression to five decimal places.

1. $\dfrac{e^2 - e^{-2}}{2}$

2. $\dfrac{e^{1.2} + e^{-1.2}}{2}$

3. $e^\pi - \pi^e$

4. $e - \left(1 + \dfrac{1}{100}\right)^{100}$

Graph each pair of functions in the same plane, either by hand or using a calculator.

5. $f(x) = (1.8)^x$, $g(x) = (1.8)^{-x}$
6. $f(x) = e^x$, $g(x) = e^x - 4$
7. $f(x) = (1.8)^x$, $g(x) = (1.8)^{x-3}$
8. $f(x) = (0.8)^x$, $g(x) = 4 \cdot (0.8)^x$

In Problems 9–12, explain without first graphing how the graphs of the given pairs of functions are related.

9. $f(x) = 3^x$, $g(x) = 2 + 3^{x-2}$

10. $f(x) = 5^x$, $g(x) = 5^{-x}$
11. $f(x) = (1.5)^x$, $g(x) = (1.6)^x$
12. $f(x) = (1.3)^x$, $g(x) = (1.3)^{2x}$
13. Match each function (a–d) with the appropriate graph in Figure 14.

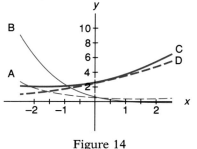

Figure 14

(a) $f(x) = (0.4)^x$
(b) $f(x) = 2 + 2^x$
(c) $f(x) = 3 \cdot (1.4)^x$
(d) $f(x) = 1 + (0.4)^{x+2}$

14. Explain why each of the graphs in Figure 15 cannot be the graph of an exponential function $f(x) = a^x$.

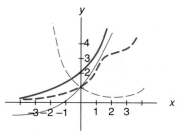

Figure 15

Without using a calculator, order each of the following from smallest to largest. Then confirm your answer with a calculator.

15. e^{2x}, 4^x, $2 \cdot 2^x$ at $x = 5$
16. e^{2x}, 4^x, $2 \cdot 2^x$ at $x = -1.3$
17. x^6, 2^x, e^x at $x = 2.3$
18. x^6, 2^x, e^x at $x = -1.3$
19. x^6, $3 \cdot 2^x$, e^x at $x = 50$
20. x^6, $x^5 + 10000$, e^x at $x = 30$

Determine the slope of the graph of each function at $x = 5$.

21. $f(x) = e^x$
22. $f(x) = 2^x$
23. $f(x) = 3^x$
24. $f(x) = e^{-x}$

Determine k so that $f(x) = e^{kx}$. Then graph $y = f(x)$ and $y = e^{kx}$ in the same plane to confirm your answer.

25. $f(x) = (12)^x$
26. $f(x) = (0.5)^x$
27. $f(x) = (2.2)^x$
28. $f(x) = (5.4)^x$

Determine the maximum point for the graph of each function, accurate to two decimal places.

29. $f(x) = 4x\, e^{-x}$ on $0 \le x \le 10$
30. $f(x) = 2x\, e^{-0.1x}$ on $0 \le x \le 20$
31. $f(x) = x^3\, e^{-0.5x}$ on $0 \le x \le 20$
32. $f(x) = \sqrt{x}\, e^{-2x}$ on $0 \le x \le 1$

B. Applications and Extensions

Order from smallest to largest, assuming that x is very very large.

33. $|x|$, x^3, $(1.1)^x$, $3x^2$
34. 2^{2x}, $x + 1000$, $50x$, e^x

By thinking about the graphs of each pair of functions, decide how many times they will intersect.

35. $f(x) = 3^x$, $g(x) = 3^x - 1$
36. $f(x) = 4^x$, $g(x) = 4^{x+1}$
37. $f(x) = 4^x$, $g(x) = 4^{-x}$

38. $f(x) = 3^x$, $g(x) = (3.2)^x$
39. $f(x) = 3^x$, $g(x) = (1.4)3^x$
40. $f(x) = x^3$, $g(x) = 3^x$
41. $f(x) = x^3 + 2$, $g(x) = 3^x$
42. $f(x) = x^2$, $g(x) = 2^x$

In Problems 43–48, determine all intersection points of the graphs of each pair of functions. Note that when the curves cross at about the same angle, it may be best to find the x-value of an intersection point by solving $f(x) - g(x) = 0$.

43. $f(x) = 2e^{2x}$, $g(x) = e^{3x}$
44. $f(x) = 0.35e^{-0.05x}$, $g(x) = 0.15e^{1.25x}$
45. $f(x) = x^5$, $g(x) = 5^x$
46. $f(x) = x^9$, $g(x) = 5^x$
47. $f(x) = xe^{1.2x}$, $g(x) = e^{1.5x}$
48. $f(x) = \dfrac{e^{-x/3}}{x + 1}$, $g(x) = e^{-x/2}$
49. Note that $f(x) = e^x$ has an inverse function. Determine $f^{-1}(0.5)$, $f^{-1}(3)$, and $f^{-1}(10)$ to three decimal places.
50. Without using a calculator, decide which grows faster for large x, $(x^x)^x$ or $x^{(x^x)}$.
51. Use your calculator to draw the graphs of $f(x) = e^x$ and

$$m(x) = \frac{e^{x+0.001} - e^{x-0.001}}{0.002}$$

in the same plane. What fact do the graphs confirm?
52. Determine the slope $m(x)$ of the graph of $f(x) = e^{3x}$ at $x = 0, 1, 2, 3,$ and 4. Compare $m(x)$ and $f(x)$ at these values and make a conjecture.
53. Use your calculator to draw the graphs of $f(x) = 3e^x$ and

$$m(x) = \frac{e^{3(x+0.001)} - e^{3(x-0.001)}}{0.002}$$

in the same plane, thus adding confirming evidence to the conjecture you should have made in Problem 52.
54. **Challenge.** Two important functions that you will meet in calculus are c and s defined by:

$$c(x) = \frac{e^x + e^{-x}}{2}$$

$$s(x) = \frac{e^x - e^{-x}}{2}$$

Graph these functions and note that c is even and that s is odd. Then use a calculator to guess at a nice relationship for each of the following. Demonstrate the result algebraically.
(a) $(c(x))^2 - (s(x))^2$
(b) $\dfrac{s(2x)}{2c(x)}$
(c) $\left(c\left(\dfrac{x}{2}\right)\right)^2 + \left(s\left(\dfrac{x}{2}\right)\right)^2$

It's easy
$\log x + \log y = \log xy$

TEASER   Let log denote logarithm to the base 10. Evaluate as a decimal.

$$\log \frac{1 \cdot 3}{2^2} + \log \frac{2 \cdot 4}{3^2} + \log \frac{3 \cdot 5}{4^2} + \cdots + \log \frac{98 \cdot 100}{99^2} + \log \frac{99 \cdot 101}{100^2}$$

3.3   LOGARITHMS AND LOGARITHMIC FUNCTIONS

John Napier (1550–1617) invented logarithms to simplify complicated calculations. His invention allowed a person to replace multiplications by additions and divisions by subtractions. He thought addition and subtraction were easier and, of course, he was right for calculations done by hand. For 350 years, logarithms (and slide rules, which depend on them) were the working tools of scientists. Today, complex calculations are done with calculators and computers. Logarithms have other uses, however, and it is still worth considering Napier's famous invention.

Recall that $b^x b^y = b^{x+y}$. On the left, we have a multiplication and on the right, an addition. If logarithms are to replace multiplications with additions, then logarithms must behave like exponents.

> **Definition.**   Assume that $b > 0$ and $b \neq 1$. The **logarithm of N to the base b** is the exponent to which b must be raised to yield N. That is,
>
> $$\log_b N = x \text{ if and only if } b^x = N$$

■ **Example 1.**   Evaluate each of the following.
(a) $\log_4 16$ (b) $\log_2 32$ (c) $\log_3(1/9)$ (d) $\log_9 27$
(e) $\log_5 1$ (f) $\log_{10} 0.01$ (g) $\log_e(e^4)$ (h) $4^{\log_4 7}$

Solution.
(a) $\log_4 16 = 2$ because $4^2 = 16$
(b) $\log_2 32 = 5$ because $2^5 = 32$
(c) $\log_3(1/9) = -2$ because $3^{-2} = 1/9$
(d) $\log_9 27 = 3/2$ because $9^{3/2} = (\sqrt{9})^3 = 27$
(e) $\log_5 1 = 0$ because $5^0 = 1$
(f) $\log_{10} 0.01 = -2$ because $10^{-2} = 0.01$
(g) $\log_e(e^4) = 4$ because $e^4 = e^4$
(h) $4^{\log_4 7} = 7$ from the definition of logarithm   ■

PROPERTIES OF LOGARITHMS

Here are the three properties of logarithms that Napier found so attractive. The number p is any real number, and M and N are positive real numbers.

We establish the first of these laws, leaving the other two to the reader. Let

$$x = \log_b M \quad \text{and} \quad y = \log_b N$$

Then, by definition,

$$M = b^x \quad \text{and} \quad N = b^y$$

so that

$$M \cdot N = b^x b^y = b^{x+y}$$

Thus, b raised to the exponent $x + y$ is $M \cdot N$, which means that

$$\log_b M \cdot N = x + y = \log_b M + \log_b N$$

■ **Example 2.** Expand each of the following using the laws of logarithms. Assume that x and y are positive.

(a) $\log_b\left(\dfrac{3x^2y^3}{x+5}\right)$ 　　　 (b) $\log_b((2\sqrt{x+1})(x^2+3)^4)$

Solution.

(a)
$$\log_b\left(\frac{3x^2y^3}{x+5}\right) = \log_b(3x^2y^3) - \log_b(x+5)$$
$$= \log_b 3 + 2\log_b x + 3\log_b y - \log_b(x+5)$$

(b) $\log_b((2\sqrt{x+1})(x^2+3)^4) = \log_b(2\sqrt{x+1}) + \log_b((x^2+3)^4)$
$$= \log_b 2 + \frac{1}{2}\log_b(x+1) + 4\log_b(x^2+3) \qquad ■$$

■ **Example 3.** Express as a single logarithm

$$\frac{1}{3}\log_b(x+3) - 2\log_b x + 3\log_b(x^2-3)$$

Solution. Using the laws of logarithms, we may rewrite this expression as

$$\log_b \frac{(x+3)^{1/3}(x^2-3)^3}{x^2} \qquad ■$$

CAUTION

A common error that students make is to replace $\log x - \log y$ by $(\log x)/(\log y)$. In fact,

$$\log x - \log y = \log\left(\frac{x}{y}\right)$$

and

$$\frac{\log x}{\log y} = \log(x^{1/\log y})$$

NATURAL LOGARITHMS

For use in calculations, logarithms with base 10 are most convenient; they are called **common logarithms** and are usually denoted simply by *log*. Your calculator probably has a key marked ⌊log⌋; however, you will seldom need to use it. Of much greater importance in mathematics and science are logarithms to the base e; they are called **natural logarithms** and are usually denoted by ln. Your calculator should have a key marked (ln); you will use it often.

Natural logarithms are all that we need; for all other logarithms can be expressed in terms of them. This fact rests on the **change of base formula,** which says

$$\log_a x = \frac{\log_b x}{\log_b a}$$

In particular

$$\boxed{\log_a x = \frac{\ln x}{\ln a}}$$

You will be asked to derive the change of base formula in Problem 74.

SOLVING EXPONENTIAL EQUATIONS

In Section 3.2, we needed to solve the equation $e^k = 2$ for k. The use of logarithms makes this a breeze. Simply take natural logarithms of both sides (if two numbers are equal then their logarithms are equal). Here are the steps.

$$e^k = 2$$

$$k \ln e = \ln 2$$

$$k = \ln 2 \approx 0.693147$$

In this solution, we used the fact that $\ln e = 1$ and we evaluated $\ln 2$ on our calculator. Whenever you need to solve an equation that has the unknown in an exponent, think of taking logarithms of both sides.

■ **Example 4.** Solve the first equation for x. Solve the second equation for T in terms of the other variables.

(a) $3 \cdot 4^{2x-5} = 20$ (b) $A = \dfrac{P}{k}(e^{kT} - 1)$

Solution. In the first case, we begin by taking logarithms, then do some algebra, postponing using a calculator until the very end. In the second case, we do some preliminary algebra to put the equation into a useful form before taking logarithms.

(a) $$\ln 3 + (2x - 5)\ln 4 = \ln 20$$

$$\ln 3 + 2x \ln 4 - 5 \ln 4 = \ln 20$$

$$x = \frac{\ln 20 + 5 \ln 4 - \ln 3}{2 \ln 4} \approx 3.184241$$

(b)

$$\frac{Ak}{P} = e^{kT} - 1$$

$$\frac{Ak}{P} + 1 = e^{kT}$$

$$\ln\left(\frac{Ak}{P} + 1\right) = kT$$

$$T = \frac{1}{k}\ln\left(\frac{Ak}{P} + 1\right)$$ ■

THE NATURAL LOGARITHMIC FUNCTION

Let $f(x) = \ln x$ and $g(x) = e^x$. Note that

$$f(g(x)) = \ln e^x = x \ln e = x$$

and

$$g(f(x)) = e^{\ln x} = e^{\log_e x} = x$$

This means that the natural logarithmic function and the natural exponential function are inverse functions. Thus, the graphs of $y = \ln x$ and $y = e^x$ are reflections across the line $y = x$ (Figure 1).

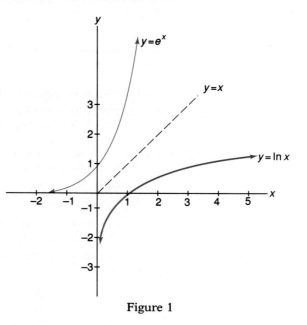

Figure 1

The domain of $f(x) = \ln x$ is the set of positive real numbers, and its range is the set of all real numbers, just the reverse of what is true for the

WE SAID SLOWLY!

To see how slowly ln grows, note that when $x = 1000$, ln x is about 7; when x reaches 1 million, ln x is nearing 14, and when x mounts to 4 trillion (the 1992 national debt), ln x is loafing along at 29. An even more slowly growing function, but still unbounded, is $f(x) =$ ln (ln x). How large would x have to be for ln (ln x) to be 10?

exponential $g(x) = e^x$. Furthermore, the y-axis is a vertical asymptote for the graph of $y = $ ln x just as the x-axis was a horizontal asymptote for $y = e^x$. Most important, observe that while ln x does grow without bound as x becomes large, it grows exceedingly slowly, in complete contrast to e^x, which grows incredibly fast. It is this slow growth that makes the log function a model for certain processes in the physical world, a matter we consider later.

■ **Example 5.** Draw the graphs of (a) $y = 5 + $ ln $(x + 2)$ and (b) $y = $ ln $|x|$.

Solution.

(a) The required graph has the same shape as the graph of $y = $ ln x but is shifted 2 units left and 5 units up (Figure 2).

(b) This graph has the domain consisting of all reals except 0, and it is symmetric with respect to the y-axis (Figure 3). ■

[−10, 10] by [−10, 10]

Figure 2

[−10, 10] by [−10, 10]

Figure 3

TEASER SOLUTION

We use the first law for logarithms to write a sum of logarithms as a single logarithm. Then we discover that most of the numbers cancel.

$$\log \frac{1 \cdot 3}{2^2} - \log \frac{2 \cdot 4}{3^2} + \log \frac{3 \cdot 5}{4^2} + \cdots + \log \frac{98 \cdot 100}{99^2} + \log \frac{99 \cdot 101}{100^2}$$

$$= \log \frac{1 \cdot 3 \cdot 2 \cdot 4 \cdot 3 \cdot 5 \cdot 4 \cdot 6 \cdots 97 \cdot 99 \cdot 98 \cdot 100 \cdot 99 \cdot 101}{2 \cdot 2 \cdot 3 \cdot 3 \cdot 4 \cdot 4 \cdot 5 \cdot 5 \cdots 98 \cdot 98 \cdot 99 \cdot 99 \cdot 100 \cdot 100}$$

$$= \log \frac{101}{200} \approx -0.296709$$

PROBLEM SET 3.3

A. Skills and Techniques

Evaluate each of the following.

1. $\log_3 27$
2. $\log_2 16$
3. $\log_{12} 144$
4. $\log_9 1$
5. $\log_7(49^2)$
6. $\log_\pi(\pi^9)$
7. $\log_{10}(0.00001)$
8. $\log_{10} 1000$
9. $\log_8 2$
10. $\log_8 4$
11. $\log_8(1/8)$
12. $\log_8(1/4)$

13. $\log_2(1/8)$
14. $\log_5(0.04)$
15. $2^{\log_2 24}$
16. $\log_7((7^7)^7)$

Expand each of the following logarithms. Assume that x and y are positive.

17. $\log (x^2 y)$
18. $\log \left(\dfrac{3x}{y}\right)$
19. $\log \left(\dfrac{x+1}{y}\right)$
20. $\log \left(\dfrac{5x^2}{y^3}\right)$
21. $\log \left(\dfrac{7x^3 y^4}{y+2}\right)$
22. $\log \left(\dfrac{x^2+1}{5xy^5}\right)$
23. $\log ((3y^2)(4x^4))^{1/3}$
24. $\log ((\sqrt{5x})(3xy^2)^3)$
25. $\log \left(\sqrt{\dfrac{x+y}{3y^2}}\right)$
26. $\log ((xy)^2(2x+y)^{-3})$
27. $\ln (e^x \sqrt{x^2+1})$
28. $\ln (e^{x+4}(x^4+2)^{1/3})$

Rewrite each expression as a single logarithm.

29. $\log x + 3 \log y$
30. $3 \log 3 + 3 \log y - 3 \log x$
31. $\dfrac{1}{2} \log (2y + x) - 2 \log y$
32. $2 \log x - 3 \log y + \log (x + y)$
33. $\dfrac{3}{2} \log (7 - x) + 4 \log y - \dfrac{1}{2} \log (3x + y)$
34. $4 \log x + 4 \log 3 - 2 \log (2x)$

Use your calculator and the change of base formula to evaluate each of the following to five decimal places.

35. $\log_2 27$
36. $\log_5 10$
37. $\log_9(13e^2)$
38. $\log_6(15e^{-3})$

Use logarithms to solve each equation for x to five decimal places.

39. $2^x = 12$
40. $5^x = 17$
41. $17^{2x-1} = 11$
42. $3^{3x+5} = 14$
43. $4e^{-3x} = 5$
44. $2e^{-2x+1} = 8$
45. $3 \cdot 6^{(x^2)} = 8$
46. $e^{(x^2-x)} = 2^x$

47. $5^{2x+2} = 4^{x-1}$
48. $7^{3x} = 2^{4x-3}$

Solve for the indicated letter in terms of the others.

49. $A = A_0 e^{-kt}$ for t
50. $B = e^{-k(t+m)}$ for m
51. $A = R\dfrac{(1+i)^n - 1}{i}$ for n
52. $R = \dfrac{iP}{1 - (1+i)^{-n}}$ for n
53. $y = \ln (2x)$ for x
54. $y = \ln (3x) - \ln A$ for x
55. $3 \ln x + \ln B = \ln A$ for x
56. $1 + \ln (AB) = \ln D$ for A

B. Applications and Extensions

Determine the domain and range of each function f. Then graph this function to see that it has an inverse function. Find the formula for $f^{-1}(x)$.

57. $f(x) = \ln (3x - 10)$
58. $f(x) = \ln \sqrt{x + 4}$
59. $f(x) = \ln (x^3 + 8)$
60. $f(x) = \ln (\ln x)$

Determine x by algebraic reasoning.

61. $\ln (x + 2) = 1 + \ln (2x - 2)$
62. $\ln x = 3 + \ln (x - 3)$
63. $e^{\ln (x^2-x)} = 6 e^{\ln 1}$
64. $\ln (e^{-2x+1}) = -e^{2 \ln x}$
65. $x^{\ln x} = 10$
66. $x^{\ln x} = x$
67. $(\ln x)^{\ln x} = x$
68. $(\ln x)^{\ln x} = \ln x$

Use a graphics calculator to solve for x in Problems 69–72.

69. $e^{-x} = \ln x$
70. $e^{x/3} = 2 \ln (2x)$
71. $(x^2 + 5)^{\ln x} = \ln (x + 5)$
72. $5 - \ln (x + 6) = 6 \ln x$
73. Establish that the following are true.

    (a) $(\sqrt{3})^{3\sqrt{3}} = (3\sqrt{3})^{\sqrt{3}}$ (b) $\left(\dfrac{9}{4}\right)^{27/8} = \left(\dfrac{27}{8}\right)^{9/4}$

74. Establish the *change of base formula*

    $$\log_a x = \dfrac{\log_b x}{\log_b a}$$

    (*Hint:* Write $x = a^{\log_a x}$ and take $\log_b$ of both sides.)
75. Prove that $\log_s t \log_t s = 1$.
76. **Challenge.** If one could write six digits to the inch, about how many miles long would the number $9^{(9^9)}$ be when written out in decimal notation? (*Hint:* First decide how the number of digits in the positive integer N is related to [[log N]].)

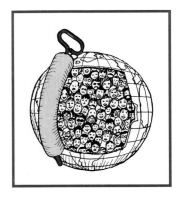

TEASER Assume that: (1) the world's population was 5.25 billion in 1990; (2) the world's population is growing at the rate of 1.6% per year; (3) it takes 1 acre of land to produce food for one person; and (4) there are 8 billion acres of arable land on the earth's surface. When will the world reach maximum sustainable population?

3.4 SCIENTIFIC APPLICATIONS

The phrase "growing exponentially" appears regularly in newspapers and on television. It is said that oil consumption, industrial pollution, and world population are growing exponentially with dire consequences for all of us. But what does this phrase really mean? For most people, it perhaps means little more than "growing rapidly," but for a scientist it has a precise definition.

Let N denote the size of a quantity at time t and let N_0 be its value at $t = 0$. If

$$N = N_0 b^t$$

for some $b > 1$, we say that N is **growing exponentially,** and, if the same equation holds for some b with $0 < b < 1$, we say N is **decaying exponentially.** We learned earlier that we can always replace b by e^k for an appropriate choice of k; so we can also describe exponential growth and decay by the equation

$$N = N_0 e^{kt}$$

with $k > 0$ for growth and $k < 0$ for decay. Figure 1 illustrates exponential growth with $k = 0.5$ and $N_0 = 4$. Figure 2 does the same for exponential decay with $k = -0.5$ and $N_0 = 40$.

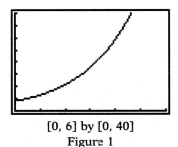

[0, 6] by [0, 40]
Figure 1

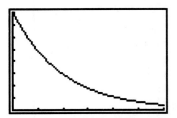

[0, 6] by [0, 40]
Figure 2

EXPONENTIAL GROWTH

In the presence of an abundant food supply and no predators, a biological population tends to grow exponentially.

■ **Example 1.** A researcher counted 100 bacteria in a colony at noon and 150 bacteria in the same colony 2 hours later. Assuming exponential growth: (a) how many bacteria will the colony have at 6:00 P.M., and (b) after what time will the colony number more than 6000?

Solution. Let $N = N(t)$ denote the number of bacteria in the colony t hours after noon. The assumption of exponential growth means that $N = 100\, e^{kt}$, where k is to be determined from the information that $N = 150$ at $t = 2$. This leads to the following calculation.

$$150 = 100\, e^{k(2)}$$

$$1.5 = e^{2k}$$

$$\ln 1.5 = \ln e^{2k} = 2k$$

$$k = \frac{1}{2} \ln 1.5 \approx 0.2027$$

Thus,

$$N = N(t) \approx 100\, e^{0.2027t}$$

(a) Substituting $t = 6$ gives

$$N(6) \approx 100\, e^{0.2027(6)} \approx 337$$

Because this model for growth is at best an approximation, we suggest reporting that there will be about 340 bacteria at 6:00 P.M.

(b) We must solve the inequality

$$100\, e^{0.2027t} > 6000$$

First we solve the equation

$$e^{0.2027t} = 60$$

which gives

$$0.2027t = \ln 60$$

$$t = \frac{\ln 60}{0.2027} \approx 20.2$$

This solution leads us to the conclusion that the colony will exceed 6000 beginning a little after 8:00 A.M. the next day. ■

> **A RULE OF THUMB**
>
> A commonly stated rule is that you should keep at least one more digit of accuracy in intermediate calculations than you plan to give in the answer. An even better rule is to avoid rounding until the final step in a calculation. Thus, in Example 1, we could have written
>
> $$N(t) = 100\, e^{(0.5 \ln 1.5)t}$$
>
> and then used this formula to evaluate $N(6)$. Because this latter rule can lead to very complicated expressions, we sometimes ignore it, even though it is an excellent rule.

■ **Example 2.** The population N of a certain city is 1200 today and is said to be growing at a rate of 2% per year (meaning that there are 2% more people at the end of a year than at the beginning). (a) Determine the formula for N, thereby showing that the growth is exponential. (b) Estimate the population 15 years from now. (c) How long will it take this population to double?

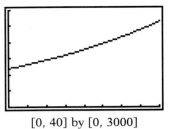

[0, 40] by [0, 3000]

Figure 3

Solution.

(a) Growing at 2% per year means that N will be 1200(1.02) a year from now, that N will be $1200(1.02)(1.02) = 1200(1.02)^2$ 2 years from now, and that in general

$$N = N(t) = 1200(1.02)^t$$

t years from now. We have graphed this equation in Figure 3.

(b) $$N(15) = 1200(1.02)^{15} \approx 1615$$

(c) We must solve the exponential equation $N(t) = 2400$, which leads to the following solution.

$$2400 = 1200(1.02)^t$$

$$2 = (1.02)^t$$

$$\ln 2 = t \ln 1.02$$

$$t = \frac{\ln 2}{\ln 1.02} \approx 35$$

The population will double in about 35 years. ∎

Example 2 introduces the notion of doubling time. As the phrase implies, the **doubling time** is the length of the time interval in which a population doubles in size. For most models of population growth, the doubling time will vary depending on when it is measured (see Problem 41). Under exponential growth, however, a population that doubles in an initial time interval of length T will double in *any* time interval of length T. To show this fact, we note that the initial doubling time T satisfies $2N_0 = N_0 e^{kT}$, that is, $e^{kT} = 2$. It follows that

$$N_0 e^{k(t+T)} = N_0 e^{kt+kT} = N_0 e^{kt} e^{kT} = 2N_0 e^{kt}$$

which is the algebraic way of saying that the population will double T units after an arbitrary time t.

∎ **Example 3.** The population of Example 1 satisfied the equation $N = N(t) = 100 \, e^{0.2027t}$. (a) Show that its doubling time is approximately 3.42 hours. (b) Calculate $N(2)$ and $N(5.42)$ to confirm the answer.

Solution.

(a) We must solve the exponential equation $200 = 100 e^{0.2027t}$, which we do by first dividing by 100 and then taking logarithms. The result is

$$t = \frac{\ln 2}{0.2027} \approx 3.42$$

(b) $$N(2) = 100 \, e^{0.2027(2)} \approx 150$$

$$N(5.42) = 100 \, e^{0.2027(5.42)} \approx 300$$ ∎

EXPONENTIAL DECAY

Fortunately, there are some things that do not grow. Many quantities decay, and some actually decay exponentially. In particular, scientists have observed that all radioactive elements decay exponentially. Recall that a quantity is decaying exponentially over time if

$$N = N_0 e^{kt}$$

where k is a negative constant.

For exponential decay, a natural concept to consider is the **half-life,** the amount of time for one-half of an initial amount to disappear. The half-life plays a role for exponential decay exactly analogous to the doubling time for exponential growth.

■ **Example 4.** The element strontium-90, a dangerous byproduct of nuclear fission, has a half-life of 28 years. Determine the time required for an initial amount to decay to $\frac{1}{10}$ its original amount.

Solution. If we assume an amount of 1 unit initially, then at time t, there will be an amount N given by $N = e^{kt}$. We can determine k from the half-life as follows.

$$0.5 = e^{k(28)}$$

$$\ln 0.5 = \ln (e^{28k}) = 28k$$

$$k = \frac{1}{28} \ln 0.5 \approx -0.02476$$

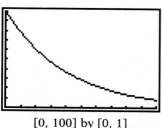

[0, 100] by [0, 1]

Figure 4

Thus,

$$N = N(t) = e^{-0.02476t}$$

an equation graphed in Figure 4. Finally, we solve for t in the equation

$$0.1 = e^{-0.02476t}$$

obtaining

$$t = \frac{\ln 0.1}{-0.02476} \approx 93 \text{ years}$$ ■

The decay of radioactive elements is used to date old objects. For example, scientists determine the age of dead organic materials on the basis of the fact that living bodies maintain a constant ratio of the radioactive isotope carbon-14 to the stable isotope carbon-12. At death, a body ceases to absorb carbon-14 from the atmosphere, and so this ratio decreases as the carbon-14 decays exponentially.

■ **Example 5.** Bones from a skeleton found in Mexico proved to have only 65.0% of the carbon-14 found in living tissue. Assuming that carbon-14 has a half-life of 5730 years, determine when the body died.

Solution. The half-life of carbon-14 determines the constant k.

$$0.5 = e^{k(5730)}$$

$$\ln 0.5 = 5730k$$

$$k = \frac{\ln 0.5}{5730} \approx -0.000121$$

To determine when 65% of the carbon-14 remains, we must solve the exponential equation

$$0.65 = e^{-0.000121t}$$

The solution is

$$t = \frac{\ln 0.65}{-0.000121} \approx 3560 \text{ years} \qquad \blacksquare$$

LOGISTIC GROWTH

Living populations cannot maintain exponential growth indefinitely because lack of food resources, disease, and predators inevitably put an upper bound on the size of a population. A more realistic model for population growth over the long run is a model called **logistic growth,** defined by

$$N = N(t) = \frac{L}{1 + Be^{-qt}}$$

Here L denotes the maximum sustainable population, $B = (L - N_0)/N_0$, and q is a positive constant.

■ **Example 6.** A fruit fly population, initially numbering 10, is growing in a container that can support at most 230 flies. After 3 days, there are 42 flies. Assuming that the logistic model is appropriate, (a) determine the constant q, (b) draw the growth curve, and (c) predict the population after 12 days.

Solution. (a) Since $B = (L - N_0)/N_0 = (230 - 10)/10 = 22$,

$$42 = \frac{230}{1 + 22e^{-3q}}$$

This equation can be solved for q, giving

$$q = -\frac{1}{3} \ln \frac{188}{(42)(22)} \approx 0.5308$$

(b) The graph of the logistic growth curve

$$N = N(t) = \frac{230}{1 + 22e^{-0.5308t}}$$

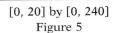

[0, 20] by [0, 240]
Figure 5

is shown in Figure 5. Note its elongated S-shape with an initial exponential growth appearance but eventual leveling off against an asymptote of $N = 230$.

(c)
$$N(12) = \frac{230}{1 + 22e^{-0.5308(12)}} \approx 222 \qquad \blacksquare$$

LOGARITHMIC GROWTH

If one quantity grows extremely slowly relative to another, we might consider modeling the growth process with a logarithmic curve. Two examples where this approach has proved useful are the Richter scale for measuring the perceived magnitude of earthquakes and the decibel scale for measuring the loudness of sounds.

MAGNITUDES OF SOME EARTHQUAKES	
1906 San Francisco	8.3
1933 Tokyo	8.9
1985 Mexico City	8.1
1989 San Francisco	7.1

Figure 6

■ **Example 7.** Figure 6 shows the magnitudes, based on the Richter scale, of four famous earthquakes. This scale relates the magnitude M of an earthquake to the amplitude A of the largest seismic wave at a distance 100 kilometers from the epicenter by the formula

$$M = \log \frac{A}{C}$$

where C is a specified constant that need not concern us and log denotes $\log_{10}$.
(a) Solve for A in the formula.
(b) The ratio of the amplitudes A of two earthquakes is considered to be a measure of their true relative strengths. On this basis, how much stronger was the 1933 Tokyo earthquake than the 1989 San Francisco earthquake?

Solution. (a) $A = C \cdot 10^M$.
(b) The ratio of the two amplitudes is

$$\frac{C \cdot 10^{8.9}}{C \cdot 10^{7.1}} = 10^{1.8} \approx 63$$

The 1933 earthquake was actually about 63 times as strong as the 1989 earthquake. ■

TEASER SOLUTION

From the given assumptions, we can say that the world population P satisfies

$$P = 5.25(1.016)^t$$

where P is measured in billions and t is the number of years after 1990. We must solve the equation

$$8 = 5.25(1.016)^t$$

which we do by taking logarithms. The result is

$$t = \frac{\ln 8 - \ln 5.25}{\ln 1.016} \approx 27$$

On the basis of the stated assumptions, the world will reach maximum sustainable population in 2017.

PROBLEM SET 3.4

A. Skills and Techniques

Identify those equations that show y growing exponentially with t, decaying exponentially with t, or neither. You may need to apply properties of exponents before you can decide. For example,

$$2^{3t+4} = 2^{3t}2^4 = 16(2^3)^t = 16(8)^t$$

1. $y = 64\left(\frac{3}{4}\right)^t$

2. $y = 9\left(\frac{4}{3}\right)^t$

3. $y = 64\left(\frac{3}{4}\right)^{2t}$

4. $y = 9\left(\dfrac{4}{3}\right)^{4t}$

5. $y = 64\left(\dfrac{3}{4}\right)^{t^2}$

6. $y = 9\left(\dfrac{4}{3}\right)^{3t+1}$

7. $y = 0.2(1.2)^{2t-1}$

8. $y = 3^t\left(\dfrac{1}{5}\right)^t$

9. $y = e^{2t-3}$

10. $y = (4t)^t$

11. $y = e^{-t+10}$

12. $y = (1 + e)^t$

In Problems 13–18, write a formula that gives the size of the indicated quantity at the end of t years.

13. The value of N if it was initially 50 and is growing at the rate of 3% per year.

14. The size N of a colony that was 100 initially and that doubles every 6 months.

15. The value V of a car that cost $15,000 initially but is losing 20% of its value each year.

16. The size A of a quantity growing exponentially that was 40 initially and 80 two years later.

17. The size A of a quantity decaying exponentially that was 50 initially and has a half-life of 10 years.

18. The size A of a quantity growing exponentially that was 50 initially and 75 at the end of 3 years.

19. A nest of ants was found to contain 100 ants when first counted and 600 ants 30 days later. Assuming exponential growth, how many ants will there be 200 days after that first count?

20. Boomtown's population was 128 on January 1, 1980, and numbered 448 exactly 10 years later. Assuming exponential growth, what will its population be on January 1, 2010?

21. A fishery pond has been stocked with 6000 trout minnows, and this fish population is expected to grow exponentially at r% per month. (a) Write a formula for the trout population P at the end of t months. (b) If $r = 2$, when will P hit 10,000?

22. If an exponentially growing population numbered 2000 initially and numbered 3000 one month later, how many will it number 12 months after that first count?

23. The population of California was 5.7 million in 1930 and 10.6 million in 1950. What did the exponential growth model predict for its population in 1980? (*Note:* The actual population in 1980 was 23.7 million.)

24. Refer to Problem 23 and use the exponential growth model and the actual population figures for 1930 and 1980 to predict California's population in the year 2000.

25. How long does it take a population to double if its size at the end of a year is 4% more than at the beginning of the year?

26. Does doubling the rate of growth halve the doubling time? (*Suggestion:* Rework Problem 25 with a rate of 8%.)

27. A certain exponentially growing population triples in 50 years. What is its doubling time?

28. An exponentially decaying substance loses two-thirds of its size in 5 years. Determine its half-life.

29. Radioactive iodine-131, used in medicine, has a half-life of 8 days. How much of an initial amount of 10 grams will remain after 20 days?

30. The half-life of radium is 1590 years. How much of an initial amount of 10 grams will remain after 500 years?

31. A radioactive element decayed from 10 grams to 7 grams in 60 days. Determine its half-life.

32. How long will it take an amount of radium (see Problem 30) to decay to 1/9 its original amount?

33. Bones from a skeleton found near an Aztec city contain 89.8% of the carbon-14 found in living tissue. How old is the skeleton (see Example 5)?

34. The Phoenician society is known to have flourished around 1000 B.C. What percentage of carbon-14 (compared with that of living tissue) would you expect to find in skeletons found in Phoenician ruins today (the end of the twentieth century)?

35. If a population is growing according to the logistic model $N = 400/(1 + 3e^{-0.2t})$, determine (a) the initial population, (b) the maximum sustainable population, and (c) the population at $t = 20$.

36. Suppose a population is growing according to the model $N = 400 - 300e^{-0.4t}$. (a) Draw the graph of this curve. (b) Does it represent logistic growth? (c) What is the maximum sustainable population?

37. A small pond can support a maximum population of 800 catfish. The pond was initially stocked with 20 catfish weighing at least 1 pound, and at the end of 3 months 64 such fish were counted. Assuming logistic growth, (a) write the appropriate growth equation, (b) draw its graph, and (c) predict the population at the end of 12 months.

38. An extremely contagious type of flu for which there is no immunization is spreading through a city of 120,000. When first diagnosed 30 people had the virus and 5 days later a total of 1000 had been infected. How long will it take for one-half the population to contract the virus? Assume logistic growth.

39. Refer to Example 7. How much stronger was the 1985 Mexico City earthquake than the 1989 San Francisco earthquake (as measured by the ratio of amplitudes)?

40. The loudness L in decibels of a sound of intensity I in watts per square meter is given by

$$L = 10 \log (I \times 10^{12})$$

What does tripling the intensity do to the loudness level?

B. Applications and Extensions

41. The text showed that the doubling time T is independent of t when it is measured for exponential growth but hinted that it will vary for other types of growth. Determine the two doubling times in each case.
  (a) $N = 100 + 2t$, starting at $t = 0$ and at $t = 20$
  (b) $N = 100 + t^2$, starting at $t = 0$ and at $t = 5$
  (c) $N = 100 - 90e^{-0.1t}$, starting at $t = 0$ and at $t = 5$

42. City A is growing linearly so that after t years, its population N_A satisfies $N_A = 1000(1 + 0.1t)$. City B is growing exponentially according to the equation $N_B = 1000e^{0.02t}$. Draw both graphs in the same plane.
  (a) Determine the time interval during which N_A will exceed N_B.
  (b) Estimate as best you can the time interval during which A will grow faster than B.

43. Follow the instructions of Problem 42 but assume now that $N_B = 1000(1 + 2 \ln (t + 1))$.

44. The pH scale for measuring acidity A of a chemical solution says that $A = -\log(H)$, where H measures the hydrogen ion concentration in moles per liter. If the acidity of human blood ranges between 7.37 and 7.44, determine the corresponding range for H.

45. Refer to Example 5. Suppose that the ratio of carbon-14 in an ancient bone to that found in living tissue cannot be reliably measured below 0.01 (that is, 1%). What is the age of the oldest bone that can be reliably dated?

46. Nocturnal brown tree snakes arrived on Guam as shipboard stowaways in 1950. With no natural predators, their population grew from perhaps 2 to an estimated 2 million by 1990. Assume logistic growth with a maximum sustainable population of 10 million.
  (a) Estimate the population in the year 2000.
  (b) How long will it take the population to double from 2 million to 4 million?

47. It is estimated that a certain manmade lake can support up to 40,000 adult walleyes. The lake was initially stocked with 500 adult walleyes. No fishing will be allowed until the adult walleye population reaches 10,000. When can fishing begin if a survey done at the end of 1 year showed 2000 adult walleyes? Assume logistic growth.

48. Assuming that the world population was 4 billion in 1975 and 5.25 billion in 1990 and that the world can sustain a maximum population of 8 billion, use the logistic model to predict the world population in 2017 (compare with the Teaser).

49. Recall the sequence of prime numbers 2, 3, 5, 7, 11, 13, . . . Let $P(x)$ denote the number of primes not exceeding x and let $f(x) = x/\ln x$.
  (a) Calculate $P(20)$ and $P(50)$. We mention that $P(10^4) = 1229$ and $P(10^6) = 78,498$.
  (b) Calculate the ratio $P(x)/f(x)$ at $x = 20, 50, 10^4$, and 10^6.
  (c) Make a conjecture. (*Note:* You might make the same conjecture that Carl Gauss made in 1792. It was proved as the famous Prime Number Theorem in 1896.)

50. **Challenge.** If one water lily can cover the surface of a pond in 50 days, how long will it take 50 water lilies to cover the same pond? Make a guess and then work it out. Assume that a water lily grows exponentially and initially covers 1/500 of the pond.

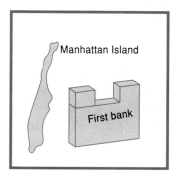
Manhattan Island

First bank

TEASER   It is said that Peter Minuit bought Manhattan Island from Native Americans in 1626 for $24. How much would this money be worth in 1996 if Minuit had invested it at 6% compounded annually?

3.5 BUSINESS APPLICATIONS

People expect money that is invested to grow in value. One way of achieving this growth is to put money in a savings account at a bank. Suppose that a **principal** of P dollars is put in a savings account today and left there for t

years. At periodic times, the bank will add amounts called **interest** to the account, causing its value to grow. Banks follow many different rules in calculating the interest, but all are based on some simple mathematical principles.

SIMPLE INTEREST

If only the original principal P draws interest, we say the bank is paying **simple interest.** For example, if Jennifer puts $100 in a bank today at 7% simple interest, then at the end of 1 year the bank will add $0.07(100) = \$7$ to her account; at the end of the second year, it will add another $7, and so on. At the end of 9 years, the value of her account will have grown to $100 + 9(7) = \$163$.

In general, if P dollars are invested at the rate r (written as a decimal), then after t years the account will be worth an amount A given by the following formula

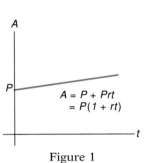

A = P + Prt
  = P(1 + rt)

Figure 1

$$\boxed{\text{Simple Interest: } A = P + Prt}$$

Since P and r are constants, the graph of this equation is a straight line (Figure 1). Simple interest produces linear growth.

■ **Example 1.**   Miguel put $550 in a savings account bearing 6.5% simple interest. How much will be in his account at the end of 12 years?

Solution. Applying the boxed formula with $P = 550$, $r = 0.065$, and $t = 12$, we obtain

$$A = 550 + 550(0.065)(12) = \$979.00 \qquad ■$$

Although simple interest is a useful investment concept, it is for practical purposes a historical oddity. Today all banks pay compound interest, our next topic.

COMPOUND INTEREST (ANNUAL COMPOUNDING)

If interest is periodically added to an account *and converted to principal* so that previous interest also earns interest, then the account is earning **compound interest.** Suppose, for example, that Jennifer puts $100 in a savings account earning 7% interest, compounded annually. Then at the end of 1 year, her account will be worth $100 + 0.07(100) = 100(1.07) = \$107$. This amount then becomes the principal during the second year so that at the end of that year, her account will be worth

$$107 + 0.07(107) = 107(1.07) = 100(1.07)^2 = \$114.49$$

At the end of 9 years, Jennifer's account will be worth

$$100(1.07)^9 = \$183.85$$

In general, if P dollars are invested in an account with interest rate r (written as a decimal) compounded annually, then after t years the account will be worth an amount A given by the following formula.

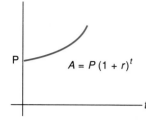

Figure 2

You should recognize this as a formula for exponential growth (Figure 2), a concept we studied in Section 3.4.

■ **Example 2.** Suppose that Miguel of Example 1 was able to invest his $550 in a savings account bearing 6.5% interest, compounded annually. How much will his account be worth at the end of 12 years?

Solution. Applying the annual compounding formula with $P = 550$, $r = 0.065$, and $t = 12$ gives

$$A = 550(1.065)^{12} = \$1171.00$$

As we expected, this is considerably more than the $979 we calculated under simple interest. ∎ ■

COMPOUNDING K TIMES PER YEAR

Interest rates are always quoted as annual rates, from now on called **nominal rates.** Compounding, however, may occur annually, quarterly, monthly, or at any stated time intervals (called periods) of equal length. If interest is compounded k times per year, then a nominal rate of r corresponds to a rate of $i = r/k$ per period. For example, suppose Jennifer invests $100 in an account bearing interest at 6%, compounded monthly. This corresponds to an interest rate of 1/2% per month; thus Jennifer's account will be worth 100(1.005) at the end of 1 month, $100(1.005)^2$ at the end of 2 months, and so on. In general a principal of P dollars, invested at rate r compounded k times per year, will be worth the following amount A at the end of t years.

$$\text{Compounding } k \text{ Times per Year:}$$
$$A = P\left(1 + \frac{r}{k}\right)^{kt} = P(1 + i)^n$$

Here $i = r/k$ is the interest rate per period and $n = kt$ is the number of periods.

■ **Example 3.** Suppose Benjamin invests $1000 in a savings account bearing 8% interest. Determine the value of the account at the end of 5 years if interest is compounded: (a) annually, (b) quarterly, (c) monthly, (d) daily.

Solution.
(a) $A = 1000(1.08)^5 = \$1469.33$
(b) $A = 1000(1.02)^{20} = \$1485.95$
(c) $A = 1000\left(1 + \dfrac{0.08}{12}\right)^{60} = \$1489.85$
(d) $A = 1000\left(1 + \dfrac{0.08}{365}\right)^{1825} = \$1491.76$

As you can see, it is to the investor's advantage to have compounding occur as often as possible. ■

CONTINUOUS COMPOUNDING

In order to attract investors' money, some banks may offer interest compounded continuously. What does this mean? Look at Example 3 again. Think about what would happen if interest were compounded every hour, or every minute, or every second. Will the corresponding amount A converge to a definite number as k, the number of times compounding occurs in a year, increases without bound?

Let's look at the general problem and consider what happens to $A = P(1 + r/k)^{kt}$ as $k \to \infty$. Using the law of exponents and replacing k/r by x, we can manipulate this expression as follows.

$$A = P(1 + r/k)^{kt} = P\left(\left(1 + \frac{r}{k}\right)^{k/r}\right)^{rt}$$

$$= P\left(\left(1 + \frac{1}{x}\right)^{x}\right)^{rt}$$

Now when k becomes large so does x; thus, the behavior of A when k becomes large depends on what happens to $(1 + 1/x)^x$ as x becomes large. We treated this latter question back in Section 3.2, where we learned that

$$\left(1 + \frac{1}{x}\right)^x \to e \text{ as } x \to \infty$$

We conclude that the amount A is converging, as k increases without bound, to the following value.

> **Continuous Compounding:** $A = Pe^{rt}$

We call this type of compounding **continuous compounding** of interest at rate r.

■ **Example 4.** Refer to Example 3. How much will Benjamin have in his account at the end of 5 years if his $1000 earns 8% interest, compounded continuously?

Solution. $A = 1000e^{0.08(5)} = \$1491.82$, just 6¢ more than he would earn under daily compounding. ■

EFFECTIVE RATES

Which is a better interest rate—8% compounded quarterly or 7.9% compounded daily? This question can be answered by computing the corresponding effective rates of interest. The **effective rate** of interest, sometimes called the annual yield, is the rate that, when compounded annually, yields the same return as the given rate. Equivalently, the effective rate is the interest earned on $1 during 1 year at the given rate. Thus, the effective rate r_e corresponding to a nominal rate r, compounded k times per year, is

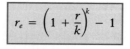

$$r_e = \left(1 + \frac{r}{k}\right)^k - 1$$

■ **Example 5.** Will Tina be better off investing money at 8% compounded quarterly or 7.9% compounded daily?

Solution. The corresponding effective rates are

$$r_e = (1 + 0.02)^4 - 1 \approx 0.082432$$

and

$$r_e = \left(1 + \frac{0.079}{365}\right)^{365} - 1 \approx 0.082195$$

Tina will earn more by investing her money at 8% compounded quarterly. ■

DOUBLING TIMES

Compound interest problems are exponential growth problems; so it makes sense to ask how long it will take for money to double at various rates (the doubling time problem of Section 3.4). The question amounts to asking how long it will take \$1 to grow to \$2 at these rates. We answer the question just as earlier by solving an exponential equation using logarithms.

■ **Example 6.** Determine the doubling time for money invested at: (a) 9% compounded monthly, (b) 7% compounded daily.

Solution.

(a)
$$\left(1 + \frac{0.09}{12}\right)^n = 2$$

$$n \ln\left(1 + \frac{0.09}{12}\right) = \ln 2$$

$$n = \frac{\ln 2}{\ln(1 + 0.09/12)} \approx 93 \text{ months}$$

(b)
$$\left(1 + \frac{0.07}{365}\right)^n = 2$$

$$n \ln\left(1 + \frac{0.07}{365}\right) = \ln 2$$

$$n = \frac{\ln 2}{\ln(1 + 0.07/365)} \approx 3615 \text{ days}$$

It takes 7¾ years in the first case and almost 10 years in the second. ■

■ **Example 7.** At what interest rate x, compounded monthly, will money double in value in 8 years?

Solution. On the basis of Example 6, we should expect the rate to be somewhat less than 9%. Do you see why? The rate x is determined by the equation

$$\left(1 + \frac{x}{12}\right)^{96} = 2$$

One way to solve this equation is to use a graphics calculator to find the x-coordinate of the intersection of $y = (1 + x/12)^{96}$ and $y = 2$ (Figure 3). But we can also solve the equation using logarithms and exponentials as follows.

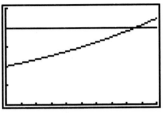

[0, 0.1] by [0, 2.5]

Figure 3

$$96 \ln\left(1 + \frac{x}{12}\right) = \ln 2$$

$$\ln\left(1 + \frac{x}{12}\right) = \frac{\ln 2}{96}$$

$$1 + \frac{x}{12} = e^{(\ln 2)/96}$$

$$x = 12\left(e^{(\ln 2)/96} - 1\right) \approx 0.086957$$

An interest rate of 8.6957%, compounded monthly, is required. ∎

A THIRD WAY

Still another way to solve

$$\left(1 + \frac{x}{12}\right)^{96} = 2$$

is by taking the 96th root of each side to obtain

$$1 + \frac{x}{12} = 2^{1/96}$$

$$x = 12(2^{1/96} - 1)$$

$$\approx 0.086957$$

TEASER SOLUTION

$A = 24(1.06)^{370} \approx \$55$ billion. Exponential growth over the long run is spectacular.

PROBLEM SET 3.5

A. Skills and Techniques

Determine the value of the account at the end of n years, assuming the principal P is invested at simple interest and rate r. Then do the same problem assuming that the interest is compounded annually.

1. $n = 10, P = \$350, r = 4.8\%$
2. $n = 5, P = \$1300, r = 5.9\%$
3. $n = 6, P = \$5000, r = 8.5\%$
4. $n = 20, P = \$5000, r = 8.5\%$
5. $n = 10, P = \$1000, r = 9\%$

6. $n = 10, P = \$1000, r = 9.5\%$
7. $n = 10, P = \$1000, r = 10\%$
8. $n = 10, P = \$1000, r = 12\%$

Determine the value of $1000 at the end of 10 years if interest is at rate r, compounded as indicated.

9. (a) $r = 6\%$, annually
      (b) $r = 6\%$, quarterly
      (c) $r = 6\%$, monthly
      (d) $r = 6\%$, daily ($k = 365$)
      (e) $r = 6\%$, continuously

10. (a) $r = 9\%$, annually
    (b) $r = 9\%$, quarterly
    (c) $r = 9\%$, monthly
    (d) $r = 9\%$, daily ($k = 365$)
    (e) $r = 9\%$, continuously
11. (a) $r = 6\%$, monthly
    (b) $r = 7\%$, monthly
    (c) $r = 8\%$, monthly
    (d) $r = 9\%$, monthly
    (e) $r = 10\%$, monthly
12. (a) $r = 9\%$, daily
    (b) $r = 10\%$, daily
    (c) $r = 11\%$, daily
    (d) $r = 12\%$, daily
    (e) $r = 12\%$, continuously

Determine the effective rate in each case.

13. 10% compounded quarterly
14. 5% compounded daily
15. 9.5% compounded daily
16. 5.2% compounded monthly
17. 9% compounded continuously
18. 5% compounded continuously

Which interest rate is better for a saver? You will have to calculate the corresponding effective rates to make a sure decision.

19. 8.1% compounded daily or 8.2% compounded quarterly
20. 6.8% compounded daily or 6.83% compounded monthly
21. 12.1% compounded daily or 12.3% compounded quarterly
22. 9.8% compounded daily or 9.83% compounded monthly
23. 8.1% compounded continuously or 8.2% compounded quarterly
24. 9.8% compounded continuously or 9.84% compounded monthly

Determine the doubling time if interest is as given.

25. (a) 8.3% compounded annually
    (b) 8.3% compounded monthly
    (c) 8.3% compounded daily
26. (a) 4.5% compounded annually
    (b) 4.5% compounded monthly
    (c) 4.5% compounded continuously

Determine the interest rate if money doubles in value in the given number of years with the indicated compounding.

27. 12 years, annually
28. 7 years, annually
29. 8 years, daily
30. 10 years, monthly

B. Applications and Extensions

31. A rule of thumb, called the Rule of 70, says that a rate of $r\%$, compounded annually, produces a doubling time of approximately $70/r$ years. Compare the answers given by this formula with those you obtained in Problems 25(a) and 26(a).
32. Develop the mathematical basis for the Rule of 70 (Problem 31) by using the approximations: $\ln 2 \approx 0.7$ and $\ln (1 + x) \approx x$ for small x.
33. Determine r if a rate of $r\%$, compounded continuously, produces an effective rate of 7.35%.
34. What principal P will grow to $1000 in 8 years if the interest rate is 9%, compounded daily?
35. At what rate r, compounded monthly, will $100 grow to $150 in 6 years?
36. How many years does it take for $200 to grow to $900 if the interest rate is 8%, compounded quarterly?
37. How long will it take for the value of $1 invested at 5%, compounded continuously, to exceed the value of $100 invested at 5% simple interest?
38. How long will it take for the value of $1 invested at 5%, compounded daily, to exceed the value of $2 invested at 4%, compounded annually?
39. John's $1000 earned interest at the rate of 7% for 5 years; then the rate changed to $r\%$ for the next 5 years. Interest was compounded annually in both cases. If the value increased to $2200 over the 10-year period, what was r?
40. Two years ago, Li put a large sum of money in bank A's savings account bearing simple interest at 9%. Today she learned that bank B will guarantee interest at the rate of 5.8%, compounded continuously. How long would she need to leave her money in bank B to justify transferring it from bank A?
41. How large does the principal P have to be to make its value under continuous compounding at the end of 2 years exceed by more than $10 what it would be under monthly compounding? Assume that both interest rates are 9%.
42. **Challenge.** Although it is true that money invested in a savings account will grow because of interest, part of the increase will be eroded away by inflation. If the annual interest rate is r, compounded annually, and the annual inflation rate is s, it is common to report the true growth rate of money j as $j = r - s$, a result that is only approximately correct. (a) Obtain the exact value of j in terms of r and s. (b) What will be the actual purchasing power (in today's money) of $1000 ten years from now if invested at 9.7%, compounded annually, and inflation remains constant at 3.5% per year? (*Hint:* If the annual inflation rate is s, then an item costing $1 at the beginning of a year will cost $$(1 + s)$ at the end of the year, which means that $1 at the end of a year is worth only $$(1 + s)^{-1}$ in today's dollars.)

Planet	Distance x	Years y
Mercury	36.0	0.241
Venus	67.3	0.615
Earth	93.0	1.000
Mars	142	1.88
Jupiter	484	11.9
Saturn	885	29.6
Uranus	1782	84.0
Neptune	2790	164
Pluto	3671	249

TEASER For each of the nine planets in our solar system, the table at the left gives the mean distance from the sun in millions of miles and the time of revolution around the sun, called the sidereal year, in earth years. In August 1992, two astronomers from the University of Hawaii discovered a tenth planet, tentatively named Smiley. Its orbit has been calculated to have a mean distance of 3700 million miles. Predict how long it takes Smiley to revolve around the sun.

3.6   NONLINEAR REGRESSION

In Section 1.7 we discussed how to fit a linear equation to experimental data using a process called linear regression. There, we assumed a linear relationship $y = a + bx$ between a dependent variable y and an independent variable x.

But we have seen throughout this chapter that not all relationships between variables are linear. For example, the number N of bacteria in a colony with unlimited food resources tends to grow exponentially with time t according to the equation $N = N_0 e^{kt}$. In this situation, we would like to consider fitting an exponential equation to the data. The fitting of nonlinear equations to experimental data is called **nonlinear regression.**

We will discuss three kinds of nonlinear regression, namely, **logarithmic, exponential,** and **power regression.** By a transformation of variables, we can change each of these nonlinear regression problems into a linear regression problem, as is illustrated in Figure 1. Thus, if the xy-data lie along a logarithmic curve, plotting y against $\ln x$ should produce a straight line; if the xy-data lie along an exponential curve, plotting $\ln y$ against x will give a straight line; and if the xy data lie along a power curve, plotting $\ln y$ against $\ln x$ will give a straight line. In each of the three cases, the least squares criterion of Section 1.7 can be used on the transformed variables to determine the parameters a and b that correspond to the best-fitting curve of the desired type. The mathematical details are a bit tricky but, fortunately, many calculators (including the TI-81 and TI-82) are programmed to do the work for us automatically.

Model	Equation	Linearization	Restrictions
Linear	$y = a + bx$	$y = a + bx$	None
Logarithmic	$y = a + b \ln x$	$y = a + b \ln x$	$x > 0$
Exponential	$y = ab^x$	$\ln y = \ln a + x \ln b$	$y > 0$
Power	$y = ax^b$	$\ln y = \ln a + b \ln x$	$x, y > 0$

Figure 1

EXPONENTIAL REGRESSION

Our first example involves a typical case of exponential growth.

Time (hr)	Population
0	37
1	100
2	262
3	747
4	3120
5	5500
6	14800

Figure 2

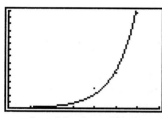

[0, 7] by [0, 15000]
Figure 3

CALCULATOR HINTS

To graph the equation and data together as in Figure 3 on the TI-81 or TI-82, first perform an exponential regression (**ExpReg** from the [STAT] menu), then copy the regression equation into **Y₁ =** (**RegEQ** from the [VARS] menu), and finally graph it together with the data. To complete the last step on the TI-82, turn on **Plot 1** from the [STATPLOT] menu and press [GRAPH]. On the TI-81, simply choose **Scatter** from the [STAT] menu.

■ **Example 1.** A researcher measured the population of a colony of bacteria every hour, as shown in Figure 2. Fit an exponential curve to these population data. Then graph the resulting regression equation and a scatter plot of the data in the same plane. Finally, write the exponential regression equation in the form $N = N_0 e^{kt}$.

Solution. We enter the data in our calculator with x denoting time and y the corresponding population size. Asked for an exponential fit to the data, our calculator produced the following numbers.

$$a = 37.185$$
$$b = 2.764$$
$$r = 0.997$$

The regression equation is

$$y = ab^x = (37.185)(2.764)^x$$

The value of r very close to 1 indicates that the fit is very good indeed. The data and the regression curve are shown in Figure 3. To write the regression equation in the form $N = N_0 e^{kt}$, we choose $N_0 = 37.185$ (close to the experimental value of 37) and solve $e^k = 2.764$ for k, obtaining $k = 1.0167$. Thus, the regression equation can be written as

$$N = N_0 e^{kt} = 37.185 e^{1.0167t}$$ ■

We can use the regression equation to predict a value of y for any desired value of x.

■ **Example 2.** Estimate the bacteria population of Example 1 at $t = 4.5$, 5.5, and 7 hours. When will the population reach 1 million bacteria?

Solution. We simply need to evaluate $N = 37.185(2.764)^t$ for the desired values of t. This gives $N \approx 3610$, 9970, and 45,800 at the three respective times.

We predict that the population will reach 1 million when

$$37.185(2.764)^t = 1,000,000$$

We can solve this equation by use of logarithms, or we can solve it graphically using a calculator and zooming. Either way, we obtain $t \approx 10$ hours. ■

LOGARITHMIC REGRESSION

Not all growth is exponential. If one quantity grows very slowly with respect to another, a logarithmic growth equation may be appropriate.

■ **Example 3.** The Catchall Catfish Company believes that the average size S (in pounds) of the catfish it harvests depends logarithmically on the amount of feed F (in pounds per day) it places in each pond. (a) Fit a logarithmic curve

Pond	Feed (lb/ day)	Average size (lb)
1	52	6.1
2	74	8.4
3	126	10.0
4	48	4.9
5	212	12.2
6	190	11.4
7	100	9.7
8	102	11.0
9	190	12.1
10	262	12.4

Figure 4

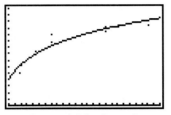

[30, 280] by [0, 15]

Figure 5

to the company's data (Figure 4) from 10 ponds and graph it together with the data. (b) How large are the catfish predicted to be if 150 pounds per day of feed are used at each pond? (c) How much feed should the company use to grow catfish to the optimal market size of 8.5 pounds?

Solution. (a) The general form of a logarithmic equation relating catfish size S to feed F is

$$S = a + b \ln F$$

where a and b are constants. For best fit, our calculator gives the following results.

$$a = -10.051$$

$$b = 4.174$$

$$r = 0.949$$

Again the closeness of the correlation coefficient r to 1 suggests that the data are fitted quite well by a logarithmic equation. The regression equation is

$$S = -10.051 + 4.174 \ln F$$

Figure 5 shows the graph of this equation together with the data.
(b) Substituting $F = 150$ in the regression equation gives an average harvest size of 10.9 pounds.
(c) To determine the amount of feed required to grow 8.5-pound catfish, we solve the regression equation for F as follows.

$$S = -10.051 + 4.174 \ln F$$

$$\ln F = \frac{S + 10.051}{4.174}$$

$$F = \exp\left(\frac{S + 10.051}{4.174}\right) = \exp\left(\frac{8.5 + 10.051}{4.174}\right) \approx 85.2$$

Thus, a pond feeding program of 85.2 pounds per day is required. ∎

POWER REGRESSION

The third type of nonlinear regression to be considered is power regression, in which the dependent variable is proportional to some power of the independent variable. The problem of falling bodies provides a good example of such a relationship in the real world. According to the law of gravity, the distance d an object falls toward the earth (in a vacuum) is related to the time of flight t by

$$d = \frac{1}{2} a_g t^2$$

or equivalently,

$$t = \sqrt{\frac{2}{a_g}} \, d^{1/2}$$

Here, a_g is a constant, called the acceleration of gravity.

Height (ft)	Fall time (s)
13.1	0.92
24.2	1.24
35.3	1.50
46.4	1.68
57.6	1.86
79.9	2.30

Figure 6

■ **Example 4.** Two students have measured the time it takes a billiard ball to hit the ground from various floors of a tall building. (a) Fit a power curve $t = ah^b$ to the students' data (Figure 6) using h as the independent variable. (b) Use this curve to determine a value for the acceleration due to gravity and compare it with the accepted value $a_g = 32$ feet per second per second. (c) Estimate the height of the top floor if the fall time is measured as 2.66 seconds.

Solution. (a) Our calculator gives $a = 0.256$ and $b = 0.495$ as the constants that provide the best fit; thus the regression equation is

$$t = 0.256\, h^{0.495}$$

Note that since $b \approx 0.5$, the data do support the expected square root dependence of t on h.

(b) Since $a = \sqrt{2/a_g}$, we find $a_g = 2/a^2 \approx 30.5$, which is reasonably close to the theoretical value $a_g = 32$.

(c) To find the height of the top floor, we solve the regression equation for h and substitute $t = 2.66$

$$h = (t/0.256)^{(1/0.495)} = (2.66/0.256)^{(1/0.495)} \approx 113$$

The top floor is about 113 feet above the ground. ■

CHOOSING THE BEST TYPE OF REGRESSION EQUATION

We have seen how to fit four types of curves to a set of data. But how do we know which to use? Sometimes, as in the last example, we have theoretical reasons for expecting a certain type of relationship between the dependent and independent variables. In other cases, however, we do not know beforehand what, if any, type of relationship exists. In these cases, a scatter plot of the data may suggest something we might reasonably try. We may, in fact, want to try fitting more than one type of equation to the data, comparing the values of the corresponding correlation coefficients. In general, the equation giving a value of $|r|$ closest to 1 fits the data best.

■ **Example 5.** To optimize production schedules, Bismarck Bakery wants to bake its bread for exactly 10.0 minutes and needs to know at what temperature to set its ovens to achieve this baking time. Bismarck has measured the correct baking time t at seven temperatures T, as shown in Figure 7. From among the four types we have considered, determine the regression equation that best fits their data and advise them on the correct 10-minute baking temperature. Use temperature as the independent variable.

Temperature T (°F)	275	300	325	350	375	400	425
Time t (min)	14.2	12.1	10.6	9.3	8.1	7.1	6.2

Figure 7

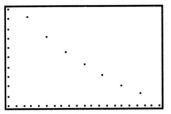

[250, 450] by [5, 15]

Figure 8

Solution. A scatter plot of the bakery's data is shown in Figure 8. The relationship between time and temperature appears to be slightly nonlinear, but it is not obvious which nonlinear equation is most appropriate. It might be best to try all three and compare the correlation coefficients. A graphics calculator gives the following.

Logarithmic	*Exponential*	*Power*
$a = 115.59$	$a = 62.878$	$a = 560906$
$b = -18.116$	$b = 0.99456$	$b = -1.8823$
$r = -0.9968$	$r = -0.9997$	$r = -0.9988$

The values of r are so close together and so near -1 that we could well use any of the three equations to model this problem. This will often happen when we attempt to fit a small, closely packed set of data. In this case, we choose the exponential equation, obtaining

$$t = (62.878)(0.99456)^T$$

Solving this equation for T in terms of t

$$T = \frac{\ln (t/62.878)}{\ln 0.99456}$$

and evaluating for $t = 10.0$ minutes, we find that the optimal oven temperature is 337°F. ∎

TEASER SOLUTION

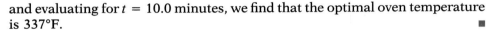

[0, 4000] by [0, 300]

Figure 9

A scatter plot of the planetary data using distance as the independent variable is shown in Figure 9 (the first four data points are overlapped by the x-axis). This plot suggests fitting by either an exponential or a power equation. Our calculator gives the following results.

Exponential	*Power*
$a = 1.3276$	$a = 0.001111$
$b = 1.0017$	$b = 1.5008$
$r = 0.893$	$r = 0.999999$

This suggests the power equation relationship

$$y = 0.001111x^{1.5008}$$

between distance x and sidereal year y. (The astronomer Johann Kepler first discovered this empirical relationship in 1618.) Using this equation with $x = 3700$, we predict that Smiley will orbit the sun every 252 earth years or so.

PROBLEM SET 3.6

A. Skills and Techniques

Use the data in Figure 10 to do the following in each case.
(a) Make a scatter plot of the data. (b) Tell whether fitting
the exponential equation $y = ab^x$ to the data will result in
a value of b greater or less than 1. (c) Use your calculator
to determine the best exponential regression equation and
confirm your answer to (b). (In the following, y vs. x means
to use y as the dependent variable and x as the indepen-
dent variable.)

x	p	q	r	s	t	u	v	w
−5	2.5	30	17	0.1	0.5	67	62	28
−2	3.3	10	12	0.2	0.9	36	31	59
−1	3.9	8	11	0.4	1.2	30	25	77
1	5.7	3	9	1.0	1.6	20	16	129
3	7.4	1.5	7	2.1	2.2	14	10	222
6	11.0	0.6	5	6.8	3.9	8	5	492
10	18.4	0.1	4	35.5	8.2	3	2	1390

Figure 10

1. *p* vs. *x*
2. *q* vs. *x*
3. *r* vs. *x*
4. *s* vs. *x*
5. *t* vs. *x*
6. *u* vs. *x*
7. *v* vs. *x*
8. *w* vs. *x*

For each given set of data (see Figure 11) do the following.
(a) Make a scatter plot of the data. (b) Determine the loga-
rithmic equation $y = a + b \ln x$ that best fits these data.
(c) Graph this equation together with the scatter plot using
the same set of axes.

a	f	g	h	x	w	y	z
2	0.60	15	−30	2	17	−1.0	69
4	1.32	23	−66	5	24	−3.7	38
6	1.85	24	−78	10	26	−4.1	5
8	2.03	29	−100	20	28	−6.3	−17
10	2.30	28	−105	50	31	−7.4	−55
12	2.56	30	−121	100	33	−8.8	−87

Figure 11

9. *f* vs. *a*
10. *g* vs. *a*
11. *h* vs. *a*
12. *w* vs. *x*
13. *y* vs. *x*
14. *z* vs. *x*

Use the data of Figure 12 to determine the power equation
$y = ax^b$ that best fits these data in each case. Then graph
this equation together with the data.

x	f	g	h	k	v	w	y	z
0.5	0.2	66	27.2	34.0	380	1.2	7.9	70
1.0	2.0	100	16.1	5.3	50	3.5	4.6	100
1.5	5.4	128	7.8	2.4	13	3.6	3.8	128
2.0	9.1	150	8.3	1.3	6	11.4	2.4	138
2.5	15.9	174	6.2	0.9	3	17.7	2.3	154
3.0	22.2	192	5.7	0.7	2	24.4	1.7	177

Figure 12

15. *f* vs. *x*
16. *g* vs. *x*
17. *h* vs. *x*
18. *k* vs. *x*
19. *v* vs. *x*
20. *w* vs. *x*
21. *y* vs. *x*
22. *z* vs. *x*

Make a scatter plot of each data set and predict which of
the four regression equations (one linear and three nonlin-
ear) will produce the line or curve that best fits the data.
Then test your prediction by determining which fitting
equation gives |r| closest to 1.

23.

x	1	2	3	5	6	8	10
y	0.4	0.5	0.7	1.4	3.3	5.8	11.2

24.

x	0.25	0.5	0.75	1.0	1.5	2.0	4.0
y	3.1	6.6	8.4	10.5	12.0	13.3	16.9

25.

x	2.2	6.4	11.3	25.8	56.2	60.3	71.8
y	22.5	41.0	66.2	136	288	323	380

26.

x	2	4	6	8	10	15	20
y	599	284	147	89	68	55	51

27.

x	0.1	0.2	0.4	0.8	1.2	1.6	2.0
y	44	38	32	26	23	21	20

28.

x	1	3	4	5	6	8	10
y	45.5	38.8	38.6	33.2	32.3	26.6	20.0

B. Applications and Extensions

29. The distance D required to bring a car to a complete stop increases with the car's speed S. (a) Determine which of our four models best fits the data in Figure 13, using S as the independent variable. (b) Write this best regression equation. (c) Estimate the stopping distances at 55 and 80 miles per hour.

Speed S (mph)	20	30	40	50	60	70
Stopping distance D (ft)	52	88	135	201	284	377

Figure 13

30. In thermodynamics, when a gas such as helium undergoes what is called a reversible adiabatic expansion, it is found that the gas pressure P and volume V are related by the equation $P = CV^{-k}$, where C and k are constants. Determine the appropriate regression equation for the data on helium given in Figure 14 and compare it with the theoretical value of $k = 5/3$.

Volume V (in^3)	26.3	30.5	36.8	44.2	58.9	97.0
Pressure P (lb/in^2)	52.8	40.5	30.0	21.8	13.9	5.8

Figure 14

31. Figure 15 presents U.S. population data for the years 1860–1970 (Source: U.S. Bureau of the Census). (a) Determine the best exponential model of population growth. (b) Draw a scatter plot of the data along with a graph of the fitted exponential equation. (c) Predict the U.S. population for the year 1980 and compare your answer with the actual Census Bureau estimate of 226.5 million. (*Hint:* Treat 1800 as year 0, 1860 as year 60, and so on.)

Year	U.S. population (millions)	U.S. population over age 65 (millions)
1860	31.4	–
1870	39.8	–
1880	50.2	–
1890	62.9	3.84
1900	76.0	4.05
1910	92.0	4.29
1920	105.7	4.67
1930	122.8	5.40
1940	131.7	6.85
1950	151.1	8.12
1960	179.3	9.30
1970	203.3	9.89

Figure 15

32. Repeat Problem 31 using a power model for the population growth. Which model more closely predicts the actual 1980 census?

33. Use both exponential and power models to fit the data in Figure 15 for the U.S. population over age 65. (a) Judging from the correlation coefficients, which model best fits the data? (b) Which model more closely predicts the 1980 census figure of 11.35 million people over age 65? (Again, let 1800 correspond to year 0.)

34. Figure 16 lists the average annual medical costs per capita in the United States on a relative scale, with 1967 corresponding to year 0 with costs of $100 (Source: U.S. Bureau of Labor Statistics). (a) Use an exponential model to predict the relative cost of medical care in the years 1975 and 1985 and compare your answers with the actual costs of 168.6 and 396.1, respectively. (b) Predict the relative cost of medical care in the year 2000.

Years	1976	1977	1978	1979	1980
Medical costs (dollars)	184.7	202.4	219.4	239.7	265.4

Years	1981	1982	1983	1984
Medical costs (dollars)	294.5	328.7	357.3	378.0

Figure 16

35. Two biology students have measured the average length of blades of grass as a function of time since sprouting. They believe the blades will grow logarithmically as a function of time. Fit a logarithmic equation to their data (Figure 17) and use it to predict the lengths of 10- and 25-day old grass blades.

Time since sprouting (days)	2	4	7	12	15	17	20
Length (in.)	1.9	3.4	4.2	5.7	6.0	6.5	6.8

Figure 17

36. In many scientific experiments, it is found that the signal-to-noise ratio S increases as the square root of the sampling time t (implying that the data become more reliable as the scientist collects them for a longer period of time). (a) On the basis of this finding, conjecture an appropriate model for S in terms of t and then fit it to the data of Figure 18. (b) Do these data support the square root dependence of S on t? (c) For how long should the scientist sample if a signal-to-noise ratio of 30 is desired?

Time t (s):	1	5	10	30	60	120
S	1.9	4.8	6.1	8.7	13.2	18.5

Figure 18

37. Figure 19 lists the average speeds of Indianapolis 500 auto race winners in 10-year intervals over the period 1920–1990. Make a scatter plot of the data and note the position of Johnny Rutherford's 1980 average speed in particular. Points such as this one that fail to follow the general trend of the data are called **outliers.** Try fitting the data both with and without the 1980 outlier to the best linear regression equations, and note how the correlation coefficient is affected. Then predict (in both cases) the year in which the 200-mile average speed barrier will be broken. Let 1900 correspond to year 0.

Year	Indianapolis 500 Winner	Average speed (mph)
1920	Gaston Chevrolet	88.16
1930	Billy Arnold	100.448
1940	Wilbur Shaw	114.277
1950	Jonny Parsons	124.002
1960	Jim Rathmann	138.767
1970	Al Unser	155.749
1980	Johnny Rutherford	142.862
1990	Arie Luyendyk	185.984

Figure 19

38. Kale has measured the mass and average diameter of 10 McIntosh apples (see Figure 20). (a) Assuming apples are perfect spheres of uniform density (that is, their masses are proportional to their volumes), write down the equation that relates their mass m to their diameter d. (b) Check on this equation by determining an appropriate nonlinear model and fitting it to Kale's data.

Diameter d (cm)	8.9	8.5	7.4	9.5	6.8
Mass m (g)	227	198	133	277	98

Diameter d (cm)	8.5	9.9	7.7	7.9	8.2
Mass m (g)	207	314	149	162	170

Figure 20

39. A decaying radioactive substance can be measured with a Geiger counter; the number of observed counts is proportional to the amount of radioactive matter remaining in the decaying sample. The readings in Figure 21 were taken over a 30-second period once an hour. (a) Determine the best exponential model to fit these data. (b) What is the half-life of this radioactive substance?

Time (hrs)	0	1	2	3	4	5	6	7	8
Counts	993	542	241	131	78	41	24	12	5

Figure 21

40. **Challenge.** The annual "Leading Money Winners" of the men's PGA and women's LPGA golf associations and their earnings are listed in Figure 22 over 5-year intervals from 1950 to 1990. (a) Separately fit exponential regression equations to the PGA and LPGA data sets and graph the resulting equations together on your calculator. (b) Predict the first year in which the women's Leading Money Winner will out-earn her male counterpart. (c) Will the first golfer to earn more than $5 million in a year be a man or a woman?

Year	PGA	Earnings	LPGA	Earnings
1950	Sam Snead	$35,758	–	–
1955	Julius Boros	65,121	Patty Berg	$16,492
1960	Aronld Palmer	75,262	Louise Suggs	16,892
1965	Jack Nicklaus	140,752	Kathy Whitworth	28,658
1970	Lee Trevino	157,037	Kathy Whitworth	30,235
1975	Jack Nicklaus	323,149	Sandra Palmer	95,805
1980	Tom Watson	530,808	Beth Daniel	231,000
1985	Curtis Strange	542,321	Nancy Lopez	416,472
1990	Greg Norman	1,165,477	Beth Daniel	863,578

Figure 22

CHAPTER 3 REVIEW PROBLEM SET

In Problems 1–10, write True or False in the blank. If false, tell why.

_____ 1. $(\sqrt[15]{3})^3 = \sqrt[5]{3}$

_____ 2. $\sqrt[4]{16} = \pm 2$

_____ 3. $\sqrt{5}\,\sqrt[3]{5} = \sqrt[6]{5}$

_____ 4. $(4/9)^{-3/2} = 27/8$

_____ 5. $\log_5 64 = 3 \log_5 4$

_____ 6. If a, b, and c are all positive, then $\ln a + 3 \ln b - (1/2) \ln c = \ln (ab^3/\sqrt{c})$.

_____ 7. If $c^{17} = 2$, then $c^{68} = 16$.

_____ 8. The number -2 is not in the domain of $\log_2 (x^2 - 2)$.

_____ 9. If a radioactive substance has a half-life of T days, then one-eighth of the original amount will be left after $3T$ days.

_____ 10. $f(x) = \ln (x^2)$ and $g(x) = 2 \ln x$ are identical functions.

In Problems 11–16, simplify, writing your answer in exponential form but with no negative exponents.

11. $\sqrt[5]{x^3}\, x^{7/5}$

12. $(\sqrt{a}\,\sqrt{ab})^2$

13. $(x^2 y^{-4} z^6)^{-1/2}$

14. $\sqrt[3]{27\sqrt{64x^4}}$

15. $(3a^{1/3} a^{-1/4})^3$

16. $\sqrt[4]{32x^{-2}/(x^6 y^{-3})}$

17. Graph the functions $f(x) = (5/4)^x$ and $g(x) = (4/5)^x$ in the same plane and describe how these graphs are related.

18. Which ultimately grows faster, x^{50} or 5^x?

19. Arrange 3^x, x^3, e^x, and $(x^5 - 10x^2)/(x + 100)$ from left to right according to the speed at which they ultimately grow, starting with the slowest growing.

20. Solve $x^5 - 3 = 3^x$, accurate to two decimal places, by graphing and zooming.

21. Find the maximum value of $f(x) = x^3 - e^x$, accurate to two decimal places.

22. Find the maximum point on the graph of $g(x) = 3 - 2xe^x$, accurate to two decimal places.

23. Evaluate each of the following.
(a) $\log_2 32$
(b) $\log_\pi 1$
(c) $\log_7 (1/49)$
(d) $\log_3 (3^\pi)$
(e) $\log_8 32$
(f) $3^{\log_3 12}$

24. Evaluate x.
(a) $\log_4 x = 3$
(b) $\log_{25} x = -\dfrac{3}{2}$
(c) $2^{2 \log_2 3} = x$
(d) $\log_6 6^x = \pi$
(e) $x = \log_{10}\sqrt{10,000}$
(f) $\log_x 64 = \dfrac{3}{2}$

25. Write $3 \log_4(x^2 + 1) - 2 \log_4(x + 2)$ as a single logarithm in the form $\log_4(\ldots)$.

26. Solve for x.
(a) $\log_2(x^2 - 1) = 3$
(b) $\log_2 x + \log_2(x + 2) = 3$

27. How is the graph of $y = \ln (x - 1) + 2$ related to the graph of $y = \ln x$?

28. If $\log_a b = 2/3$, find $\log_b a$.

29. Use your calculator to evaluate $\log_7 16$.

30. Use logarithms to solve $3^{2x+2} = 20$ for x.

31. Solve $5/(2 + e^{-2x}) = 2$ for x by first manipulating and then using logarithms.

32. Solve $3x + \ln \sqrt{2x - 1} = x^2 - 5$ accurate to two decimal places.

33. How are the graphs of $y = \log_4 x$ and $y = 4^x$ related to each other?

34. Arrange $x/2$, $\ln(x + 5)$, $\log_2 x$, and $\ln(x^2 - 5)$ from left to right according to the speed at which they ultimately grow, beginning with the slowest growing.

35. A radioactive substance decays according to the formula $y = y_0 e^{-0.055t}$ where t is measured in years.

    (a) Find its half-life.

    (b) How long does it take to decay to 1/10 its original amount?

36. A radioactive substance initially weighing 40 grams decayed exponentially so that after 20 days only 10 grams were left. Determine the formula in the form $y = y_0 e^{kt}$ for the amount y left after t days.

37. The doubling time for a quantity growing exponentially is 500 years. How long will it take to grow to five times its initial size?

38. The half-life of a quantity decaying exponentially is 2 hours. How long will it take to reach one-seventh its initial size?

39. Assume that the amount y (in grams) of a substance grows according to the formula $y = 50(1 - e^{-0.03t})$ where t is in days.

    (a) When will the amount reach 47 grams?

    (b) What is the limit on the amount y as $t \to \infty$?

40. How long will it take $100 to accumulate to $160 if it is invested at 8.5% (a) compounded annually? (b) Compounded continuously?

41. Determine the value of $1000 at the end of 8 years if interest is at 8% and the interest is (a) simple interest, (b) compounded quarterly, (c) compounded daily.

42. First Bank offers 8.3% interest compounded quarterly while Second Bank offers 8.25% interest compounded monthly. Determine which bank is offering the better deal to a saver.

43. Determine how long it will take for the value of $1000 invested at 6% compounded annually to surpass that of $1200 invested at 9% simple interest.

44. Determine the interest rate if $500 grew to $700 in 5 years assuming interest was compounded annually.

45. Figure 1 presents data for the U.S. population density in number of people per square mile.

    (a) Determine the exponential equation that best fits these data, using time t as the independent variable and assuming that $t = 0$ corresponds to the year 1800.

    (b) Graph this equation together with a scatter plot of the data.

    (c) Determine the doubling time for the U.S. population density.

    (d) According to this model, when will the population density reach 100 people per square mile?

Year	U.S. Population Density
1800	6.1
1820	5.5
1840	9.8
1860	10.6
1880	14.2
1900	21.5
1920	29.9
1940	37.2
1960	50.6
1980	64.0

Figure 1

46. The data in Figure 2 show how the world record time for the men's 1-mile race has progressed since Roger Bannister broke the 4-minute barrier in 1954. Note that times are given in minutes, seconds, and tenths of seconds.

    (a) Determine the exponential equation that best fits the record time y as a function of year x, letting $x = 0$ correspond to 1954.

    (b) Plot the data together with this equation.

    (c) Predict what the world record time will be in the year 2010.

    (d) When does this model predict that a 3½-minute mile will be attained?

Year	New World Record Time
1954	3:59.4
1954	3:58.0
1957	3:57.2
1958	3:54.5
1962	3:54.4
1964	3:54.1
1965	3:53.6
1966	3:51.3
1967	3:51.1
1975	3:51.0
1975	3:49.4
1979	3:49.0
1980	3:48.8
1981	3:48.5
1981	3:48.4
1981	3:47.3
1985	3:46.3

Figure 2

47. The share values of a certain well known stock fund were reported as shown in Figure 3. (a) Using $t =$

0 to correspond to 1950, determine which of the four models we have studied fits these data best and use it to write an equation giving the share value y in terms of t. (b) On the basis of your answer to (a), predict the value of a share in the year 2000. (c) Express your opinion as to the worth of this model in predicting share values 50 years from now.

Year	Share Value ($)
1952	1.0196
1962	3.4398
1972	8.6803
1982	14.0567
1991	57.7715

Figure 3

4

THE TRIGONOMETRIC FUNCTIONS

The trigonometric functions are useful in the study of triangles, but that is not their major use either in mathematics or science. What is most significant about these functions is the repeating nature of their graphs. Repetitive behavior is typical of many processes in nature as well as in business and industry. A familiar example is the electrocardiogram (illustrated on the tape below), which records the electric impulses produced by a beating heart.

Any function that repeats itself over and over in a continuing manner is said to be periodic. The premier examples of periodic functions are the sine and cosine. Their wavelike graphs will soon become very familiar to you (see Section 4.4).

Although Leonhard Euler had earlier

Joseph Fourier
(1768–1830)

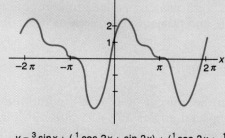

expanded some special functions in an infinite series of sines and cosines, it was Joseph Fourier who suggested that any periodic function could be so represented. In 1807, he submitted a paper to the Academy of Sciences in Paris on heat flow. In it, he expanded an arbitrary function in a Fourier series, that is, in a sum of possibly infinitely many terms of the form $a_k \cos kx + b_k \sin kx$. Here k runs through the positive integers 1, 2, 3, . . . and a_k and b_k are numbers depending on k. The example shown below has just three terms but already shows the possibility of representing quite complicated behavior. Three prominent mathematicians read Fourier's paper and rejected it. The paper was ahead of its time and it lacked rigor. Gradually, Fourier's ideas won acceptance and today are considered to be among the biggest steps in the development of mathematics during the nineteenth century.

$$y = \tfrac{3}{2}\sin x + (\tfrac{1}{2}\cos 2x + \sin 2x) + (\tfrac{1}{2}\cos 3x + \tfrac{1}{4}\sin 3x)$$

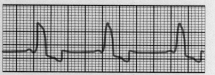

Electrocardiogram

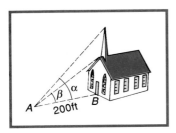

TEASER   From a point on the ground 200 feet away from a point directly beneath the church steeple, Andrew measures the angles of elevation of the top and bottom of the steeple to be $\alpha = 35.3°$ and $\beta = 26.0°$, respectively. Help Andrew find the height of the steeple.

4.1   RIGHT TRIANGLE TRIGONOMETRY

We introduce trigonometry by going back to its roots. The word *trigonometry* is derived from two Greek words, *trigonon*, meaning triangle, and *metria*, meaning measurement. For more than 1500 years, trigonometry was a study of the relationship between the sides and angles of a triangle together with its application to surveying, navigation, and astronomy. These notions are still important. Moreover, they provide a simple background for the more abstract concepts of modern trigonometry to be treated later.

Consider a right triangle, by which we mean a triangle with a 90° angle. The other two angles are acute (less than 90°), since the sum of all three angles is 180°. Let θ (the Greek letter *theta*) denote one of these acute angles. We may label the three sides relative to θ as adjacent side, opposite side, and hypotenuse (Figure 1). In terms of the lengths of these sides, we introduce the three fundamental ratios of trigonometry, namely, sine of θ, cosine of θ, and tangent of θ. Using obvious abbreviations, we define these ratios as follows.

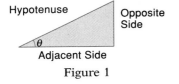

Figure 1

$$\sin \theta = \frac{\text{opp}}{\text{hyp}} \qquad \cos \theta = \frac{\text{adj}}{\text{hyp}} \qquad \tan \theta = \frac{\text{opp}}{\text{adj}}$$

Thus, with every acute angle θ, we associate three numbers, $\sin \theta$, $\cos \theta$, and $\tan \theta$. A careful reader might wonder whether these numbers depend only on the size of θ; or do they also depend on the size of the right triangle that contains θ? The answer to the latter question is no. Consider two right triangles each with the angle θ (Figure 2). These triangles are similar (which means they have equal angles), and so by a theorem from geometry, their ratios of corresponding sides are equal. We conclude that for a given θ, $\sin \theta$ has the same value no matter what right triangle is used to compute it. So do $\cos \theta$ and $\tan \theta$. Thus, sine, cosine, and tangent are functions; we call them trigonometric functions.

Figure 2

VALUES OF SINE, COSINE, AND TANGENT

We can use the Pythagorean theorem ($a^2 + b^2 = c^2$) to find the values of sine, cosine, and tangent of the special angles 30°, 45°, and 60°. First, a 45° right triangle with equal sides of length $\sqrt{2}$ has hypotenuse of length 2. Second, a

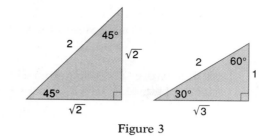

Figure 3

30° right triangle of hypotenuse 2 has side opposite the 30° angle of length 1; therefore the other side has length √3. Both triangles are shown in Figure 3. From these triangles, we obtain the following exact values.

$$\sin 45° = \frac{\sqrt{2}}{2} \qquad \cos 45° = \frac{\sqrt{2}}{2} \qquad \tan 45° = 1$$

$$\sin 30° = \frac{1}{2} \qquad \cos 30° = \frac{\sqrt{3}}{2} \qquad \tan 30° = \frac{1}{\sqrt{3}} = \frac{\sqrt{3}}{3}$$

$$\sin 60° = \frac{\sqrt{3}}{2} \qquad \cos 60° = \frac{1}{2} \qquad \tan 60° = \sqrt{3}$$

What about other angles? In the past, every trigonometry text had an extensive table giving decimal approximations to sin θ, cos θ, and tan θ. Today, we find these approximations with a calculator.

■ **Example 1.** Find the values of (a) sin 71°, (b) cos 19°, and (c) tan 57.256°.

Solution. We should first check that our calculator is in its degree mode (rather than radian mode, which is to be discussed later). Then, we simply press the obvious keys and obtain the results shown on the screen in Figure 4. (Note that the degree symbol does not appear on the screen. If you wish the degree symbol to appear, see the box titled The Degree Symbol.) ■

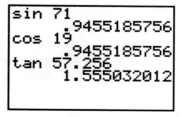

Figure 4

THE DEGREE SYMBOL °

You can display the degree symbol (as in 30°) on either the TI-81 or the TI-82. After entering an angle, press

MATH 6 (TI-81)

or

ANGLE 1 (TI-82)

In either case, doing this forces the calculator to evaluate in degree mode even if the calculator is set to radian mode.

The first two answers in Example 1 are the same. Do you see why this is so? Note that 71° and 19° are complementary angles; that is, they sum to 90°. Now use your calculator to find the values of sin 42° and cos 48°; these also are complementary angles. Make a conjecture. We will have more to say about complementary angles later.

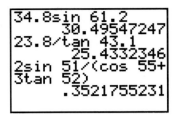

Figure 5

The need to calculate various expressions that involve the three trigonometric functions arises often; we must be proficient in doing this with a calculator.

■ **Example 2.** Use your calculator to find the value of each expression: (a) 34.8 sin 61.2°, (b) 23.8/tan 43.1°, (c) 2 sin 51°/(cos 55° + 3 tan 52°).

Solution. The answers are shown on the screen in Figure 5. ■

SOLVING TRIANGLES

A triangle has six key elements: three sides and three angles. If we know three of these elements including at least one side, we are able to determine the others. Doing this is called **solving a triangle.** We adopt the practice of naming the three sides (and their lengths) of a right triangle with the letters a, b, and c, with c being the hypotenuse. We label the angles opposite these sides with the Greek letters α, β, and γ. If we need to refer to the vertices (corner points), we will name them A, B, and C. These conventions are illustrated in Figure 6. Note that since $\gamma = 90°$, two additional pieces of information including at least one side suffice to determine the triangle.

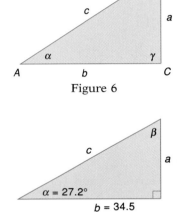

Figure 6

■ **Example 3 (One side and one angle given).** Solve the right triangle for which $\alpha = 27.2°$ and $b = 34.5$ (Figure 7).

Solution. Three elements (namely, β, a, and c) are to be determined:
Step 1: The two acute angles must sum to 90°; thus,

$$\beta = 90° - 27.2° = 62.8°$$

Step 2: Since tan 27.2° = opp/adj = a/34.5, we conclude that

$$a = 34.5 \tan 27.2° \approx 17.7$$

Step 3: We could use the Pythagorean theorem to determine c. Alternatively, cos 27.2° = adj/hyp = 34.5/c, and so

$$c = \frac{34.5}{\cos 27.2°} \approx 38.8$$ ■

Figure 7

■ **Example 4 (Two sides given).** Suppose the two legs of a right triangle have lengths $a = 34.56$ and $b = 75.23$ (Figure 8). Solve the triangle.

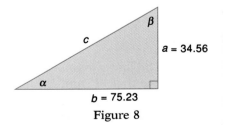

Figure 8

168 • The Trigonometric Functions

Solution. Step 1:

$$\tan \alpha = \text{opp/adj} = 34.56/75.23 = 0.45939$$

We know tan α, but we want to know α; for this we need the inverse function $\tan^{-1}$. Your calculator has a key for this and should give

$$\tan^{-1} 0.45939 \approx 24.67°$$

Rather than using the two-step process in Step 1, it is better to calculate $\tan^{-1}(34.56/75.23)$ directly.

Step 2:

$$\beta = 90° - \alpha \approx 90° - 24.67° = 65.33°$$

Step 3: To determine c, we could use the fact that $\sin \alpha = a/c$, but this time we will use the Pythagorean theorem.

$$c = \sqrt{34.56^2 + 75.23^2} \approx 82.79 \quad \blacksquare$$

Note that in both Examples 3 and 4 we gave our final answers with the same accuracy as the data. It is good to follow this practice when it seems likely that the data were obtained as a result of measurements.

APPLICATIONS

Our first application is to the type of problem that surveyors face.

■ **Example 5.** George needs to determine the width of a river and doesn't want to get his feet wet. Directly across from a tree B on the opposite bank, he drives a stake at C (Figure 9). Then he measures a distance of 200 feet along the shore from C to a point A. Finally, he measures the angle $\alpha = \angle CAB$ to be 51.2°. Use this information to help George determine the distance from C to B.

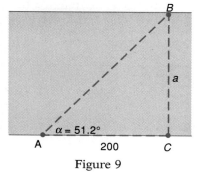

Figure 9

Solution. Since $\tan 51.2° = a/200$,

$$a = 200 \tan 51.2° \approx 249$$

The river is approximately 249 feet wide. $\quad \blacksquare$

■ **Example 6.** A speed boat is on its way to rescue the crew of an overturned sailboat. A helicopter in the vertical plane of the two boats and 550 feet above the water reports the angle of depression (angle down from the horizontal) of the two boats to be 22.3° and 17.2° as shown in Figure 10. Determine the distance between the boats.

Figure 10

Solution. We need to determine $x + y$ in Figure 10. Note that $\alpha = 72.8°$, that $\beta = 67.7°$, and that $\tan 72.8° = x/550$ and $\tan 67.7° = y/550$. Thus,

$$x + y = 550 \tan 72.8° + 550 \tan 67.7° \approx 3120 \text{ feet} \quad ■$$

TEASER SOLUTION

Referring to Figure 11, we observe that

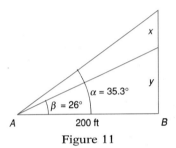

Figure 11

$$\tan 35.3° = \frac{x + y}{200} \quad \text{and} \quad \tan 26.0° = \frac{y}{200}$$

We solve for y in the second equation, substitute in the first, and then manipulate the result to obtain the following sequence of equivalent equations.

$$\tan 35.3° = \frac{x + 200 \tan 26°}{200}$$

$$200 \tan 35.3° = x + 200 \tan 26°$$

$$x = 200 \tan 35.3° - 200 \tan 26° \approx 44.1 \text{ feet}$$

PROBLEM SET 4.1

A. Skills and Techniques

Use your calculator to evaluate each of the following. Put your calculator in degree mode.

1. $\sin 47.32°$
2. $\tan 31.56°$
3. $\cos 17.432°$
4. $\sin 87.564°$
5. $\tan 56.321°$
6. $\cos 28.678°$
7. $\sin 25.2° - \cos 34.3°$
8. $\tan 51.2° + \sin 12.9°$
9. $(\cos 13.8°)(\tan 43.2°)$

10. $\dfrac{\sin 48.9°}{\cos 54.2°}$
11. $\dfrac{\tan 23.1° - \cos 46.8°}{\sin 36.2°}$
12. $\dfrac{\sin 53.1° + \tan 22.2°}{\cos 42.1°}$
13. $\dfrac{(\cos 32.2°)^2 + 1.234}{\tan 54.3°}$
14. $\dfrac{(\tan 14.7°)^3 - \sin 11.7°}{\sin 39.4°}$
15. $\dfrac{(\sin 21.1° - \cos 32.2°)^2}{\tan 12.3^{°2}} + \sin 32.7°$
16. $\dfrac{(\tan 32.2°)^2}{\sin 15.5° + \cos 13.7°} - \cos 51.2°$

Evaluate θ accurate to the nearest thousandth of a degree. For example, to find θ if sin θ = 0.2364, press sin⁻¹ 0.2364. *You should obtain 13.67416235 and report the answer as 13.674°.*

17. $\sin \theta = 0.86544$
18. $\cos \theta = 0.78325$
19. $\tan \theta = 1.34581$
20. $\sin \theta = 0.23456$
21. $\cos \theta = 0.23765$
22. $\tan \theta = 0.65409$

Solve for x in each of the following diagrams.

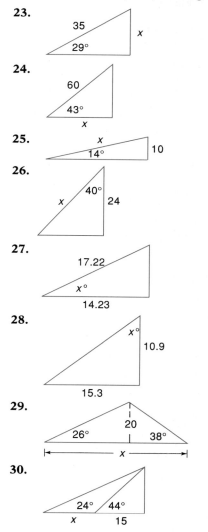

23.

24.

25.

26.

27.

28.

29.

30.

In Problems 31–40, solve the right triangles with the given data. Begin by sketching the triangle. Give your answers to the same accuracy as the least accurate of the data.

31. $\alpha = 42.3°, c = 35.1$
32. $\beta = 29.3°, c = 50.2$
33. $\beta = 56.25°, c = 91.34$
34. $\alpha = 69.98°, c = 10.87$
35. $\alpha = 39.45°, a = 120.3$
36. $\alpha = 40.45°, b = 165.3$
37. $a = 10.67, b = 12.34$
38. $a = 24.51, b = 10.34$
39. $a = 40.3, c = 50.5$
40. $a = 40.42, c = 41.32$
41. A 20-foot ladder leans against a vertical wall, making an angle of 76° with the level ground. How high up the wall is the upper end of the ladder?
42. Find the angle of elevation of the sun if a woman 5 feet 9 inches tall casts a shadow on the ground 45.7 feet long. (The **angle of elevation** is the upward angle made with the horizontal.)
43. The angle of elevation of the sun is 33.4°. How long a shadow does a girl 4 feet 6 inches tall cast on the ground?
44. A guy wire to a vertical pole makes an angle of 69° with the level ground and is 14 feet from the pole at ground level. How long is the guy wire?

B. Applications and Extensions

45. From the top of a lighthouse 120 feet above sea level, the **angle of depression** (the downward angle from the horizontal) to a boat adrift on the sea is 9.4° (Figure 12). How far from the foot of the lighthouse is the boat?

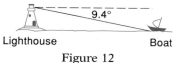

Figure 12

46. With her hands 5 feet above the ground, Sally is pulling a kite. If the kite is 200 feet above the ground and the taut kite string makes an angle of 32.4° with the horizontal, how many feet of string are out?
47. Determine *x* in Figure 13.

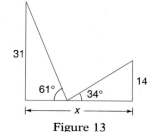

Figure 13

48. A plane is flying directly away from a ground observer at a constant rate, maintaining an elevation of 15,000 feet. At a certain instant, the observer measures the angle of elevation as 44° and 15 seconds later as 31°. How fast is the plane flying in miles per hour?

49. From a window in an office building, I am looking at a television tower that is 600 meters away (horizontally). The angle of elevation of the top of the tower is 19.6° and the angle of depression of the base of the tower is 21.3°. How tall is the tower?

50. The vertical distance from the first to the second floor of a certain department store is 28 feet. The escalator, which has a horizontal reach of 96 feet, takes 25 seconds to carry a person between floors. How fast does the escalator travel?

51. The Great Pyramid is about 480 feet high and its square base measures 760 feet on a side. Find the angle of elevation of one of its edges, that is, find β in Figure 14.

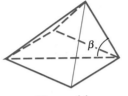

Figure 14

52. Find the angle between a principal diagonal and a face diagonal of a cube.

53. A ship S_1 left the harbor O at noon sailing in the direction S45°E at 24 miles per hour. At 1:00 P.M. a second ship S_2 left the same harbor sailing in the direction N39°E at 28 miles per hour (see Figure 15).

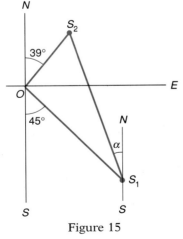

Figure 15

(a) Determine the distance between the ships at 2:30 P.M.
(b) At 2:30 P.M. the bearing from S_1 to S_2 is NαW. Determine α.

54. Determine the angle α between the chord AC and the diameter AB for the circle in Figure 16.

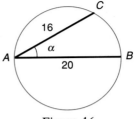

Figure 16

55. A regular hexagon (six equal sides) is inscribed in a circle of radius 4. Find the perimeter P and area A of this hexagon.

56. A regular decagon (10 equal sides) is inscribed in a circle of radius 12. What percent of the area of the circle is the area of the decagon?

57. Find the exact area of the regular six-pointed Star of David that is inscribed in a circle of radius 1 (Figure 17).

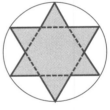

Figure 17

58. **Challenge.** Find the exact area of the regular five-pointed star (the pentagram) that is inscribed in a circle of radius 1 (Figure 18).

Figure 18

TEASER The pedal sprocket of Lee's bicycle has a radius of 8 centimeters, the rear wheel sprocket has a radius of 4 centimeters, and the tires have a radius of 35 centimeters. How far did Lee travel if he pedaled for 60 revolutions of the pedal sprocket?

4.2 GENERAL ANGLES AND ARCS

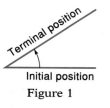

Figure 1

To prepare for the modern treatment of the trigonometric functions that will come in the next section, we need to generalize the notion of angle and relate it to an arc on a circle. We consider a ray that rotates about its endpoint from an initial position to a terminal position (Figure 1). The angle is defined to be positive if the rotation is counterclockwise and negative if the rotation is clockwise. Moreover, this rotation may measure any real number of degrees. Figure 2 illustrates these ideas for six different angles.

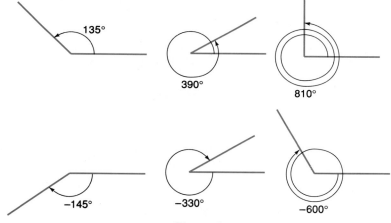

Figure 2

Degree measure is attributable to the ancient Babylonians, who divided the circle into 360 parts, each of size 1 degree, then divided an angle of 1 degree into 60 parts, each measuring 1 minute, and finally divided an angle of 1 minute into 60 parts of size 1 second. While we will occasionally use degree measure in this book, we will not use minutes and seconds. If we need to measure an angle to finer accuracy than a degree, we will use decimal parts (as we did in Section 4.1). Thus, we will write 40.5° rather than 40°30′.

RADIAN MEASURE

The best way to measure angles is in radians. Conside a circle of radius r. The familiar formula $C = 2\pi r$ tells us that the circumference has 2π arcs (about

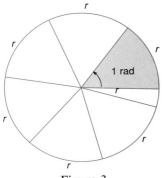

Figure 3

6.28 arcs) of length r around it (Figure 3). A central angle, meaning an angle with vertex at the center of the circle, measures one **radian** (abbreviated as *rad*) if it cuts off an arc of length r. Thus, an angle of size 360° measures 2π radians, and an angle of size 180° measures π radians. We abbreviate the latter by writing

$$180° = \pi \text{ radians}$$

Moreover, we have these equivalents for special angles.

$$30° = \frac{\pi}{6} \text{ radians} \qquad 45° = \frac{\pi}{4} \text{ radians}$$

$$60° = \frac{\pi}{3} \text{ radians} \qquad 90° = \frac{\pi}{2} \text{ radians}$$

Since

$$1° = \frac{\pi}{180} \approx 0.01745 \text{ radians}$$

and

$$1 \text{ radian} = \frac{180°}{\pi} \approx 57.296°$$

we have the following conversion rules.

To change from degrees to radians, multiply by $\pi/180$.

To change from radians to degrees, multiply by $180/\pi$.

■ **Example 1.** (a) Change 313.92° to radians. (b) Change 1.245 radians to degrees.

Solution.

(a) $313.92° = 313.92 \left(\dfrac{\pi}{180}\right) \approx 5.4789$ radians

(b) $1.245 \text{ radians} = 1.245 \left(\dfrac{180}{\pi}\right) \approx 71.33°$ ■

ARC LENGTH AND AREA

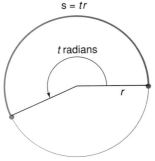

$s = tr$

t radians

r

Figure 4

Radian measure is almost always used in calculus and advanced mathematics. Division of a circle into 360 parts was quite arbitrary; its division into parts of radius length (2π parts) is more natural. Because of this circumstance, formulas using radian measure tend to be simple, whereas those using degree measure are often complicated. As an example, consider arc length on a circle. Let t be the radian measure of a central angle θ of a circle of radius r (Figure 4). This angle cuts off an arc of length s satisfying the simple formula

$$s = rt$$

A second nice formula is that for the area of a sector cut off from a circle by a central angle of t radians (Figure 5). Note that the ratio of the area A of this sector to the area of the whole circle is the ratio of t to 2π; that is, $A/\pi r^2 = t/2\pi$. Thus,

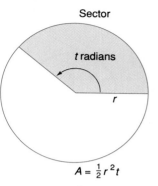

Sector

t radians

r

$A = \frac{1}{2}r^2t$

Figure 5

$$A = \frac{1}{2}r^2t$$

■ **Example 2.** Consider a central angle of $137.24°$ on a circle of radius 21.23 centimeters (Figure 6). (a) Find the length s of the arc that is cut off. (b) Calculate the area A of the sector that is cut off. (c) Determine the area A_0 of the shaded region.

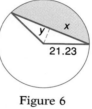

y x

21.23

Figure 6

Solution. We must measure the angle in radians.

$$137.24° \approx 2.39529 \text{ radians}$$

(a) $s \approx 21.23(2.39529) \approx 50.85$ centimeters.

(b) $A \approx \frac{1}{2}(21.23)^2(2.39529) \approx 539.8$ square centimeters.

(c) Let x denote the half-width and y the height of the central triangle. Since half the given angle measures $68.62°$, we see that $\sin 68.62° = x/21.23$ and $\cos 68.62° = y/21.23$. This gives $x \approx 19.769$ centimeters and $y \approx 7.739$ centimeters. Thus, the area of the central triangle is $A_1 = bh/2 = 19.769(7.739) \approx 153.0$ square centimeters. We conclude that the area of the shaded region is $A_0 = A - A_1 \approx 539.8 - 153.0 = 386.8$ square centimeters. ■

THE UNIT CIRCLE

The formula for arc length takes a particularly simple form when $r = 1$; namely, $s = t$. We emphasize its meaning.

On the unit circle, the radian measure of a central angle and the length of the arc it determines are identical.

How do we interpret this statement when the radian measure of the angle is more than 2π or when it is negative? Imagine an infinitely long string on which the real number scale has been marked. Wrap this string around the unit circle with the zero point of the scale at the point (1, 0) as shown in Figure 7.

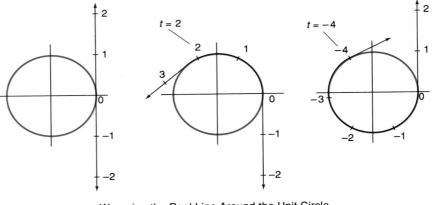

Wrapping the Real Line Around the Unit Circle

Figure 7

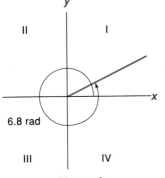

Figure 8

Consider an angle in **standard position,** which means that its vertex is at the origin and its initial side is along the *x*-axis. The radian measure of the angle is the (directed) length of the piece of string cut off by the angle. For example, an angle of 8 radians cuts off a piece of string of length 8 units, and an angle of -21 radians (a clockwise angle) cuts off a piece of string of directed length -21 units.

Let θ be an angle in standard position. It will often be important for us to know the quadrant in which the terminal side of θ lies. For example, the terminal side of an angle of 6.8 (a little more than 2π) radians lies in quadrant I (Figure 8). Similarly, the terminal side of an angle of -2.4 radians lies in quadrant III. For short, we say that the first angle is a first-quadrant angle and the second one is a third-quadrant angle.

■ **Example 3.** Determine the quadrant of the terminal side for each of the following angles. (a) 16 radians, (b) 57 radians, (c) -23 radians.

Solution. The idea is to first reduce the given number by removing as many multiples of $2\pi \approx 6.2832$ as possible, since this reduction does not affect the position of the terminal side of the angle.

(a) $16 - 2(2\pi) = 3.4336$, so the given angle is a third-quadrant angle.

(b) $57 - 9(2\pi) = 0.4513$, so the given angle is a first-quadrant angle.

(c) $23 - 3(2\pi) = 4.1504$, so an angle that measures -23 radians has the same terminal side as one that measures -4.1504 radians. It is a second-quadrant angle. ■

ANGULAR VELOCITY

Our world runs on wheels. Imagine a rotating wheel with its center at the origin. The rate at which the wheel is turning, called its **angular velocity,** is the rate at which the angle from the *x*-axis to a radial line on the wheel is changing. Angular velocity is often measured in revolutions per unit of time (for example, tachometers measure the speed of a motor in revolutions per minute). In mathematics, however, angular velocities are best measured in radians per unit of time. A quantity related to the angular velocity of a spinning wheel is the **linear velocity** of a point on the rim of the wheel (the actual distance the point travels per unit of time). The relationship is simple. A point on the rim of a wheel of radius r, spinning with an angular velocity of ω radians per unit of time, has a linear velocity of v length units per unit of time where

$$v = r\omega$$

■ **Example 4.** A fly sits on the rim of a wheel of radius 23.25 centimeters. If the wheel is turning at 100 revolutions per minute, what is the linear velocity of the fly? In other words, how fast would the fly be moving if it fell off the spinning wheel?

Solution. An angular velocity of 100 revolutions per minute corresponds to $100(2\pi) \approx 628.32$ radians per minute. This means the linear velocity is $v = r\omega \approx 23.25(628.32) \approx 14{,}610$ centimeters per minute. ■

■ **Example 5.** Assume that the earth is a sphere of radius 3960 miles. Author Dale Varberg lives in St. Paul exactly on the 45th parallel, that is, midway between the North Pole and the equator. How fast in miles per hour is he moving because of the spinning of the earth on its axis?

Solution. We must first determine the radius r of the "wheel" on whose rim Varberg is spinning (Figure 9). Note that $r^2 + r^2 = 3960^2$, which implies that $r \approx 2800$ miles. The angular velocity of this wheel is $2\pi/24$ radians per hour, and so Varberg's linear velocity is $v = r\omega \approx 2800(2\pi/24) \approx 733$ miles per hour. ■

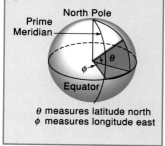

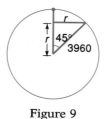

Figure 9

TEASER SOLUTION

Sixty revolutions of the 8-centimeter-radius pedal sprocket makes 120 revolutions of the 4-centimeter-radius rear sprocket. Thus, the tires turn through 120 revolutions, which is $120(2\pi)$ radians. It follows that these 35-centimeter-radius tires will travel a linear distance of

$$s = rt = 35(120)(2\pi) \approx 26{,}400 \text{ centimeters}$$
$$\approx 264 \text{ meters}$$

---

177 • 4.2 General Angles and Arcs

PROBLEM SET 4.2

A. Skills and Techniques

Convert all degree measures to radian measure and convert all radian measures to degree measure. Radian measures should be given in terms of π and also as a decimal rounded to four decimal places (for example, 30° = π/6 radians = 0.5236 radians).

1. 120°
2. 225°
3. 210°
4. 330°
5. 540°
6. 450°
7. 34°
8. 92°
9. 54.5°
10. $\left(\dfrac{150}{\pi}\right)^{\circ}$
11. $\dfrac{4\pi}{3}$ rad
12. $\dfrac{5\pi}{6}$ rad
13. $\dfrac{-2\pi}{3}$ rad
14. 3π rad
15. $\dfrac{11\pi}{4}$ rad
16. $\dfrac{11}{4}$ rad
17. 3 rad
18. 4.23 rad

Find the radian measure of the central angle that cuts off on a circle of radius 6 an arc of the indicated length.

19. 5
20. 12
21. 16.23
22. 35.35

In Problems 23–26, find the length of arc and area of the sector cut off on a circle of radius 9 by a central angle of the indicated measure.

23. 2.6 rad
24. 5.5 rad
25. $\dfrac{4\pi}{3}$ rad
26. $\dfrac{7\pi}{4}$ rad
27. A central angle of measure 120° cuts off an arc of length 32.35 inches on a circle. Determine the radius of the circle.
28. A sector of a circle, determined by a central angle of 43.4°, has area 64.2 square inches. Determine the radius of the circle.
29. Through how many radians does the minute hand of a clock turn in 3.5 hours? The hour hand in 2 hours?
30. Find the area of the region swept out by the minute hand of a clock between 1:15 and 1:52, assuming this hand is 8 centimeters long.

Sketch the angle with the indicated measure in standard position, and name the quadrant in which its terminal side resides.

31. 340°
32. 761°
33. 8 rad
34. 18 rad
35. −500°
36. −910°
37. −22.3 rad
38. −50 rad

In Problems 39–42, determine the linear velocity of a point on the rim of a wheel of radius 10 inches that is turning at the indicated rate.

39. 600° per second
40. 5 rad per minute
41. 40 revolutions per minute
42. 1200 revolutions per minute
43. Martina is pedaling her tricycle so that the front wheel of radius 8 inches turns at 2 revolutions per second. How fast is she traveling down the sidewalk (in miles per hour)?
44. Dale is riding a bicycle with tires of radius 30 centimeters so that the wheels turn through 70 revolutions per minute. How fast is he traveling down the road (in kilometers per hour)?
45. Suppose the tire on a truck has an outer diameter of 2.5 feet. How many revolutions per minute does it make when traveling at 60 miles per hour?
46. The rear wheels of a farm tractor are 6 feet tall. How many revolutions per minute do they make when the tractor is going 20 miles per hour?
47. New York City is located at 40.5° latitude north. How far is it from the equator (assuming the earth is a sphere of radius 3960 miles)?
48. If a radar site is only 500 miles from the North Pole, what is its latitude?

B. Applications and Extensions

49. A string of length 100 is wrapped counterclockwise around the unit circle with its initial end at (1, 0). In what quadrant does its terminal end lie? Estimate the coordinates of the terminal end.

50. Determine the perimeter and the area of the sector shown in Figure 10. Its radius is 10.5.

Figure 10

51. Determine the perimeter and the area of the shaded region in Figure 11.

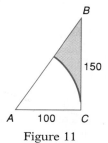

Figure 11

52. Determine the perimeter and the area of the shaded region in Figure 12.

Figure 12

53. Figure 13 shows a circle with diameter *AB* of length 20. Find (a) the length *AC* and (b) the area of the shaded region.

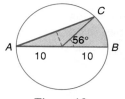

Figure 13

54. A cone has radius of base *R* and slant height *L*. Obtain the formula for its lateral surface area. (*Hint:* Imagine the cone to be made of paper, slit it up the side, and lay it flat in the plane.)

55. The minute hand and the hour hand of a clock are both 6 inches long and reach to the edge of the dial. Find the area of the pie-shaped region between the two hands at 5:40.

56. The angle subtended by the sun at the earth (93 million miles away) is 0.0093 radians. Find the diameter of the sun.

57. A dead fly is stuck to a belt that passes over two pulleys 6 and 8 inches in radius (Figure 14). The smaller pulley turns at 20 revolutions per minute.

(a) How fast is the fly moving? (b) How fast is the larger pulley turning?

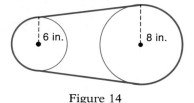

Figure 14

58. How much area will a radius on the larger pulley of Figure 14 sweep out when the small pulley turns through 3 revolutions?

59. Assume that the earth is a sphere of radius 3960 miles. How fast (in miles per hour) is a point on the equator moving as a result of the earth's rotation about its axis?

60. A belt traveling at the rate of 60 feet per second drives a pulley at the rate of 900 revolutions per minute. Find the radius of the pulley.

61. The pedal sprocket of Julio's bicycle has radius 9 centimeters, the rear sprocket has radius 3 centimeters, and the tires have radius 35 centimeters. How far did Julio travel if he pedaled 50 revolutions of the pedal sprocket?

62. The orbit of the earth about the sun is an ellipse that is nearly circular with radius 93 million miles. Approximately, what is the earth's speed (in miles per hour) in its path around the sun? Assume that a complete orbit takes 365.25 days.

63. A nautical mile is the length of 1 minute (1/60 of a degree) of arc on a great circle of the earth (with radius 3960 ordinary miles). How many ordinary miles are there in a nautical mile?

64. Oslo, Norway, and St. Petersburg, Russia, are both located at 60° latitude north. Oslo is at longitude 6° east (of the prime meridian), whereas St. Petersburg is at 30° east. How far apart are these two cities along the 60° parallel?

65. A pilot flew one-fourth of the way around the earth along the 30° parallel traveling east, then flew straight south to the equator. How far did she fly?

66. **Challenge.** Each of two circles, both of radius *r*, has its center on the rim of the other (Figure 15). Find the formula for the area of their intersection.

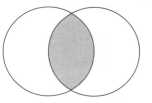

Figure 15

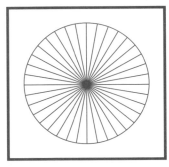

TEASER   Evaluate

$$\sin 1° + \sin 2° + \sin 3° + \ldots + \sin 358° + \sin 359°$$

4.3   THE SINE AND COSINE FUNCTIONS

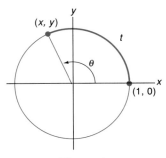

Figure 1

The way is prepared for the modern definition of the sine and cosine functions. Consider an arbitrary angle θ with radian measure t. Place θ in **standard position;** that is, put θ in the coordinate plane so that its vertex is at the origin and its initial side is along the positive x-axis. Let (x, y) be the coordinates of the point where the terminal side of θ intersects with the unit circle (Figure 1). Then

$$\sin \theta = \sin t = y \qquad \cos \theta = \cos t = x$$

Notice that we have defined the sine and cosine for any angle θ and also for the corresponding number t. Both concepts are important. In geometric situations, angles play a central role, and we are likely to need the sines and cosines of angles. In most of pure mathematics and in many scientific applications, however, it is the trigonometric functions of numbers that are important.

Related to this dual usage are two notational conventions. If we write sin 12°, it is clear that we are indicating the sine of an angle of size 12°. But we do not write sin (12 radians). Rather, we write sin 12, and you cannot tell from the notation whether we mean the sine of an angle of size 12 radians or the sine of the number 12. The context will usually make the choice clear, and, in any case, no confusion arises because both understandings result in the same value.

■ **Example 1.**   Use a calculator to evaluate each of the following.

(a) sin 9.81°   (b) sin 9.81   (c) cos (−403°)
(d) cos 12.3ᶜ   (e) cos 12.3   (f) sin (−6.78)

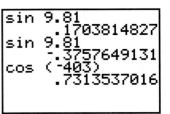

Figure 2

Solution. We begin by putting our calculator in the correct mode—degree or radian—for the problem at hand. Most calculators do not automatically display a degree symbol, so there is some potential for error from using the wrong mode. The results shown on the screen in Figure 2 are confusing because the calculator was in radian mode for the second entry but in degree mode for the other two. Make sure that you can reproduce the following answers on your calculator.

(a) $\sin 9.81° = 0.1703814827$
(b) $\sin 9.81 = -0.3757649131$
(c) $\cos(-403°) = 0.7313537016$
(d) $\cos 12.3° = 0.9770455744$
(e) $\cos 12.3 = 0.9647326179$
(f) $\sin(-6.78) = -0.4766277411$ ∎

CONSISTENCY WITH EARLIER DEFINITIONS

Do the right triangle definitions for sine and cosine given in Section 4.1 agree with those given here? Yes. Consider a right triangle ABC with an acute angle θ. Orient the triangle so that θ is in standard position, thus determining the point $B'(x, y)$ where the hypotenuse intersects the unit circle and the point $C'(x, 0)$ directly below it on the x-axis (Figure 3).

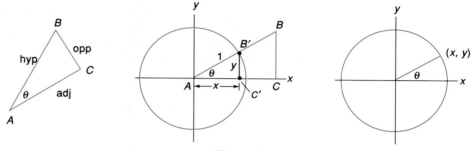

Figure 3

Notice that triangles ABC and $AB'C'$ are similar. It follows that

$$\frac{\text{opp}}{\text{hyp}} = \frac{BC}{AB} = \frac{B'C'}{AB'} = \frac{y}{1} = y$$

$$\frac{\text{adj}}{\text{hyp}} = \frac{AC}{AB} = \frac{AC'}{AB'} = \frac{x}{1} = x$$

On the left are the old definitions of $\sin \theta$ and $\cos \theta$; on the right are the new ones. They are consistent.

SPECIAL ANGLES

sin > 0 cos < 0	sin > 0 cos > 0
sin < 0 cos < 0	sin < 0 cos > 0

Figure 4

Consider first the question of the sign (plus or minus) of the sine and cosine functions in the four quadrants, that is, for angles with terminal sides in the four quadrants. Since $\sin \theta$ is the y-coordinate of a point, the sine function is positive in quadrants I and II and negative in quadrants III and IV. Similarly $\cos \theta = x$, so the cosine function is positive in quadrants I and IV and negative in quadrants II and III (Figure 4).

We learned the sines and cosines of the special angles 30°, 45°, and 60° in Section 4.1. Here we want to consider a more extensive set of angles that can be related to these earlier angles. And we will measure these angles in radians.

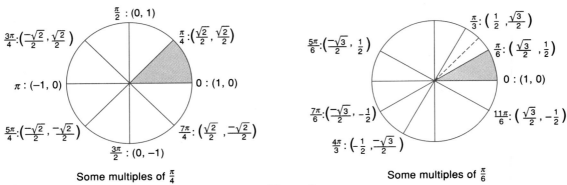

Some multiples of $\frac{\pi}{4}$ Some multiples of $\frac{\pi}{6}$

Figure 5

Study the diagrams in Figure 5. Look first at the first quadrants and note the coordinates of the points on the unit circles corresponding to $\pi/4$, $\pi/6$, and $\pi/3$. From symmetries and knowledge of signs, we can determine the sines and cosines of many other angles as the diagrams indicate. Notice, for example, how the coordinates of the points corresponding to $t = 5\pi/6$, $7\pi/6$, and $11\pi/6$ are related to those corresponding to $t = \pi/6$. The table in Figure 6 has results useful enough that you may wish to memorize them.

t	0	$\frac{\pi}{6}$	$\frac{\pi}{4}$	$\frac{\pi}{3}$	$\frac{\pi}{2}$	π	$\frac{3\pi}{2}$
$\cos t$	1	$\frac{\sqrt{3}}{2}$	$\frac{\sqrt{2}}{2}$	$\frac{1}{2}$	0	-1	0
$\sin t$	0	$\frac{1}{2}$	$\frac{\sqrt{2}}{2}$	$\frac{\sqrt{3}}{2}$	1	0	-1

Figure 6

■ **Example 2.** Determine each value by relating the given angle to a special angle. Do not use a calculator.

(a) $\cos \dfrac{5\pi}{3}$ (b) $\sin \dfrac{5\pi}{6}$ (c) $\sin \dfrac{5\pi}{3}$

(d) $\cos 27\pi$ (e) $\sin \left(-\dfrac{13\pi}{6} \right)$ (f) $\cos \dfrac{13\pi}{4}$

(g) $\sin 300°$ (h) $\cos (-450°)$ (i) $\sin 840°$

Solution. You should draw a diagram of the angle in each case to help you relate the angle to a special angle. For example, Figure 7 shows an appropriate diagram for (a). When an angle is larger than 2π (or $360°$), you should first reduce the angle by removing a multiple of 2π.

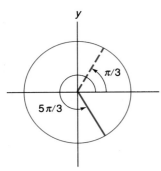

Figure 7

(a) $\cos \dfrac{5\pi}{3} = \cos \dfrac{\pi}{3} = \dfrac{1}{2}$

(b) $\sin \dfrac{5\pi}{6} = \sin \dfrac{\pi}{6} = \dfrac{1}{2}$

(c) $\sin \dfrac{5\pi}{3} = -\sin \dfrac{\pi}{3} = -\dfrac{\sqrt{3}}{2}$

(d) $\cos 27\pi = \cos \pi = -1$

(e) $\sin \left(-\dfrac{13\pi}{6} \right) = \sin \left(-\dfrac{\pi}{6} \right) = -\sin \dfrac{\pi}{6} = -\dfrac{1}{2}$

(f) $\cos \dfrac{13\pi}{4} = \cos \dfrac{5\pi}{4} = -\cos \dfrac{\pi}{4} = -\dfrac{\sqrt{2}}{2}$

(g) $\sin 300° = -\sin 60° = -\dfrac{\sqrt{3}}{2}$

(h) $\cos(-450°) = \cos(-90°) = \cos 90° = 0$

(i) $\sin 840° = \sin 120° = \sin 60° = \dfrac{\sqrt{3}}{2}$ ∎

■ **Example 3.** Use a calculator to confirm the answers in Example 2. Give answers to five-decimal-place accuracy.

Solution. Our calculator gives these answers.
(a) 0.5
(b) 0.5
(c) −0.86603
(d) −1
(e) −0.5
(f) −0.70711
(g) −0.86603
(h) 0
(i) 0.86603 ∎

PROPERTIES OF SINES AND COSINES

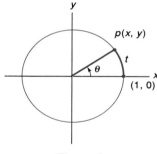

Figure 8

Think of what happens to x and y as t increases from 0 to 2π in Figure 8, that is, as the point $P(x, y)$ travels around the unit circle. For example, x starts at 1 and steadily decreases until it reaches its smallest value of -1 at $t = \pi$; then it starts to increase until it is back to 1 at $t = 2\pi$. We have just described the behavior of $\cos t$ as t increases from 0 to 2π. You should give a similar description of the behavior of $\sin t$.

Since x and y are coordinates of points on the unit circle, it follows that

$$-1 \le \sin t \le 1$$
$$-1 \le \cos t \le 1$$

Also $x^2 + y^2 = 1$ on the unit circle, and, since $x = \cos t$ and $y = \sin t$, it follows that

$$(\sin t)^2 + (\cos t)^2 = 1$$

It is conventional to write $\sin^2 t$ instead of $(\sin t)^2$ and $\cos^2 t$ instead of $(\cos t)^2$. Thus, we have the first of many fundamental trigonometric identities (equations that are true for all t).

$$\sin^2 t + \cos^2 t = 1$$

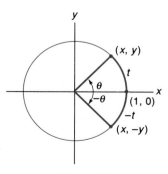

Figure 9

Using symmetry with respect to the x-axis (Figure 9), it is easy to see two more identities.

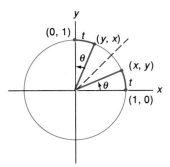

Figure 10

$$\sin(-t) = -\sin t$$
$$\cos(-t) = \cos t$$

Therefore, *sine is an odd function, whereas cosine is an even function.* Using symmetry in the line $y = x$ (Figure 10), we obtain a pair of identities for complementary angles that we first met in Section 4.1.

$$\sin\left(\frac{\pi}{2} - t\right) = \cos t$$
$$\cos\left(\frac{\pi}{2} - t\right) = \sin t$$

Finally, we note that both sine and cosine are periodic functions with period 2π, that is,

$$\sin(t + 2\pi) = \sin t$$
$$\cos(t + 2\pi) = \cos t$$

This follows from the fact that the points on the unit circle corresponding to t and $t + 2\pi$ are identical.

We have stated these identities for the number t. Of course, they are also valid if the number t is replaced by the angle θ. Also we may replace t by $\pi/2 - t$ (as we will do in part (d) of Example 5) or by $3t^2 - 5t$ or by any other expression. Trigonometric identities will play an increasingly important role as we continue our study. Here are two simple examples of how we use them.

■ **Example 4.** Given that $\pi/2 \le t \le \pi$ and $\sin t = 3/5$, find the exact value of $\cos t$.

Solution. We use the identity $\sin^2 t + \cos^2 t = 1$ to write

$$\cos^2 t = 1 - \sin^2 t$$
$$= 1 - \frac{9}{25} = \frac{16}{25}$$

Since cosine is negative in the second quadrant, we infer that

$$\cos t = -\sqrt{\frac{16}{25}} = -\frac{4}{5} \qquad ■$$

■ **Example 5.** Given that $\cos t = -1/\sqrt{5}$ and $\pi \le t \le 3\pi/2$, determine the exact value of each of the following.

(a) $\sin t$ (b) $\cos\left(\frac{\pi}{2} - t\right)$ (c) $\cos(t + 4\pi)$ (d) $\sin\left(t - \frac{\pi}{2}\right)$

Solution.
(a) $\sin^2 t = 1 - \cos^2 t$ and so

$$\sin t = -\sqrt{1 - (-1/\sqrt{5})^2} = -\sqrt{\frac{4}{5}} = \frac{-2}{\sqrt{5}}$$

(b) $\cos\left(\dfrac{\pi}{2} - t\right) = \sin t = -\dfrac{2}{\sqrt{5}}$

(c) $\cos(t + 4\pi) = \cos(t + 2\pi) = \cos t = \dfrac{-1}{\sqrt{5}}$

(d) $\sin\left(t - \dfrac{\pi}{2}\right) = \sin\left(-\left(\dfrac{\pi}{2} - t\right)\right) = -\sin\left(\dfrac{\pi}{2} - t\right) = -\cos t = \dfrac{1}{\sqrt{5}}$

Note that in the second step we used the identity $\sin(-t) = -\sin t$ with t replaced by $\pi/2 - t$. ∎

TEASER SOLUTION

We pair the terms as indicated below.

$$\sin 1° + \sin 2° + \sin 3° + \ldots + \sin 357° + \sin 358° + \sin 359°$$

From thinking about Figure 11, we deduce that the sum of each pair is 0; the middle term ($\sin 180°$) is also 0. We conclude that the sum of all the terms is 0.

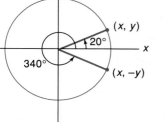

Figure 11

PROBLEM SET 4.3

A. Skills and Techniques

Use a calculator to evaluate each of the following, correct to five decimal places.

1. $\sin 13.7°$
2. $\cos 17.6°$
3. $\cos 7.6°$
4. $\sin 20.4°$
5. $\sin 13.7$
6. $\cos 17.6$
7. $\cos 7.6$
8. $\sin 20.4$
9. $\cos(-78.3°)$
10. $\sin(-123.3°)$
11. $\sin(649.3°)$
12. $\cos 3.57$
13. $\cos(-20.2)$
14. $\sin(-15.3)$

Use symmetry and your knowledge of special angles to find the exact value of each of the following. Confirm your answer with a calculator.

15. $\cos\left(\dfrac{2\pi}{3}\right)$

16. $\sin\left(\dfrac{2\pi}{3}\right)$

17. $\sin\left(-\dfrac{5\pi}{3}\right)$

18. $\cos\left(-\dfrac{9\pi}{4}\right)$

19. $\sin\left(\dfrac{15\pi}{4}\right)$

20. $\cos(19\pi)$

21. $\cos(-100\pi)$

22. $\sin\left(-\dfrac{11\pi}{4}\right)$

23. $\sin(-540°)$

24. $\cos(-480°)$

25. $\cos(780°)$

26. $\sin(855°)$

Without using a calculator, determine the sign (plus or minus) of each of the following

27. $\cos 357°$
28. $\sin 183°$
29. $\sin 550°$

30. cos 10
31. cos (−6.5)
32. sin (−12)

In each case, indicate which of the two expressions has the larger value. You should be able to do this without a calculator by thinking about the unit circle and the definitions of sine and cosine.

33. cos 1 or cos 1.2
34. sin 2 or sin 3
35. sin 3 or sin 3.2
36. cos 2 or cos 3
37. sin 1 or cos 1
38. sin 2 or cos 2
39. sin 3 or sin 3 cos 3
40. cos 3 or sin 3 cos 3

Determine the exact value of the first expression, given the information that follows it.

41. $\sin t; \cos t = \dfrac{5}{13}, 0 \le t \le \dfrac{\pi}{2}$

42. $\cos t; \sin t = \dfrac{5}{13}, \dfrac{\pi}{2} \le t \le \pi$

43. $\cos t; \sin t = -\dfrac{2}{5}, \dfrac{3\pi}{2} \le t \le 2\pi$

44. $\sin t; \cos t = -\dfrac{1}{3}, \pi \le t \le \dfrac{3\pi}{2}$

45. $\sin\left(t - \dfrac{\pi}{2}\right); \cos t = \dfrac{3}{4}$

46. $\cos\left(t - \dfrac{\pi}{2}\right); \cos t = \dfrac{3}{5}, 0 \le t \le \dfrac{\pi}{2}$

47. $\cos(t + 6\pi); \sin t = \dfrac{3}{5}, \dfrac{\pi}{2} \le t \le \pi$

48. $\sin(-t); \cos(t + 4\pi) = \dfrac{4}{5}, \dfrac{3\pi}{2} \le t \le 2\pi$

49. $(\sin t + \cos t)^2 - 2 \sin t \cos t; 0 \le t \le 2\pi$

50. $(\cos t - \sin t)^2; \sin t = \dfrac{3}{5}, -\dfrac{\pi}{2} \le t \le \dfrac{\pi}{2}$

B. Applications and Extensions

51. Let $P(t)$ denote the terminal point of a string of length $|t|$ having initial point (1, 0) and wrapped around the unit circle (counterclockwise if $t > 0$, clockwise if $t < 0$). Find the coordinates of $P(t)$.

   (a) $P\left(\dfrac{5\pi}{2}\right)$

   (b) $P(-19\pi)$

   (c) $P(7.15)$

   (d) $P(-15.32)$

52. The string of Problem 51 has terminal point $P(t) = (x, -1/2)$ in the third quadrant. Determine x and the value for t between 0 and 2π.

53. Determine the length of the line segment (Figure 12) connecting $P(\pi/6)$ and $P(-3\pi/4)$.

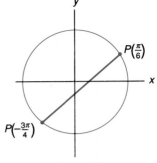

Figure 12

54. Find a value of t between 0 and $\pi/2$ where $\sin t = 2 \cos t$.

55. For what values of t satisfying $0 \le t \le 2\pi$ are the following true?

   (a) $\sin t = \cos t$

   (b) $\dfrac{1}{2} < \sin t < \dfrac{\sqrt{3}}{2}$

   (c) $\cos^2 t \ge 0.25$

   (d) $\cos^2 t > \sin^2 t$

56. Find the four smallest positive solutions to the following equations.

   (a) $\sin t = 1$

   (b) $|\cos t| = \dfrac{1}{2}$

   (c) $\cos t = -\dfrac{\sqrt{2}}{2}$

   (d) $\sin t = -\dfrac{\sqrt{3}}{2}$

57. Use Figure 13 to discover identities for $\sin(\pi + t)$ and $\cos(\pi + t)$.

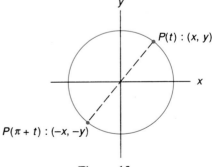

Figure 13

58. Draw a figure to convince yourself that $P(t)$ and $P(\pi - t)$ are symmetric with respect to the y-axis. State identities for $\sin(\pi - t)$ and $\cos(\pi - t)$.

59. If $P(t)$ has coordinates $(1/3, -2\sqrt{2}/3)$, find exact values for each of the following. You may need the results of Problems 57 and 58 for the later parts.

(a) $\sin(-t)$

(b) $\sin\left(\dfrac{\pi}{2} - t\right)$

(c) $\cos(2\pi + t)$

(d) $\cos(2\pi - t)$

(e) $\cos(\pi + t)$

(f) $\sin(\pi + t)$

(g) $\cos(\pi - t)$

(h) $\sin(\pi - t)$

60. Fill in all the blanks in Figure 14 with exact values.

61. Evaluate

$$\cos 1° + \cos 2° + \cos 3° + \ldots$$
$$+ \cos 357° + \cos 358° + \cos 359°$$

62. Challenge. Evaluate

$$\sin^2 1° + \sin^2 2° + \sin^2 3° + \ldots$$
$$+ \sin^2 357° + \sin^2 358° + \sin^2 359°$$

$\sin t$	$\cos t$	$\sin(t+\pi)$	$\cos(t+\pi)$	$\sin(\pi-t)$	$\sin(2\pi-t)$	Least positive value of t
$\sqrt{3}/2$	$-\frac{1}{2}$					
	$\sqrt{2}/2$	$\sqrt{2}/2$				
$-\frac{1}{2}$			$-\sqrt{3}/2$			
-1						
			$\sqrt{3}/2$		$\frac{1}{2}$	
	0				-1	

Figure 14

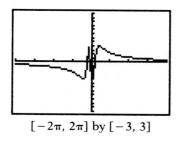

$[-2\pi, 2\pi]$ by $[-3, 3]$

TEASER   One of the strangest of graphs is that of $f(x) = \sin(1/x)$. Where does this graph cross the x-axis? Try graphing this function on your calculator using different domains. Then draw a hand sketch.

4.4 GRAPHS OF SINE AND COSINE FUNCTIONS

We want now to study sine and cosine as functions, functions whose domain is the set $\mathbb{R}$ of all real numbers. We intend to look at their basic graphs and then learn how to modify them by changing certain parameters. These two functions are the premier examples of periodic functions, that is, functions that continually repeat themselves. Therefore we want to study periodicity—both as an abstract concept and as a phenomenon of nature.

To begin, let us clarify a point of possible confusion. So far in this chapter, we have used t as the domain variable and have defined $\sin t$ and $\cos t$. Moreover, we have defined these functions in terms of the coordinates (x, y) of a point

TEST YOUR INTUITION

You are riding a Ferris wheel of radius 30 feet that is turning counter-clockwise at 1 radian per second. The center of the wheel is at the origin of the coordinate system and you are at the point (30, 0) at time $t = 0$. (a) Sketch a rough graph that represents the y-coordinate of your nose at time t. (b) Do the same for the case where the wheel stopped for 10 seconds to let off a passenger exactly 2 revolutions after you were at (30, 0).

After studying this lesson, you should find these tasks rather easy. See Problem 41 for related material.

on the unit circle. It would seem logical therefore to draw the graph of $u = \sin t$ in the tu-plane. But tradition is very strong; it is conventional to draw the graph of $y = \sin x$ in the xy-plane. Of course, the equations $u = \sin t$ and $y = \sin x$ determine the same ordered pairs; thus, their graphs are identical except for the names we use. But we must warn you: Do not confuse the usage of x and y in this section with their usage in Section 4.3.

THE GRAPHS OF $Y = \sin X$ AND $Y = \cos X$

Before we draw the graphs of these equations on a calculator, we should think about how we would construct hand-drawn graphs. We would first create a table of values (Figure 1). Then we would plot the corresponding points and connect them with a smooth curve (Figure 2). Finally we would extend the graph as far as desired in either direction using periodicity. Note that it is convenient to use π as the basic unit on the x-axis.

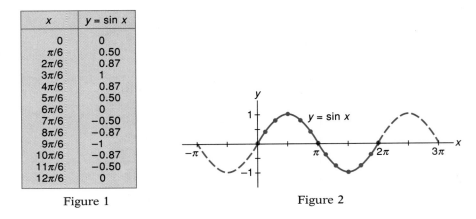

x	$y = \sin x$
0	0
$\pi/6$	0.50
$2\pi/6$	0.87
$3\pi/6$	1
$4\pi/6$	0.87
$5\pi/6$	0.50
$6\pi/6$	0
$7\pi/6$	-0.50
$8\pi/6$	-0.87
$9\pi/6$	-1
$10\pi/6$	-0.87
$11\pi/6$	-0.50
$12\pi/6$	0

Figure 1

Figure 2

Of course, it is much easier to let a calculator do the work for us. Your calculator probably has a special procedure for setting range values appropriate for the trigonometric functions. (On the TI-81 or TI-82, pressing ZOOM and selecting 7 sets the horizontal window size to decimal equivalents of $[-2\pi, 2\pi]$.)

■ **Example 1.** Use a calculator to graph $y = \sin x$ and $y = \cos x$ for $-2\pi \leq x \leq 2\pi$.

Solution. The graphs are shown in Figures 3 and 4, respectively. ■

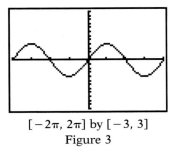

$[-2\pi, 2\pi]$ by $[-3, 3]$
Figure 3

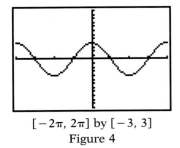

$[-2\pi, 2\pi]$ by $[-3, 3]$
Figure 4

Let us observe some things from the graphs in Figures 2, 3, and 4.

1. The values of both functions vary between −1 and 1.
2. Sine is an odd function (the graph is symmetric with respect to the origin); cosine is an even function (the graph is symmetric with respect to the y-axis).
3. The graph of $y = \sin x$ has the same shape as that of $y = \cos x$ but is shifted $\pi/2$ units right.
4. Both functions are periodic, with period 2π.

Actually, we encountered each of these facts in Section 4.3, where they appeared as boxed results. Using one of our trigonometric identities allows us to demonstrate statement 3 algebraically.

$$\cos\left(t - \frac{\pi}{2}\right) = \cos\left(-\left(\frac{\pi}{2} - t\right)\right) = \cos\left(\frac{\pi}{2} - t\right) = \sin t$$

THE GRAPHS OF $Y = A \sin (Bx + C)$ AND $Y = A \cos (Bx + C)$

Once we have mastered the basic graphs of $y = \sin x$ and $y = \cos x$, we can consider various transformations of these graphs. We will do this in stages.

Consider first the graph of $y = A \sin x$. The period of this graph is still 2π, but the graph is stretched vertically by a factor of $|A|$. Changing A from positive to negative reflects the graph across the x-axis.

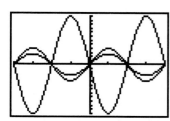

$[-2\pi, 2\pi]$ by $[-3, 3]$

Figure 5

■ **Example 2.** Draw the graphs of $y = \sin x$, $y = (2/3) \sin x$, and $y = -3 \sin x$ in the same plane.

Solution. The three graphs are shown in Figure 5. ■

We define the **amplitude** of a periodic function to be the magnitude of its variation from its median position; that is, the amplitude is $(M - m)/2$ where M and m are the largest and smallest values of the function. Thus, we see that the function $f(x) = A \sin x$ has amplitude $|A|$. What we have said for the sine function holds as well for $g(x) = A \cos x$.

■ **Example 3.** Determine the amplitudes of each of the following periodic functions.

(a) $f(x) = 4 \cos x$ (b) $g(x) = 2 + 4 \cos x$

Solution. Both have amplitude 4. The graph of the second has the same shape as that of the first, but is shifted upward 2 units. You may want to check this by using a calculator to draw the two graphs. ■

Next we consider the graph of $y = \sin Bx$, where B is a positive number. Changing B shrinks or expands the graph horizontally.

■ **Example 4.** Use a calculator to draw the graphs of $y = \sin x$, $y = \sin 2x$, and $y = \sin (x/2)$.

Solution. The three graphs, drawn to the same scale, are shown in Figures 6, 7, and 8. Note that the periods are 2π, π, and 4π, respectively. ∎

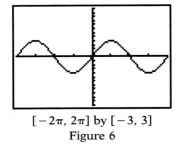

$[-2\pi, 2\pi]$ by $[-3, 3]$

Figure 6

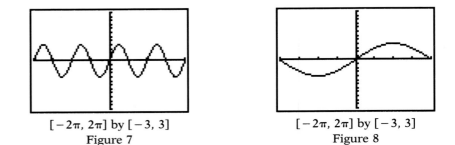

$[-2\pi, 2\pi]$ by $[-3, 3]$

Figure 7

$[-2\pi, 2\pi]$ by $[-3, 3]$

Figure 8

The **period** p of a periodic function f is the smallest positive number p such that $f(x + p) = f(x)$ for all x in the domain of f. After thinking about the problems in Example 4, we assert that $f(x) = \sin Bx$ with $B > 0$ has period $2\pi/B$. The same assertion holds for $g(x) = \cos Bx$.

Next consider $g(x) = \sin (Bx + C) = \sin (B(x + C/B))$ where again $B > 0$. From what we learned about shifts in Section 2.8, we know that the graph of g has the same shape as that of $f(x) = \sin Bx$ but is shifted $-C/B$ units (left if $C > 0$ and right if $C < 0$). The number $-C/B$ is called the **phase shift** of g. What is true for the sine is also true for the cosine.

■ **Example 5.** Explain how the graph of $y = \cos (x/2 + \pi/4)$ is related to the graph of $y = \cos (x/2)$. Draw both graphs in the same plane to confirm the answer.

Solution. The phase shift of the first graph is

$$-\frac{\pi/4}{1/2} = -\frac{\pi}{2}$$

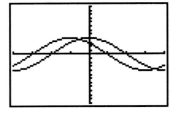

$[-2\pi, 2\pi]$ by $[-3, 3]$

Figure 9

This means that the first graph is shifted $\pi/2$ units left as compared with the second graph. This shift is confirmed in Figure 9. ∎

Finally, we are ready to put these ideas together.

■ **Example 6.** Determine the amplitude, period, and phase shift for $y = 3 \sin (x/2 - \pi/4)$. Make a hand-drawn graph of this equation and then confirm it by drawing the graph with a calculator.

Solution. The three key numbers are as follows:

Amplitude: 3

Period: $\dfrac{2\pi}{B} = \dfrac{2\pi}{1/2} = 4\pi$

Phase shift: $-\dfrac{C}{B} = \dfrac{\pi/4}{1/2} = \dfrac{\pi}{2}$

To draw the graph by hand, we first draw the graph of $y = 3 \sin (x/2)$; then we shift it $\pi/2$ units right (Figure 10). The corresponding graph drawn on a calculator is shown in Figure 11. ∎

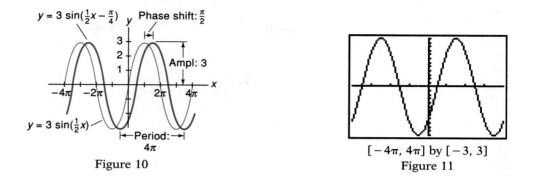

Figure 10

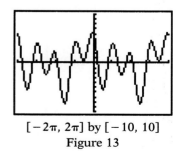

$[-4\pi, 4\pi]$ by $[-3, 3]$

Figure 11

Many phenomena that occur in the physical world display the behavior typified by the graph of $y = A \sin (Bx + C)$ in which x measures time. We mention in particular the action of a vibrating spring, the voltage of an alternating current, and the vertical motion of a point on a spinning wheel. The problem set will explore some of these ideas.

OTHER SINE AND COSINE GRAPHS

There is no end to the experimenting we can do with various combinations of sines and cosines, using the capabilities of a graphing calculator.

∎ **Example 7.** Draw the graphs of $y = 4 \sin 3x - 2 \cos 2x$ and $y = 2 \cos x - 3 \sin 2x + 4 \cos 4x$. Use these graphs to determine the amplitude and period in each case.

Solution. The graphs are shown in Figures 12 and 13. The first has amplitude $(6.00 + 5.15)/2 = 5.58$ and period 2π; the second has amplitude $(6.27 + 8.43)/2 = 7.35$ and period 2π. ∎

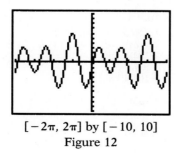

$[-2\pi, 2\pi]$ by $[-10, 10]$

Figure 12

$[-2\pi, 2\pi]$ by $[-10, 10]$

Figure 13

191 • 4.4 Graphs of Sine and Cosine Functions

Recall that $\sin t = 0$ for $t = n\pi$, where n is any integer. Thus the graph of $y = \sin (1/x)$ crosses the x-axis whenever $1/x$ is an integer multiple of π, that is, at

$$x = \pm\frac{1}{\pi}, \pm\frac{1}{2\pi}, \pm\frac{1}{3\pi}, \pm\frac{1}{4\pi}, \ldots$$

A calculator can give only a poor imitation of the graph (Figure 14). The hand-drawn graph in Figure 15 gives a better but still inadequate picture of this very complicated graph. It has infinitely many wiggles in any neighborhood of the origin.

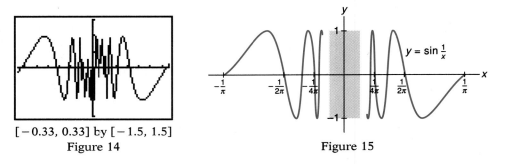

$[-0.33, 0.33]$ by $[-1.5, 1.5]$
Figure 14

Figure 15

PROBLEM SET 4.4

A. Skills and Techniques

In each case, determine the period, amplitude, and phase shift. Then sketch the graph on the interval $-2\pi \leq x \leq 2\pi$ by hand. Finally confirm your graph by using a graphics calculator.

1. $y = 2 \cos 2x$

2. $y = \dfrac{2}{3} \sin 3x$

3. $y = \dfrac{3}{2} \sin \dfrac{1}{3}x$

4. $y = 4 \sin \dfrac{2}{3}x$

5. $y = -1.5 \sin 3x$

6. $y = -2.5 \cos 4x$

7. $y = \sin \left(4x + \dfrac{\pi}{2}\right)$

8. $y = 3 \cos \left(x + \dfrac{\pi}{4}\right)$

9. $y = 2 \cos \left(2x - \dfrac{\pi}{3}\right)$

10. $y = 2 \sin \left(\dfrac{1}{2}x - \dfrac{\pi}{6}\right)$

11. $y = 2 + 2 \cos (3x)$

12. $y = -3 + \cos \left(3x - \dfrac{\pi}{2}\right)$

Determine A, B, and D so that each of the following is the graph of $y = D + A \sin Bx$. Assume that $B > 0$.

13.

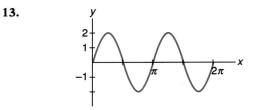

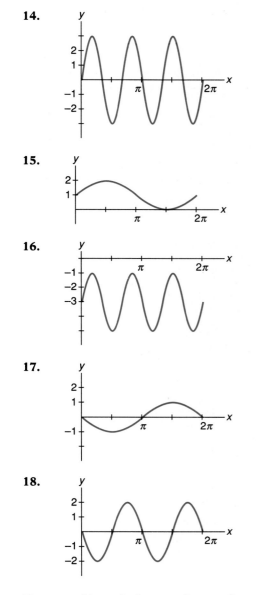

14.

15.

16.

17.

18.

Use a graphics calculator to draw each graph. From this graph, determine the period and amplitude of the given periodic function.

19. $y = 3 \sin x + \cos 2x$

20. $y = -2 \cos 2x + 3 \sin 4x$

21. $y = 0.7 \sin \left(\frac{1}{2}x\right) - 0.3 \sin x + 1.3 \cos 2x$

22. $y = -4 + 4 \sin \left(\frac{1}{3}x\right) + 2 \cos x$

23. $y = 3 \cos^2 x$
24. $y = 2 \sin^2 2x + \cos 4x$
25. $y = 3 \sin 2x \cos x$
26. $y = 1 - \cos 3x$

Use a graphics calculator to find all solutions of the given equation on $0 \le x \le 2\pi$, accurate to three decimal places.

27. $2 \sin x = x$
28. $\sin x = 2 \cos x$
29. $4 \cos \frac{1}{2}x = -3 + \sin x$
30. $2 \sin \frac{1}{2}x = -3x + 5$

B. Applications and Extensions

31. A weight attached to a spring (Figure 16) is bobbing up and down so that $y = 8 + 4 \cos (1.5t + 0.9)$ where y and t are measured in feet and seconds, respectively. It is said to be exhibiting **simple harmonic motion.** What is the farthest the weight gets from the ceiling, and when does this first happen for $t > 0$?

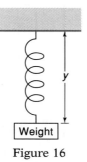

Figure 16

32. A cork is bobbing up and down in a lake so that its height y (in centimeters) at time t (in seconds) is given by $y = 10 \sin (3.1t - 0.5)$. Determine the lowest level of the cork and the time at which it first reaches this level after $t = 0$.

33. Friction in the spring and air resistance will cause the amplitude of the motion of the weight of Problem 31 to decrease with time **(damped harmonic motion).** A typical formula for y in this case is

$$y = 8 + 4e^{-0.16t} \cos (1.5t + 0.9)$$

Graph this motion on $0 \le t \le 25$. Then determine during $t \ge 1$ the farthest the weight gets from the ceiling and when this happens.

34. Draw the graph of $y = 2^{-0.1t} \cos 2t$ on $0 \le t \le 25$ and use it to determine the least t_0 so that $y \le 0.25$ for $t \ge t_0$.

35. The motion of a certain <u>clock</u> pendulum (Figure 17) satisfies $\theta = 0.2 \cos(t\sqrt{32/L})$, where t is measured in seconds and L is in feet. (a) Write a formula for T, the time for a complete oscillation. (b) Determine L so that this oscillation will require exactly 1 second.

Figure 17

36. In predator-prey systems, the number of predators and the number of prey tend to vary periodically. In a certain region the number of coyotes was given by $C = 200 + 50 \sin(2t - 0.7)$ and the number of rabbits by $R = 1000 + 150 \sin 2t$, where t is measured in years. (a) Graph C and R together on the interval $0 \le t \le 6$. (b) Determine the length of the time interval after the rabbit population reaches its maximum until the coyote population reaches its maximum.

37. The voltage drop E across the terminals in a certain alternating current circuit is approximately $E = 156 \sin(110\pi t)$, where t is in seconds. Determine the **frequency**, that is, the number of complete cycles per second.

38. The carrier wave for the radio wave of a certain FM station has the form $y = A \sin(2\pi \cdot 10^8 t)$, where t is measured in seconds. What is the frequency for this wave?

39. Try graphing $y = \sin 140x$ for the range values $[-2\pi, 2\pi]$ by $[-3, 3]$. Then use the range values $[-1, 1]$ by $[-3, 3]$. Determine range values that will model the behavior of this function correctly. (*Note:* Let this problem serve to warn you against unthinking reliance on your calculator.)

40. A wheel with center at the origin is rotating counterclockwise at 4 radians per second. A free-hanging shaft 8 centimeters long is attached to the wheel at a point 5 centimeters from the center, and this point is initially at (5, 0) as shown in Figure 18. Determine the coordinates of P at time t, assuming that the shaft remains vertical.

41. A Ferris wheel with radius 30 feet and center at the origin is turning counterclockwise at 1 radian per second. Determine the y-coordinate of a seat at time t seconds if that seat was at (30, 0) at time $t = 0$.

42. Rework Problem 41 assuming that the wheel turns at 2 radians per second and that the seat in question was at $(0, -30)$ at $t = 0$.

43. Consider the wheel-piston device shown in Figure 19 (which is analogous to the crankshaft and

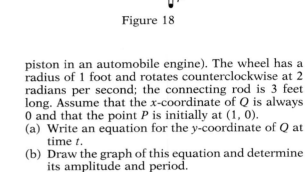

Figure 18

piston in an automobile engine). The wheel has a radius of 1 foot and rotates counterclockwise at 2 radians per second; the connecting rod is 3 feet long. Assume that the x-coordinate of Q is always 0 and that the point P is initially at (1, 0).
(a) Write an equation for the y-coordinate of Q at time t.
(b) Draw the graph of this equation and determine its amplitude and period.

44. Repeat Problem 43 assuming that the wheel has radius 1.5, rotates at 60 revolutions per second, and point P is at (1.5, 0) initially.

45. Use a calculator to draw the graph of $y = x - [\![x]\!]$ on $0 \le x \le 4$ and determine its period and amplitude. Here $[\![\ ]\!]$ denotes the greatest integer function (see Section 2.9).

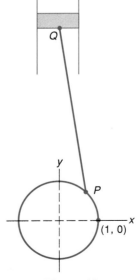

Figure 19

46. On the basis of Problem 45 and without graphing, determine the period and amplitude of the graph of $y = 2(3x - [\![3x]\!])$. Confirm your answer by drawing its graph.

47. Draw the graph of $e^{\cos 2x}$ and determine its period and amplitude.

48. How many solutions does $\sin(e^x) = 0.5$ have on $0 \le x \le 4$?

49. Draw the graph of $y = (\sin 3x)/x$ on $-2\pi \le x \le 2\pi$ and determine what happens to y as x approaches 0. Experiment with various values of B and thereby conjecture what happens to y as x approaches 0 in $y = (\sin Bx)/x$.

50. Find the slope of the tangent line to $y = \sin x$ at $x = 1$ and at $x = 2$ (see Section 2.3).

51. Draw the graph of the slope function

$$m(x) = \frac{\sin(x + 0.01) - \sin(x - 0.01)}{0.02}$$

Then make a conjecture about this slope function.

52. **Challenge.** How many solutions does $\sin(1/x) = x$ have on the interval $0.02 \le x \le 4$? On the interval $x > 0$?

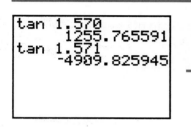

TEASER   Look at the screen at the left, which shows the values of tan 1.570 and tan 1.571 to be of opposite sign and far apart. Is something wrong with the calculator? Explain.

4.5   FOUR MORE TRIGONOMETRIC FUNCTIONS

Certainly, the sine and the cosine are the most important trigonometric functions. If you master their graphs and their properties, you will be in good shape for the uses of trigonometry that arise in science and more advanced mathematics. In fact, the four trigonometric functions to be introduced next are simply special quotients of sine and cosine that arise often enough to be given special names. We refer to the tangent, cotangent, secant, and cosecant functions, which we now define using obvious abbreviations.

DEFINITION

$$\tan x = \frac{\sin x}{\cos x} \qquad \cot x = \frac{\cos x}{\sin x} = \frac{1}{\tan x}$$

$$\sec x = \frac{1}{\cos x} \qquad \csc x = \frac{1}{\sin x}$$

Of course, we must omit from the domains of these new functions any values of x that make a denominator 0.

Most calculators have a key for the tangent function, but cotangent, secant, and cosecant must be calculated from their definitions.

■ **Example 1.**   Use a calculator to evaluate each of the following.

(a) sec 2.34   (b) csc 3.01   (c) tan 188°   (d) cot 856°

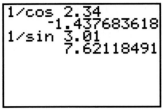

```
1/cos 2.34
       -1.437683618
1/sin 3.01
        7.62118491
```

Figure 1

Solution. Note the screen in Figure 1, which shows how to calculate the first two values.

(a) sec 2.34 = −1.437683618
(b) csc 3.01 = 7.62118491
(c) tan 188° = 0.1405408347
(d) cot 856° = −1.035530314 ■

SOME IMPORTANT IDENTITIES

First, we note two Pythagorean-type relationships.

$$1 + \tan^2 x = \sec^2 x$$
$$1 + \cot^2 x = \csc^2 x$$

These identities are easy to justify using a simple chain of equalities. To prove the first, we write

$$1 + \tan^2 x = 1 + \left(\frac{\sin x}{\cos x}\right)^2 = 1 + \frac{\sin^2 x}{\cos^2 x} = \frac{\cos^2 x + \sin^2 x}{\cos^2 x}$$

$$= \frac{1}{\cos^2 x} = \left(\frac{1}{\cos x}\right)^2 = \sec^2 x$$

The second is proved similarly.

■ **Example 2.** Given that $\tan x = 5/12$ and $\pi < x \le 3\pi/2$, determine exact values for each of the other trigonometric functions at x.

Solution. We make use of identities that we have learned and keep in mind that sin, cos, sec, and csc are all negative in the third quadrant.

(a) $\cot x = \dfrac{1}{\tan x} = \dfrac{12}{5}$

(b) $\sec x = -\sqrt{1 + \tan^2 x} = -\sqrt{1 + 25/144} = -\sqrt{169/144} = -\dfrac{13}{12}$

(c) $\cos x = \dfrac{1}{\sec x} = -\dfrac{12}{13}$

(d) $\sin x = -\sqrt{1 - \cos^2 x} = -\sqrt{1 - 144/169} = -\dfrac{5}{13}$

(e) $\csc x = \dfrac{1}{\sin x} = -\dfrac{13}{5}$ ■

Here are more cofunction identities to go with those that we learned earlier for sin and cos.

$$\tan\left(\frac{\pi}{2} - x\right) = \cot x \qquad \cot\left(\frac{\pi}{2} - x\right) = \tan x$$

$$\sec\left(\frac{\pi}{2} - x\right) = \csc x \qquad \csc\left(\frac{\pi}{2} - x\right) = \sec x$$

AN ALTERNATIVE

Some people may find it easier to remember the two Pythagorean-type identities for tangent and cotangent if they derive them as follows. Start with the familiar identity

$$\sin^2 x + \cos^2 x = 1$$

Divide both sides by $\cos^2 x$ to get the tangent identity and divide both sides by $\sin^2 x$ to get the cotangent identity.

We may summarize these identities in one general identity for cofunctions. If F and COF are cofunctions, then

$$F\left(\frac{\pi}{2} - x\right) = COF(x), \qquad F(90° - \theta) = COF(\theta)$$

To demonstrate the cofunction identity for the tangent, proceed as follows.

$$\tan\left(\frac{\pi}{2} - x\right) = \frac{\sin\left(\dfrac{\pi}{2} - x\right)}{\cos\left(\dfrac{\pi}{2} - x\right)} = \frac{\cos x}{\sin x} = \cot x$$

Each of the others is proved in a similar manner.

■ **Example 3.** Given that $\sin 22° = a$ and $\cos 74° = b$, express

$$K = \sin^2 74° - \sin^2 376° \sec^2 74° - \cos^2 68°$$

in terms of a and b and simplify.

Solution. Note first that $\cos 68° = \sin 22° = a$ and $\sin 16° = \cos 74° = b$. Thus,

$$K = 1 - \cos^2 74° - \frac{\sin^2 16°}{\cos^2 74°} - \sin^2 22°$$

$$= 1 - b^2 - \frac{b^2}{b^2} - a^2 = -b^2 - a^2 \qquad ■$$

Next, we claim that tan, cot, and csc are odd functions, whereas sec is an even function; that is,

$\tan(-x) = -\tan x$	$\cot(-x) = -\cot x$
$\sec(-x) = \sec x$	$\csc(-x) = -\csc x$

The justification is straightforward. For example,

$$\cot(-x) = \frac{\cos(-x)}{\sin(-x)} = \frac{\cos x}{-\sin x} = -\cot x$$

GRAPHS OF TANGENT AND COTANGENT FUNCTIONS

We expect that the graph of $y = \tan x = (\sin x)/\cos x$ will have vertical asymptotes at x-values where $\cos x = 0$, that is, at $x = \pm\pi/2, \pm 3\pi/2$, and so on. To make a hand-drawn graph, we first draw in dotted vertical lines at these places. Then we prepare a table (Figure 2) of ordered pairs $(x, \tan x)$ and plot the corresponding points. Finally, we connect these points with a smooth curve. The result is shown in Figure 3.

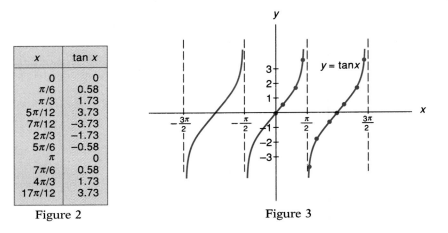

x	$\tan x$
0	0
$\pi/6$	0.58
$\pi/3$	1.73
$5\pi/12$	3.73
$7\pi/12$	-3.73
$2\pi/3$	-1.73
$5\pi/6$	-0.58
π	0
$7\pi/6$	0.58
$4\pi/3$	1.73
$17\pi/12$	3.73

Figure 2

Figure 3

Of course, we can obtain the graph of $y = \tan x$ much more easily by using a calculator (in radian mode). We show the result both in dot mode and in connected mode (Figures 4 and 5). In connected mode, solid vertical lines appear where the asymptotes belong. They represent the calculator's attempt to produce a connected graph. Keep in mind that asymptotes are not part of the graph of $y = \tan x$; they are just guidelines for the graph. For this reason, we prefer to have our calculator draw graphs that have jumps in dot mode.

$[-2\pi, 2\pi]$ by $[-3, 3]$

Figure 4

$[-2\pi, 2\pi]$ by $[-3, 3]$

Figure 5

We observe several facts about the tangent function from these graphs.

1. The domain is $\{x: x \neq \pm\pi/2, \pm 3\pi/2, \pm 5\pi/2, \ldots\}$.
2. The range is the set $\mathbb{R}$ of all real numbers.
3. The period is π, because $\tan (x + \pi) = \tan x$ and π is the smallest positive number with this property. We will demonstrate this algebraically later.

■ **Example 4.** Draw the graph of $y = \tan (x/2)$ and determine its period.

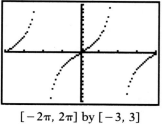

[-2π, 2π] by [-3, 3]

Figure 6

Solution. Figure 6 displays the graph in dot mode. The period is 2π.　■

■ **Example 5.**　Draw the graph of $y = \cot x = (\cos x)/\sin x$ in both dot mode and connected mode.

Solution. The required graphs are in Figures 7 and 8. Note that there are asymptotes at $x = n\pi$ when n is an integer. This is because $\sin x = 0$ at these values.　■

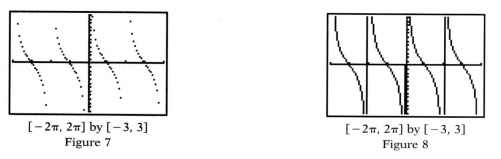

[-2π, 2π] by [-3, 3]
Figure 7

[-2π, 2π] by [-3, 3]
Figure 8

GRAPHS OF SECANT AND COSECANT

Since $\sec x = 1/\cos x$ the graph of $y = \sec x$ will have vertical asymptotes at odd multiples of $\pi/2$, just as the graph of the tangent did. We use a calculator to produce the graph in both dot mode and connected mode (Figures 9 and 10).

From these graphs, we note the following facts about the secant function.

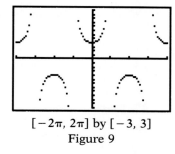

[-2π, 2π] by [-3, 3]
Figure 9

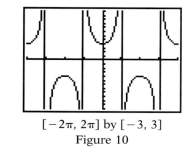

[-2π, 2π] by [-3, 3]
Figure 10

1.　The domain is $\{x: x \neq \pm\pi/2, \pm 3\pi/2, \pm 5\pi/2, \ldots\}$.
2.　The range is $\{y: |y| \geq 1\}$.
3.　The period is 2π.

■ **Example 6.**　Draw the graph of $y = \csc x$ and determine its domain, range, and period.

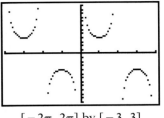

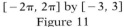

$[-2\pi, 2\pi]$ by $[-3, 3]$

Figure 11

Solution. Figure 11 shows the graph in dot mode.

1. The domain is $\{x: x \neq n\pi, n \text{ an integer}\}$.
2. The range is $\{y: |y| \geq 1\}$.
3. The period is 2π. ■

TEASER SOLUTION

Now that we have seen the graph of the tangent function, there is nothing mysterious about the fact that $\tan 1.570 \approx 1256$ and $\tan 1.571 \approx -4910$. For when x goes from 1.570 to 1.571, x passes over the point $x = \pi/2 \approx 1.570796$, at which there is a vertical asymptote for the graph. In fact, $\tan x \to \infty$ as $x \to \pi/2^-$ and $\tan x \to -\infty$ as $x \to \pi/2^+$.

PROBLEM SET 4.5

A. Skills and Techniques

Evaluate each of the following, giving your answers accurate to five decimal places.

1. $\cot 46.2°$
2. $\tan 34.3°$
3. $\sec 453°$
4. $\csc 512°$
5. $\tan 14.2$
6. $\cot (-2.545)$
7. $\csc (-3.123)$
8. $\sec (-1.59)$
9. $\sec (\tan 1.5)$
10. $\csc (\csc 4.3)$
11. $\dfrac{\sec 2.7 + \tan 1.4}{\csc 2.2}$
12. $\dfrac{\csc 5.2 - \cot 1.2}{\sec 0.9}$
13. $\csc^2 2.56 - \cot^2 1.23$
14. $\left(\dfrac{\sec 2.3}{\tan 4.1}\right)^3$

Use the given information and identities to obtain the exact values of the other five trigonometric functions at the number x.

15. $\sin x = -\dfrac{3}{5}$ and $\pi \leq x \leq \dfrac{3\pi}{2}$
16. $\cos x = -\dfrac{4}{5}$ and $\dfrac{\pi}{2} \leq x \leq \pi$
17. $\cot x = \dfrac{5}{12}$ and $0 \leq x \leq \dfrac{\pi}{2}$
18. $\sec x = \dfrac{13}{5}$ and $\dfrac{3\pi}{2} \leq x \leq 2\pi$

Assume that $\cos 74° = a$ and $\sin 23° = b$. Use identities to express K in terms of a and b.

19. $K = \sin 16° \sin 383° \sin^2 74°$
20. $K = \sec 74° \sin (-23°) \tan^2 23°$
21. $K = 1 + \tan^2 74° + \csc^2 23°$
22. $K = \csc^2 74° - \tan^2 74°$
23. $K = \sin 74° \tan 74° \sec 74°$
24. $K = \cos 23° \cot 23° \csc 23°$

Use algebra and identities to simplify each of the following.

25. $(\csc x + \cot x)(\csc x - \cot x)$
26. $\dfrac{\tan^2 x}{\sec^2 x} + \dfrac{\cot^2 x}{\csc^2 x}$
27. $\tan x + \cot x - \sec x \csc x$
28. $\dfrac{\tan x}{\sec x + 1} + \dfrac{\tan x}{\sec x - 1}$
29. $\dfrac{\cot^4 x - \csc^4 x}{\cot^2 x + \csc^2 x}$
30. $\dfrac{3 \cot x + \csc x}{\csc x} - 1$

In each case, use a graphics calculator to draw the graph of the given function for at least one full period. Use either the connected or the dot mode, whichever you think gives the best representation. Then determine the range and period.

31. $f(x) = \tan 2x$

32. $f(x) = \sec 3x$

33. $f(x) = \csc \dfrac{1}{2} x$

34. $f(x) = \cot \dfrac{1}{3} x$

35. $f(x) = \sec \left(x - \dfrac{\pi}{6} \right)$

36. $f(x) = \tan \left(\dfrac{1}{2} x - \dfrac{\pi}{6} \right)$

37. $f(x) = 3 \sec 2x$

38. $f(x) = \dfrac{3}{2} \csc \dfrac{1}{2} x$

39. $f(x) = 2 \tan \pi x$

40. $f(x) = -\cot 2\pi x$

41. $f(x) = \cot (\cos x)$

42. $f(x) = \tan (\sin x)$

43. $f(x) = \tan^2 x + 2 \tan x - 1$

44. $f(x) = 3 \sec x - \tan x$

45. $f(x) = \sec \dfrac{1}{2} x - 2 \sin 3x$

46. $f(x) = \sec^2 x - 3 \sec x$

B. Applications and Extensions

47. Write each of the following in terms of sines and cosines and simplify.

(a) $\dfrac{\sec \theta \csc \theta}{\tan \theta + \cot \theta}$

(b) $(\tan \theta)(\cos \theta - \csc \theta)$

(c) $\dfrac{(1 + \tan \theta)^2}{\sec^2 \theta}$

(d) $\dfrac{\sec \theta \cot \theta}{\sec^2\theta - \tan^2\theta}$

(e) $\dfrac{\cot \theta - \tan \theta}{\csc \theta - \sec \theta}$

(f) $\tan^4\theta - \sec^4\theta$

48. Draw the graph of each of the following and use it to conjecture an identity. Then prove your conjecture.

(a) $\sec x - \sin x \tan x$ (b) $\dfrac{\sin x + \tan x}{1 + \sec x}$

49. If $\tan \theta = 12/5$ and $\csc \theta < 0$, evaluate $\csc \theta - \sec \theta$.

50. If $\cot \theta = -4/3$ and $\sec \theta < 0$, evaluate $\sec^3\theta + \sin^3\theta$.

51. Use the identities $\sin (t + \pi) = -\sin t$ and $\cos (t + \pi) = -\cos t$ (taken from **Problem 57 of Section 4.3**) to find corresponding identities for

(a) $\tan (t + \pi)$

(b) $\cot (t + \pi)$

(c) $\sec (t + \pi)$

(d) $\csc (t + \pi)$

What do you conclude from (a) and (b) about the periods of tangent and cotangent?

52. Draw the graphs of $y = \tan x$ and $y = \sec x$ for $0 \leq x < \pi/2$. Do these graphs cross? Justify your answer.

53. Solve the equation $\csc(x/2) = \sec x$ for $0 < x < \pi/2$.

54. Solve the equation $\csc x = 5 \csc\sqrt{x}$ for $0 < x < 2\pi$.

55. Find the minimum value of $\csc^2 x + \cos x$.

56. Determine the range of $f(x) = 64 \csc x + 27 \sec x$ for $0 < x < \pi/2$.

57. How many solutions does $\sin (\tan x) = 0.5$ have for $0 < x < \pi/2$?

58. Let a line (Figure 12) make a counterclockwise angle θ with the positive x-axis (θ is called the **angle of inclination** of the line). How are the slope m and angle of inclination related?

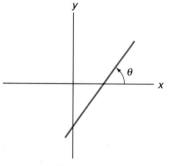

Figure 12

59. Find the equation of the line segment in Figure 13 (see Problem 58).

60. Find the coordinates of P in Figure 13.

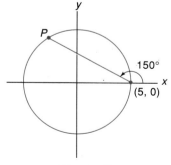

Figure 13

61. The face of a clock is in the *xy*-plane with center at the origin and 12 on the positive *y*-axis. Both hands of the clock are 5 units long.
(a) Find the equation of the line through the minute hand at 2:24.
(b) Find the equation of the line through the tips of the two hands at 12:50.

62. Consider the center of the earth, a sphere of radius 3960 miles, to be at the origin as shown in Figure 14. A pilot *h* units above the surface of the earth at $(3960 + h, 0)$ sights a bright light *P* on the horizon with the angle of depression 2.1°. Determine (a) *h*, (b) the equation of the pilot's line of sight, and (c) the distance between the pilot and the light *P*.

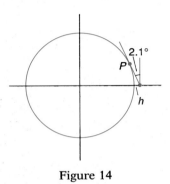

Figure 14

63. Express the length of the line segment *L* shown in Figure 15 in terms of the angle *θ*. Then find the minimum value of *L*. (*Note:* This minimum *L* is the length of the longest pipe that can be carried horizontally around the corner shown in Figure 15.)

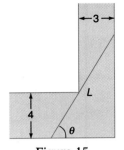

Figure 15

64. **Challenge.** Express the length *L* of the crossed belt that intersects in angle 2α and fits around wheels of radius *r* and *R* (Figure 16) in terms of *r*, *R*, and α.

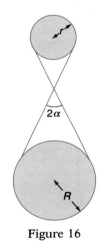

Figure 16

TEASER   Nancy is studying a painting that hangs flat against a museum wall. The painting is 4 feet high and its base is 8 feet above the floor. How far from the wall should Nancy stand for a maximum viewing angle *θ* if her eyes are 5 feet above the floor?

It would be a good idea to review Section 2.8 (Inverse Functions) before jumping into this section. Recall that only one-to-one functions have inverse functions. For a function with equation $y = f(x)$ to be one-to-one, its graph must meet every horizontal line in at most one point. The trigonometric functions are about as far from meeting this criterion as possible. Review their graphs and you will see that in each case, a typical horizontal line meets their graphs in infinitely many points. To make them have inverses, we must drastically restrict their domains.

THE INVERSE SINE

The graph of the sine function is shown in Figure 1. We want to restrict its domain in such a way that it still assumes its full range of values but takes on each of those values only once. There are many ways to do this, but one way has won universal acceptance. We restrict the domain to the interval $-\pi/2 \le x \le \pi/2$. The corresponding part of the graph is shown in solid color in Figure 1. When we write $\sin^{-1}$, it is always assumed that we have restricted the domain of the sine in this way.

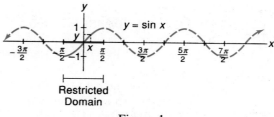

Restricted
Domain

Figure 1

A little thought (or looking at the graph of $y = \sin x$) should convince you that

$$\sin^{-1}\left(\frac{1}{2}\right) = \frac{\pi}{6} \approx 0.5236$$

$$\sin^{-1}(1) = \frac{\pi}{2} \approx 1.5708$$

$$\sin^{-1}\left(-\frac{1}{2}\right) = -\frac{\pi}{6} \approx -0.5236$$

$$\sin^{-1}(0) = 0$$

In general,

$$x = \sin^{-1} y \text{ if and only if } y = \sin x \text{ and } -\frac{\pi}{2} \le x \le \frac{\pi}{2}$$

Moreover,

$$\sin(\sin^{-1} y) = y \text{ for } -1 \le y \le 1$$

$$\sin^{-1}(\sin x) = x \text{ for } -\frac{\pi}{2} \le x \le \frac{\pi}{2}$$

Fortunately, most calculators have a $\sin^{-1}$ key that is programmed to be consistent with the definition just given. You should use your calculator to check the calculations above. Then try it on the following example.

■ **Example 1.** Evaluate each of the following, using radian measure.
(a) $\sin^{-1}(0.1234)$
(b) $\sin^{-1}(-0.8975)$
(c) $\sin^{-1}(\cos 2.12)$
(d) $\sin^{-1}(2.12)$
(e) $\sin^{-1}(\sin 1.012)$
(f) $\sin(\sin^{-1} 0.965)$

Solution.
(a) $\sin^{-1}(0.1234) = 0.1237153458$
(b) $\sin^{-1}(-0.8975) = -1.114067654$
(c) $\sin^{-1}(\cos 2.12) = -0.5492036732$
(d) $\sin^{-1}(2.12)$ gives an error message.
(e) $\sin^{-1}(\sin 1.012) = 1.012$
(f) $\sin(\sin^{-1} 0.965) = 0.965$

If these last three results surprise you, you have not been paying attention. ■

An alternative notation for $\sin^{-1} y$ is arcsin y (the arc on the unit circle whose sine is y). Thus arcsin $1 = \pi/2$.

We are interested in the graph of $y = \sin^{-1} x$. This is, of course, the graph of $y = \sin x$, restricted to $-\pi/2 \le x \le \pi/2$ and reflected across the line $y = x$. We show this in Figure 2. A corresponding calculator graph of $y = \sin^{-1} x$ appears in Figure 3.

Figure 2

$[-2, 2]$ by $[-2, 2]$
Figure 3

THE INVERSE COSINE

One look at the graph of the cosine should convince you that we should not restrict its domain to the same interval as that for the sine. Rather, we choose to restrict its domain to the interval $0 \leq x \leq \pi$, on which you will note that cosine is one-to-one (Figure 4).

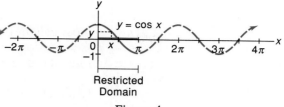

Figure 4

Consequently,

$$x = \cos^{-1} y \text{ if and only if } y = \cos x \text{ and } 0 \leq x \leq \pi$$

Thus,

$$\cos^{-1}\left(\frac{1}{2}\right) = \frac{\pi}{3} \quad \text{and} \quad \cos^{-1}(-1) = \pi$$

■ **Example 2.** Use a calculator to evaluate each of the following.

(a) $\cos^{-1}(0.567)$
(b) $\cos^{-1}(-0.987)$
(c) $\cos(\cos^{-1}(0.453))$
(d) $\cos^{-1}(\cos 6)$

Solution.

(a) $\cos^{-1}(0.567) = 0.9679370804$
(b) $\cos^{-1}(-0.987) = 2.980172303$
(c) $\cos(\cos^{-1}(0.453)) = 0.453$
(d) $\cos^{-1}(\cos 6) = 0.2831853072$

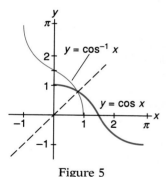

Figure 5

Does the answer to (d) surprise you? Note that 6 is not in the interval $0 \leq x \leq \pi$, on which $\cos^{-1}$ undoes cos; in other words, 6 is not in the domain of the restricted cosine. ■

The graphs of $y = \cos x$ (restricted) and $y = \cos^{-1} x$ are shown in Figure 5. They are reflections in the line $y = x$.

THE INVERSE TANGENT

The procedure that we use should be clear by now. The standard restriction for the domain of the tangent is the interval $-\pi/2 < x < \pi/2$. Thus

$$x = \tan^{-1} y \text{ if and only if } y = \tan x \text{ and } -\frac{\pi}{2} < x < \frac{\pi}{2}$$

This restriction and the corresponding graph of $y = \tan^{-1} x$ are shown in Figures 6 and 7.

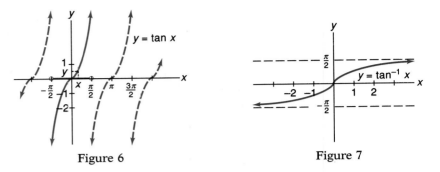

Figure 6   Figure 7

You can confirm the graph in Figure 7 on a calculator (Figure 8).

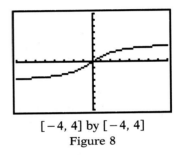

$[-4, 4]$ by $[-4, 4]$
Figure 8

THE INVERSE SECANT

Everything we have discussed so far regarding inverse trigonometric functions is standardized and agreed to by everyone. But authors disagree on how to form an inverse for the secant function. We choose to restrict the domain of the secant to $\{x: 0 \leq x \leq \pi, x \neq \pi/2\}$. The graph of this restricted secant function and its inverse are shown in Figures 9 and 10.

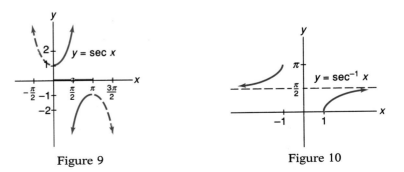

Figure 9   Figure 10

One of the advantages of our choice of restriction is the relation

$$\sec^{-1} x = \cos^{-1}\left(\frac{1}{x}\right)$$

It is this relation that we use to calculate inverse secants on a calculator, because calculators typically do not have a $\sec^{-1}$ key.

■ **Example 3.** Calculate (a) $\sec^{-1}(2.23)$ (b) $\sec^{-1}(-4)$.

Solution.

(a) $\sec^{-1}(2.23) = \cos^{-1}\left(\frac{1}{2.23}\right) = 1.105787721$

(b) $\sec^{-1}(-4) = \cos^{-1}\left(\frac{-1}{4}\right) = 1.823476582$ ■

The inverse cotangent and inverse cosecant play no role in calculus so we will not bother to define them.

INVERSE TRIGONOMETRIC FUNCTIONS AND TRIANGLES

For some applications, it is important to be able to relate the inverse trigonometric functions to right triangles.

■ **Example 4.** In each of the following right triangles, write θ explicitly in terms of x.

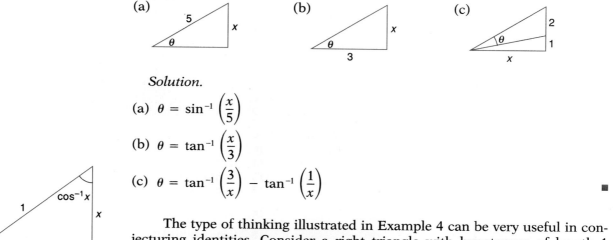

Solution.

(a) $\theta = \sin^{-1}\left(\frac{x}{5}\right)$

(b) $\theta = \tan^{-1}\left(\frac{x}{3}\right)$

(c) $\theta = \tan^{-1}\left(\frac{3}{x}\right) - \tan^{-1}\left(\frac{1}{x}\right)$ ■

The type of thinking illustrated in Example 4 can be very useful in conjecturing identities. Consider a right triangle with hypotenuse of length 1 (Figure 11). Call the length of one leg x; then the other leg has length $\sqrt{1 - x^2}$, by the Pythagorean theorem. The two acute angles are $\sin^{-1} x$ and $\cos^{-1} x$. Thus,

Figure 11

1. $\cos(\sin^{-1} x) = \sqrt{1 - x^2}$
2. $\sin(\cos^{-1} x) = \sqrt{1 - x^2}$
3. $\sin^{-1} x + \cos^{-1} x = \dfrac{\pi}{2}$

One question remains. Since our derivations of these relations were based on the narrow trigonometry of right triangles, are these relations genuine identities? For example, are they valid for negative values of x? The answer is yes. We demonstrate statement 1 algebraically.

Let $\theta = \sin^{-1} x$ and note that $-\pi/2 \leq \theta \leq \pi/2$ so that $\cos \theta \geq 0$. Thus, using the identity $\cos^2 \theta = 1 - \sin^2 \theta$, we obtain

$$\cos \theta = \sqrt{1 - \sin^2 \theta}$$

that is,

$$\cos(\sin^{-1} x) = \sqrt{1 - \sin^2 (\sin^{-1} x)}$$
$$= \sqrt{1 - x^2}$$

TEASER SOLUTION

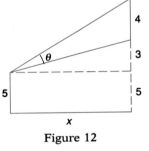

Figure 12

Refer to Figure 12 and note that

$$\theta = \tan^{-1}\left(\frac{7}{x}\right) - \tan^{-1}\left(\frac{3}{x}\right)$$

We wish to determine the value of x that maximizes θ, which we do by methods discussed in Section 2.4. Our conclusion is that θ reaches a maximum value of 0.41 radians (about 24°) when $x \approx 4.6$ feet.

PROBLEM SET 4.6

A. Skills and Techniques

Without using a calculator, evaluate each of the following exactly. If the given expression is undefined, say so. Keep in mind that $\sec^{-1}x = \cos^{-1}(1/x)$.

1. $\cos^{-1}\left(\dfrac{1}{2}\right)$

2. $\sin^{-1}\left(\dfrac{1}{2}\right)$

3. $\cos^{-1}\left(-\dfrac{1}{2}\right)$

4. $\sin^{-1}\left(-\dfrac{1}{2}\right)$

5. $\sin^{-1}(-1)$

6. $\cos^{-1}(1)$
7. $\cos^{-1}(-1)$
8. $\sin^{-1}(1)$
9. $\sin^{-1}(1.5)$
10. $\tan^{-1}(1)$
11. $\tan^{-1}(-1)$
12. $\tan^{-1}(\sqrt{3})$
13. $\sec^{-1}(2)$

14. $\sec^{-1}\left(\dfrac{1}{2}\right)$

15. $\sin(\sin^{-1}(0.3))$

16. $\sin\left(\cos^{-1}\left(\dfrac{\sqrt{3}}{2}\right)\right)$

17. $\tan^{-1}(\tan(0.30))$
18. $\cos^{-1}(\cos 2\pi)$
19. $\tan^{-1}(\tan 2\pi)$
20. $\sec^{-1}(\cos \pi)$

Use your calculator to evaluate each of the following. If it gives you an error message, be sure you understand why.

21. $\sin^{-1}(0.2346)$
22. $\cos^{-1}(0.7791)$
23. $\sin^{-1}(-0.2346)$
24. $\cos^{-1}(-0.7791)$
25. $\tan^{-1}(3.2451)$
26. $\tan^{-1}(-3.2451)$
27. $\sec^{-1}(-2.1234)$
28. $\sec^{-1}(2.1234)$
29. $\cos^{-1}(1.2012)$
30. $\sin^{-1}(-0.9988)$
31. $\tan^{-1}\pi$
32. $\sin^{-1}\pi$
33. $\sin(\cos^{-1}(0.23))$
34. $\tan(\sin^{-1}(0.97))$
35. $\sin^{-1}(\cos(8.27))$
36. $\cos^{-1}(\tan(0.54))$
37. $\cos^{-1}(\tan(1.57))$
38. $\sec^{-1}(\tan(0.54))$
39. $\tan^{-1}(\tan^{-1}\pi)$
40. $\sec^{-1}(\sec^{-1}\pi)$

Use your calculator to draw the graph of each function. Specify its domain and range.

41. $f(x) = \sin^{-1}\left(\frac{1}{2}x\right)$

42. $f(x) = \cos^{-1}\left(\frac{1}{3}x\right)$

43. $f(x) = \tan^{-1}(8x)$
44. $f(x) = \sec^{-1}(0.3x)$
45. $f(x) = \cos(\cos^{-1}x)$
46. $f(x) = \cos(\tan^{-1}x)$

In each of the following right triangles, write θ explicitly in terms of x, using inverse trigonometric functions.

47.

48.

49.

50.

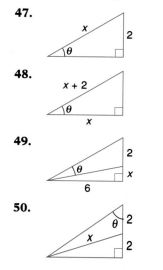

Specify the domain and range of each of the following.

51. $\sin^{-1}x + \cos^{-1}x$
52. $\cos^{-1}x + \sec^{-1}x$
53. $\tan^{-1}x + \tan x$
54. $\sin^{-1}x + \tan^{-1}x$
55. $2\sin^{-1}3x + 2\pi$
56. $3\tan^{-1}\left(\frac{x}{4}\right) + \frac{3\pi}{2}$

B. Applications and Extensions

57. Let $f(x) = 2\sin^{-1}x + \sin 16x$. Determine the domain and the range of f.
58. A man 5 feet 10 inches tall casts a shadow 32 feet long. Determine the angle of elevation of the sun.
59. Find all positive solutions to $\sin^{-1}x = 1/\sin x$.
60. Find any solutions to $\tan^{-1}x = 1/\tan x$ between 0 and $\pi/2$.
61. Determine the maximum point on the graph of $y = \sin x + \cos(\tan^{-1}x)$ for $-2\pi \leq x \leq 2\pi$.
62. Mentally, decide what the graph of $f(x) = \sin^{-1}(\sin x)$ should look like. Then confirm your answer by drawing the graph using a graphics calculator.
63. Follow the directions of Problem 62 for $f(x) = \cos^{-1}(\cos x)$. Then write a nice trigonometric formula of the form $A\cos(\cos^{-1}Bx)$ for $\ll x \gg$, the distance to the nearest integer (see Problem 55 of Section 2.9).
64. If $\tan\theta = 2x + 1$ with $x > -1/2$ and $0 < \theta < \pi/2$, write an explicit expression for $T = \theta - \sec\theta$ in terms of x. Then evaluate T when $x = 1.8$.
65. Graph $y = \sin^{-1}x$ and $y = \tan^{-1}(x/\sqrt{1 - x^2})$ and conjecture an identity. Draw a right triangle picture that shows why this identity holds.
66. Graph $y = \sec(\tan^{-1}x)$ and $y = \sqrt{1 + x^2}$ and conjecture an identity. Prove this identity.
67. A motorist is driving down a straight road looking at a billboard that is 50 feet wide, perpendicular to the road, and whose nearest edge is 30 feet from A, the point where she will pass the billboard (Figure 13). At what point does she see the billboard at the best viewing angle (that is, the largest angle θ)?

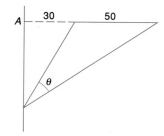

Figure 13

68. **Challenge.** A goat is tethered to a stake at the edge of a circular pond of radius r (Figure 14) by means of a rope of length kr, $0 < k \le 2$. Find an explicit formula for the goat's grazing area A in terms of r and k. Then determine A when $r = 100$ feet and $k = 1.5$.

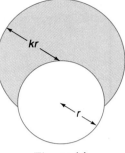

Figure 14

CHAPTER 4 REVIEW PROBLEM SET

In Problems 1–10, write True or False in the blank. If false, tell why.

_____ **1.** $\sin\left(\dfrac{\pi}{3}\right) = \cos\left(\dfrac{\pi}{6}\right)$

_____ **2.** An angle measuring 135° measures $3\pi/4$ radians.

_____ **3.** If $\sin\theta = 0$, then $\cos\theta = 1$.

_____ **4.** $\cos(-t) = -\cos t$ for all t.

_____ **5.** If $\beta = \alpha + \pi$, then $\sin\beta = \sin\alpha$.

_____ **6.** If $\cot t = a$, then $|\csc t| = \sqrt{1 + a^2}$.

_____ **7.** $\sin^{-1}x = \dfrac{1}{\sin x}$

_____ **8.** 1000° measures a fourth-quadrant angle.

_____ **9.** The amplitude of the periodic function $f(t) = 1 + 2\sin t$ is 3.

_____ **10.** The period of the periodic function $f(t) = 4\cos(3t/2)$ is $4\pi/3$.

Solve the right triangles in Problems 11 and 12, assuming standard notation.

11. $a = 9$, $c = 15$
12. $\alpha = 72.4°$, $b = 29.6$

Find the exact value of each of the following (without the use of a calculator) by relating the given angle to a special angle.

13. $\sin\left(\dfrac{3\pi}{2}\right)$

14. $\tan\left(\dfrac{\pi}{6}\right)$

15. $\sec\left(\dfrac{5\pi}{3}\right)$

16. $\cot 315°$

17. $\sin(-225°)$

18. $\cos 7\pi$

19. $\tan\left(\dfrac{3\pi}{4}\right)$

20. $\csc\left(\dfrac{7\pi}{2}\right)$

In Problems 21–24, find the two exact values of t between 0 and 2π for which the following is true.

21. $\cos t = -\dfrac{1}{2}$

22. $\tan t = -1$

23. $\sin t = \sqrt{2}/2$

24. $\sec t = 2$

25. If $\sin t = 3/7$ and $\pi/2 < t < \pi$, find $\tan t$ exactly.

26. Find θ if $\cos\theta = 1/2$ and $-90° < \theta < 90°$.

27. If $\cos\theta = 3/5$, find
   (a) $\cos(-\theta)$
   (b) $\sin(\theta - 90°)$
   (c) $\cos(\theta + 180°)$
   (d) $|\sin\theta|$

28. If θ is a second-quadrant angle, express each of the following in terms of $\sin\theta$.
   (a) $\cos\theta$
   (b) $\csc\theta$
   (c) $\tan\theta$

29. If $(-5, 12)$ is on the terminal side of angle θ in standard position, find each of the following exactly.
   (a) $\tan\theta$
   (b) $\csc\theta$

30. For what values of t between 0 and 2π is
   (a) $\sin t > 0$
   (b) $\sin 2t > 0$

31. Determine the range of the function $f(t) = 3\cos t - 2$.

32. Use the fact that the sine function is odd and the cosine function is even to show that the tangent function is odd.

33. Find a and b so that the curve $y = a\sin x + b\cos x$ passes through the points $(\pi/2, 2)$ and $(\pi/3, 4)$.

34. Let $f(t) = 2 + 4\sin t$ for $0 \le t \le 2\pi$.
   (a) Find the maximum and minimum values of $f(t)$.
   (b) Where does the graph cross the t-axis?

35. Write in terms of cos t.
    (a) $\cos(t + 4\pi)$
    (b) $\cos(t + \pi)$
    (c) $\sin^2 t$
    (d) $\sin\left(t - \dfrac{\pi}{2}\right)$

36. If $\tan\theta = 3/4$ and $\sin\theta < 0$, evaluate $\sec^2\theta - \sin^3\theta$ exactly.

37. Show that the area A of a right triangle is $(a^2/2)\cot\alpha$, assuming the standard notation.

38. Calculate.
    (a) $\dfrac{\sin^2 137° - \tan 51°}{\cot 114°}$
    (b) $\left(\dfrac{\cos 5.13 + \sin 2.54}{\sin 2.11 - \cos 3.63}\right)^2$

In Problems 39–42, determine the period, amplitude, and phase shift of each function. Then sketch its graph on the interval $-2\pi \le x \le 2\pi$ by hand. Confirm your graph with a graphics calculator.

39. $f(x) = 3\sin 2x$

40. $f(x) = -2.5\cos\left(0.5x + \dfrac{\pi}{8}\right)$

41. $f(x) = \cos\left(4x + \dfrac{\pi}{2}\right)$

42. $f(x) = 5 - 5\sin\left(2x - \dfrac{\pi}{2}\right)$

43. Use a graphics calculator to determine the period and amplitude of $f(x) = \sin x + 2\sin 3x$.

44. Graph each function on your graphics calculator and determine its domain, range, and period.
    (a) $3\cot\pi x$
    (b) $\cos 2x + \tan x$
    (c) $3/(1 + \cos^2 x)$
    (d) $0.1\csc 0.5x$

45. Find all solutions of the given equation on the interval $0 \le x \le \pi$ by using a graphics calculator.
    (a) $\sin\left(\dfrac{1}{2}x\right) = 2|\cos x|$
    (b) $4\sin 2x = \cos\left(x + \dfrac{\pi}{2}\right)$

46. Determine the maximum and minimum value of $f(x) = x\sin(x^2)$ on $0 \le x \le \pi$, accurate to two decimal places.

47. Determine the exact domain and range of $f(x) = (3/x)\tan^{-1}x$.

48. Solve the equation $3x = \cos^{-1}x$, accurate to two decimal places.

In Problems 49 and 50, calculate each value exactly (without use of a calculator).

49. (a) $\sin^{-1}\left(\dfrac{\sqrt{2}}{2}\right)$
    (b) $\cos^{-1}\left(-\dfrac{1}{2}\right)$

    (c) $\tan^{-1}(-1)$
    (d) $\sec^{-1}(\sqrt{2})$

50. (a) $\sec(\sec^{-1}(2.5))$
    (b) $\sin^{-1}\left(\sin\dfrac{3\pi}{4}\right)$
    (c) $\csc(\sin^{-1}(0.6))$
    (d) $\sec(\cos^{-1}(-0.2))$

51. Find two exact values of t between 0 and 2π at which $\sin 2t$ takes on its maximum value.

52. Angie is pedaling her tricycle so that the front wheel of radius 10 inches is turning at 300° per second.
    (a) Determine her speed in feet per second.
    (b) How far down the road will she travel in 2 minutes?

53. Find the area of the shaded region in Figure 1. The arc is part of a circle of radius 120.

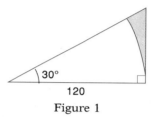

30°

120

Figure 1

54. A 16-foot ladder leans against a vertical wall. If the ladder makes an angle of 55° with the ground, how high up the wall does it reach?

55. If the bottom of the ladder in Problem 54 is pulled one foot farther from the wall, how far does the top of the ladder slide down the wall?

56. Find x and y in Figure 2.

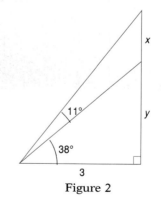

x

y

11°

38°

3

Figure 2

57. The curve on a highway is an arc of a circle of radius 500 meters and subtends an angle of 18° at the center of the circle. Find
    (a) the length of this arc
    (b) the straight-line distance between the two ends of the arc

58. At how many points do the graphs of $y = \sin 2t$ and $y = -\cos t$ intersect as t varies from 0 to 4π?

59. A regular octagon (eight sides) is inscribed in a circle of radius 12 centimeters. Determine the perimeter of this octagon.

60. The Pentagon building in Washington, D.C., has the shape of a regular pentagon with five sides each of length 921 feet. Find the area of the region within the perimeter of this building.

61. Determine an appropriate restricted domain so that $f(x) = \sin 2x$ will have an inverse and then give a formula for $f^{-1}(x)$.

62. Show that cot $(\sin^{-1} x) = \sqrt{1 - x^2}/x$ for $-1 \leq x \leq 1$, $x \neq 0$.

63. Find all positive solutions to $\cos^{-1} x = 1/\cos x$, accurate to four decimal places.

64. In Figure 2, replace the angle 38° by a general angle θ. Express x in terms of θ. Then determine θ, accurate to the nearest 100th of a degree, so that $x = 20$.

5

TRIGONOMETRIC IDENTITIES, EQUATIONS, AND LAWS

The notion of number (scalar) is very old; that of vector is quite recent and its invention owes more to physicists, chemists, and engineers than it does to mathematicians. Many of the quantities that scientists deal with—forces, velocities, torques, and so on—cannot be specified by simply giving a number. These quantities are determined rather by both a number and a direction and are best symbolized by an arrow. While the roots of this idea go back further, it was the founder of electromagnetic theory, James Clerk Maxwell (1831–79), who saw the need for vector quantities in physics and who separated them into components along the coordinate axes (see the diagram below).

But the recognition of vector analysis as a subject worthy of study on its own merits is due to the physical chemist,

Josiah Gibbs, and the engineer, Oliver Heaviside. Gibbs was born in New Haven, Connecticut, graduated from Yale University, and studied in Paris, Berlin, and Heidelberg. He was offered the professorship of mathematical physics at Yale, the first such position in America, and accepted it even though it carried no salary for the first 10 years. Beginning about 1880, he developed the symbolism and algebra of vectors. In 1901, the full treatment of his ideas was presented by one of his students, E. B. Wilson, in an influential book titled Vector Analysis. To honor Gibbs, the American Mathematical Society sponsors the Gibbs Lecture at its annual meeting each year. This lecture features a recent application of mathematics to a problem of science. Our introduction to the topic of vectors is in Section 5.8.

Josiah Willard Gibbs
(1839–1903)

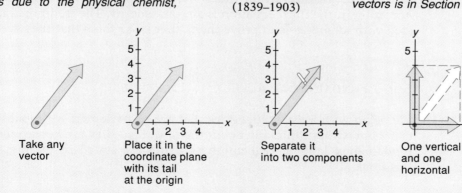

Take any vector

Place it in the coordinate plane with its tail at the origin

Separate it into two components

One vertical and one horizontal

TEASER   If $a + b = 1$ and $a^2 + b^2 = 2$, determine the value of $a^3 + b^3$.

5.1   ALGEBRA REVIEW (OPTIONAL)

A prominent part of this chapter deals with the algebra of the trigonometric functions. This subject depends strongly on the more fundamental algebra of polynomials, which might be called basic algebra. Of course, we have been using basic algebra all along, though it was not our main focus. Here we will need to flex our algebraic muscles, and it seems wise to prepare. Readers with strong algebraic skills may treat this section as a review of things learned long ago.

We begin with some terminology. Any algebraic expression that contains an equal sign is called an **equation.** For example,

$$x^2 - 1 = 0, \quad x^2 - 1 = (x - 1)(x + 1), \quad \frac{x^2 - y^2}{x - y} = x + y$$

are all equations. The first is called a **conditional equation** because it is true only for some values of the variable x, namely, -1 and 1. The traditional task associated with a conditional equation is to solve it, that is, to find those values of x that make the equation true. The second equation above is an identity; it is true for all values of x. In fact, any equation that is true for all the values of its variables that are in the domain of both sides is called an **identity.** Accordingly, the third equation is also an identity. The traditional task associated with identities is to prove them, that is, to show that they are identities.

SOME FUNDAMENTAL ALGEBRAIC IDENTITIES

We begin with a list of identities that are so basic that we should be able to use them without thinking about them. In fact, they are consequences of the fundamental laws of real numbers that were stated in the first section of this book.

ALGEBRAIC IDENTITIES	
1.	$a(x + y + z) = ax + ay + az$
2.	$(x + a)(y + b) = xy + xb + ay + ab$
3.	$(x + y)^2 = x^2 + 2xy + y^2$
4.	$(x - y)^2 = x^2 - 2xy + y^2$
5.	$(x - y)(x + y) = x^2 - y^2$
6.	$(x + y)^3 = x^3 + 3x^2y + 3xy^2 + y^3$
7.	$(x - y)^3 = x^3 - 3x^2y + 3xy^2 - y^3$
8.	$(x + y)(x^2 - xy + y^2) = x^3 + y^3$
9.	$(x - y)(x^2 + xy + y^2) = x^3 - y^3$

These identities allow us to expand algebraic expressions into sums, as we now illustrate.

■ **Example 1.** Expand (a) $(2s - t)(s + 3t)$, (b) $(2u - v)^2$, (c) $(3c + 2d)^3$

Solution. One of the key skills in algebra is the ability to mentally replace one variable by another. We will need that skill in each part of this example. (a) We apply identity 2 with $x = 2s$, $a = -t$, $y = s$, and $b = 3t$.

$$(2s + -t)(s + 3t) = (2s)(s) + (2s)(3t) + (-t)(s) + (-t)(3t)$$
$$= 2s^2 + 6st - ts - 3t^2$$
$$= 2s^2 + 5st - 3t^2$$

A good way to think of this is to recognize that there are four terms in the sum: the products of the firsts, the outers, the inners, and the lasts (abbreviated by the acronym *FOIL*). Often the outers and inners can be combined.
(b) Apply identity 4 with $2u$ as x and v as y.

$$(2u - v)^2 = (2u)^2 - 2(2u)(v) + v^2 = 4u^2 - 4uv + v^2$$

(c) Apply identity 6 with $3c$ as x and $2d$ as y.

$$(3c + 2d)^3 = (3c)^3 + 3(3c)^2(2d) + 3(3c)(2d)^2 + (2d)^3$$
$$= 27c^3 + 54c^2d + 36cd^2 + 8d^3 \qquad ■$$

FACTORING

The identities in our list should be read backward as well as forward. Read from right to left, they are factoring identities, a subject we consider now. To factor an expression is to write it as a product. In each case, our goal is to factor as much as possible.

■ **Example 2 (Taking out a common factor).** Factor $4xy^3 - 8x^2y^2z$.

Solution. We apply identity 1. We could take out the common factor 4, or we could take out $4x$, but the best choice is to take out $4xy^2$.

$$4xy^3 - 8x^2y^2z = 4xy^2(y - 2xz) \qquad \blacksquare$$

■ **Example 3 (Factoring a trinomial).** Factor (a) $x^2 - 4x - 12$, (b) $2x^2 + 9x + 4$.

Solution. Factoring a trinomial is largely a matter of experimenting, using identity 2.
(a) We expect to factor this trinomial as $(x - a)(x + b)$ with $ab = 12$. As possible values for a and b, we mentally try the various factors of 12 (meaning 3 and 4, 2 and 6, 1 and 12), looking for a combination that gives a difference of 4. We are led to

$$x^2 - 4x - 12 = (x - 6)(x + 2)$$

(b) Here we expect to factor as $(2x + a)(x + b)$ with $ab = 4$. Experimenting eventually leads to

$$2x^2 + 9x + 4 = (2x + 1)(x + 4) \qquad \blacksquare$$

■ **Example 4 (Perfect squares).** Factor $4x^2y^4 + 4xy^2 + 1$.

Solution. We note that the first and last terms are squares, which suggests that this trinomial might be a perfect square (identity 3). But does the middle term check? Yes, it does.

$$4x^2y^4 + 4xy^2 + 1 = (2xy^2 + 1)^2 \qquad \blacksquare$$

■ **Example 5 (Difference of squares).** Factor (a) $9u^2 - 4v^2$, (b) $u^2 + 6u + 9 - v^2$, (c) $u^4 - v^4$.

Solution. Refer to identity 5.
(a) $9u^2 - 4v^2 = (3u - 2v)(3u + 2v)$
(b) $u^2 + 6u + 9 - v^2 = (u + 3)^2 - v^2$
$$= (u + 3 - v)(u + 3 + v)$$
(c) $u^4 - v^4 = (u^2 - v^2)(u^2 + v^2)$
$$= (u - v)(u + v)(u^2 + v^2)$$

This last example reminds us that we may be able to factor and then factor again. $\qquad \blacksquare$

■ **Example 6 (Sum and difference of cubes).** Factor (a) $z^6 + 27$, (b) $2u^3w^2 - 16w^2$.

Solution. Use identities 8 and 9.
(a) $z^6 + 27 = (z^2)^3 + (3)^3 = (z^2 + 3)(z^4 - 3z^2 + 9)$
(b) $2u^3w^2 - 16w^2 = 2w^2(u^3 - 8)$
$$= 2w^2(u - 2)(u^2 + 2u + 4) \qquad \blacksquare$$

HANDLING FRACTIONS

First, we review the basic properties of fractions.

RULES FOR FRACTIONS

1. (Cancellation) $\dfrac{ac}{bc} = \dfrac{a}{b}$ provided $c \neq 0$

2. (Signs) $\dfrac{-a}{b} = \dfrac{a}{-b} = -\dfrac{a}{b} = -\dfrac{-a}{-b}$

3. (Addition) $\dfrac{a}{c} + \dfrac{b}{c} = \dfrac{a+b}{c}$

4. (Multiplication) $\dfrac{a}{b} \cdot \dfrac{c}{d} = \dfrac{ac}{bd}$

5. (Division) $\dfrac{a/b}{c/d} = \dfrac{ad}{bc}$

■ **Example 7.** Simplify (a) $(x^2 - 4y^2)/(x - 2y)$, (b) $(x^2 - x - 20)/(x^2 + 5x + 4)$.

Solution.

(a) $\dfrac{x^2 - 4y^2}{x - 2y} = \dfrac{(x - 2y)(x + 2y)}{x - 2y} = x + 2y$

(b) $\dfrac{x^2 - x - 20}{x^2 + 5x + 4} = \dfrac{(x - 5)(x + 4)}{(x + 1)(x + 4)} = \dfrac{x - 5}{x + 1}$ ■

■ **Example 8.** Perform the indicated operation and simplify.

(a) $\dfrac{3x - 2}{x + 2} \cdot \dfrac{x^2 - 4}{6xy - 4y}$

(b) $\dfrac{(x^2 - 5x + 4)/(2x + 6)}{(2x^2 - x - 1)/(x^2 + 5x + 6)}$

Solution.

(a) $\dfrac{3x - 2}{x + 2} \cdot \dfrac{x^2 - 4}{6xy - 4y} = \dfrac{(3x - 2)(x - 2)(x + 2)}{(x + 2)(2y)(3x - 2)} = \dfrac{x - 2}{2y}$

(b) $\dfrac{(x^2 - 5x + 4)/(2x + 6)}{(2x^2 - x - 1)/(x^2 + 5x + 6)} = \dfrac{x^2 - 5x + 4}{2x + 6} \cdot \dfrac{x^2 + 5x + 6}{2x^2 - x - 1}$

$\qquad\qquad = \dfrac{(x - 4)(x - 1)(x + 2)(x + 3)}{2(x + 3)(x - 1)(2x + 1)}$

$\qquad\qquad = \dfrac{(x - 4)(x + 2)}{2(2x + 1)}$ ■

■ **Example 9.** Combine and simplify.

(a) $\dfrac{4}{2x - 1} + \dfrac{x - 2}{1 - 2x}$

(b) $\dfrac{x}{4x^2 - 1} + \dfrac{6x}{4x - 2} - \dfrac{3x + 1}{2x + 1}$

(c) $x^2 + \dfrac{x}{x + 1} + \dfrac{2}{x - 1}$

Solution. Our goal is to write all the terms of the expression with the same denominator (the lowest common denominator) and then add using rule 3.

(a) $\dfrac{4}{2x-1} + \dfrac{x-2}{1-2x} = \dfrac{4}{2x-1} + \dfrac{-x+2}{2x-1} = \dfrac{6-x}{2x-1}$

(b) $\dfrac{x}{4x^2-1} + \dfrac{6x}{4x-2} - \dfrac{3x+1}{2x+1} = \dfrac{x}{(2x-1)(2x+1)} + \dfrac{6x}{2(2x-1)} - \dfrac{3x+1}{2x+1}$

$\qquad = \dfrac{x}{(2x-1)(2x+1)} + \dfrac{3x(2x+1)}{(2x-1)(2x+1)} - \dfrac{(3x+1)(2x-1)}{(2x-1)(2x+1)}$

$\qquad = \dfrac{x + 6x^2 + 3x - (6x^2 - x - 1)}{(2x-1)(2x+1)}$

$\qquad = \dfrac{5x+1}{(2x-1)(2x+1)}$

(c) $x^2 + \dfrac{x}{x+1} + \dfrac{2}{x-1} = \dfrac{x^2(x+1)(x-1)}{(x+1)(x-1)} + \dfrac{x(x-1)}{(x+1)(x-1)} + \dfrac{2(x+1)}{(x+1)(x-1)}$

$\qquad = \dfrac{x^4 - x^2 + x^2 - x + 2x + 2}{(x+1)(x-1)}$

$\qquad = \dfrac{x^4 + x + 2}{(x+1)(x-1)}$ ∎

TEASER SOLUTION

We are given that $a + b = 1$ and $a^2 + b^2 = 2$. We could solve these two equations simultaneously to determine a and b and then substitute into the expression $a^3 + b^3$ to obtain its value. That is the hard way to do the problem. Let's look for an easier way. Note first that

$$1 = (a + b)^2 = a^2 + 2ab + b^2 = 2 + 2ab$$

from which we conclude that $ab = -1/2$. Next

$$1 = (a + b)^3 = a^3 + 3a^2b + 3ab^2 + b^3$$
$$= a^3 + b^3 + 3ab(a + b)$$
$$= a^3 + b^3 + 3\left(-\dfrac{1}{2}\right)(1)$$

This implies that $a^3 + b^3 = 1 + 3/2 = 5/2$.

PROBLEM SET 5.1

A. Skills and Techniques

Perform the indicated operations and simplify.

1. $x^2 - 3x + x(2x - 4) - 4x^2$
2. $3x^2 - 2x(x + 4) - 5(-1 - 2x)$
3. $2t(t^2 - 3t + 1) - t^3 + (t - 1)(t + 1)$
4. $u(-u^2 + 3u - 2) + (u + 2)^2$
5. $(3x - 4y)^2$
6. $\left(2a + \dfrac{1}{2}b\right)^2$
7. $(2x - 4y^2)(2x + 4y^2)$
8. $(2t^3 - 3s)(2t^3 + 3s)$
9. $(3a^2 + bc)^2$
10. $(3u - v^3)^2$
11. $(3u + 2v)(u - 2v)$
12. $(2s - 3t)(4s - t)$

13. $(2x^2 - yz)(3x^2 + 2yz) - x^2yz$
14. $(3rs - t)(2rs + 5t) - 6r^2s^2 + 5t^2$
15. $(2s - t)^3$
16. $(u^2 - 3v)^3$
17. $xy(x^2 - 2y)(x^2 + 2y)$
18. $x^2(2x - 3y^2)^2 + 12x^3y^2$
19. $(x + y - 4)(x + y + 4)$
20. $(u - v + w)(u + v - w)$

Factor each of the following as much as possible.

21. $3a^2b^3c - 6a^3bc^2$
22. $2uv^2w^3 - 4u^3vw^2 + 6uvw^2$
23. $x^4y^2 - x^2z^2$
24. $a^3 - 4a^2b + 4ab^2$
25. $y^2 - 4y - 12$
26. $x^2 - 3x - 40$
27. $9x^2 + 24x + 16$
28. $4z^2 - 4z - 3$
29. $8a^3 - 27b^3$
30. $x^3 + 8y^3$
31. $16xu^3 + 2xv^3$
32. $9 - 9t^6$
33. $a^3b^3 - 3a^2b^2c + 3abc^2 - c^3$
34. $x^3 + 9x^2y + 27xy^2 + 27y^3$

Perform the indicated operations and simplify.

35. $\dfrac{x + 5}{x^2 - 25}$

36. $\dfrac{8x + 16}{x^2 - 4}$

37. $\dfrac{x^2 - 9}{x + 2} \cdot \dfrac{x^2 - 4}{x - 3}$

38. $\left(1 + \dfrac{1}{x + 2}\right)\left(\dfrac{4}{3x + 9}\right)$

39. $\dfrac{x^2y^2}{x - y}\left(\dfrac{x}{y^2} - \dfrac{y}{x^2}\right)$

40. $\left(\dfrac{x^2 + 5x}{x^2 - 16}\right)\left(\dfrac{x^2 - 2x - 24}{x^2 - x - 30}\right)$

41. $\dfrac{x + 2}{(x^2 - 4)/x}$

42. $\dfrac{(x + 2)/(x^2 - 3x)}{(x^2 - 4)/x}$

43. $\dfrac{4}{2x - 1} + \dfrac{8x}{1 - 2x}$

44. $\dfrac{x}{6x - 2} - \dfrac{3}{1 - 3x}$

45. $\dfrac{2}{6y - 2} + \dfrac{y}{9y^2 - 1} - \dfrac{2y + 1}{1 - 3y}$

46. $\dfrac{2x}{x^2 - y^2} + \dfrac{1}{x + y} + \dfrac{1}{y - x}$

B. Applications and Extensions

47. Expand and simplify.
    (a) $(x - \sqrt{2}y)(x + \sqrt{2}y) - (x - \sqrt{2}y)^2$
    (b) $(2x - y)^3 + 12x^2y - 6xy^2$
48. Expand $(a + b + c)^2$. Then find the coefficient of x^5 in the expansion of $(x^4 + 2x^3 + 3x^2)^2$.
49. Factor.
    (a) $(x + 2y)^2 + 6(x + 2y) + 9$
    (b) $(m - n)^2 + 5(m - n) + 4$
50. Factor.
    (a) $(a + 2)^3(b + 1)^2 + (a + 2)^2(b + 1)^3$
    (b) $x^{n+3} + 4x^n + x^3 + 4$

In Problems 51–58, simplify the given expression.

51. $\dfrac{x^3 + a^3}{x^3 - a^3} \cdot \dfrac{x - a}{x + a}$

52. $\dfrac{1 + 2/b}{1 - 4/b^2}$

53. $\dfrac{1/(x + 2) - 3/(x^2 - 4)}{3/(x - 2)}$

54. $\dfrac{(a - b)/(a + b) - (a + b)/(a - b)}{ab/(a - b)}$

55. $1 - \dfrac{x - 1/x}{1 - 1/x}$

56. $\dfrac{a^2/b^2 - b^2/a^2}{a/b - b/a}$

57. $\dfrac{n - n^2/(n - m)}{1 + m^2/(n^2 - m^2)}$

58. $\dfrac{x^3 - 8y^3}{x^2 - 4y^2} + \dfrac{2xy}{x + 2y}$

59. A triple (a, b, c) of positive integers is called a *Pythagorean triple* if $a^2 + b^2 = c^2$. For example, $(3, 4, 5)$ and $(5, 12, 13)$ are Pythagorean triples. Show that $(2m, m^2 - 1, m^2 + 1)$ is a Pythagorean triple if m is an integer greater than 1.
60. Under what conditions is $(m^2 - n^2, 2mn, m^2 + n^2)$ a Pythagorean triple?
61. Use algebraic tricks to find the exact value of

$$\left(1 - \dfrac{1}{2^2}\right)\left(1 - \dfrac{1}{3^2}\right)\left(1 - \dfrac{1}{4^2}\right)$$

$$\cdots \left(1 - \dfrac{1}{98^2}\right)\left(1 - \dfrac{1}{99^2}\right)$$

62. If $(x + y)^2 = 15$ and $xy = 6$, evaluate $x^2 + y^2$.
63. If $x + y = \sqrt{11}$ and $x^2 + y^2 = 16$, find the values of xy and $x^4 + y^4$. (*Hint:* It will be helpful to know that $(x + y)^4 = x^4 + 4x^3y + 6x^2y^2 + 4xy^3 + y^4$.)
64. **Challenge.** If $a + b + c = 1$, $a^2 + b^2 + c^2 = 2$, and $a^3 + b^3 + c^3 = 3$, evaluate $a^4 + b^4 + c^4$.

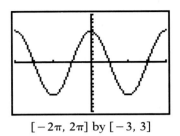

$[-2\pi, 2\pi]$ by $[-3, 3]$

TEASER At the left, we show the graph of $f(x) = (\cos^3 x)(1 - \tan^4 x + \sec^4 x)$ on the interval $-2\pi \le x \le 2\pi$. Conjecture an identity. Prove it algebraically.

5.2 BASIC TRIGONOMETRIC IDENTITIES

Complicated combinations of the six trigonometric functions occur often in mathematics. It is important that we be able to take such combinations and simplify them. To do this, we must be skilled at algebraic manipulations and we must know the fundamental identities of trigonometry.

FUNDAMENTAL TRIGONOMETRIC IDENTITIES

1. $\tan x = \dfrac{\sin x}{\cos x}$

2. $\cot x = \dfrac{\cos x}{\sin x} = \dfrac{1}{\tan x}$

3. $\sec x = \dfrac{1}{\cos x}$

4. $\csc x = \dfrac{1}{\sin x}$

5. $\sin^2 x + \cos^2 x = 1$

6. $1 + \tan^2 x = \sec^2 x$

7. $1 + \cot^2 x = \csc^2 x$

8. $\sin\left(\dfrac{\pi}{2} - x\right) = \cos x$

9. $\cos\left(\dfrac{\pi}{2} - x\right) = \sin x$

10. $\sin(-x) = -\sin x$

11. $\cos(-x) = \cos x$

We have seen all of these identities before (the first four are actually definitions); now would be a good time to commit them to memory.

SIMPLIFYING TRIGONOMETRIC EXPRESSIONS

If you cannot think of anything else to do, a good start is to express everything in terms of sines and cosines. Keep in mind that since $\sin^2 x + \cos^2 x = 1$, it is also true that $\sin^2 x = 1 - \cos^2 x$.

■ **Example 1.** Simplify $(\csc x + \cot x)(1 - \cos x)$.

Solution.

$$(\csc x + \cot x)(1 - \cos x) = \left(\frac{1}{\sin x} + \frac{\cos x}{\sin x}\right)(1 - \cos x)$$

$$= \frac{1 + \cos x}{\sin x} \cdot \frac{1 - \cos x}{1} = \frac{1 - \cos^2 x}{\sin x}$$

$$= \frac{\sin^2 x}{\sin x} = \sin x \qquad ■$$

Sometimes factoring can be the key to a simplification. Recall that

$$a^4 - b^4 = (a^2 - b^2)(a^2 + b^2) = (a - b)(a + b)(a^2 + b^2)$$

■ **Example 2.** Simplify $(\sin^4 x - \cos^4 x)/(\sin x - \cos x)$.

Solution.

$$\frac{\sin^4 x - \cos^4 x}{\sin x - \cos x} = \frac{(\sin^2 x - \cos^2 x)(\sin^2 x + \cos^2 x)}{\sin x - \cos x}$$
$$= \frac{(\sin x - \cos x)(\sin x + \cos x)(1)}{\sin x - \cos x}$$
$$= \sin x + \cos x \qquad\blacksquare$$

Often a simple manipulation with fractions is the key.

■ **Example 3.** Simplify $\cot u + (1 - 2\cos^2 u)/(\sin u \cos u)$.

Solution.

$$\cot u + \frac{1 - 2\cos^2 u}{\sin u \cos u} = \frac{\cos u}{\sin u} + \frac{1 - 2\cos^2 u}{\sin u \cos u}$$
$$= \frac{\cos^2 u}{\sin u \cos u} + \frac{1 - 2\cos^2 u}{\sin u \cos u} = \frac{1 - \cos^2 u}{\sin u \cos u}$$
$$= \frac{\sin^2 u}{\sin u \cos u} = \frac{\sin u}{\cos u} = \tan u \qquad\blacksquare$$

Occasionally, it is important to be able to express all of the trigonometric functions in terms of one of them. We can do this in an unambiguous way provided the variable is restricted to one quadrant.

■ **Example 4.** Given that $\pi/2 < u < \pi$, express $\sin u$, $\cos u$, $\cot u$, $\sec u$, and $\csc u$ in terms of $\tan u$.

Solution. Since $\sec^2 u = 1 + \tan^2 u$ and $\sec u$ is negative in the second quadrant,

$$\sec u = -\sqrt{1 + \tan^2 u}$$

Also

$$\cos u = \frac{1}{\sec u} = -\frac{1}{\sqrt{1 + \tan^2 u}}$$
$$\cot u = \frac{1}{\tan u}$$
$$\sin u = \frac{(\sin u)(\cos u)}{\cos u} = \tan u \cos u = -\frac{\tan u}{\sqrt{1 + \tan^2 u}}$$
$$\csc u = \frac{1}{\sin u} = -\frac{\sqrt{1 + \tan^2 u}}{\tan u} \qquad\blacksquare$$

PROVING IDENTITIES

A traditional task for trigonometry students is to prove that certain equations are identities. This is generally a matter of choosing one side of the equation and showing by a chain of equalities that it is equal to the other side. We illustrate several useful techniques.

■ **Example 5 (Simplify the complicated side).**   Prove that the equation $\sin t + \cos t \cot t = \csc t$ is an identity.

Solution.

$$\sin t + \cos t \cot t = \sin t + \cos t \frac{\cos t}{\sin t}$$

$$= \frac{\sin^2 t}{\sin t} + \frac{\cos^2 t}{\sin t} = \frac{\sin^2 t + \cos^2 t}{\sin t} = \frac{1}{\sin t} = \csc t \quad ■$$

■ **Example 6 (Change to sines and cosines).**   Prove that

$$\frac{\tan u + \tan u \sin^2 u - \cos u \sin u}{\sin u \tan u} = 2 \sin u$$

is an identity.

Solution.

$$\frac{\tan u + \tan u \sin^2 u - \cos u \sin u}{\sin u \tan u}$$

$$= \frac{\sin u/\cos u + \sin^3 u/\cos u - \cos u \sin u}{\sin^2 u/\cos u}$$

$$= \frac{\sin u/\cos u + \sin^3 u/\cos u - \cos^2 u \sin u/\cos u}{\sin^2 u/\cos u} \cdot \frac{\cos u}{\cos u}$$

$$= \frac{\sin u + \sin^3 u - \cos^2 u \sin u}{\sin^2 u}$$

$$= \frac{\sin u (1 + \sin^2 u - \cos^2 u)}{\sin^2 u}$$

$$= \frac{\sin^2 u + \sin^2 u}{\sin u} = 2 \sin u \quad ■$$

When you are proving that an equation is an identity, it pays to look before you leap. Changing the more complicated side to sines and cosines, as in the above example, is often the best thing to do. But not always. Sometimes, the simpler side gives us a clue as to how we should reshape the other side.

■ **Example 7 (Note the character of the simple side).**   Prove that

$$\frac{\sec^2 u + 2 \tan u}{2 - \sec^2 u} = \frac{1 + \tan u}{1 - \tan u}$$

is an identity.

Solution. The simpler right side involves only tangents, suggesting that we should try to express the left side in terms of tangents.

$$\frac{\sec^2 u + 2 \tan u}{2 - \sec^2 u} = \frac{1 + \tan^2 u + 2 \tan u}{2 - (1 + \tan^2 u)}$$

$$= \frac{1 + 2 \tan u + \tan^2 u}{1 - \tan^2 u} = \frac{(1 + \tan u)^2}{(1 - \tan u)(1 + \tan u)} = \frac{1 + \tan u}{1 - \tan u}$$

$$■$$

■ **Example 8 (Multiply by g(x)/g(x)).** Prove that

$$\frac{\sin(-x)}{\sin(\pi/2 - x) - 1} = \frac{1 + \cos x}{\sin x}$$

is an identity.

Solution. First we simplify the left side. Then we note the expression $1 + \cos x$ in the numerator on the right. This suggests that we multiply the left side by 1, written as $(1 + \cos x)/(1 + \cos x)$.

$$\frac{\sin(-x)}{\sin(\pi/2 - x) - 1} = \frac{-\sin x}{\cos x - 1} = \frac{\sin x}{1 - \cos x}$$

$$= \frac{\sin x}{1 - \cos x} \cdot \frac{1 + \cos x}{1 + \cos x}$$

$$= \frac{\sin x\,(1 + \cos x)}{1 - \cos^2 x} = \frac{\sin x\,(1 + \cos x)}{\sin^2 x}$$

$$= \frac{1 + \cos x}{\sin x} \qquad ■$$

■ **Example 9 (Factor one side).** Prove that

$$\csc^4 x - \cot^4 x = \frac{1 + \cos^2 x}{1 - \cos^2 x}$$

Solution. We are attracted to the left side because we notice that it can be factored. After we have factored, we use fundamental identities in a straightforward way.

$$\csc^4 x - \cot^4 x = (\csc^2 x - \cot^2 x)(\csc^2 x + \cot^2 x)$$

$$= (1 + \cot^2 x - \cot^2 x)(\csc^2 x + \cot^2 x)$$

$$= (1)\!\left(\frac{1}{\sin^2 x} + \frac{\cos^2 x}{\sin^2 x}\right) = \frac{1 + \cos^2 x}{\sin^2 x}$$

$$= \frac{1 + \cos^2 x}{1 - \cos^2 x} \qquad ■$$

USING A GRAPHICS CALCULATOR

A graphics calculator can help us discover both equalities and inequalities. Our Teaser problem illustrates the former. Our next example illustrates using a calculator to discover an inequality.

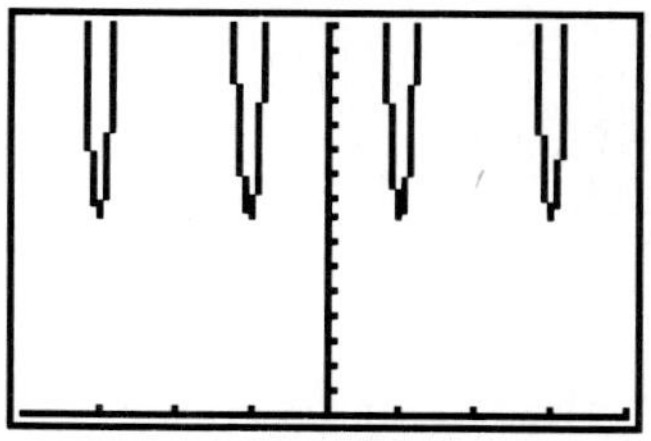

■ **Example 10.** Draw the graph of $f(x) = \tan^4 x + \cot^4 x$. Use this graph to conjecture an inequality that is true for all x in the domain of f. Demonstrate this result algebraically.

Solution. The graph of f is shown in Figure 1. It suggests that $\tan^4 x + \cot^4 x \geq 2$. To prove this algebraically, we write

$$\tan^4 x + \cot^4 x = \tan^4 x - 2 + \cot^4 x + 2$$
$$= (\tan^2 x - \cot^2 x)^2 + 2 \geq 2 \qquad ■$$

$[-\pi, \pi]$ by $[0, 4]$
Figure 1

The graph of $f(x) = (\cos^3 x)(1 - \tan^4 x + \sec^4 x)$ is shown in Figure 2. This graph looks suspiciously like a cosine curve. After noting the units on the x- and y-axes, we conjecture that $f(x) = 2 \cos x$.

An algebraic proof of this conjecture goes as follows.

$$(\cos^3 x)(1 - \tan^4 x + \sec^4 x) = \cos^3 x - \cos^3 x \frac{\sin^4 x}{\cos^4 x} + \cos^3 x \frac{1}{\cos^4 x}$$

$$= \cos^3 x - \frac{\sin^4 x}{\cos x} + \frac{1}{\cos x} = \frac{\cos^4 x - \sin^4 x + 1}{\cos x}$$

$$= \frac{(\cos^2 x - \sin^2 x)(\cos^2 x + \sin^2 x) + 1}{\cos x}$$

$$= \frac{\cos^2 x - (1 - \cos^2 x) + 1}{\cos x}$$

$$= \frac{2 \cos^2 x}{\cos x} = 2 \cos x$$

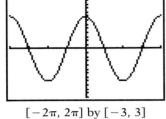

$[-2\pi, 2\pi]$ by $[-3, 3]$
Figure 2

PROBLEM SET 5.2

A. Skills and Techniques

Simplify each of the following expressions.

1. $(\sec x + \tan x)(1 - \sin x)$
2. $\dfrac{\sin u}{\csc u} + \dfrac{\cos u}{\sec u}$
3. $(1 - \sin^2 t)(\sec^2 t - 1)$
4. $\sin t \,(\csc t - \sin t)$
5. $(\sec u + \tan u)(\csc u - 1)$
6. $\dfrac{1 + \sec x}{\sin x + \tan x}$
7. $\dfrac{1 - \tan^2 u}{1 + \tan^2 u} + \sin^2 u$
8. $\dfrac{\cos x - \sec x}{1 + \cos x} + \sec x$
9. $\dfrac{\tan u}{\sec u - 1} - \dfrac{\tan u}{\sec u + 1}$
10. $(\csc x - \cot x)^2(\csc x + \cot x)^2$
11. $\dfrac{(\sec^4 x - \tan^4 x)(1 + \sin x)}{\sec x + \tan x}$
12. $\dfrac{\sin u}{1 - \cot u} - \dfrac{\cos u}{\tan u - 1}$
13. $\sin^2\!\left(\dfrac{\pi}{2} - u\right) \csc u - \tan^2\!\left(\dfrac{\pi}{2} - u\right) \sin u$
14. $\dfrac{\cos^3(\pi/2 - x)}{\cos x - \cos^3 x}$

Express the five remaining trigonometric functions in terms of the given one.

15. $\sin x; \ \pi < x < \dfrac{3\pi}{2}$
16. $\csc x; \ \dfrac{3\pi}{2} < x < 2\pi$
17. $\sec x; \ \dfrac{\pi}{2} < x < \pi$
18. $\cos x; \ 0 < x < \dfrac{\pi}{2}$

Prove that each of the following is an identity.

19. $\dfrac{\sec^2 t - 1}{\sec^2 t} = \sin^2 t$
20. $\cos t \,(\tan t + \cot t) = \csc t$
21. $\dfrac{\sec t - 1}{\tan t} = \dfrac{\tan t}{\sec t + 1}$
22. $\dfrac{1 - \tan u}{1 + \tan u} = \dfrac{\cot u - 1}{\cot u + 1}$
23. $\dfrac{\tan^2 u}{\sec u + 1} = \dfrac{1 - \cos u}{\cos u}$
24. $\dfrac{\cot x}{\csc x + 1} = \dfrac{\csc x - 1}{\cot x}$
25. $\dfrac{\sin t + \cos t}{\tan^2 t - 1} = \dfrac{\cos^2 t}{\sin t - \cos t}$
26. $\dfrac{\sec u - \cos u}{1 + \cos u} = \sec u - 1$

27. $(1 + \tan^2 t)(\cos t + \sin t) = (1 + \tan t) \sec t$

28. $1 - (\cos t + \sin t)(\cos t - \sin t) = 2 \sin^2 t$

29. $2 \sec^2 y - 1 = \dfrac{1 + \sin^2 y}{\cos^2 y}$

30. $(\sin x + \cos x)(\sec x + \csc x) = 2 + \tan x + \cot x$

31. $\dfrac{\cos z}{1 + \cos z} = \dfrac{\sin z}{\sin z + \tan z}$

32. $2 \sin^2 t + 3 \cos^2 t + \sec^2 t = (\sec t + \cos t)^2$

Graph each function and then conjecture an identity. Prove your conjecture algebraically.

33. $f(x) = (\sin x + \csc x)^2 + (\cos x + \sec x)^2 - (\tan x + \cot x)^2$

34. $f(x) = \dfrac{1}{1 + \sin^2 x} + \dfrac{1}{1 + \cos^2 x}$
    $\qquad + \dfrac{1}{1 + \sec^2 x} + \dfrac{1}{1 + \csc^2 x}$

Graph each function and conjecture an inequality. Then prove your conjecture.

35. $f(x) = \dfrac{1}{1 - \cos x} + \dfrac{1}{1 + \cos x}$

36. $f(x) = \cos^2 x + 4 \sec^2 x$

B. Applications and Extensions

37. Simplify
$$\frac{1 - \sin x \cos x}{\cos x \, (\sec x - \csc x)} \cdot \frac{\sin^2 x - \cos^2 x}{\sin^3 x + \cos^3 x}$$

to one of the six trigonometric functions.

38. By graphing, find the smallest $a > 0$ so that
$$-a \le \frac{1 - \tan^2 x}{1 + \tan^2 x} \le a$$

Then find an algebraic way to demonstrate this double inequality.

In Problems 39–52, demonstrate that each equality is an identity.

39. $(\csc t + \cot t)^2 = \dfrac{1 + \cos t}{1 - \cos t}$

40. $\sec^4 y - \tan^4 y = \dfrac{1 + \sin^2 y}{\cos^2 y}$

41. $\dfrac{\cos x + \sin x}{\cos x - \sin x} = \dfrac{1 + \tan x}{1 - \tan x}$

42. $\dfrac{1 + \cos x}{1 - \cos x} - \dfrac{1 - \cos x}{1 + \cos x} = 4 \cot x \csc x$

43. $\dfrac{2 \tan u}{1 - \tan^2 u} + \dfrac{1}{\cos^2 u - \sin^2 u} = \dfrac{\cos u + \sin u}{\cos u - \sin u}$

44. $\sec t + \cos t = \sin t \tan t + 2 \cos t$

45. $\dfrac{\cos^3 t + \sin^3 t}{\cos t + \sin t} = 1 - \sin t \cos t$

46. $\dfrac{\tan x}{1 + \tan x} + \dfrac{\cot x}{1 - \cot x} = \dfrac{\tan x + \cot x}{\tan x - \cot x}$

47. $\dfrac{1 - \cos \theta}{1 + \cos \theta} = \left(\dfrac{1 - \cos \theta}{\sin \theta}\right)^2$

48. $\dfrac{(\sec^2 \theta + \tan^2 \theta)^2}{\sec^4 \theta - \tan^4 \theta} = \sec^2 \theta + \tan^2 \theta$

49. $(\csc t - \cot t)^4 (\csc t + \cot t)^4 = 1$

50. $(\sec t + \tan t)^5 (\sec t - \tan t)^6 = \dfrac{1 - \sin t}{\cos t}$

51. $\sin^6 u + \cos^6 u = 1 - 3 \sin^2 u \cos^2 u$

52. $\dfrac{\cos^2 x - \cos^2 y}{\cot^2 x - \cot^2 y} = \sin^2 x \sin^2 y$

53. In a later section, we will learn that
$$\tan 3x = \frac{3 \tan x - \tan^3 x}{1 - 3 \tan^2 x}$$

Taking this for granted, show that
$$\cot 3x = \frac{3 \cot x - \cot^3 x}{1 - 3 \cot^2 x}$$

Note the similarity in form of these two identities.

54. **Challenge.** Generalize Problem 53 by showing that, if $\tan kx = f(\tan x)$ and if k is an odd number, then $\cot kx = f(\cot x)$. (*Hint:* Let $x = \pi/2 - y$.)

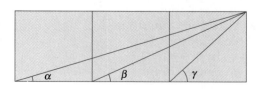

TEASER The figure shows three identical squares side by side. Prove that $\alpha + \beta = \gamma$. *Hint:* It is enough to show that
$$\tan (\alpha + \beta) = \tan \gamma$$

5.3 ADDITION IDENTITIES

When you study calculus, you will meet expressions such as sin $(\alpha + \beta)$ and cos $(\alpha - \beta)$. If you are like some students, you will be tempted to replace these expressions by sin α + sin β and cos α − cos β, respectively. If you do, you will make a terrible mistake. Note, for example, that

$$\sin (30° + 60°) = \sin 90° = 1$$
$$\sin 30° + \sin 60° \approx 0.5 + 0.87 = 1.37$$

and that

$$\cos \left(\frac{\pi}{4} - \frac{\pi}{4} \right) = \cos 0 = 1$$

$$\cos \frac{\pi}{4} - \cos \frac{\pi}{4} = 0$$

This shows that sine and cosine do not operate on $\alpha + \beta$ in a simple way; our task is to discover the more complicated relationships that exist.

A KEY IDENTITY

Consider the two unit circles shown in Figure 1. Each displays a central angle of $\alpha - \beta$ together with a corresponding chord. The orientation of the angle $\alpha - \beta$ surely does not affect the length L of the chord it determines. Moreover, we know the coordinates of the endpoints of these two chords (in terms of sines and cosines) and so the distance formula gives us two different expressions for L^2. When we equate these expressions, a remarkable formula falls out.

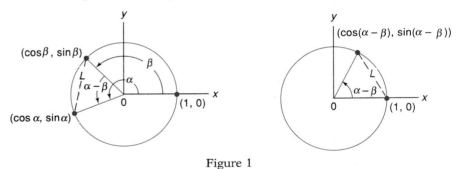

Figure 1

Using the left figure, we obtain

$$L^2 = (\cos \alpha - \cos \beta)^2 + (\sin \alpha - \sin \beta)^2$$
$$= \cos^2 \alpha - 2 \cos \alpha \cos \beta + \cos^2 \beta$$
$$\quad + \sin^2 \alpha - 2 \sin \alpha \sin \beta + \sin^2 \beta$$
$$= 2 - 2(\cos \alpha \cos \beta + \sin \alpha \sin \beta)$$

On the other hand, from the right figure we get

$$L^2 = (\cos(\alpha - \beta) - 1)^2 + (\sin(\alpha - \beta) - 0)^2$$
$$= \cos^2(\alpha - \beta) - 2 \cos(\alpha - \beta) + 1 + \sin^2(\alpha - \beta)$$
$$= 2 - 2 \cos(\alpha - \beta)$$

Thus,

$$\cos(\alpha - \beta) = \cos\alpha\cos\beta + \sin\alpha\sin\beta$$

This identity is important enough to be memorized, and it is best to do so in words.

The cosine of a difference is the cosine of the first times the cosine of the second plus the sine of the first times the sine of the second.

Of course, this identity is true whether α and β are angles measured in degrees, angles measured in radians, or simply numbers.

■ **Example 1.** Use the above identity to evaluate (a) $\cos(\pi/2 - \pi/6)$ and (b) $\cos(15°)$ exactly.

Solution. Since $\pi/2 - \pi/6 = \pi/3$, we know the first answer has to be 1/2; and it is.

(a) $\cos\left(\dfrac{\pi}{2} - \dfrac{\pi}{6}\right) = \cos\dfrac{\pi}{2}\cos\dfrac{\pi}{6} + \sin\dfrac{\pi}{2}\sin\dfrac{\pi}{6}$

$$= 0 \cdot \dfrac{\sqrt{3}}{2} + 1 \cdot \dfrac{1}{2} = \dfrac{1}{2}$$

(b) $\cos(15°) = \cos(45° - 30°)$

$\qquad\qquad = \cos 45°\cos 30° + \sin 45°\sin 30°$

$$= \dfrac{\sqrt{2}}{2}\cdot\dfrac{\sqrt{3}}{2} + \dfrac{\sqrt{2}}{2}\cdot\dfrac{1}{2} = \dfrac{\sqrt{6} + \sqrt{2}}{4} \qquad\qquad ■$$

RELATED IDENTITIES

If we replace β by $-\beta$ in the boxed result; we obtain

$$\cos(\alpha - (-\beta)) = \cos\alpha\cos(-\beta) + \sin\alpha\sin(-\beta)$$
$$= \cos\alpha\cos\beta + \sin\alpha(-\sin\beta)$$

Thus,

$$\cos(\alpha + \beta) = \cos\alpha\cos\beta - \sin\alpha\sin\beta$$

To derive the corresponding result for $\sin(\alpha + \beta)$, we write

$$\sin(\alpha + \beta) = \cos\left(\dfrac{\pi}{2} - (\alpha + \beta)\right) = \cos\left(\left(\dfrac{\pi}{2} - \alpha\right) - \beta\right)$$

$$= \cos\left(\dfrac{\pi}{2} - \alpha\right)\cos\beta + \sin\left(\dfrac{\pi}{2} - \alpha\right)\sin\beta$$

Therefore,

$$\sin(\alpha + \beta) = \sin\alpha\cos\beta + \cos\alpha\sin\beta$$

Finally, replacing β by $-\beta$ in this latter formula yields the identity

$$\boxed{\sin(\alpha - \beta) = \sin\alpha\cos\beta - \cos\alpha\sin\beta}$$

All four of these important addition identities should be memorized (in words).

■ **Example 2.** Use the boxed identities to evaluate exactly (a) $\cos 105°$ and (b) $\sin(-5\pi/12)$.

Solution.

(a) $\cos 105° = \cos(45° + 60°)$
$$= \cos 45° \cos 60° - \sin 45° \sin 60°$$
$$= \frac{\sqrt{2}}{2} \cdot \frac{1}{2} - \frac{\sqrt{2}}{2} \cdot \frac{\sqrt{3}}{2} = \frac{\sqrt{2} - \sqrt{6}}{4}$$

(b) $\sin\left(-\frac{5\pi}{12}\right) = -\sin\left(\frac{5\pi}{12}\right) = -\sin\left(\frac{\pi}{4} + \frac{\pi}{6}\right)$
$$= -\sin\left(\frac{\pi}{4}\right)\cos\left(\frac{\pi}{6}\right) - \cos\left(\frac{\pi}{4}\right)\sin\left(\frac{\pi}{6}\right)$$
$$= -\frac{\sqrt{2}}{2} \cdot \frac{\sqrt{3}}{2} - \frac{\sqrt{2}}{2} \cdot \frac{1}{2} = -\frac{\sqrt{6} + \sqrt{2}}{4} \qquad ■$$

■ **Example 3.** Simplify each of the following by relating them to the addition identities.

(a) $\sin 39° \cos 51° + \cos 39° \sin 51°$
(b) $\cos 1.55 \cos 0.55 + \sin 1.55 \sin 0.55$
(c) $\sin(x + h) \cos h - \cos(x + h) \sin h$

Solution.

(a) $\sin(39° + 51°) = \sin 90° = 1$
(b) $\cos(1.55 - 0.55) = \cos 1$
(c) $\sin(x + h - h) = \sin x \qquad ■$

■ **Example 4.** Suppose $3\pi/2 \le u \le 2\pi$, $\pi \le v \le 3\pi/2$, $\cos u = 4/5$, and $\sin v = -12/13$. Determine $\sin(u - v)$.

Solution.

$$\sin u = -\sqrt{1 - \cos^2 u} = -\sqrt{1 - 16/25} = -\frac{3}{5}$$

$$\cos v = -\sqrt{1 - \sin^2 v} = -\sqrt{1 - 144/169} = -\frac{5}{13}$$

Thus,

$$\sin(u - v) = \sin u \cos v - \cos u \sin v$$
$$= \left(-\frac{3}{5}\right)\left(-\frac{5}{13}\right) - \left(\frac{4}{5}\right)\left(-\frac{12}{13}\right) = \frac{63}{65} \qquad ■$$

■ **Example 5.** Find simple formulas for each of the following.

(a) $\cos\left(t + \dfrac{\pi}{2}\right)$,   (b) $\sin\left(t + \dfrac{\pi}{2}\right)$,   (c) $\cos(t + \pi)$,   (d) $\sin(t + \pi)$

Solution.

(a) $\cos\left(t + \dfrac{\pi}{2}\right) = \cos t \cos\dfrac{\pi}{2} - \sin t \sin\dfrac{\pi}{2}$

$\qquad = (\cos t)(0) - (\sin t)(1) = -\sin t$

(b) $\sin\left(t + \dfrac{\pi}{2}\right) = \sin t \cos\dfrac{\pi}{2} + \cos t \sin\dfrac{\pi}{2}$

$\qquad = (\sin t)(0) + (\cos t)(1) = \cos t$

(c) $\cos(t + \pi) = \cos t \cos\pi - \sin t \sin\pi$

$\qquad = (\cos t)(-1) - (\sin t)(0) = -\cos t$

(d) $\sin(t + \pi) = \sin t \cos\pi + \cos t \sin\pi$

$\qquad = (\sin t)(-1) + (\cos t)(0) = -\sin t$ ■

THE ADDITION IDENTITY FOR THE TANGENT

To obtain an identity for $\tan(\alpha + \beta)$, we begin with a very natural step.

$$\tan(\alpha + \beta) = \frac{\sin(\alpha + \beta)}{\cos(\alpha + \beta)} = \frac{\sin\alpha\cos\beta + \cos\alpha\sin\beta}{\cos\alpha\cos\beta - \sin\alpha\sin\beta}$$

Next, divide numerator and denominator by $\cos\alpha\cos\beta$ to obtain

$$\frac{(\sin\alpha\cos\beta)/(\cos\alpha\cos\beta) + (\cos\alpha\sin\beta)/(\cos\alpha\cos\beta)}{(\cos\alpha\cos\beta)/(\cos\alpha\cos\beta) - (\sin\alpha\sin\beta)/(\cos\alpha\cos\beta)}$$

Thus,

$$\tan(\alpha + \beta) = \frac{\tan\alpha + \tan\beta}{1 - \tan\alpha\tan\beta}$$

Replacing β by $-\beta$ gives the corresponding result for a difference.

$$\tan(\alpha - \beta) = \frac{\tan\alpha - \tan\beta}{1 + \tan\alpha\tan\beta}$$

This last identity has an important application to finding the angle between two lines. Let β be the smallest counterclockwise angle from the positive x-axis to the line $y = m_1x + q_1$. We call β the **angle of inclination** of this line. Similarly, let $\alpha > \beta$ be the angle of inclination of the line $y = m_2x + q_2$. Finally, let θ be the counterclockwise angle from the first line to the second

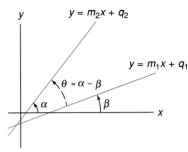

$y = m_2x + q_2$

$y = m_1x + q_1$

$\theta = \alpha - \beta$

α β

Figure 2

one (Figure 2). Note that $\theta = \alpha - \beta$ and that $m_1 = \tan \beta$ and $m_2 = \tan \alpha$. We conclude that

$$\tan \theta = \frac{m_2 - m_1}{1 + m_1 m_2}$$

■ **Example 6.** Determine the angle θ from the line $y = 0.5x + 1$ to the line $y = 4x - 3$.

Solution. According to the formula above,

$$\tan \theta = \frac{4 - 0.5}{1 + 2} = \frac{3.5}{3}$$

Thus,

$$\theta = \tan^{-1}\left(\frac{3.5}{3}\right) \approx 0.86217 \text{ radians} \approx 49.399°$$

Our next example is a little trickier.

■ **Example 7.** Determine the smallest positive angle θ from the line $y = 0.6x + 2$ to the line $y = -x + 1$ (Figure 3).

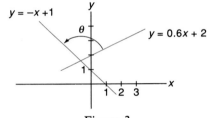

Figure 3

Solution.

$$\tan \theta = \frac{-1 - 0.6}{1 + (0.6)(-1)} = -4$$

Now $\phi = \tan^{-1}(-4) \approx -1.3258$ radians (about $-76°$), but this is not the angle we are looking for. Rather θ is the second-quadrant angle whose tangent is -4. Figure 4 shows the relation between ϕ and θ. We conclude that $\theta \approx 1.8158$ radians, or about $104.04°$. ■

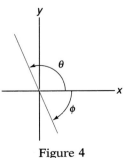

Figure 4

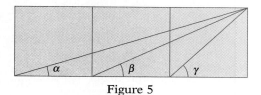

Figure 5

Refer to Figure 5 and note that $\tan \alpha = 1/3$, that $\tan \beta = 1/2$, and that $\tan \gamma = 1$. Moreover,

$$\tan (\alpha + \beta) = \frac{\tan \alpha + \tan \beta}{1 - \tan \alpha \tan \beta} = \frac{\frac{1}{3} + \frac{1}{2}}{1 - \frac{1}{3} \cdot \frac{1}{2}} = 1$$

Thus, $\tan (\alpha + \beta) = \tan \gamma$.

PROBLEM SET 5.3

A. Skills and Techniques

Calculate each pair exactly.

1. (a) $\sin \dfrac{\pi}{4} + \sin \dfrac{\pi}{6}$

   (b) $\sin \left(\dfrac{\pi}{4} + \dfrac{\pi}{6} \right)$

2. (a) $\cos \dfrac{5\pi}{6} + \cos \dfrac{\pi}{6}$

   (b) $\cos \left(\dfrac{5\pi}{6} + \dfrac{\pi}{6} \right)$

3. (a) $\cos \dfrac{3\pi}{4} - \cos \dfrac{\pi}{6}$

   (b) $\cos \left(\dfrac{3\pi}{4} - \dfrac{\pi}{6} \right)$

4. (a) $\sin \dfrac{\pi}{3} - \sin \dfrac{\pi}{4}$

   (b) $\sin \left(\dfrac{\pi}{3} - \dfrac{\pi}{4} \right)$

Use the identities of this section to find exact values for each of the following.

5. $\sin 105°$
6. $\sin 15°$
7. $\cos 75°$
8. $\cos 165°$
9. $\sin 13° \cos 17° + \cos 13° \sin 17°$

10. $\cos 66° \cos 6° + \sin 66° \sin 6°$

11. $\cos \left(\dfrac{\pi}{4} - 2 \right) \cos 2 - \sin \left(\dfrac{\pi}{4} - 2 \right) \sin 2$

12. $\sin \left(\dfrac{\pi}{3} + \dfrac{\pi}{5} \right) \cos \dfrac{\pi}{5} - \cos \left(\dfrac{\pi}{3} + \dfrac{\pi}{5} \right) \sin \dfrac{\pi}{5}$

13. $\cos^2 \left(\dfrac{\pi}{12} \right) - \sin^2 \left(\dfrac{\pi}{12} \right)$

14. $\sin \left(\dfrac{5\pi}{12} + t \right) \cos \left(\dfrac{\pi}{3} - t \right)$
    $$+ \cos \left(\dfrac{5\pi}{12} + t \right) \sin \left(\dfrac{\pi}{3} - t \right)$$

In Problems 15–24, simplify the given expression.

15. $\cos 5 \cos 4 + \sin 5 \sin 4$
16. $\sin 12 \cos 10 - \cos 12 \sin 10$
17. $\sin (\alpha + \beta) \cos \beta - \cos (\alpha + \beta) \sin \beta$
18. $\cos (\alpha + \beta) \cos (\alpha - \beta) - \sin (\alpha + \beta) \sin (\alpha - \beta)$
19. $\sin \left(t - \dfrac{3\pi}{2} \right)$

20. $\cos \left(t + \dfrac{3\pi}{2} \right)$

21. $\cos \left(\dfrac{\pi}{3} - u \right) - \cos \left(\dfrac{\pi}{3} + u \right)$

22. $\sin \left(t + \dfrac{\pi}{6} \right) - \sin \left(t - \dfrac{\pi}{6} \right)$

23. $\sin 4\theta \cos \theta + \cos 4\theta \sin \theta$

24. $\cos (5t - \pi) \cos 4t + \sin (5t - \pi) \sin 4t$

25. Suppose α and β are second-quadrant angles with $\sin \alpha = 3/5$ and $\cos \beta = -3/5$. Determine (a) $\sin (\alpha - \beta)$ and (b) $\cos (\alpha + \beta)$, exactly.

26. Suppose α and β are fourth-quadrant angles with $\sin \alpha = -12/13$ and $\cos \beta = 5/13$. Determine (a) $\sin (\alpha + \beta)$ and (b) $\cos (\alpha - \beta)$, exactly.

27. Suppose α is a second-quadrant angle with $\sin \alpha = 3/5$, and β is a third-quadrant angle with $\cos \beta = -12/13$. Determine (a) $\tan (\alpha - \beta)$ and (b) $\tan (\alpha + \beta)$, exactly.

28. Let α and β be second- and third-quadrant angles, respectively, with $\cos \alpha = \cos \beta = -3/7$. Determine the quadrant for $\alpha - \beta$.

In Problems 29–34, determine the smallest positive angle from the first line to the second line, accurate to the nearest 100th of a degree.

29. $y = 0.6x + 4, y = 3x + 1$

30. $y = -5x + 2, y = -1.5x - 2$

31. $2y - 3x = 1, 2y + 5x = 1$

32. $3y - 4x = 5, 3y + 2x = 4$

33. $y = 0.7x - 1, y = -0.3x + 1$

34. $y = 1.2x + 3, y = -0.5x + 2$

B. Applications and Extensions

In Problems 35–44, prove that each equality is an identity.

35. $\tan (s + \pi) = \tan s$

36. $\tan \left(t + \dfrac{\pi}{4} \right) = \dfrac{1 + \tan t}{1 - \tan t}$

37. $\cot (u + v) = \dfrac{\cot u \cot v - 1}{\cot u + \cot v}$

38. $\dfrac{\sin (u + v)}{\sin (u - v)} = \dfrac{\tan u + \tan v}{\tan u - \tan v}$

39. $\dfrac{\cos 2t}{\sin t} + \dfrac{\sin 2t}{\cos t} = \csc t$

40. $\dfrac{\cos 5t}{\sin t} - \dfrac{\sin 5t}{\cos t} = \cos 6t \sec t \csc t$

41. $\dfrac{\sin (x + h) - \sin x}{h} = \sin x \dfrac{\cos h - 1}{h} + \cos x \dfrac{\sin h}{h}$

42. $\dfrac{\sin (u - v)}{\cos u \cos v} + \dfrac{\sin (v - w)}{\cos v \cos w} + \dfrac{\sin (w - u)}{\cos w \cos u} = 0$

43. $\sin (x + y) \sin (x - y) = \sin^2 x - \sin^2 y$

44. $\cos (x + y) \cos (x - y) = \cos^2 x - \sin^2 y$

45. Graph $f(t) = \sin^2 (t - 2\pi/3) + \sin^2 t + \sin^2 (t + 2\pi/3)$. Conjecture an identity. Prove it.

46. Graph $f(t) = \sin t + \cos t$. From the graph, conjecture values of A and c so that $\sin t + \cos t = A \sin (t - c)$ is an identity. Prove your conjecture.

47. In Figure 6, $BC = 2AB = 2CD$. Determine $\alpha + \beta$ exactly.

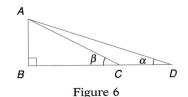

Figure 6

48. Given that $\sin \alpha = 4/5$, $\sin \beta = 5/13$, and $\sin \gamma = 7/25$ where α, β, and γ are first-quadrant angles, determine $\sin (\alpha + \beta + \gamma)$ exactly.

49. Demonstrate that if α, β, and γ are the three angles of a triangle, then $\sin (\alpha + \beta) = \sin \gamma$ and $\cos (\alpha + \beta) = -\cos \gamma$. (Hint: $\alpha + \beta + \gamma = 180°$.)

50. Prove that
$$\cos (u + v)(\cos (u + v) - 2 \cos u \cos v)$$
$$= -\cos (u + v) \cos (u - v) = \sin^2 v - \cos^2 u$$

51. Show that if α, β, and γ are the angles of a triangle, then
$$\cos^2 \alpha + \cos^2 \beta + \cos^2 \gamma + 2 \cos \alpha \cos \beta \cos \gamma = 1$$

52. **Challenge.** Show that if α, β, and γ are the angles of a (nonright) triangle, then
$$\tan \alpha + \tan \beta + \tan \gamma = \tan \alpha \tan \beta \tan \gamma$$

TEASER   Draw the graph of $y = \sin (x/2) + \sqrt{3} \cos (x/2)$ for $-2\pi \le x \le 2\pi$. Then superimpose the graph of $y = 2 \sin (x/2 + \pi/3)$. Make a conjecture. Prove it.

5.4 MORE IDENTITIES

The trigonometric functions occur throughout mathematics and science but often in complicated combinations that must be manipulated or simplified to be useful. Among the tools for doing this are a host of identities that follow easily from the addition identities of Section 5.3. These new identities, spawned by the addition identities, are the subject of this section.

DOUBLE-ANGLE IDENTITIES

The addition identities tell us that

$$\sin (t + t) = \sin t \cos t + \cos t \sin t$$
$$\cos (t + t) = \cos t \cos t - \sin t \sin t$$

By simply writing $2t = t + t$, we are lead to the important double-angle identities.

DOUBLE-ANGLE IDENTITIES

$$\sin 2t = 2 \sin t \cos t$$
$$\cos 2t = \cos^2 t - \sin^2 t = 2 \cos^2 t - 1 = 1 - 2 \sin^2 t$$

Note that we list three equivalent forms for the cosine identity; each of them is useful.

■ **Example 1.** Suppose that $\cos t = 3/5$ and $3\pi/2 \le t \le 2\pi$. Evaluate $\sin 2t$ and $\cos 2t$.

Solution. First, we note that $\sin t = -\sqrt{1 - (3/5)^2} = -4/5$. Thus,

$$\sin 2t = 2 \sin t \cos t = 2\left(-\frac{4}{5}\right)\left(\frac{3}{5}\right) = -\frac{24}{25}$$

$$\cos 2t = 2 \cos^2 t - 1 = 2\left(\frac{3}{5}\right)^2 - 1 = -\frac{7}{25}$$ ■

Occasionally, we need to use the double-angle identities in the reverse direction as in our next example.

■ **Example 2.** Rewrite each expression in a more compact way.
(a) $2 \sin 4u \cos 4u$, (b) $\sin^2 3t - \cos^2 3t$, (c) $6 \cos^2 (3t/2) - 3$.

Solution.
(a) $$2 \sin 4u \cos 4u = \sin (2 \cdot 4u) = \sin 8u$$

(b) $$\sin^2 3t - \cos^2 3t = -(\cos^2 3t - \sin^2 3t)$$
$$= -\cos (2 \cdot 3t) = -\cos 6t$$

(c)
$$6 \cos^2\left(\frac{3}{2}t\right) - 3 = 3\left(2 \cos^2\left(\frac{3}{2}t\right) - 1\right)$$
$$= 3 \cos\left(2 \cdot \frac{3}{2}t\right) = 3 \cos 3t \qquad \blacksquare$$

■ **Example 3.** Express $\sin 3t$ in terms of $\sin t$.

Solution.
$$\sin 3t = \sin(2t + t) = \sin 2t \cos t + \cos 2t \sin t$$
$$= 2 \sin t \cos^2 t + (1 - 2 \sin^2 t) \sin t$$
$$= 2 \sin t (1 - \sin^2 t) + \sin t - 2 \sin^3 t$$
$$= 2 \sin t - 2 \sin^3 t + \sin t - 2 \sin^3 t$$
$$= 3 \sin t - 4 \sin^3 t \qquad \blacksquare$$

HALF-ANGLE IDENTITIES

Consider the identities

$$\cos t = 1 - 2 \sin^2\left(\frac{t}{2}\right) \qquad \cos t = 2 \cos^2\left(\frac{t}{2}\right) - 1$$

In the first, we solve for $\sin(t/2)$ and, in the second, we solve for $\cos(t/2)$. We obtain two more important identities.

HALF-ANGLE IDENTITIES

$$\sin \frac{t}{2} = \pm\sqrt{\frac{1 - \cos t}{2}} \qquad \cos \frac{t}{2} = \pm\sqrt{\frac{1 + \cos t}{2}}$$

Unfortunately, there is no way to remove the ambiguity of the signs, other than writing these identities in their squared forms. Of course, for a specific t we can determine the appropriate sign by noting the quadrant corresponding to $t/2$. Thus, in Example 4, we use the plus sign for $\cos 15°$.

■ **Example 4.** Use a half-angle formula to find $\cos 15°$ exactly. Reconcile the answer with the value $(\sqrt{6} + \sqrt{2})/4$ that we obtained by a different method in Example 1 of Section 5.3.

Solution.

$$\cos 15° = \sqrt{\frac{1 + \cos 30°}{2}} = \sqrt{\frac{1 + \sqrt{3}/2}{2}} = \sqrt{\frac{2 + \sqrt{3}}{4}}$$

The earlier answer $(\sqrt{6} + \sqrt{2})/4$ looks quite different, but let's look at its square.

$$\left(\frac{\sqrt{6} + \sqrt{2}}{4}\right)^2 = \frac{6 + 2\sqrt{12} + 2}{16} = \frac{8 + 4\sqrt{3}}{16} = \frac{2 + \sqrt{3}}{4}$$

We conclude that the two answers are equivalent. $\qquad \blacksquare$

■ **Example 5.** Prove the following half-angle identity for the tangent.

$$\tan \frac{t}{2} = \frac{\sin t}{1 + \cos t}$$

Solution.

$$\frac{\sin t}{1 + \cos t} = \frac{2 \sin (t/2) \cos (t/2)}{1 + 2 \cos^2(t/2) - 1} = \frac{\sin (t/2)}{\cos (t/2)} = \tan \frac{t}{2} \qquad ■$$

PRODUCT-TO-SUM IDENTITIES

Recall the following identities.

$$\cos x \cos y - \sin x \sin y = \cos (x + y)$$
$$\cos x \cos y + \sin x \sin y = \cos (x - y)$$

When we add them together and divide by 2, we obtain the first of the four identities below. The other three can be demonstrated in a similar fashion.

PRODUCT-TO-SUM IDENTITIES

$$\cos x \cos y = \frac{1}{2}(\cos (x + y) + \cos (x - y))$$

$$\sin x \sin y = -\frac{1}{2}(\cos (x + y) - \cos (x - y))$$

$$\sin x \cos y = \frac{1}{2}(\sin (x + y) + \sin (x - y))$$

$$\cos x \sin y = \frac{1}{2}(\sin (x + y) - \sin (x - y))$$

■ **Example 6.** Prove that $\sin (u + v) \sin (u - v) = \sin^2 u - \sin^2 v$ is an identity.

Solution. We appeal to the second of the product-to-sum identities.

$$\sin (u + v) \sin (u - v) = -\frac{1}{2}(\cos 2u - \cos 2v)$$
$$= -\frac{1}{2}(1 - 2 \sin^2 u - (1 - 2 \sin^2 v))$$
$$= \sin^2 u - \sin^2 v \qquad ■$$

SUM-TO-PRODUCT IDENTITIES

Let $u = x + y$ and $v = x - y$ so that $x = (u + v)/2$ and $y = (u - v)/2$. When we substitute these expressions in the product-to-sum identities, we obtain another list of four identities.

■ **Example 7.** Prove that

$$\frac{\cos 9t + \cos 3t}{\sin 9t - \sin 3t} = \cot 3t$$

is an identity.

Solution. We apply the sum-to-product identities to the left side.

$$\frac{\cos 9t + \cos 3t}{\sin 9t - \sin 3t} = \frac{2 \cos 6t \cos 3t}{2 \cos 6t \sin 3t} = \frac{\cos 3t}{\sin 3t} = \cot 3t \qquad ■$$

SIMPLE HARMONIC MOTION

If the value of a quantity can be modeled by the equation $y = A \sin (Bx + C)$, with $B > 0$, the quantity is said to undergo **simple harmonic motion.** We first met this phenomenon in Section 4.4 where we also discussed the related concepts of amplitude, period, and phase shift. Recall the following relationships.

$|A|$:   amplitude

$\dfrac{2\pi}{B}$:   period

$\dfrac{-C}{B}$:   phase shift

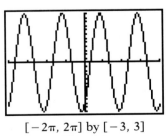

$[-2\pi, 2\pi]$ by $[-3, 3]$
Figure 1

Thus, $y = 3 \sin (2x - \pi/3)$ has amplitude 3, period π, and phase shift $\pi/6$ (Figure 1).

When we expand $A \sin (Bx + C)$ using an addition identity, we obtain

$$A \sin (Bx + C) = A \sin Bx \cos C + A \cos Bx \sin C,$$

which has the form $A_1 \sin Bx + A_2 \cos Bx$. A natural question to ask is whether this statement can be reversed. Can we always write

$$A_1 \sin Bx + A_2 \cos Bx = A \sin (Bx + C)$$

for appropriate choices of A and C? The answer is yes. To see this, let $A =$

$\sqrt{(A_1)^2 + (A_2)^2}$ and let C be an angle in standard position whose terminal side passes through the point (A_1, A_2), as shown in Figure 2. Then

$$A_1 \sin Bx + A_2 \cos Bx = A\left(\frac{A_1}{A}\right) \sin Bx + A\left(\frac{A_2}{A}\right) \cos Bx$$
$$= A \cos C \sin Bx + A \sin C \cos Bx$$
$$= A \sin (Bx + C)$$

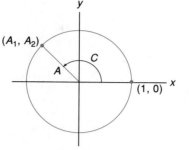

Figure 2

■ **Example 8.** Express $4 \sin 3x - 4 \cos 3x$ in the form $A \sin (3x + C)$.

Solution. Let $A = \sqrt{4^2 + (-4)^2} = \sqrt{32} = 4\sqrt{2}$ and let C be a standard angle passing through $(4, -4)$; that is, let $C = 7\pi/4$. Then

$$4 \sin 3x - 4 \cos 3x = 4\sqrt{2} \sin \left(3x + \frac{7\pi}{4}\right)$$

You can check that this is correct by expanding the right side.   ■

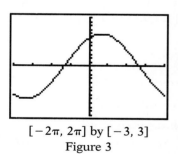

$[-2\pi, 2\pi]$ by $[-3, 3]$

Figure 3

TEASER SOLUTION

The graphs of $y = \sin (x/2) + \sqrt{3} \cos (x/2)$ and $y = 2 \sin (x/2 + \pi/3)$ appear to be identical when superimposed (Figure 3). We conjecture that

$$\sin \frac{1}{2}x + \sqrt{3} \cos \frac{1}{2}x = 2 \sin \left(\frac{1}{2}x + \frac{\pi}{3}\right)$$

This can be demonstrated by calculating A and C as shown above. Alternatively, we may expand the right side (using an addition identity) and thereby show that it is equal to the left side.

$$2 \sin \left(\frac{1}{2}x + \frac{\pi}{3}\right) = 2\left(\sin \frac{1}{2}x \cos \frac{\pi}{3} + \cos \frac{1}{2}x \sin \frac{\pi}{3}\right)$$
$$= 2 \cdot \frac{1}{2} \sin \frac{1}{2}x + 2 \frac{\sqrt{3}}{2} \cos \frac{1}{2}x$$
$$= \sin \frac{1}{2}x + \sqrt{3} \cos \frac{1}{2}x$$

PROBLEM SET 5.4

Refer to Figure 4. Then determine the exact value of each of the following.

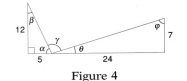

Figure 4

1. $\sin 2\alpha$
2. $\cos 2\theta$
3. $\tan 2\beta$
4. $\csc 2\phi$
5. $\cos\left(\dfrac{\theta}{2}\right)$
6. $\sin\left(\dfrac{\alpha}{2}\right)$
7. $\tan\left(\dfrac{\beta}{2}\right)$
8. $\cot\left(\dfrac{\phi}{2}\right)$
9. $\sin(2\gamma)$
10. $\sin\left(\dfrac{\gamma}{2}\right)$

One or two pieces of information are given. Determine the exact value of the last expression.

11. $\sin t = \dfrac{4}{5}, \dfrac{\pi}{2} < t < \pi; \sin 2t$
12. $\cos t = \dfrac{5}{13}, \dfrac{3\pi}{2} < t < 2\pi; \sin 2t$
13. $\csc t = \dfrac{13}{5}, \cos t < 0; \tan 2t$
14. $\sec t = \dfrac{3}{2}, \sin t > 0; \tan 2t$
15. $\cos t = \dfrac{4}{5}, 0 < t < \dfrac{\pi}{2}; \cos\left(\dfrac{t}{2}\right)$
16. $\sin t = -\dfrac{5}{13}, \pi < t < \dfrac{3\pi}{2}; \sin\left(\dfrac{t}{2}\right)$
17. $\tan t = -\dfrac{3}{4}, \cos t < 0; \tan\left(\dfrac{t}{2}\right)$
18. $\cos t = -\dfrac{1}{3}, \dfrac{\pi}{2} < t < \pi; \cos\left(\dfrac{t}{2}\right)$
19. $\sin t = \dfrac{2}{3}, \tan t > 0; \sin 4t$
20. $\cos t = \dfrac{1}{3}; \cos 4t$
21. $\cos t = -\dfrac{3}{5}; \cos^2\left(\dfrac{t}{2}\right)\cos^2 t \sec 2t$

22. $\sin t = \dfrac{4}{5}, \cos t < 0; \tan\left(\dfrac{t}{2}\right)\tan t \tan 2t$

Use half-angle identities to obtain the exact value of each of the following.

23. $\sin\left(\dfrac{\pi}{12}\right)$
24. $\cos\left(\dfrac{\pi}{8}\right)$
25. $\tan\left(\dfrac{\pi}{8}\right)$
26. $\sin 105°$
27. $\cos 22.5°$
28. $\tan 67.5°$

Use identities from this section to write each expression in a more compact form.

29. $4 \sin 4t \cos 4t$
30. $\cos^2\left(\dfrac{t}{2}\right) - \sin^2\left(\dfrac{t}{2}\right)$
31. $\dfrac{1 + \cos 2x}{2}$
32. $\dfrac{1 - \cos 2u}{2}$
33. $2 \sin^2 4t - 1$
34. $1 - 2\cos^2\left(\dfrac{t}{4}\right)$
35. $\dfrac{\cos 5t - \cos t}{\sin 5t + \sin t}$
36. $\dfrac{\cos 5t + \cos 3t}{\sin 5t + \sin 3t}$

Write each expression in terms of $\cos\theta$ alone.

37. $\cos 3\theta$　　　　38. $\cos 5\theta$

Prove that each of the following is an identity.

39. $(\sin t + \cos t)^2 = 1 + \sin 2t$
40. $\csc 2t + \cot 2t = \cot t$
41. $\sin^2 t \cos^2 t = \dfrac{1}{8}(1 - \cos 4t)$
42. $\cot\left(\dfrac{t}{2}\right) = \dfrac{\sin t}{1 - \cos t}$
43. $\tan 2u = \dfrac{2 \tan u}{1 - \tan^2 u}$
44. $\cot 2u = \dfrac{\cot^2 u - 1}{2 \cot u}$
45. $\dfrac{2 \tan x}{1 + \tan^2 x} = \sin 2x$
46. $\dfrac{1 - \tan^2 x}{1 + \tan^2 x} = \cos 2x$

47. $\cos 4\theta = 8 \cos^4 \theta - 8 \cos^2 \theta + 1$
48. $\sin 4\theta = 4 \sin \theta (2 \cos^3 \theta - \cos \theta)$

Determine A, B, and C so that each expression is equal to A sin (Bx + C).

49. $5 \sin 2t + 5\sqrt{3} \cos 2t$
50. $-2 \sin 3t + 2 \cos 3t$
51. $-2\sqrt{3} \sin t + 2 \cos t$
52. $3\sqrt{3} \sin 5t - 3 \cos 5t$

B. Applications and Extensions

In Problems 53–68, establish that the given equality is an identity.

53. $\cos^4 z - \sin^4 z = \cos 2z$

54. $\dfrac{1 - \cos 4x}{\tan^2 2x} = 2 \cos^2 2x$

55. $1 + \dfrac{1 - \cos 8t}{1 + \cos 8t} = \sec^2 4t$

56. $\sec 2t = \dfrac{\sec^2 t}{2 - \sec^2 t}$

57. $\tan\left(\dfrac{\theta}{2}\right) - \sin \theta = \dfrac{-\sin \theta}{1 + \sec \theta}$

58. $(2 - \sec^2 2\theta) \tan 4\theta = 2 \tan 2\theta$
59. $3 \cos 2t + 4 \sin 2t = (3 \cos t - \sin t)(\cos t + 3 \sin t)$
60. $\csc 2x - \cot 2x = \tan x$

61. $2(\cos 3x \cos x + \sin 3x \sin x)^2 = 1 + \cos 4x$

62. $\dfrac{1 + \sin 2x + \cos 2x}{1 + \sin 2x - \cos 2x} = \cot x$

63. $\tan 3t = \dfrac{3 \tan t - \tan^3 t}{1 - 3 \tan^2 t}$

64. $\cos^4 u = \dfrac{3}{8} + \dfrac{1}{2} \cos 2u + \dfrac{1}{8} \cos 4u$

65. $\sin^4 u + \cos^4 u = \dfrac{3}{4} + \dfrac{1}{4} \cos 4u$

66. $\cos^6 u - \sin^6 u = \cos 2u - \dfrac{1}{4} \sin^2 2u \cos 2u$

67. $\cos^2 x + \cos^2 2x + \cos^2 3x$
$= 1 + 2 \cos x \cos 2x \cos 3x$

68. $\cos x \cos 2x \cos 4x \cos 8x \cos 16x = \dfrac{\sin 32x}{32 \sin x}$

69. Let α and β be the acute angles of a right triangle. Demonstrate that $\sin 2\alpha - \sin 2\beta = 0$ and that $\cos 2\alpha + \cos 2\beta = 0$.

70. Let α, β, and γ be the three angles of a triangle. Prove that

$$\sin 2\alpha + \sin 2\beta + \sin 2\gamma = 4 \sin \alpha \sin \beta \sin \gamma$$

71. Draw the graph of $y = 2 \cos x - 4 \sin x \sin 2x$. Conjecture an identity. Prove it.
72. Draw the graph of $y = (\sin 2t)(3 - 16 \cos^2 t \sin^2 t) - \sin 6t$. Conjecture an identity. Prove it.
73. Draw the graph of $y = 32 \cos^6 t - 48 \cos^4 t + 18 \cos^2 t - 1$. Conjecture an identity. Prove it.
74. **Challenge.** Determine the exact value of $\sin 1°$ $\sin 3° \sin 5° \ldots \sin 175° \sin 177° \sin 179°$.

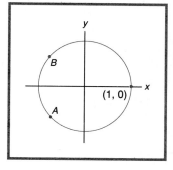

TEASER Two bugs, A and B, crawl around a vertical unit circle, both starting at (1, 0). Bug A moves at 3 units per minute and bug B at 2 units per minute. Determine the first time when bug B will be directly above bug A. At what time do they first become 1.5 units apart?

5.5 TRIGONOMETRIC EQUATIONS

To solve an algebraic equation $f(x) = 0$, we typically factor $f(x)$, set each factor equal to 0, and then solve the resulting linear equations. For example, to solve $x^3 - x^2 - 6x = 0$, we write $x(x - 3)(x + 2) = 0$ and set each factor equal to 0, thereby identifying the three solutions 0, 3, and -2. If we are unable to factor $f(x)$, we may graph $y = f(x)$ and use the zooming process to find the x-coordinates of the points where the graph crosses the x-axis.

Similar methods apply to solving trigonometric equations. First, we explore the factoring method. Consider, for example, the equation

$$3 \sin^2 x + 7 \sin x + 2 = 0$$

which in factored form becomes

$$(\sin x + 2)(3 \sin x + 1) = 0$$

Setting the first factor to 0 yields no solutions (remember that $-1 \leq \sin x \leq 1$); setting the second factor to zero gives

$$\sin x = -\frac{1}{3}$$

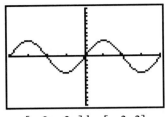

[$-2\pi, 2\pi$] by [$-3, 3$]

Figure 1

Our first impulse is to say that $x = \sin^{-1}(-1/3)$, press the appropriate keys on a calculator, and report the answer as -0.3398. One look at the graph of the sine function (Figure 1), however, tells us that $\sin x = -1/3$ has infinitely many solutions. Suppose we restrict attention to the interval $0 \leq x \leq 2\pi$. Then the graph tells us that there are two solutions but neither of them is -0.3398. How do we find the two solutions we want? The answer has to do with the concept of reference angles.

REFERENCE ANGLES AND REFERENCE NUMBERS

Let θ be an angle in standard position and let t be its radian measure. Associated with θ is an angle θ_0, called its **reference angle,** and defined to be the smallest nonnegative angle between the terminal side of θ and the x-axis. The radian measure t_0 of θ_0 is called the **reference number** corresponding to t. Note that all four of the angles shown in Figure 2 have the same reference angle θ_0 and the same reference number t_0.

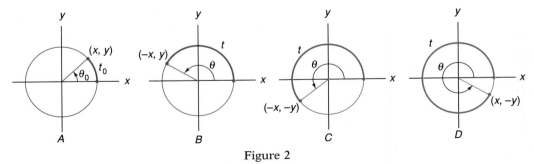

Figure 2

Here is the crucial fact, a fact that can be seen from the symmetries in Figure 2. For any of the six trigonometric functions T,

$$T(\theta) = \pm T(\theta_0) \quad \text{and} \quad T(t) = \pm T(t_0)$$

■ **Example 1.** Express each of the following in terms of the same trigonometric function of a reference angle (number).
(a) $\sin 154°$ (b) $\cos 193°$ (c) $\tan 5.5411$ (d) $\sin 3.6789$

Solution. Figure 3 shows the appropriate reference angles (numbers). The plus or minus sign is determined from our knowledge of the trigonometric functions in the four quadrants (for example, sine is positive in quadrants I and II and negative in quadrants III and IV).

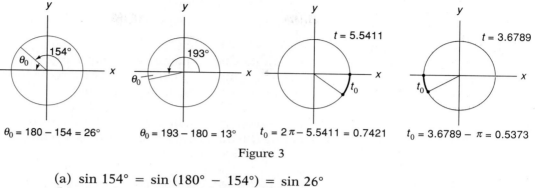

$$\theta_0 = 180 - 154 = 26° \qquad \theta_0 = 193 - 180 = 13° \qquad t_0 = 2\pi - 5.5411 = 0.7421 \qquad t_0 = 3.6789 - \pi = 0.5373$$

Figure 3

(a) $\sin 154° = \sin (180° - 154°) = \sin 26°$
(b) $\cos 193° = -\cos (193° - 180°) = -\cos 13°$
(c) $\tan 5.5411 = -\tan (2\pi - 5.5411) = -\tan 0.7421$
(d) $\sin 3.6789 = -\sin (3.6789 - \pi) = -\sin 0.5373$ ∎

Returning to the matter of solving an equation of the form $\sin x = c$ where c is a constant, we first find the reference angle (number) for x by calculating $x_0 = \sin^{-1} |c|$, which can be done using a calculator. Then we determine those angles (numbers) on the given interval that have x_0 as reference angle (number).

■ Example 2. Find all solutions of $\sin x = -1/3$ on the interval $0 \le x \le 2\pi$. See Figure 1.

Solution. We find $x_0 = \sin^{-1}(1/3) \approx 0.3398$. The required solutions are $\pi + 0.3398$ and $2\pi - 0.3398$, that is, 3.4814 and 5.9434. ∎

■ Example 3. Find all solutions of $\tan x = 0.7891$ on the interval $0 \le x \le 2\pi$. See Figure 4.

Solution. A calculator gives $x_0 = \tan^{-1}(0.7891) \approx 0.6681$. Recalling that the tangent is positive in the first and third quadrants, we see that the required solutions are 0.6681 and $\pi + 0.6681 \approx 3.8097$. ∎

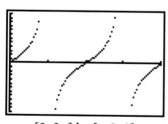

$[0, 2\pi]$ by $[-3, 3]$
Figure 4

QUADRATIC TYPE EQUATIONS

The equation $x^2 = 3/4$ has the two solutions $x = \pm\sqrt{3}/2$; they are obtained by taking square roots.

■ Example 4. Find all solutions of $\cos^2 x = 3/4$ on the interval $0 \le x \le 2\pi$.

Solution. By taking square roots, we obtain $\cos x = \sqrt{3}/2$ and $\cos x = -\sqrt{3}/2$. The reference number (which we should be able to determine mentally) is $x_0 = \cos^{-1}(\sqrt{3}/2) = \pi/6$. There are four solutions to the given equation; they are $\pi/6$, $5\pi/6$, $7\pi/6$, and $11\pi/6$. ∎

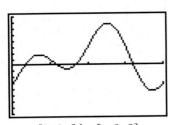

[0, 2π] by [−8, 8]

Figure 5

■ **Example 5.** Find all solutions of $2 \sin^2 x - \sin x - 1 = 0$ on $0 \le x \le 2\pi$.

Solution. We may factor the given equation to obtain $(2 \sin x + 1)(\sin x - 1) = 0$. Thus, we must solve $\sin x = -1/2$ and $\sin x = 1$. The first has the solutions $7\pi/6$ and $11\pi/6$; the second has the single solution $\pi/2$. ■

■ **Example 6.** Find the zeros of the function $f(x) = (2 \sin x - 1)(\sin x + 3 \cos x)$ for x in the interval $0 \le x \le 2\pi$.

Solution. To help our thinking, we graph $y = f(x)$ and note that it has four zeros in the given interval (Figure 5). They are to be found by solving $\sin x = 1/2$ and $\sin x = -3 \cos x$ (equivalently, $\tan x = -3$). The first is easy; it has the solutions $\pi/6$ and $5\pi/6$. To solve the second, note that $x_0 = \tan^{-1}(3) \approx 1.2490$. From this, we obtain the two additional solutions $\pi - 1.2490$ and $2\pi - 1.2490$. Written in decimal form, the four required solutions are 0.5236, 2.6180, 1.8926, and 5.0342. ■

GRAPHICAL METHODS

What if we are given a complicated equation that we cannot factor? Then we may be forced to go to a graphical method, where a graphics calculator becomes an indispensable aid.

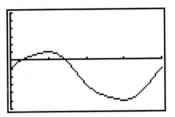

[0, 2π] by [−6, 6]

Figure 6

■ **Example 7.** Solve $\cos^3 x + 3 \sin x - 2 = 0$ for $0 \le x \le 2\pi$.

Solution. We begin by graphing $y = \cos^3 x + 3 \sin x - 2$ on the required interval (Figure 6) and note that there are two solutions, one near 0.5 and the other near 2. Zooming allows us to determine these solutions to any desired degree of accuracy. To three-decimal-place accuracy, we obtain 0.429 and 2.280. ■

AN APPLICATION

Before introducing an application involving simple harmonic motion, we solve another simple equation, which, however, has a new twist.

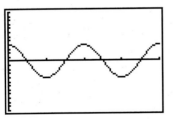

[0, 2π] by [−3, 3]

Figure 7

■ **Example 8.** Find all solutions of $\cos 2x = 0.34$ on the interval $0 \le x \le 2\pi$.

Solution. One glance at the graph of $\cos 2x$ (Figure 7) shows us that we are looking for four answers. A calculator gives $\cos^{-1}(0.34) \approx 1.2239$, which is a value for $2x$. Now to make x range between 0 and 2π, we must make $2x$ range between 0 and 4π. Thus, three other values for $2x$ are $2\pi - 1.2239$, $2\pi + 1.2239$, and $4\pi - 1.2239$. Dividing each of the four solutions for $2x$ by 2 gives us the solutions we want: 0.6119, 2.5297, 3.7535, and 5.6712. ■

■ **Example 9.** In Lake City, the length L of daylight, in minutes, on a day that comes t days after the beginning of the year is given by

$$L = L(t) = 720 + 240 \sin\left(\frac{2\pi}{365}(t - 81)\right)$$

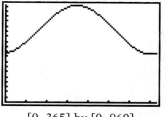

[0, 365] by [0, 960]

Figure 8

(a) On what day does L reach a maximum?

(b) On what days does L exceed 15 hours (900 minutes)?

Solution. The graph of L is shown in Figure 8.

(a) Since sin ($\pi/2$) = 1, L reaches its maximum when $(2\pi/365)(t - 81) = \pi/2$, that is, when $t \approx 172$ (June 21). On that day, there will be 720 + 240 = 960 minutes (16 hours of daylight).

(b) We must solve the inequality

$$720 + 240 \sin\left(\frac{2\pi}{365}(t - 81)\right) \geq 900$$

which means that we need to find the two solutions of

$$\sin\left(\frac{2\pi}{365}(t - 81)\right) = 0.75$$

on the interval $0 \leq t \leq 365$. Now $\sin^{-1}(0.75) = 0.8481$, so the two solutions of $\sin x = 0.75$ on $0 \leq x \leq 2\pi$ are 0.8481 and $\pi - 0.8481 \approx 2.2935$. Setting $(2\pi/365)(t - 81)$ equal to these two values and solving for t gives 130 and 214. Thus, Lake City has more than 15 hours of daylight from day 130 to day 214, that is, from May 10 to August 2. ∎

TEASER SOLUTION

Figure 9 shows the position of the two bugs after t minutes. Note that the coordinates of the bugs A and B are (cos $3t$, sin $3t$) and (cos $2t$, sin $2t$) after t minutes. Thus, bug B will be directly over bug A when cos $2t$ = cos $3t$. To solve this equation, we graph $f(t) = \cos 2t - \cos 3t$ (Figure 10) and use the zooming process to determine that the graph first crosses the t-axis after $t = 0$ at $t \approx 1.26$ minutes.

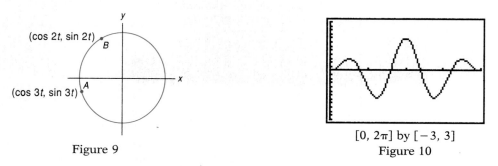

Figure 9

[0, 2π] by [−3, 3]

Figure 10

The two bugs first reach 1.5 units apart when

$$(\cos 3t - \cos 2t)^2 + (\sin 3t - \sin 2t)^2 = 1.5^2$$

which looks like an algebraic mess. But this equation simplifies after expanding and collecting terms to

$$2 - 2(\cos 3t \cos 2t + \sin 3t \sin 2t) = 2.25$$

and then, by use of the addition identity for cos $(3t - 2t)$, to

$$\cos t = \frac{2.25 - 2}{-2} = -0.125$$

This last equation has the solution $t \approx 1.70$. The two bugs will first reach 1.5 units apart at 1.70 minutes.

PROBLEM SET 5.5

A. Skills and Techniques

Express the given function in terms of the same function at the reference angle (number). For example, cos 121° = −cos 59°.

1. sin 332°
2. tan 192°
3. cos 296°
4. sin 213°
5. tan (−146°)
6. cos (−37°)
7. cot (478°)
8. sec (567°)
9. sin 3
10. cos 4
11. tan 3.33
12. tan 5.21
13. cos 7.2832
14. sin 8.4531

Find all solutions to the given equation on the interval $0 \le t \le 2\pi$. You should be able to do many of them mentally and give exact answers. For others you will need a calculator and then you should give answers accurate to three decimal places.

15. $\sin t = 0$
16. $\sin t = 1$
17. $\cos t = 0.5$
18. $\cos t = -\sqrt{3}/2$
19. $\sin t = -0.2345$
20. $\cos t = -0.5432$
21. $\tan t = -1.4567$
22. $\tan t = 5.3212$
23. $\cos t = 1.3212$
24. $\csc t = 0.8765$
25. $\tan^2 t = 1$
26. $4 \sin^2 t - 2\sqrt{3} = 0$
27. $(\sin t - 2)(2 \cos t - 1) = 0$
28. $\sec t \, (3 \tan t - \sqrt{3}) = 0$
29. $2 \sin^2 t + \sin t = 0$
30. $\sec^2 t = 1 + \tan^2 t$
31. $\tan^2 t - \tan t - 2 = 0$
32. $\sec^2 t - \sec t - 12 = 0$

33. $\cot^2 t + 7 \cot t + 12 = 0$
34. $2 \sin^2 t + 3 \sin t = -1$
35. $\cos^3 t - 4 \cos^2 t + 4 \cos t = 0$
36. $\tan^2 t \sin t + 4 \sin t = 0$
37. $\cos 2t = 3 \sin t - 1$
38. $\sin t = 4 \cos t$
39. $\sin 3t = 0.5$
40. $\tan 2t = -\sqrt{3}$
41. $3 \cos^3 t + 2 \cos^2 t + 2 \cos t - 1 = 0$
42. $4 \tan^3 t + 3 \tan t - 2 = 0$
43. $\cos^3 2t - \sin t + 1 = 0$
44. $\tan^4 (0.3t) + 3 \cos 2t - 4 = 0$

Solve each inequality on the interval $0 \le x \le 2\pi$, accurate to three decimal places.

45. $2 \cos x \le 3 \sin x$
46. $2 \cos x \ge \sin \left(\dfrac{x}{2}\right)$
47. $\sin x \ge 0.5x + 0.1$
48. $\tan^{-1} x \ge \tan \left(\dfrac{x}{4}\right)$

B. Applications and Extensions

In Problems 49–58, find the exact values of all solutions in the interval $0 \le t \le 2\pi$. You may need to do some algebra or use identities as a first step. For example, in Problem 49, begin by squaring both sides and then write everything in terms of cosines. Be sure to check your answers, because squaring can introduce extraneous solutions.

49. $1 - \cos t = \sqrt{3} \sin t$
50. $\sin t + \cos t = 1$
51. $\tan 2t = 3 \tan t$
52. $\cos \left(\dfrac{x}{2}\right) = 1 + \cos x$
53. $4 \sin t - 4 \sin^3 t + \cos t = 0$
54. $\cos^8 t - \sin^8 t = 0$
55. $\sin t \cos t \, (\sin^2 t - \cos^2 t) = -1$
56. $\sin t \cos t = -\dfrac{1}{2}$

57. $\sin(1 + \cos t) = 0$

58. $\cos(\sin^2 t + 1) = 1$

59. A ray of light from the lamp L in Figure 11 reflects off a mirror to the object O.
  (a) Find the distance x.
  (b) Write an equation for θ.
  (c) Solve this equation.

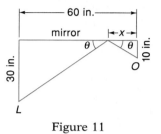

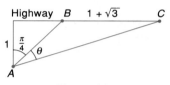

Figure 11

60. Lin and Fred are lost in a desert 1 mile from a highway, at point A in Figure 12. Each strikes out in a different direction to get to the highway. Lin gets to the highway at point B, and Fred arrives at point C, $1 + \sqrt{3}$ miles farther down the road. Write an equation for θ and solve it.

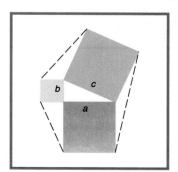

Figure 12

61. In a certain city, the number T of hours of daylight in any given day is approximated by

$$T = 12 + 4\sin\left(\frac{2\pi}{365}(t - 92)\right)$$

where t is the number of days after midnight on December 31.
  (a) How many hours of daylight are there on August 1?
  (b) How many days of the year have more than 14 hours of daylight?

62. The range r in feet (Figure 13) of a projectile shot from ground level with an angle of elevation θ and initial velocity v is given by $r = r(\theta) = (v^2/16)\sin\theta\cos\theta$ where v is in feet per second.
  (a) What value of θ gives the maximum range?
  (b) Suppose $v = 500$ and a target is 7000 feet downrange. Determine two values for θ that allow the projectile to hit the target.

Figure 13

63. Solve $\sin 2x \geq \cos x$ exactly on the interval $0 \leq x \leq 2\pi$.

64. Find the exact solutions to $\sin x \sin 2x \sin 3x = \cos x$ on $0 \leq x \leq \pi$.

65. Find the exact solutions to $\cos^8 u + \sin^8 u = 41/128$ on $0 \leq u \leq \pi$. (*Hint:* Begin by using half-angle identities.)

66. **Challenge.** Show that $t = \pi/4$ is the only solution to

$$\frac{a + b\cos t}{b + a\sin t} = \frac{a + b\sin t}{b + a\cos t}$$

on the interval $0 \leq t \leq \pi$. (*Hint:* $a^2 + b^2 \geq 2|ab|$.)

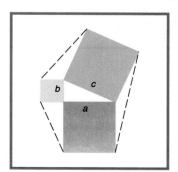

TEASER   A rubber band is stretched around the Pythagorean figure shown at the left ($a^2 + b^2 = c^2$). Find a simple formula for the total area of the region thus enclosed.

We suggest that you review Section 4.1, which introduced the sine and cosine functions as they are related to right triangles. We remind you of two important facts. If θ denotes one of the acute angles in a right triangle, then

$$\sin \theta = \frac{\text{opp}}{\text{hyp}} \qquad \cos \theta = \frac{\text{adj}}{\text{hyp}}$$

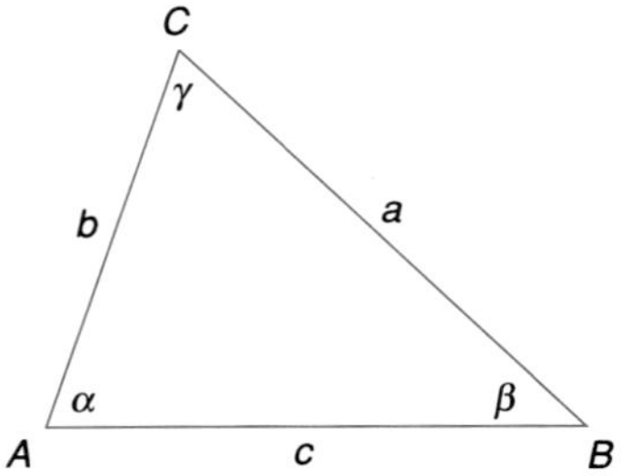

where opp, adj, and hyp are abbreviations for the lengths of the opposite side, adjacent side, and hypotenuse.

Our goal in this and the next section is to relate the sine and cosine functions to the lengths of the sides in a general triangle, that is, a triangle in which there may or may not be a right angle. We will denote the vertices of such a triangle by A, B, and C, the corresponding angles by α, β, and γ, and the lengths of the sides opposite these angles by a, b, and c (Figure 1). We remind you of an identity that will play a critical role in what follows, namely,

$$\sin \theta = \sin (180° - \theta)$$

Figure 1

THE LAW OF SINES PROVED

The angle α, being an angle in a general triangle, can be either acute or obtuse (or right). The two situations to be considered are shown in Figure 2 with appropriate notation.

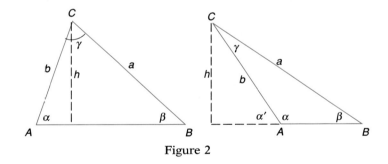

Figure 2

Consider the first triangle in Figure 2 and note that we have two expressions for h (the length of the perpendicular from C to AB), namely,

$$h = b \sin \alpha \qquad h = a \sin \beta$$

Equating these expressions gives

$$b \sin \alpha = a \sin \beta$$

and after dividing both sides by ab

$$\frac{\sin \alpha}{a} = \frac{\sin \beta}{b}$$

Next, consider the second triangle in Figure 2. Since $\sin \alpha' = \sin \alpha$, the relations derived for the first triangle hold also for the second. Furthermore, in both cases the roles of α and γ can be interchanged. Thus, we can state the following very general result.

LAW OF SINES

In any triangle with angles α, β, γ and corresponding lengths of opposite sides a, b, c,

$$\frac{\sin \alpha}{a} = \frac{\sin \beta}{b} = \frac{\sin \gamma}{c}$$

SOLVING TRIANGLES

In general, we can solve a triangle (determine its unknown parts) if we know three pieces of information, one of which is a side. The law of sines is particularly useful in two cases: **AAS** (two angles and a side are given) and **SSA** (two sides and the angle opposite one of them are given).

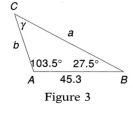

Figure 3

■ **Example 1 (AAS).** Suppose that in triangle ABC, $\alpha = 103.5°$, $\beta = 27.5°$, and $c = 45.3$ (Figure 3). Find γ, a, and b.

Solution.
Step 1: Since $\alpha + \beta + \gamma = 180°$,

$$\gamma = 180° - (103.5° + 27.5°) = 49°$$

Step 2: By the law of sines,

$$\frac{a}{\sin 103.5°} = \frac{45.3}{\sin 49°}$$

Thus,

$$a = \frac{45.3 \sin 103.5°}{\sin 49°} \approx 58.4$$

Step 3: Also by the law of sines,

$$\frac{b}{\sin 27.5°} = \frac{45.3}{\sin 49°}$$

and so

$$b = \frac{45.3 \sin 27.5°}{\sin 49°} \approx 27.7 \qquad\blacksquare$$

The next case (SSA) is called the **ambiguous case** because the given information does not always determine a unique triangle. We suppose that α, a, and b are given. If $a \geq b$, a unique triangle is determined. If $a < b$, there are three cases. Figure 4 shows the various possibilities and Examples 2, 3, and 4 offer illustrations.

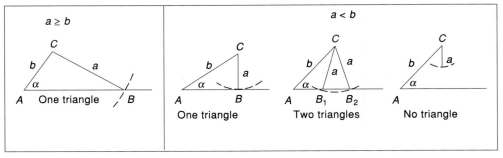

Figure 4

■ **Example 2 (SSA with $a \geq b$).** Suppose that in triangle ABC, $\alpha = 110.6°$, $a = 48.3$, and $b = 32.4$. Solve the triangle (Figure 5).

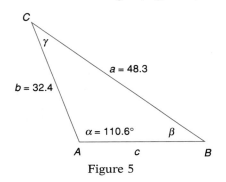

Figure 5

Solution.

Step 1: By the law of sines

$$\frac{\sin \beta}{32.4} = \frac{\sin 110.6°}{48.3}$$

Therefore

$$\sin \beta = \frac{32.4 \sin 110.6°}{48.3} \approx 0.62792$$

and $\beta \approx 38.9°$.

Step 2: $\gamma = 180° - (\alpha + \beta) \approx 180° - (110.6° + 38.9°) = 30.5°$

Step 3: $\dfrac{c}{\sin 30.5°} = \dfrac{48.3}{\sin 110.6°}$

and so

$$c = \frac{48.3 \sin 30.5°}{\sin 110.6°} \approx 26.2$$

■

■ **Example 3 (SSA with *a* < *b*).** Suppose that in triangle *ABC*, $\alpha = 36.21°$, $a = 10.42$, and $b = 13.14$ (Figure 6). Solve the two triangles determined by this information.

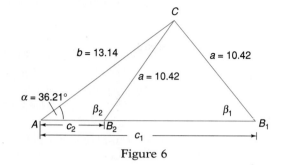

Figure 6

Solution.

Step 1:
$$\frac{\sin \beta}{13.14} = \frac{\sin 36.21°}{10.42}$$

$$\sin \beta = \frac{13.14 \sin 36.21°}{10.42} \approx 0.74495$$

Now $\sin^{-1}(0.74495) \approx 48.16°$, so the two angles we want are

$$\beta_1 \approx 48.16° \qquad \beta_2 \approx 180° - 48.16° = 131.84°$$

Step 2:
$$\gamma_1 \approx 180° - (36.21° + 48.16°) = 95.63°$$
$$\gamma_2 \approx 180° - (36.21° + 131.84°) = 11.95°$$

Step 3:
$$c_1 \approx \frac{10.42 \sin 95.63°}{\sin 36.21°} \approx 17.55$$

$$c_2 \approx \frac{10.42 \sin 11.95°}{\sin 36.21°} \approx 3.65$$

Note that we do not give our answers with greater accuracy than the data. ■

■ **Example 4 (SSA with *a* < *b*).** Suppose we are told that triangle *ABC* has $\alpha = 72.2°$, $a = 14.3$, and $b = 16.9$. Solve the triangle (if possible).

Solution. We begin by attempting to draw the triangle (Figure 7), but we are immediately suspicious that side *a* may not be long enough to reach to side *c*. We forge ahead, however, and apply the law of sines.

$$\sin \beta = \frac{16.9 \sin 72.2°}{14.3} \approx 1.1252$$

Of course, no angle β has a sine greater than 1. We conclude that the given information is inconsistent; no such triangle exists.

What if a similar calculation shows that $\sin \beta$ is exactly 1? Then there is a unique triangle consistent with the given information; it is a right triangle. ■

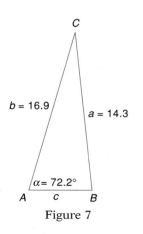

Figure 7

■ **Example 5.** Cities A and B are 403 miles apart as the crow flies. But the shortest highway route goes from A to C to B as shown in Figure 8. Ira can fly from A to B for $210.50 or take a taxi at 45¢ per mile. Which route is cheaper and by how much?

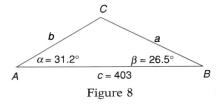

Figure 8

Solution. First we must determine the distances a and b, which we do using the law of sines.

$$a = \frac{403 \sin 31.2°}{\sin 122.3°} \approx 247 \text{ miles}$$

$$b = \frac{403 \sin 26.5°}{\sin 122.3°} \approx 213 \text{ miles}$$

Thus, the cost by taxi is $(247 + 213)(0.45) = \$207.00$, which is $3.50 cheaper than by air. ■

AN IMPORTANT AREA FORMULA

Consider the two triangles shown in Figure 9. Because the area A of a triangle is given by the formula $A = bh/2$ and for either triangle, $h = c \sin \alpha$, we conclude that

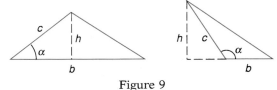

Figure 9

$$A = \frac{1}{2} bc \sin \alpha$$

For example, the area of the triangle with $\alpha = 38.7°$, $b = 34.5$ meters, and $c = 41.2$ meters is

$$A = \frac{1}{2}(34.5)(41.2) \sin 38.7° \approx 444 \text{ square meters}$$

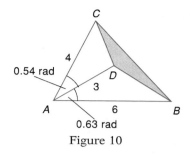

Figure 10

■ **Example 6.** Find the area A of the shaded triangle BCD in Figure 10.

Solution. The required area can be found by subtracting the areas of triangles ACD and ABD from that of ABC. Thus,

$$A = \frac{1}{2}(4)(6)\sin 1.17 - \frac{1}{2}(3)(4)\sin 0.54 - \frac{1}{2}(3)(6)\sin 0.63$$

$$\approx 2.66 \text{ square units}$$ ■

■ **Example 7.** Spokes of lengths 4, 3, and 6 emanate from the same point O and are separated by angles of size t radians, $0 \le t \le \pi$, as shown in three configurations in Figure 11. The tips of these spokes determine the shaded triangle ABC. Write a formula for the area A of this triangle and determine its maximum value.

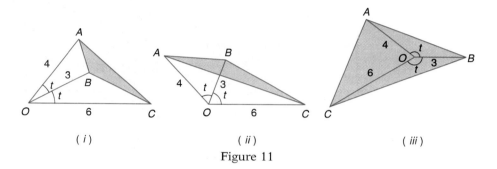

(i)   (ii)   (iii)

Figure 11

Solution. From left to right, the areas A in the three cases of Figure 11 are as follows.

(i) $A = \dfrac{1}{2}(24\sin 2t - 12\sin t - 18\sin t)$

(ii) $A = \dfrac{1}{2}(12\sin t + 18\sin t - 24\sin 2t)$

(iii) $A = \dfrac{1}{2}(12\sin t + 18\sin t + 24\sin(2\pi - 2t))$

$ = \dfrac{1}{2}(12\sin t + 18\sin t - 24\sin 2t)$

In each case, the formula simplifies to either $12\sin 2t - 15\sin t$ or its negative and, since A is positive, we may summarize in the single formula

$$A = |\,12\sin 2t - 15\sin t\,|$$

The graph of this equation is shown in Figure 12. Using a calculator, we find that A is maximized when $t = 2.175$ radians (about 124.6°), with a maximum value of $A = 23.56$. ■

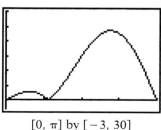

$[0, \pi]$ by $[-3, 30]$

Figure 12

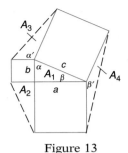

Figure 13

TEASER SOLUTION

The configuration whose area A we are to find is reproduced in Figure 13. Note that since $\alpha + \alpha' = 180°$, $\sin \alpha' = \sin \alpha$; similarly, $\sin \beta' = \sin \beta$. Thus

$$A = a^2 + b^2 + c^2 + A_1 + A_2 + A_3 + A_4$$

$$= a^2 + b^2 + a^2 + b^2 + \frac{1}{2} ab + \frac{1}{2} ab + \frac{1}{2} bc \sin \alpha + \frac{1}{2} ac \sin \beta$$

$$= 2a^2 + 2b^2 + ab + \frac{1}{2} bc \cdot \frac{a}{c} + \frac{1}{2} ac \cdot \frac{b}{c} = 2(a^2 + b^2 + ab)$$

PROBLEM SET 5.6

A. Skills and Techniques

Solve the triangles with the indicated parts. Report your answers with the same accuracy as the data.

1. $\alpha = 43.34°$, $\beta = 80.85°$, $a = 14.36$
2. $\beta = 122.23°$, $\gamma = 13.78°$, $a = 295.34$
3. $\alpha = \gamma = 63.24°$, $b = 51.23$
4. $\alpha = \beta = 14.561°$, $c = 31.234$
5. $\alpha = 114.78°$, $a = 46.57$, $b = 34.32$
6. $\beta = 145.23°$, $a = 46.57$, $b = 85.32$
7. $\alpha = 30.267°$, $a = 8.459$, $b = 5.112$
8. $\beta = 59.76°$, $a = 11.44$, $b = 12.03$
9. $\alpha = 30.267°$, $a = 5.112$, $b = 8.459$
10. $\beta = 59.76°$, $a = 12.03$, $b = 11.44$
11. $\alpha = 44.23°$, $a = 6.91$, $b = 10.23$
12. $\alpha = 30.00°$, $a = 12.59$, $b = 25.18$

In Problems 13–18, find the area of the triangle with the given data.

13. $\alpha = 40.24°$, $b = 20.23$, $c = 30.43$
14. $\gamma = 130.24°$, $a = 13.98$, $b = 32.45$
15. $\alpha = 25.32°$, $\beta = 112.56°$, $c = 30.12$
16. $\alpha = 29.34°$, $\gamma = 45.78°$, $a = 20.98$
17. $\alpha = 1.12$, $\beta = 1.37$, $a = 5.61$
18. $\alpha = 1.01$, $a = 9.23$, $b = 5.31$
19. Two observers stationed 110 meters apart at A and B on the bank of a river are looking at a post situated at a point C on the opposite bank. They measure angles CAB and CBA to be 43.2° and 57.3°, respectively. How far is the observer at B from the post at C?
20. A telegraph pole leans away from the sun at an angle of 11° from the vertical (Figure 14). The pole casts a shadow 96 feet long on the horizontal ground when the angle of elevation of the sun is 23°. Find the length of the pole.

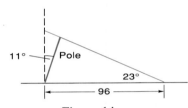

Figure 14

21. A vertical pole 60 feet long is standing on an incline that makes an angle θ with the horizontal (Figure 15). When the angle of elevation of the sun is 58°, the pole casts a shadow 138 feet long directly down the incline. Determine θ.

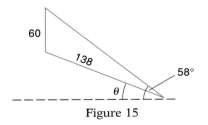

Figure 15

22. Two forest rangers 15 miles apart at points A and B along a straight road observe a fire at point C. The ranger at A measures angle CAB as 43.6° and the one at B measures angle CBA as 79.3°. (a) How far is the ranger at A from the fire? (b) How far is it from the road to the fire?

B. Applications and Extensions

23. The children's slide at the park is 30 feet long and inclines 36° from the horizontal. The ladder to the top is 18 feet long. How steep is the ladder; that is, what angle does it make with the horizontal? Assume that the slide is straight and that the bottom end of the slide is at the same level as the bottom end of the ladder.

24. Prevailing winds have caused an old tree to incline 11° eastward from the vertical. The sun in the west is 32° above the horizontal. How long a shadow is cast by the tree if the tree measures 114 feet from top to bottom?

25. A rectangular room, 16 feet by 30 feet, has an open beam ceiling. The two parts of the ceiling make angles of 65° and 32° with the horizontal (an end view is shown in Figure 16). Find the total area of the ceiling.

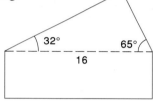

Figure 16

26. Sheila Sather, traveling north on a straight road at a constant rate of 60 miles per hour, sighted flames shooting up into the air at a point 20° west of north. Exactly 1 hour later, the fire was 59° west of south. Determine the shortest distance from the road to the fire.

27. A lighthouse stands at a certain distance out from a straight shoreline. It throws a beam of light that revolves at a constant rate of 1 revolution per minute. A short time after shining on the nearest point on the shore, the beam reaches a point on the shore that is 2640 feet from the lighthouse, and 3 seconds later it reaches a point 2000 feet farther along the shore. How far is the lighthouse from the shoreline?

28. In Figure 17, AC is 10 meters longer than CB. Determine the length of CD.

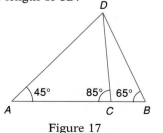

Figure 17

29. Two points A and B are on level ground and in line with the base C of a tower. The angles of elevation of the top of the tower at A and B are 21° and 35°, respectively. How tall is the tower if A and B are 300 feet apart?

30. Tang and Wang left point O together, Tang heading in the direction N58.6°E and Wang in the direction N33.2°E. After 20 seconds, they were 200 feet apart. If Tang ran at 12 feet per second, how fast did Wang run?

31. Spokes OD, OF, and OE of lengths 12, 6, and 10 radiate from a common point O. The angles DOF and FOE are each 20°. Find the area of triangle DEF.

32. The dial of a clock has a radius of 5 inches, and the minute and hour hands are 4 and 3 inches long, respectively (Figure 18). D is a fixed point on the rim of the dial at the 12 mark, E is the tip of the minute hand, and F is the tip of the hour hand. The points D, E, and F determine a triangle that changes with time. Let t denote the number of minutes after 12:00 and let $A(t)$ be the area of triangle DEF at time t. Show that

$$A(t) = \frac{1}{2} \left| 20 \sin \frac{\pi t}{30} - 15 \sin \frac{\pi t}{360} - 12 \sin \frac{11\pi t}{360} \right|$$

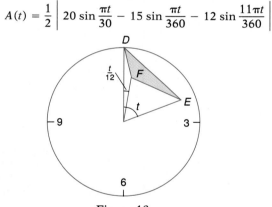

Figure 18

33. Graph the function $A(t)$ of Problem 32 for $0 \le t \le 60$. Determine the time between 12:00 and 1:00 when the area is a maximum, and calculate this maximum area.

34. Determine the first time after 12:00 when the points D, E, and F of Problem 32 are collinear.

35. Four line segments of lengths 3, 4, 5, and 6 radiate like spokes from a common point. Their outer ends are the vertices of a quadrilateral Q. Determine the maximum possible area of Q.

36. Lines $y = 0.8x - 1$, $y = -0.8x + 7$, and $y = -2.8x - 1$ bound a triangle. Determine its area. (*Hint:* Section 5.3 gives a formula for the angle from one line to another.)

37. Let 2ϕ denote the angle at a point of the regular six-pointed star shown in Figure 19. Express the area A of this star in terms of ϕ and the edge length r.

38. **Challenge.** Refer to Figure 19 and assume that the circle has radius 1. Express the area A and perimeter P of the star in terms of ϕ alone. Be sure to check that your formulas give the right answers when $\phi = 0°$ and when $\phi = 60°$. At $\phi = 60°$ the star is a regular hexagon with $A = 3\sqrt{3}/2$ and $P = 6$.

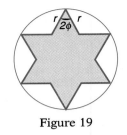

Figure 19

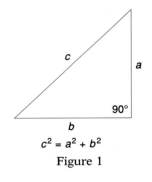

TEASER The hour and minute hands of a clock are 3 and 4 inches long, respectively. At some time between 12:00 and 12:30, the tips of these hands are 6 inches apart. What time is it then?

5.7   THE LAW OF COSINES

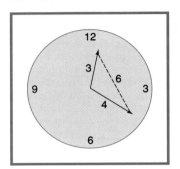

$$c^2 = a^2 + b^2$$

Figure 1

By now you are very familiar with the Pythagorean relationship $c^2 = a^2 + b^2$, which relates the lengths of the hypotenuse and legs of a right triangle (Figure 1). A question worth considering is how this relationship must be modified for an arbitrary triangle with sides a, b, and c. The answer, $c^2 = a^2 + b^2 - 2ab \cos \gamma$, is known as the law of cosines (Figure 2). Note that the correction term $-2ab \cos \gamma$ reduces to 0 when $\gamma = 90°$.

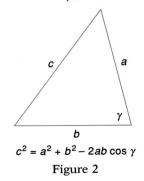

$$c^2 = a^2 + b^2 - 2ab \cos \gamma$$

Figure 2

THE LAW OF COSINES PROVED

Let ABC be a general triangle. Since α may be either acute or obtuse, we ask you to consider the two situations shown in Figure 3. Concentrate first on the left diagram in Figure 3. Note that we have two expressions for the length $\overline{CD}$, namely,

$$\overline{CD}^2 = b^2 - \overline{AD}^2 = b^2 - (b \cos \alpha)^2$$

and

$$\overline{CD}^2 = a^2 - \overline{BD}^2 = a^2 - (c - b \cos \alpha)^2$$

Thus,

$$b^2 - b^2 \cos^2 \alpha = a^2 - c^2 + 2bc \cos \alpha - b^2 \cos^2 \alpha$$

After a little algebra, this gives

$$b^2 + c^2 - 2bc \cos \alpha = a^2$$

which is one form of the law of cosines.

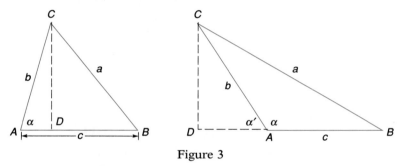

Figure 3

Next, refer to the right diagram in Figure 3. Note that $\cos \alpha' = \cos (180° - \alpha) = -\cos \alpha$, which means that $c - b \cos \alpha = c + b \cos \alpha'$. If you now go through the steps above, you will discover that every statement made for the left diagram also holds for the right diagram. Finally, we mention that what has been done for angle α could be done as well for angles β and γ. Doing so would be simply a matter of changing notation.

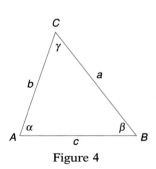

Figure 4

LAW OF COSINES

In any triangle with angles α, β, γ and corresponding opposite sides a, b, c (Figure 4),

$$a^2 = b^2 + c^2 - 2bc \cos \alpha$$
$$b^2 = a^2 + c^2 - 2ac \cos \beta$$
$$c^2 = a^2 + b^2 - 2ab \cos \gamma$$

It is wise to memorize this law in words.

The square of any side of a triangle is equal to the sum of the squares of the other two sides minus twice the product of those sides times the cosine of the angle between them.

SOLVING TRIANGLES

We mentioned in Section 5.6 that three pieces of information, of which at least one is a side, allow us to solve a triangle. In that section, we handled the cases AAS (two angles and any side are given) and SSA (two sides and the angle opposite one of them are given). Here, we will illustrate the cases SAS (two sides and the included angle are given) and SSS (three sides are given).

■ **Example 1 (SAS).** Suppose that in triangle ABC, $b = 18.1$, $c = 12.3$, and $\alpha = 115°$ (Figure 5). Determine a, β, and γ.

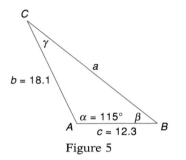

Figure 5

Solution.
Step 1: By the law of cosines,

$$a^2 = (18.1)^2 + (12.3)^2 - 2(18.1)(12.3) \cos 115°$$
$$a \approx 25.8$$

Step 2: Now we can use the law of sines.

$$\sin \beta = \frac{18.1 \sin 115°}{25.8} \approx 0.63582$$
$$\beta \approx 39.5°$$

Step 3: $\qquad\qquad \gamma \approx 180° - (115° + 39.5°) = 25.5°$ ■

■ **Example 2 (SSS).** Solve the triangle ABC in which $a = 13.15$, $b = 15.56$, and $c = 17.21$ (Figure 6).

Solution.
Step 1: By the law of cosines,

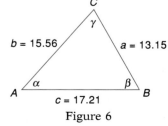

Figure 6

$$\cos \alpha = \frac{b^2 + c^2 - a^2}{2bc}$$
$$= \frac{(15.56)^2 + (17.21)^2 - (13.15)^2}{2(15.56)(17.21)} \approx 0.68221$$
$$\alpha \approx 46.98°$$

Step 2: By the law of sines,

$$\sin \beta = \frac{15.56 \sin 46.98°}{13.15} \approx 0.86511$$
$$\beta \approx 59.89°$$

Step 3: $\qquad \gamma \approx 180° - (46.98° + 59.89°) = 73.13°$ ∎

■ **Example 3.** The keepers of lighthouses A and B, which are 1.52 miles apart along a straight shoreline, sight an overturned sailboat C at angles 70.3° and 48.2° as shown in Figure 7. They notify a rescue boat, which is docked at D, 0.75 miles farther down the shore from B. How far is the rescue boat from the sailboat?

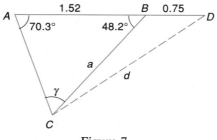

Figure 7

Solution. First, we determine angle γ to be

$$\gamma = 180° - (70.3° + 48.2°) = 61.5°$$

Then we apply the law of sines to determine a.

$$a = \frac{1.52 \sin 70.3°}{\sin 61.5°} \approx 1.6284 \approx 1.63 \text{ miles}$$

Finally, we use the law of cosines to find the required distance d.

$$d^2 = (1.6284)^2 + (0.75)^2 - 2(1.6284)(0.75) \cos 131.8°$$
$$d \approx 2.20 \text{ miles}$$ ∎

HERON'S AREA FORMULA

Inasmuch as the three sides of a triangle completely determine the triangle, they must also determine the area of the triangle. About 2000 years ago, Heron of Alexandria found the formula that illustrates this fact. Let a, b, and c denote the lengths of the three sides and let $s = (a + b + c)/2$ be its semiperimeter. Then the area A of the triangle is given by

$$A = \sqrt{s(s - a)(s - b)(s - c)}$$

The proof of this fact is subtle, depending on the clever matching of the area formula of Section 5.6 with the law of cosines. We begin by writing the law of cosines in the form

$$2bc \cos \alpha = b^2 + c^2 - a^2$$

a formula that we will use shortly. Next, we take the area formula $A = (bc/2) \sin \alpha$ in its squared form and manipulate it very carefully.

$$A^2 = \frac{1}{4} b^2 c^2 \sin^2 \alpha = \frac{1}{4} b^2 c^2 (1 - \cos^2 \alpha)$$

$$= \frac{1}{16} (2bc)(1 + \cos \alpha)(2bc)(1 - \cos \alpha)$$

$$= \frac{1}{16} (2bc + 2bc \cos \alpha)(2bc - 2bc \cos \alpha)$$

$$= \frac{1}{16} (2bc + b^2 + c^2 - a^2)(2bc - b^2 - c^2 + a^2)$$

$$= \frac{1}{16} ((b + c)^2 - a^2)(a^2 - (b - c)^2)$$

$$= \frac{(b + c + a)(b + c - a)(a - b + c)(a + b - c)}{2 \quad\quad 2 \quad\quad 2 \quad\quad 2}$$

$$= \left(\frac{a + b + c}{2}\right)\left(\frac{a + b + c}{2} - a\right)\left(\frac{a + b + c}{2} - b\right)\left(\frac{a + b + c}{2} - c\right)$$

$$= s(s - a)(s - b)(s - c)$$

■ **Example 4.**　Find the area of the triangle with sides 21.4, 29.6, and 34.2.

Solution. The semiperimeter is

$$s = \frac{21.4 + 29.6 + 34.2}{2} = 42.6$$

Thus, by Heron's formula, the area A is

$$A = ((42.6)(42.6 - 21.4)(42.6 - 29.6)(42.6 - 34.2))^{1/2}$$
$$\approx 314 \text{ square units}　■$$

■ **Example 5.**　Spokes *OD*, *OE*, and *OF* of lengths 12, 6, and 10 radiate from a common point *O* and are separated by angles of 20° and 22°, as shown in Figure 8. Find the perimeter *P* and area *A* of triangle *DEF*.

Solution. We use the law of cosines first.

$$\overline{DE} = (12^2 + 6^2 - 2(12)(6) \cos 20°)^{1/2} \approx 6.685 \approx 6.68$$
$$\overline{EF} = (6^2 + 10^2 - 2(6)(10) \cos 22°)^{1/2} \approx 4.974 \approx 4.97$$
$$\overline{DF} = (12^2 + 10^2 - 2(12)(10) \cos 42°)^{1/2} \approx 8.102 \approx 8.10$$

Thus,

$$P \approx 6.685 + 4.974 + 8.102 \approx 19.76 \text{ units}$$

and

$$A = (9.88(9.88 - 6.68)(9.88 - 4.97)(9.88 - 8.10))^{1/2}$$
$$\approx 16.6 \text{ square units}　■$$

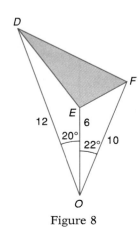

Figure 8

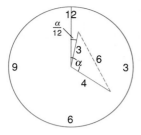

Figure 9

TEASER SOLUTION

We reproduce the clock in Figure 9. Note that when the minute hand moves through an angle α, the hour hand moves through angle $\alpha/12$. By the law of cosines,

$$36 = 9 + 16 - 24 \cos\left(\frac{11\alpha}{12}\right)$$

from which we conclude that $\cos(11\alpha/12) = -11/24$ and so $11\alpha/12 \approx 117.2796°$, or $\alpha \approx 127.9414°$. Let t denote the number of minutes after 12:00, and note that 1 minute on a clock dial measures an angle of 6°. Thus $t = \alpha/6 \approx 21.324$. The tips of the hands will be 6 inches apart at about 12:21.324.

PROBLEM SET 5.7

A. Skills and Techniques

In Problems 1–10, solve the triangle with the given data.

1. $\alpha = 60.23°, b = 14.76, c = 10.98$
2. $\beta = 56.89°, a = c = 12.34$
3. $\gamma = 120.45°, a = 12.98, b = 24.56$
4. $\alpha = 152.34°, b = 35.67, c = 41.36$
5. $\alpha = 1.75, b = 45.21, c = 34.56$
6. $\beta = 2.49, a = 12.43, c = 30.58$
7. $a = 10.23, b = 11.54, c = 12.58$
8. $a = 4.561, b = 6.345, c = 7.562$
9. $a = 52.12, b = 121.33, c = 69.05$
10. $a = 0.243, b = 0.365, c = 0.602$

In Problems 11–14, use Heron's formula to find the area of the triangle with the given data.

11. $a = 5.32, b = 6.44, c = 8.26$
12. $a = 5, b = 12, c = 13.$ (Surprise?)
13. $a = 7.32, b = 6.54,$ perimeter $= 16.48$
14. $a = 14.678, b = 12.542, c = 13.584$
15. At one corner of a triangular field, the angle measures 52.4°. The sides that meet at this corner are 100 meters and 120 meters long. How long is the third side?
16. To approximate the distance between two points A and B on opposite sides of a swamp, a surveyor selects a point C and measures it to be 140 meters from A and 260 meters from B. Then she measures the angle ACB, which turns out to be 49°. What is the calculated distance from A to B?
17. Two runners start from the same point at noon, one of them heading north at 6 miles per hour and the other heading 68° east of north at 8 miles per hour (Figure 10). What is the distance between them at 3:00 that afternoon?

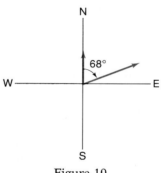

Figure 10

18. A 50-foot pole stands on top of a hill that slants 20° from the horizontal. How long must a rope be to reach from the top of the pole to a point 88 feet directly downhill (that is, on the slant) from the base of the pole?
19. A triangular garden plot has sides of length 35 meters, 40 meters, and 60 meters. Find the largest angle of the triangle.
20. A piece of wire 60 inches long is bent into the shape of a triangle. Find the angles of the triangle if two of the sides have lengths 24 inches and 20 inches.

B. Applications and Extensions

21. A triangular garden plot has sides measuring 42 meters, 50 meters, and 63 meters. Find the measure of the smallest angle.

22. A diagonal and a side of a parallelogram measure 80 centimeters and 25 centimeters, respectively, and the angle between them measures 47°. Find the length of the other diagonal. Recall that the diagonals of a parallelogram bisect each other.

23. Two cars, starting from the intersection of two straight highways, travel along the highways at speeds of 55 miles per hour and 65 miles per hour. If the angle of intersection of the highways measures 72°, how far apart are the cars after 36 minutes?

24. Buoys A, B, and C mark the vertices of a triangular racing course on a lake. Buoys A and B are 4200 feet apart, buoys A and C are 3800 feet apart, and angle CAB measures 100°. If the winning boat in a race covered the course in 6.4 minutes, what was its average speed in miles per hour?

25. The sides of a triangular garden plot with area 200 square meters are in the proportion 3:2:2. Find the length of each side to the nearest 100th of a meter.

26. A woman running at 11 feet per second ran first in the direction N67.9°E for 8 seconds and then in the direction N29.8°E for the next 12 seconds. How far was she then from the starting point?

27. Spokes OD, OF, and OE of lengths 13, 7, and 11 radiate from a common point O. The angles DOF and FOE are each 18°. Find the perimeter of triangle DEF (Figure 11).

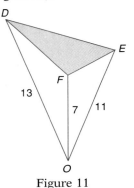

Figure 11

28. A quadrilateral Q has sides of length 1, 2, 3, and 4, respectively. The angle between the first pair of sides is 120°. Find the angle between the other pair of sides and also the exact area of Q.

29. For the triangle ABC in Figure 12, let r be the radius of the inscribed circle and let $s = (a + b + c)/2$ be its semiperimeter.

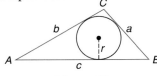

Figure 12

(a) Show that the area of the triangle is rs.

(b) Show that $r = \sqrt{(s - a)(s - b)(s - c)/s}$.
(c) Find r for a triangle with sides 5, 6, and 7.

30. Consider a triangle with sides of length 4, 5, and 6. Show that one of its angles is twice another. (*Hint:* Show that the cosine of twice one angle is equal to the cosine of another angle.)

31. In the triangle with sides of length a, b, and c, let $a_1, b_1,$ and c_1 denote the lengths of the corresponding medians to these sides from the opposite vertices. Show that

$$a_1^2 + b_1^2 + c_1^2 = \frac{3}{4}(a^2 + b^2 + c^2)$$

32. Determine the exact length of AB in Figure 13.

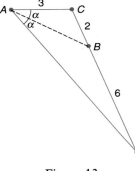

Figure 13

33. Refer to the clock illustration in Problem 32 of Section 5.6. Convince yourself that the perimeter $P(t)$ of triangle DEF at time t is given by

$$P(t) = \left(41 - 40\cos\frac{\pi t}{30}\right)^{1/2}$$
$$+ \left(34 - 30\cos\frac{\pi t}{360}\right)^{1/2} + \left(25 - 24\cos\frac{11\pi t}{360}\right)^{1/2}$$

Use the graph of $P(t)$ to determine the time between 12:00 and 1:10 when the perimeter is a maximum.

34. Determine the time between 12:10 and 1:10 when the perimeter of the triangle of Problem 33 is a minimum.

35. Three mutually tangent circles have radii 4, 5, and 6 (Figure 14). Determine the area of the curved white region between the three circles.

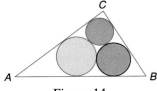

Figure 14

36. **Challenge.** Find the area of the triangle ABC (Figure 14) that circumscribes the three circles of Problem 35.

260 • Trigonometric Identities, Equations, and Laws

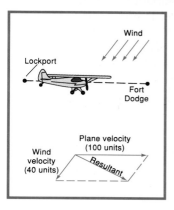

5.8 VECTORS

Tail   Head

Figure 1

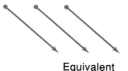

Equivalent
vectors

Figure 2

Many quantities that occur in science (for example, length, mass, volume, and electric charge) can be specified by giving a single number. We call these quantities and the numbers that measure them **scalars.** Other quantities, such as velocity, force, torque, and displacement, must be specified by giving both a magnitude and a direction. We call such quantities **vectors** and represent them by arrows (directed line segments). The length and direction of the arrow are the magnitude and direction of the vector. Thus, in the Teaser, the plane's velocity appears as an arrow 100 units long pointing eastward, and the wind velocity is shown as an arrow 40 units long pointing southwest.

Arrows that we draw, like those shot from a bow (Figure 1) have two ends: the **tail** (or initial point) and the **head** (or terminal point). Two vectors are considered to be equal if they have the same magnitude and direction (Figure 2). We symbolize vectors with boldface letters, as in **u** and **v.** (Because boldfacing is hard to accomplish in handwriting, you might use $\vec{u}$ and $\vec{v}$.) The magnitude of vector **u** is symbolized by $\|\mathbf{u}\|$.

To find the **sum,** or resultant, of **u** and **v,** move **v** without changing its magnitude or direction until its tail coincides with the head of **u.** Then **u** + **v** is the vector connecting the tail of **u** to the head of **v.** Alternatively, move **v** so that its tail matches that of **u.** Then **u** + **v** is the vector with this common tail, which coincides with the diagonal of the parallelogram determined by **u** and **v.** Both approaches are illustrated in Figure 3. It is easy to see that addition of vectors is both commutative and associative.

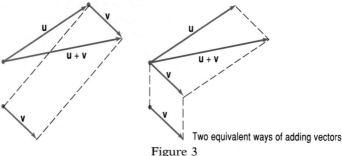

Two equivalent ways of adding vectors

Figure 3

■ **Example 1.** In navigation, directions are specified by giving an angle, called the **bearing,** with respect to a north-south line. Thus a bearing of N35°E denotes an angle whose initial side points north and whose terminal side points 35° east of north. If a ship sails 70 miles in the direction N35°E and then 90 miles straight east, what are its distance and bearing with respect to its starting point?

Solution. Displacements are vectors. Our job is to discover the length and bearing of the sum **w** of two displacements (Figure 4). We first use a little geometry to determine that $\beta = 125°$. Then by the law of cosines

$$\|\mathbf{w}\|^2 = (70)^2 + (90)^2 - 2(70)(90)\cos 125° \approx 20{,}227$$
$$\|\mathbf{w}\| \approx 142$$

By the law of sines,

$$\frac{\sin \alpha}{90} = \frac{\sin 125°}{142}$$
$$\sin \alpha = \frac{90 \sin 125°}{142} \approx 0.5192$$
$$\alpha \approx 31°$$

Thus the bearing of **w** is N66°E. ■

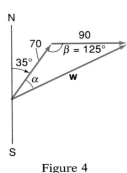

N

70   90

35° $\beta = 125°$

α **w**

S

Figure 4

■ **Example 2.** The river is flowing at 6 miles per hour, and Bob's boat travels at 20 miles per hour in still water. In what direction should he aim his boat if he wants to go straight across the river?

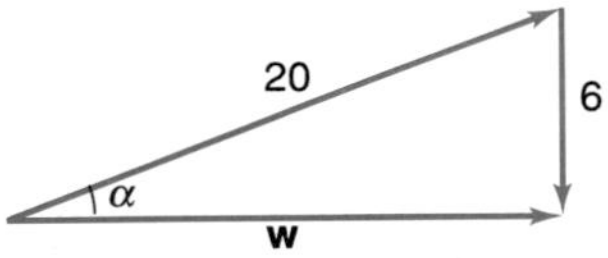

Solution. The given velocities are vectors, and we want their sum **w** to point straight across the river. Our job is to determine the angle α that **w** makes with the vector that represents the velocity of the boat (Figure 5).

$$\sin \alpha = \frac{6}{20} = 0.3000$$
$$\alpha \approx 17°$$ ■

20   6

α **w**

Figure 5

SCALAR MULTIPLICATION AND SUBTRACTION

If **u** is a vector, then 3**u** is the vector with the same direction as **u** but three times as long; $-2\mathbf{u}$ is twice as long as **u** but oppositely directed (Figure 6). More generally, $c\mathbf{u}$ has magnitude $|c|$ times that of **u** and is similarly or oppositely directed depending on whether c is positive or negative. In particular, $(-1)\mathbf{u}$, usually written $-\mathbf{u}$, has the same length as **u** but the opposite direction. It is called the negative of **u** because when we add it to **u**, the result is a vector that has shriveled to a point. This special vector (the only vector without direction) is called the **zero vector** and is denoted by **0.** It is the identity element for addition; that is, $\mathbf{u} + \mathbf{0} = \mathbf{0} + \mathbf{u} = \mathbf{u}$. You can check that scalar multiplication has the properties you would expect. For example,

$$(a + b)\mathbf{u} = a\mathbf{u} + b\mathbf{u}$$

and

$$a(\mathbf{u} + \mathbf{v}) = a\mathbf{u} + a\mathbf{v}$$

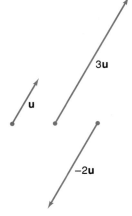

3**u**

u

−2**u**

Figure 6

Finally, subtraction is defined by

$$\mathbf{u} - \mathbf{v} = \mathbf{u} + (-\mathbf{v})$$

and so

$$a(\mathbf{u} - \mathbf{v}) = a\mathbf{u} - a\mathbf{v}$$

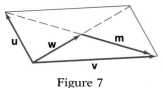

Figure 7

■ **Example 3.** Express vector **m** in Figure 7 in terms of **u** and **v**.

Solution. First, we note that $\mathbf{w} = (1/2)(\mathbf{u} + \mathbf{v})$. Second, $\mathbf{w} + \mathbf{m} = \mathbf{v}$ and so

$$\mathbf{m} = \mathbf{v} - \mathbf{w} = \mathbf{v} - \frac{1}{2}(\mathbf{u} + \mathbf{v}) = \frac{1}{2}\mathbf{v} - \frac{1}{2}\mathbf{u} \qquad ■$$

THE ALGEBRA OF VECTORS

So far our treatment of vectors has been largely geometric. To give the subject a more algebraic flavor, we first suppose that all vectors have been placed in the coordinate system with their tails at the origin. Then a vector is completely determined by the position of its head. Next we select two vectors to play permanent and special roles. The first, labeled **i,** is the vector from $(0, 0)$ to $(1, 0)$; the second, labeled **j,** is the vector from $(0, 0)$ to $(0, 1)$. Then, as Figure 8 shows, an arbitrary vector **u** with its head at (a, b) can be expressed uniquely in the form

$$\mathbf{u} = a\mathbf{i} + b\mathbf{j}$$

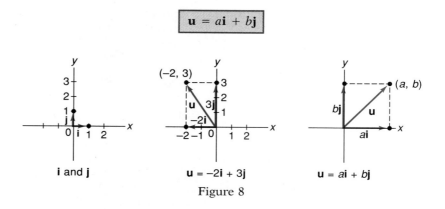

Figure 8

The vectors $a\mathbf{i}$ and $b\mathbf{j}$ are called the **vector components of u,** while a and b are called its **scalar components.** Notice that the magnitude of **u** is expressed in terms of its scalar components by

$$\|\mathbf{u}\| = \sqrt{a^2 + b^2}$$

To add two vectors $\mathbf{u} = a\mathbf{i} + b\mathbf{j}$ and $\mathbf{v} = c\mathbf{i} + d\mathbf{j}$, simply add the corresponding components; that is,

$$\mathbf{u} + \mathbf{v} = (a + c)\mathbf{i} + (b + d)\mathbf{j}$$

Similarly, to multiply a vector by a scalar k, multiply each component by k. Thus,

$$k\mathbf{u} = ka\mathbf{i} + kb\mathbf{j}$$

To see that these new algebraic rules are equivalent to the old geometric ones, study Figure 9.

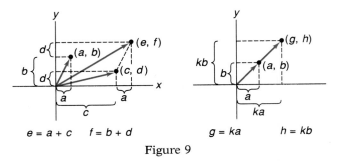

$$e = a + c \qquad f = b + d \qquad\qquad g = ka \qquad h = kb$$

Figure 9

Once the rules for addition and scalar multiplication are established, the rule for subtraction follows easily.

$$\mathbf{u} - \mathbf{v} = a\mathbf{i} + b\mathbf{j} + (-1)(c\mathbf{i} + d\mathbf{j}) = (a - c)\mathbf{i} + (b - d)\mathbf{j}$$

■ **Example 4.** Write the vector **v** represented by the arrow from $P(1, 5)$ to $Q(6, 2)$ in the **ij** form and then subtract it from $\mathbf{u} = 8\mathbf{i} - 7\mathbf{j}$.

Solution. To find the components of **v**, we move it so that its tail is at the origin and determine the coordinates of its head (Figure 10). We do this algebraically by subtracting the coordinates of P from those of Q obtaining $(6 - 1, 2 - 5) = (5, -3)$. Thus, $\mathbf{v} = 5\mathbf{i} - 3\mathbf{j}$ and

$$\mathbf{u} - \mathbf{v} = 8\mathbf{i} - 7\mathbf{j} - (5\mathbf{i} - 3\mathbf{j}) = 3\mathbf{i} - 4\mathbf{j} \qquad\qquad ■$$

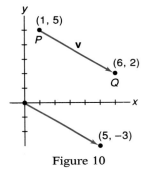

Figure 10

THE DOT PRODUCT

Is there a sensible way to multiply two vectors together? Yes; in fact, there are two kinds of products. One, called the **vector product,** requires three-dimensional space and therefore falls outside of the scope of this course. The other, called the **dot product,** or **scalar product,** can be introduced now. If $\mathbf{u} = a\mathbf{i} + b\mathbf{j}$ and $\mathbf{v} = c\mathbf{i} + d\mathbf{j}$, then the dot product of **u** and **v** is the scalar given by

$$\boxed{\mathbf{u} \cdot \mathbf{v} = ac + bd}$$

Why would anyone be interested in the dot product? To answer this, we need its geometric interpretation. Suppose that the heads of $\mathbf{u} = a\mathbf{i} + b\mathbf{j}$ and $\mathbf{v} = c\mathbf{i} + d\mathbf{j}$ are at (a, b) and (c, d), as shown in Figure 11. Then we may think of $\mathbf{v} - \mathbf{u}$ as the vector from (a, b) to (c, d). Let θ denote the smallest positive angle between **u** and **v**. By the law of cosines,

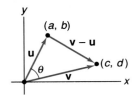

Figure 11

$$\|\mathbf{v} - \mathbf{u}\|^2 = \|\mathbf{u}\|^2 + \|\mathbf{v}\|^2 - 2\|\mathbf{u}\| \|\mathbf{v}\| \cos \theta$$
$$(a - c)^2 + (b - d)^2 = a^2 + b^2 + c^2 + d^2 - 2\|\mathbf{u}\| \|\mathbf{v}\| \cos \theta$$
$$-2ac - 2bd = -2\|\mathbf{u}\| \|\mathbf{v}\| \cos \theta$$
$$ac + bd = \|\mathbf{u}\| \|\mathbf{v}\| \cos \theta$$

The last equality gives us a geometric formula for the dot product.

$$\boxed{\mathbf{u} \cdot \mathbf{v} = \|\mathbf{u}\| \|\mathbf{v}\| \cos \theta}$$

Of what use is this formula? For one thing, it gives us an easy way to tell when two vectors are perpendicular. Since $\cos \theta$ is 0 if and only if θ is 90°, we see that:

Two vectors are perpendicular if and only if their dot product is zero.

More generally, we can use the formula to find the angle between any two vectors.

■ **Example 5.** Find the angle between the vectors $\mathbf{u} = 3\mathbf{i} + 4\mathbf{j}$ and $\mathbf{v} = -2\mathbf{i} + 3\mathbf{j}$.

Solution.

$$\cos \theta = \frac{\mathbf{u} \cdot \mathbf{v}}{\|\mathbf{u}\| \|\mathbf{v}\|} = \frac{(3)(-2) + (4)(3)}{\sqrt{9 + 16}\sqrt{4 + 9}} = \frac{6}{5\sqrt{13}} \approx 0.3328$$

We conclude that $\theta \approx 70.6°$. ■

In physics, the **work** done by a force $\mathbf{F}$ in moving an object from P to Q is defined to be the product of the magnitude of that force times the distance from P to Q. This formula assumes that the force is in the direction of the motion. In the more general case where the force $\mathbf{F}$ is at an angle to the motion, we must replace the magnitude of $\mathbf{F}$ by its scalar component in the direction of the motion. If both $\mathbf{F}$ and the displacement $\mathbf{D}$ from P to Q are treated as vectors, it can be shown that

$$\boxed{\text{Work} = \mathbf{F} \cdot \mathbf{D}}$$

■ **Example 6.** Find the work done by the force $\mathbf{F} = 8\mathbf{i} + 12\mathbf{j}$, measured in pounds, in moving an object from $P(2, 3)$ to $Q(14, 5)$, measured in feet.

Solution. Written as a vector, the displacement is

$$\mathbf{D} = (14 - 2)\mathbf{i} + (5 - 3)\mathbf{j} = 12\mathbf{i} + 2\mathbf{j}$$

Thus

$$\text{Work} = \mathbf{F} \cdot \mathbf{D} = (8)(12) + (12)(2) = 120 \text{ foot-pounds}$$ ■

TEASER SOLUTION

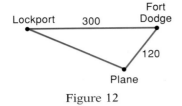

Figure 12

One way to answer Sylvia's question is to say that she is 3(40) miles southwest of where she hoped to be (Figure 12), that is, southwest of Fort Dodge. But suppose she wants to know her position relative to her starting point. How far east and how far south is she from Lockport? To answer this question, we will put Sylvia's problem into vector language. Her velocity of 100 miles per hour straight east is the vector $\mathbf{u} = 100\mathbf{i}$. The wind velocity of 40 miles per hour coming from the northeast is a vector $\mathbf{v}$ of length 40 pointing southwest (that is, with a bearing of S45°W). Using trigonometry (Figure 13), we write this vector in component form

Figure 13

$$\mathbf{v} = -40 \sin 45° \, \mathbf{i} - 40 \cos 45° \, \mathbf{j} = -20\sqrt{2}\mathbf{i} - 20\sqrt{2}\mathbf{j}$$

Thus, Sylvia's velocity with respect to the ground is

$$\mathbf{u} + \mathbf{v} = (100 - 20\sqrt{2})\mathbf{i} - 20\sqrt{2}\mathbf{j} \approx 71.716\mathbf{i} - 28.284\mathbf{j}$$

Her displacement after 3 hours is represented by the vector

$$3(\mathbf{u} + \mathbf{v}) \approx 215.147\mathbf{i} - 84.853\mathbf{j}$$

which means that Sylvia is about 215 miles east and 85 miles south of her starting point. This is a distance of

$$\sqrt{215.147^2 + 84.853^2} \approx 231 \text{ miles}$$

from the starting point.

PROBLEM SET 5.8

A. Skills and Techniques

*In Problems 1–4, draw the vector **w** so that its tail is at the heavy dot.*

1. $\mathbf{w} = \mathbf{u} + \mathbf{v}$

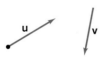

2. $\mathbf{w} = \mathbf{u} - \mathbf{v}$

3. $\mathbf{w} = -2\mathbf{u} + \frac{1}{2}\mathbf{v}$

4. $\mathbf{w} = \mathbf{u} - 3\mathbf{v}$

5. Draw the sum of the three vectors shown in Figure 14.

6. Draw $\mathbf{u} - \mathbf{v} + \frac{1}{2}\mathbf{w}$ for Figure 14.

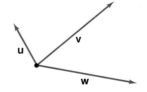

Figure 14

7. If I walk 10 miles N45°E and then 10 miles straight north, how far am I from my starting point?

8. In Problem 7, what is the bearing of my final position with respect to my starting point?

9. An airplane flew 100 kilometers in the direction S51°W and then 145 kilometers S39°W. What were the airplane's distance and bearing with respect to its starting point?

10. A ship sailed 11.2 miles straight north and then 48.3 miles N13.2°W. Find its distance and bearing with respect to the starting point.

11. A wind with velocity 58 miles per hour is blowing in the direction N20°W. An airplane that flies at 425 miles per hour in still air is supposed to fly straight north. How should the airplane be headed and how fast will it then be flying with respect to the ground?

12. A ship is sailing due south at 20 miles per hour. A man walks west (that is, at right angles to the side of the ship) across the deck at 3 miles per hour. What are the magnitude and direction of his velocity relative to the surface of the water?

13. In Figure 15, $\|\mathbf{u}\| = \|\mathbf{v}\| = 10$. Find the magnitude and direction of a force $\mathbf{w}$ needed to counterbalance $\mathbf{u}$ and $\mathbf{v}$.

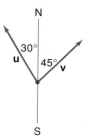

Figure 15

14. Jack pushes on a post from the direction S30°E with a force of 50 pounds. Wayne pushes on the same post from the direction S60°W with a force of 40 pounds. What is the magnitude and direction of the resultant force?

15. Refer to Figure 16. Express each of the following in terms of $\overrightarrow{AD}$ and $\overrightarrow{AB}$.
    (a) $\overrightarrow{BD}$ (b) $\overrightarrow{AF}$ (c) $\overrightarrow{DE}$ (d) $\overrightarrow{AF} - \overrightarrow{DE}$

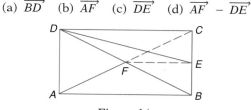

Figure 16

16. Express $\mathbf{w}$ in terms of $\mathbf{u}$ and $\mathbf{v}$ (Figure 17).

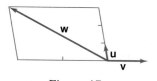

Figure 17

In Problems 17–20, find $3\mathbf{u} - \mathbf{v}$, $\mathbf{u} \cdot \mathbf{v}$, and $\cos \theta$ for the given vectors $\mathbf{u}$ and $\mathbf{v}$.

17. $\mathbf{u} = 3\mathbf{i} - 4\mathbf{j}$, $\mathbf{v} = 5\mathbf{i} + 12\mathbf{j}$
18. $\mathbf{u} = \mathbf{i} + \sqrt{3}\mathbf{j}$, $\mathbf{v} = 6\mathbf{i} - 8\mathbf{j}$
19. $\mathbf{u} = 2\mathbf{i} - \mathbf{j}$, $\mathbf{v} = 3\mathbf{i} - 4\mathbf{j}$
20. $\mathbf{u} = \mathbf{i} + \mathbf{j}$, $\mathbf{v} = \mathbf{i} - \mathbf{j}$
21. If $\mathbf{u} = 14.1\mathbf{i} + 32.7\mathbf{j}$ and $\mathbf{v} = 19.2\mathbf{i} - 13.3\mathbf{j}$, find θ, the smallest positive angle between $\mathbf{u}$ and $\mathbf{v}$.
22. Determine the length of $2\mathbf{u} - 3\mathbf{v}$, where $\mathbf{u}$ and $\mathbf{v}$ are the vectors in Problem 21.

In Problems 23–26, let $\mathbf{u}$ be the vector from P to Q and $\mathbf{v}$ be the vector from P to R. Write both vectors in the form $a\mathbf{i} + b\mathbf{j}$ and then find $\mathbf{u} \cdot \mathbf{v}$.

23. $P(1, 1)$, $Q(6, 3)$, $R(5, -2)$
24. $P(-1, 2)$, $Q(-3, 6)$, $R(0, -5)$
25. $P(1, 1)$, $Q(-3, -4)$, $R(-5, 6)$
26. $P(-1, -1)$, $Q(3, -5)$, $R(2, 4)$
27. If $\mathbf{u}$ is a vector 10 units long pointing in the direction N30°W, write $\mathbf{u}$ in the form $a\mathbf{i} + b\mathbf{j}$.
28. If $\mathbf{u}$ is a vector 9 units long pointing in the direction S21°W, write $\mathbf{u}$ in the form $a\mathbf{i} + b\mathbf{j}$.
29. Determine x so that $x\mathbf{i} + \mathbf{j}$ is perpendicular to $3\mathbf{i} - 4\mathbf{j}$.
30. Determine two vectors that are perpendicular to $2\mathbf{i} + 5\mathbf{j}$. (*Hint:* Try $x\mathbf{i} + \mathbf{j}$ and $x\mathbf{i} - \mathbf{j}$.)
31. Find a vector of unit length that has the same direction as $\mathbf{u} = 3\mathbf{i} - 4\mathbf{j}$. (*Hint:* Try $\mathbf{u}/\|\mathbf{u}\|$.)
32. Find two vectors of unit length that are perpendicular to $2\mathbf{i} + 3\mathbf{j}$.
33. Find the work done by the force $\mathbf{F} = 3\mathbf{i} + 10\mathbf{j}$ in moving an object north 10 units.
34. Find the work done by a S70°E force of 100 dynes in moving an object 50 centimeters east.
35. Find the work done by a N45°E force of 50 dynes in moving an object from $(1, 1)$ to $(6, 9)$, with the distance measured in centimeters.
36. Find the work done by $\mathbf{F} = 3\mathbf{i} + 4\mathbf{j}$ in moving an object from $(0, 0)$ to $(-6, 0)$. Interpret the negative answer.

B. Applications and Extensions

37. Two men are pushing an object along the ground. One is pushing with a force of 50 pounds in the direction N32°W and the other is pushing with a force of 100 pounds in the direction N30°E. In what direction is the object moving?
38. The vectors $\mathbf{u}$, $\mathbf{v}$, $\mathbf{w}$, $\mathbf{p}$, $\mathbf{q}$, have directions N90°E, N60°E, N45°E, N30°W, N0°E and lengths 20, 15, 20, 20, 10, respectively. Write the resultant of these five vectors in the form $a\mathbf{i} + b\mathbf{j}$.

39. An 80-pound weight is suspended from a ceiling by two cables as shown in Figure 18, producing forces **u** and **v** in the cables. Determine the tensions $\|\mathbf{u}\|$ and $\|\mathbf{v}\|$. (*Hint:* **u** + **v** = 0**i** + 80**j**.)

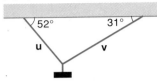

Figure 18

40. Find the work done by a force of 100 pounds directed N45°E in moving an object along the line from (1, 1) to (7, 5), distances measured in feet.

41. Three vectors form the edges of a triangle and are oriented clockwise around it. What is their sum?

42. Let **u, v,** and **w** be the vectors from the vertices of a triangle to the midpoints of the opposite edges (the medians). Show that **u** + **v** + **w** = **0**.

43. Alice and Bette left point *P* at the same time and met at point *Q* 2 hours later. To get there Alice walked a straight path, but Bette first walked 1 mile south and then 2 miles in the direction S60°E. How fast did Alice walk, assuming that she walked at a constant rate?

44. Suppose that **u, v,** and **w** of Figure 19 point in the directions N60°E, S45°E, N25°W and have lengths 60, $30\sqrt{2}$, 100, respectively. Find the length of **u** + **v** + **w**.

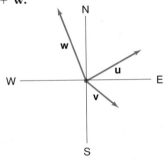

Figure 19

45. The 4-inch dial of a clock is oriented with its center at the origin and the number 12 on the positive *x*-axis (Figure 20). The minute hand is 3 inches long.

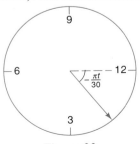

Figure 20

Let *t* denote the number of minutes after 12:00. Show that the minute hand is the vector

$$\left(3\cos\frac{\pi t}{30}\right)\mathbf{i} - \left(3\sin\frac{\pi t}{30}\right)\mathbf{j}$$

46. Let *d(t)* denote the distance from the tip of the minute hand of Problem 45 to the point on the rim of the 4-inch dial at 12. Show that

$$d(t) = \left(25 - 24\cos\frac{\pi t}{30}\right)^{1/2}$$

47. Draw the graph of *d(t)* from Problem 46 and determine the amplitude and period of this periodic function.

48. Redo Problem 47 for the 2-inch hour hand. To do this, you will first have to determine the appropriate formula for *d(t)*.

49. Show that for any two vectors **u** and **v**,

$$|\mathbf{u} \cdot \mathbf{v}| \leq \|\mathbf{u}\|\, \|\mathbf{v}\|$$

When will equality hold?

50. Prove that **u** · **v** = **v** · **u** and that **u** · (**v** + **w**) = **u** · **v** + **u** · **w**.

51. If **u** + **v** and **u** − **v** are perpendicular, what can we conclude about $\|\mathbf{u}\|$ and $\|\mathbf{v}\|$?

52. Show that $\|\mathbf{u} + \mathbf{v}\|^2 + \|\mathbf{u} - \mathbf{v}\|^2 = 2(\|\mathbf{u}\|^2 + \|\mathbf{v}\|^2)$.

53. Here is a vector problem to try on the TI-81 or TI-82. Suppose an airplane with a ground speed of 450 miles per hour is headed in the direction N58°E and the wind is blowing at 100 miles per hour in the direction S48°E. Set the calculator to parametric, simultaneous, and degree mode and enter the following in the ⒴ menu.

$$X_{1T} = 450T \cos 32$$
$$Y_{1T} = 450T \sin 32$$
$$X_{2T} = 100T \cos (-42)$$
$$Y_{2T} = 100T \sin (-42)$$
$$X_{3T} = X_{1T} + X_{2T}$$
$$Y_{3T} = Y_{1T} + Y_{2T}$$

Finally set the range values Tmin = 0, Tmax = 3, Tstep = 0.1, Xmin = 0, Xmax = 1500, Ymin = −500, and Ymax = 1000. Then press the ⒼⓇⒶⓅⒽ key. Convince yourself that the three vectors being drawn represent at time *T* the position of the plane if there were no wind, the position of an object being blown by the wind, and the actual position of the plane. Use the standard cursor and zooming to find the *xy*-coordinates of the plane after 3 hours.

54. Redo Problem 53 for a plane flying with a ground speed of 380 miles per hour and headed in the direction N41°E assuming the wind is blowing at 90 miles per hour in the direction S11°E. Use the cursor to find the *xy*-coordinates of the plane after 3.25 hours.

55. Find the exact value of sin 18°, which may be needed to complete Problem 56. (*Hint:* Let $\theta = 18°$ and note that $\cos 3\theta = \sin 2\theta = 2 \sin \theta \cos \theta$ and that $\cos 3\theta = \cos(2\theta + \theta) = (1 - 4\sin^2\theta)\cos\theta$.)

56. **Challenge.** Let **u** and **v** denote adjacent edges of a regular pentagon that is inscribed in a circle of radius 1 (Figure 21). Let $\mathbf{w} = \mathbf{u} + \mathbf{v}$.
(a) Express $\mathbf{u} \cdot \mathbf{w}$ and $\|\mathbf{u}\| \, \|\mathbf{w}\|$ in terms of cos 36°.
(b) Use your calculator to guess the exact value of $(\|\mathbf{u}\| \, \|\mathbf{w}\|)^2$ and then prove that this is the correct value.

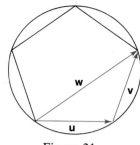

Figure 21

CHAPTER 5 REVIEW PROBLEM SET

In Problems 1–10, write True or False in the blank. If false, tell why.

———— **1.** $\sin \theta \cos \theta \tan \theta = 1 - \cos^2 \theta$.
———— **2.** A trigonometric equation can have at most two solutions on the interval 0 to 2π.
———— **3.** If a trigonometric equation has infinitely many solutions, it is an identity.
———— **4.** The equation $4 \sec^2 t - 1 = 0$ has no solution.
———— **5.** $2 \sin 1.5 \cos 1.5 = \sin 3$.
———— **6.** $\sin(\alpha + \beta) = \sin \alpha + \sin \beta$ is not true for any α and β.
———— **7.** $\cos 75° \sin 15° - \sin 75° \cos 15° = -\sqrt{3}/2$.
———— **8.** Addition of vectors is associative but not commutative.
———— **9.** For any vectors **u** and **v**, $\|\mathbf{u} + \mathbf{v}\| \le \|\mathbf{u}\| + \|\mathbf{v}\|$.
———— **10.** The Pythagorean theorem is a special case of the law of cosines.

11. Factor as much as possible.
(a) $5ab^2c^4 - 20a^3b^2c^2$
(b) $x^6 - 4x^4y + 4x^2y^2$
(c) $\tan^2 u - 5 \tan u - 6$
(d) $\sin^3 t + 8 \cos^3 t$

12. Simplify as much as possible
(a) $\dfrac{x^2 + 3x - 18}{x^2 - 9}$
(b) $\dfrac{5x}{x - 7} - \dfrac{7x}{14 - 2x}$
(c) $\dfrac{\tan^4 t - 1}{\tan^2 t - 1}$
(d) $\dfrac{\cos x}{\cos x + \sin x} - \dfrac{\sin x \cos x - \sin^2 x}{\cos^2 x - \sin^2 x}$

13. Simplify.
(a) $\dfrac{(x^2 - 2xy + y^2)/4xy}{(x^2 - y^2)/(16x^2 y - 4xy)}$
(b) $\dfrac{(\sin^3 x - \cos^3 x)/\cos 2x}{(1 + \sin x \cos x)/(\cos x + \sin x)}$

14. Rewrite each expression in terms of cos t and simplify.
(a) $2 - 3 \sin^2 t$

(b) $\dfrac{\sin^2 t \sec t}{1 + \sec t}$

(c) $\sin^2 2t \cos^2 \left(\dfrac{t}{2}\right)$

Prove that each of the following equations is an identity.

15. $\cot \theta \cos \theta = \csc \theta - \sin \theta$
16. $\sec t - \cos t = \sin t \tan t$
17. $\left(\cos \dfrac{t}{2} + \sin \dfrac{t}{2}\right)^2 = 1 + \sin t$
18. $\sec^4 \theta - \sec^2 \theta = \tan^4 \theta + \tan^2 \theta$
19. $\tan u + \cot u = \sec u \csc u$
20. $\dfrac{1 - \cos x}{\sin x} = \dfrac{\sin x}{1 + \cos x}$

In Problems 21–24 use appropriate identities to simplify each expression and then evaluate.

21. $\cos 153° \cos 33° + \sin 153° \sin 33°$
22. $\sin \dfrac{\pi}{8} \cos \dfrac{3\pi}{8} + \cos \dfrac{\pi}{8} \sin \dfrac{3\pi}{8}$
23. $2 \sin^2 112.5° - 1$
24. $\dfrac{\tan 20° + \tan 25°}{1 - \tan 20° \tan 25°}$
25. Express $\tan \theta \tan 2\theta$ in terms of $\sin \theta$.
26. Express $\cos 4t$ in terms of $\sin t$.
27. If $\sin t = -\dfrac{12}{13}$ and $\pi < t < \dfrac{3\pi}{2}$, evaluate the following.
(a) $\cos t$
(b) $\sin 2t$
(c) $\cos 2t$
(d) $\tan \left(\dfrac{t}{2}\right)$

28. If $\sin \alpha = \dfrac{2}{3}$, $\sin \beta = \dfrac{4}{5}$, $\cos \alpha < 0$, and $\cos \beta < 0$, evaluate the following.
(a) $\cos \alpha$
(b) $\cos \beta$
(c) $\sin(\alpha + \beta)$

Prove that each equation is an identity.

29. $\cos\left(u + \dfrac{\pi}{3}\right)\cos(\pi - u)$

$$- \sin\left(u + \dfrac{\pi}{3}\right)\sin(\pi - u) = -\dfrac{1}{2}$$

30. $\dfrac{\cos 5t}{\sin t} - \dfrac{\sin 5t}{\cos t} = \dfrac{2\cos 6t}{\sin 2t}$

31. $\csc 2t + \cot 2t = \cot t$

32. $\sin 3\theta = 3\sin\theta - 4\sin^3\theta$

33. $\dfrac{\sin(\alpha - \beta)}{\cos\alpha\cos\beta} = \tan\alpha - \tan\beta$

34. $\dfrac{1 - \tan^2(u/2)}{1 + \tan^2(u/2)} = \cos u$

In Problems 35–36, calculate each expression exactly without use of a calculator.

35. (a) $\sin(2\cos^{-1}(0.8))$
(b) $\cos(2\cos^{-1}(0.7))$

36. (a) $\sin(\cos^{-1}(0.6) + \cos^{-1}(0.5))$
(b) $\tan(\tan^{-1}(1) + \tan^{-1}(2))$

Solve the equations in Problems 37–44 on the interval $0 \le t \le 2\pi$, using algebraic methods.

37. $\tan t = -\sqrt{3}/3$

38. $2\cos^2 t - \cos t = 0$

39. $\sin^2 t - 2\sin t - 3 = 0$

40. $\sec^4 t - 3\sec^2 t + 2 = 0$

41. $\sin 2t = \dfrac{1}{2}$

42. $\cos 4t = -\dfrac{1}{2}$

43. $3 + \cos 2t = 5\cos t$

44. $\tan^3 t + \tan^2 t - 3\tan t = 3$

45. Draw the graph of $f(t) = 4\tan t \cos^2 t \csc(2t + \pi)$. Conjecture an identity and then prove it.

46. Use a graphics calculator to find all solutions to $\sin^3 x + \cos^2 x = 0$ on the interval $0 \le x \le 2\pi$, accurate to two decimal places.

47. It is known that $4(\tan^{-1}(2) + \tan^{-1}(3)) = n\pi$ where n is an integer. Use your calculator to guess the value of n and then prove the result algebraically.

48. Draw the graph of $\tan^{-1}(x/\sqrt{1 - x^2})$. Conjecture an identity. Prove it.

Solve each of the triangles in Problems 49–52.

49. $\alpha = 30°, \beta = 45°, c = 10$

50. $a = 2, b = 3, c = 4$

51. $\beta = 142°, b = 94, a = 67$

52. $\gamma = 37.6°, a = 11.6, b = 20.3$

53. Find the area of the triangle in Problem 52.

54. Find the area of a triangle whose sides measure 7, 8, and 9.

55. At a certain point A along a straight river, my line of sight to a tree B on the opposite bank makes an angle of 30° with the river. After I walk 100 yards farther along the river to the point C, my line of

sight to B makes an angle of 45° with the river (Figure 1). Determine
(a) the distance $\overline{AB}$
(b) the width of the river

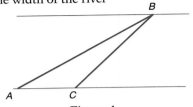

Figure 1

56. Find two triangles for which $a = 1.5, b = 2$, and $\alpha = 40°$.

57. Let O, A, and B denote the corners of the sector of a circle of radius 10 and central angle at O of 36°. Find the area of the region inside the sector AOB but outside the triangle AOB.

58. If $\mathbf{u}$ is a vector 6 units long pointing in the direction N60°W and $\mathbf{v} = -10\mathbf{i} + 15\mathbf{j}$, calculate $\mathbf{u} + (4/5)\mathbf{v}$ and write it in the form $a\mathbf{i} + b\mathbf{j}$.

59. Let $\mathbf{u} = 5\mathbf{i} - 12\mathbf{j}$ and $\mathbf{v} = 24\mathbf{i} + 7\mathbf{j}$. Calculate each of the following.
(a) $\|\mathbf{u}\|$
(b) $\|\mathbf{v}\|$
(c) $\mathbf{u} \cdot \mathbf{v}$
(d) θ, the angle between $\mathbf{u}$ and $\mathbf{v}$
(e) a unit length vector with the same direction as $\mathbf{v}$

60. For $\mathbf{u}$ and $\mathbf{v}$ as in Problem 59 determine
(a) the scalar projection of $\mathbf{u}$ on $\mathbf{v}$
(b) the vector projection of $\mathbf{u}$ on $\mathbf{v}$

61. Marge and Fred are pushing an object along the ground. Marge pushes with a force $\mathbf{u}$ of 120 pounds in the direction N arccos (4/5) E; Fred pushes with a force $\mathbf{v}$ of 90 pounds straight north. Calculate the resultant force $\mathbf{w} = \mathbf{u} + \mathbf{v}$ and write it in the form $a\mathbf{i} + b\mathbf{j}$.

62. In Problem 61, find the work done by $\mathbf{w}$ in moving the object
(a) 4 feet in the direction of $\mathbf{w}$
(b) 4 feet in the direction of $\mathbf{u}$

63. Draw the graph of $f(x) = 2\sin(1 + \cos x)$. Then determine exactly each of the following.
(a) its period
(b) its amplitude
(c) its zeros on $0 \le x \le 2\pi$

64. Determine the area A of the shaded region of Figure 2 in terms of t. Then find t so that $A = 2$, accurate to three decimal places.

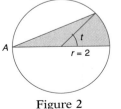

Figure 2

6

SYSTEMS OF EQUATIONS AND INEQUALITIES

Already in Section 1.1, we indicated part of the path that mathematicians took in extending the notion of number: natural numbers to integers to rational numbers to real numbers to complex numbers. Each of these number classes obeys the commutative law of multiplication: $ab = ba$. To William Rowan Hamilton, it seemed that this law was part of the concept of number; only when he reluctantly gave it up as he walked across a Dublin bridge in 1843 was he able to invent the next class of numbers—the quaterions. If you look closely, you can still find in the stone of that bridge the inscription: $ab \neq ba$.

Quaterions are of little interest today but they prepared the way for the English mathematician, Arthur Cayley, to take the next step—to matrices. A matrix is just a rectangular array of real (or complex) num-

bers. Cayley defined addition and multiplication for these number boxes (Section 6.3) only to discover that once again the commutative law of multiplication fails. Specializing to square matrices, Cayley asked if his new number boxes would have multiplicative inverses. Some do and some don't (Section 6.4) but those that have inverses can be manipulated much like ordinary numbers, so long as we avoid commutation.

The drive toward extension and abstraction reaches a climax with the work of the German mathematician, Emmy Noether. This talented woman taught us that the study of numbers and their algebra is simply the study of a set of objects with well-defined operations that are subject to more or less arbitrary laws (axioms). To her we owe the modern abstract view of algebra.

Arthur Cayley (1821–1895)

Emmy Noether (1882–1935)

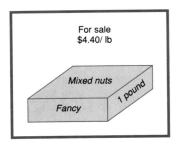

For sale
$4.40/ lb

Mixed nuts

Fancy

1 pound

TEASER   Grocer Sara Sellmor has one 60-pound crate of filberts, two 60-pound crates of cashews, and an almost unlimited supply of peanuts. She plans to use up all her filberts and cashews in packaging uniform 1-pound boxes of mixed nuts. Filberts sell for $6 per pound, cashews $5 per pound, and peanuts $3 per pound. How much of each kind should she put in a box if it is to sell for $4.40?

6.1   EQUIVALENT SYSTEMS OF EQUATIONS

In Section 1.3 (Example 7), we met our first system of equations, the pair of equations $y = 2x + 1$ and $y = x^2 - x - 3$. We discovered that this system had two ordered pairs as solutions, namely, $(-1, -1)$ and $(4, 9)$. Thus, $x = -1$ and $y = -1$ make both equations true equalities. So do $x = 4$ and $y = 9$. Here we consider the more general situation of a system of n equations in n unknowns. By a **solution** of such a system, we mean an ordered n-tuple that satisfies all n equations simultaneously. And to solve such a system means to find all its solutions.

As an example, consider the following system of three equations in three unknowns.

$$4x - 3y + 5z = 8$$
$$2x + 2y - 6z = -6$$
$$5x - 3y - 2z = -5$$

To solve this system means to find all triples (x, y, z) that satisfy each of these equations. You may check that $(1, 2, 2)$ is one such triple; in fact, that triple turns out to be the only solution.

The system just described is called a **linear system** because each of the equations in it is linear, that is, of the form $ax + by + cz = d$. The system mentioned at the beginning is nonlinear because one of its equations, $y = x^2 - x - 3$, is nonlinear. Most of our interest in this section will be in linear systems.

> **GENERAL LINEAR SYSTEMS**
>
> We are restricting attention to linear systems in which the number m of equations is equal to the number n of unknowns. Most of what we say, however, is valid also in the case where $m \neq n$.

OPERATIONS THAT LEAD TO EQUIVALENT SYSTEMS

To solve a system of equations requires tools. Our principal tool is a set of operations that allow us to go from one system to an **equivalent system,** that is, to another system with the same solutions. These operations allow us (at least in the linear case) to simplify a system until its solutions are obvious. Here are the operations we are permitted to use.

Operation 1.   *Interchanging the positions of two equations*

Operation 2.   *Multiplying an equation by a nonzero constant, that is, replacing an equation by a nonzero multiple of itself*

Operation 3.   *Adding a multiple of one equation to another, that is, replacing an equation by the sum of that equation and a multiple of another*

Let us illustrate a way to solve a system of two linear equations in two unknowns using these operations.

■ **Example 1.** Solve the system

$$2x - 2y = 4$$
$$3x + 5y = 14$$

Solution. Using operation 2, we replace the given system by

$$x - y = 2$$
$$3x + 5y = 14$$

Then, using operation 3, we add -3 times the first equation to the second one, obtaining

$$x - y = 2$$
$$8y = 8$$

In this form, the solution is obvious. The second equation says that $y = 1$, and, when this value is substituted in the first equation, it says that $x = 3$. You should check that the ordered pair $(3, 1)$ satisfies both equations of the original system. In fact, $(3, 1)$ is the only solution. ■

THE THREE POSSIBILITIES FOR A LINEAR SYSTEM

Look at a system of two linear equations in two unknowns from a geometric point of view. The two equations determine two lines. Either the lines are parallel and distinct (no solutions), or they are identical (infinitely many solutions), or they intersect in exactly one point (a single solution). Figure 1 illustrates these three cases.

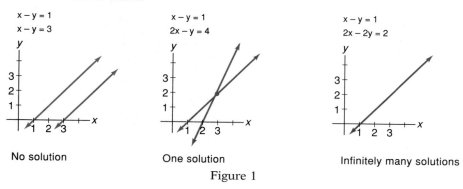

Figure 1

Although linear equations in three or more unknowns no longer represent lines (but rather planes and hyperplanes), it is still true that a linear system either has no solution, infinitely many solutions, or exactly one solution. Thus, if you discover two solutions for a linear system, you can immediately conclude that it has infinitely many solutions.

SOLVING LARGER SYSTEMS

What we need is a systematic procedure that will lead us to a solution. We begin with an extremely simple case, a so-called **triangular system,** by which we mean a system in which the unknowns below the main northwest-southeast diagonal have zero coefficients (Figure 2).

■ **Example 2.** Solve the system

$$3x - 2y - 4z = -5$$
$$y + 2z = 4$$
$$3z = 9$$

Solution. We use a process called **back substitution.** This means that we solve the last equation ($z = 3$), substitute this value in the next higher equation and solve it ($y = -2$), substitute these two values in the next higher equation and solve it ($x = 1$), and so on. In our case, this leads to the unique solution $(1, -2, 3)$. ■

Next, we illustrate how a general linear system of n equations in n unknowns can be reduced to triangular form.

■ **Example 3.** Solve the system

$$2x + 4y - 2z = -10$$
$$-3x + 4y - 2z = 5$$
$$5x + 6y + 3z = 3$$

Solution. We prefer that the first equation have ± 1 as its leading coefficient. Multiplying this equation by 1/2 (operation 2) leads to the equivalent system

$$x + 2y - z = -5$$
$$-3x + 4y - 2z = 5$$
$$5x + 6y + 3z = 3$$

Next we use operation 3 twice, adding 3 times the first equation to the second equation and -5 times the first equation to the third equation, obtaining

$$x + 2y - z = -5$$
$$10y - 5z = -10$$
$$-4y + 8z = 28$$

Multiplying the second equation by 1/5 and the third equation by 1/4 gives the equivalent system

$$x + 2y - z = -5$$
$$2y - z = -2$$
$$-y + 2z = 7$$

Operation 1 allows us to interchange the second and third equations, giving

$$x + 2y - z = -5$$
$$-y + 2z = 7$$
$$2y - z = -2$$

thereby making the second equation have leading coefficient -1. Finally, we add twice the second equation to the third and obtain the triangular system

UNIQUENESS?

Here is an important point to make. There are many different ways to go about reducing a system to triangular form. Our first step in Example 3 could have been to multiply the first equation by 3/2 and add to the second equation. Even the coefficients in the final triangular form are not unique. What is unique is the solution we get.

$$x + 2y - z = -5$$
$$-y + 2z = 7$$
$$3z = 12$$

When we solve this system by back substitution, we obtain $(-3, 1, 4)$ as the unique solution to the system. ■

SYSTEMS WITH NO OR INFINITELY MANY SOLUTIONS

When a system has a unique solution, the procedure described above determines that solution. The same procedure also establishes when a system has no solution.

■ **Example 4.** Solve the system

$$x - 2y + 3z = 10$$
$$2x - 3y - z = 4$$
$$5x - 9y + 8z = 6$$

Solution. Using operation 3, we eliminate x from the second and third equations to obtain

$$x - 2y + 3z = 10$$
$$y - 7z = -16$$
$$y - 7z = -44$$

It is already clear that the last two equations are incompatible. However, pushing on to triangular form by adding -1 times the second equation to the third, we get

$$x - 2y + 3z = 10$$
$$y - 7z = -16$$
$$0z = -28$$

which is a system with no solution. The system is **inconsistent.** ■

Finally, we use a similar process to solve a system that has infinitely many solutions.

■ **Example 5.** Find all solutions of the system

$$x - 2y + 3z = 10$$
$$2x - 3y - z = 8$$
$$4x - 7y + 5z = 28$$

Solution. After eliminating x from the second and third equations using operation 3, we have the system

$$x - 2y + 3z = 10$$
$$y - 7z = -12$$
$$y - 7z = -12$$

Adding -1 times the second equation to the third produces the triangular system

WHY Z?

In Example 5, we solved for x and y in terms of z. Alternatively, we could have solved for x and z in terms of y. Or we could have solved for y and z in terms of x. In any case, one of the variables can be given arbitrary values.

$$x - 2y + 3z = 10$$
$$y - 7z = -12$$
$$0z = 0$$

The third equation tells us that z can be any number whatever; we say that z is arbitrary. When we solve the second equation for y in terms of z, substitute this expression in the first equation, and solve for x, we get

$$y = 7z - 12$$
$$x = 2y - 3z + 10$$
$$= 2(7z - 12) - 3z + 10 = 11z - 14$$

We may report the set of solutions as

$$x = 11z - 14$$
$$y = 7z - 12$$
$$z \text{ arbitrary}$$

Alternatively, we may say that the set of solutions consists of all ordered triples of the form $(11z - 14, 7z - 12, z)$. If we set $z = 0$, we get $(-14, -12, 0)$; if we set $z = 1$, we get $(-3, -5, 1)$; and if we set $z = -1$, we get $(-25, -19, -1)$. It is clear that there are infinitely many solutions. ∎

TEASER SOLUTION

Let x, y, and z be the amounts of peanuts, cashews, and filberts, respectively, that will go into each 1-pound box of mixed nuts. Thus, $x + y + z = 1$. Since the amount of cashews to be used is twice that of filberts, $y = 2z$. Finally, the value of 1 pound of mixed nuts is the sum of the values of the three kinds of nuts that make it up; that is, $4.40 = 3x + 5y + 6z$. We conclude that we must solve the linear system

$$x + y + z = 1$$
$$y - 2z = 0$$
$$3x + 5y + 6z = 4.4$$

The triangulation process leads first to

$$x + y + z = 1$$
$$y - 2z = 0$$
$$2y + 3z = 1.4$$

and then to

$$x + y + z = 1$$
$$y - 2z = 0$$
$$7z = 1.4$$

Back substitution gives the solution $(0.4, 0.4, 0.2)$. Sara should put 0.4 pound of peanuts, 0.4 pound of cashews, and 0.2 pound of filberts in each box. Incidentally, this means that she can make $60/0.2 = 300$ boxes altogether.

PROBLEM SET 6.1

A. Skills and Techniques

Solve each system of equations.

1. $2x - 3y = 7$
   $y = -1$
2. $5x - 3y = -25$
   $y = 5$
3. $x = -2$
   $2x + 7y = 24$
4. $x = 5$
   $3x + 4y = 3$
5. $x - 3y = 7$
   $4x + y = 2$
6. $5x + 6y = 27$
   $x - y = 1$
7. $2x - y + 3z = -6$
   $2y - z = 2$
   $z = -2$
8. $x + 2y - z = -4$
   $3y + z = 2$
   $z = 5$
9. $3x - 2y + 5z = -10$
   $y - 4z = 8$
   $2y + z = 7$
10. $4x + 5y - 6z = 31$
    $y - 2z = 7$
    $5y + z = 2$
11. $x + 2y + z = 8$
    $2x - y + 3z = 15$
    $-x + 3y - 3z = -11$
12. $x + y + z = 5$
    $-4x + 2y - 3z = -9$
    $2x - 3y + 2z = 5$
13. $x - 2y + 3z = 0$
    $2x - 3y - 4z = 0$
    $x + y - 4z = 0$
14. $x + 4y - z = 0$
    $-x - 3y + 5z = 0$
    $3x + y - 2z = 0$
15. $x + y + z + w = 10$
    $y + 3z - w = 7$
    $x + y + 2z = 11$
    $x - 3y + w = -14$
16. $2x + y + z = 3$
    $y + z + w = 5$
    $4x + z + w = 0$
    $3y - z - 2w = 0$

Some of the following systems have infinitely many solutions and some are inconsistent. Solve each system or state that it is inconsistent.

17. $x - 4y + z = 18$
    $2x - 7y - 2z = 4$
    $3x - 11y - z = 22$
18. $x + y - 3z = 10$
    $2x + 5y + z = 18$
    $5x + 8y - 8z = 48$
19. $x - 2y + 3z = -2$
    $3x - 6y + 9z = -6$
    $-2x + 4y - 6z = 4$
20. $-4x + y - z = 5$
    $4x - y + z = -5$
    $-24x + 6y - 6z = 30$
21. $2x - y + 4z = 0$
    $3x + 2y - z = 0$
    $9x - y + 11z = 0$
22. $x + 3y - 2z = 0$
    $2x + y + z = 0$
    $y - z = 0$
23. $x - 4y + z = 18$
    $2x - 7y - 2z = 4$
    $3x - 11y - z = 10$
24. $x + y - 3z = 10$
    $2x + 5y + z = 18$
    $5x + 8y - 8z = 50$
25. $x + 3y - 2z = 10$
    $2x + y + z = 4$
    $5y - 5z = 16$
26. $x - 2y + 3z = -2$
    $3x - 6y + 9z = -6$
    $-2x + 4y - 6z = 0$

B. Applications and Extensions

In Problems 27–30, solve for s and t. (Hint: Begin by making a substitution. For example, in Problem 27, let $x = 1/s$ and $y = 1/t$. Solve for x and y; then solve for s and t.)

27. $2/s + 2/t = -2$
    $3/s + 4/t = -6$
28. $1/s + 3/t = 9$
    $3/s - 2/t = -0.5$
29. $4 \ln s - 2 \ln t = 0$
    $\ln s + 3 \ln t = 14$
30. $e^s + e^t = 2$
    $6e^s - 4e^t = -3$
31. If the system $x + 2y = 4$ and $ax + 3y = b$ has infinitely many solutions, what are a and b?
32. Helena claims that she has $4.40 in nickels, dimes, and quarters, that she has four times as many dimes as quarters, and that she has 40 coins in all. Is this possible? If so, determine how many coins of each kind she has.
33. A three-digit number equals 19 times the sum of its digits. If the digits are reversed, the resulting number is greater than the given number by 297. The tens digit exceeds the units digit by 3. Find the number.
34. Find the equation of the parabola $y = ax^2 + bx + c$ that goes through $(-1, 6)$, $(1, 0)$, and $(2, 3)$.

35. Find the equation of the cubic curve $y = ax^3 + bx^2 + cx + d$ that goes through $(-2, -6)$, $(-1, 5)$, $(1, 3)$, and $(2, 14)$.

36. A chemist has three hydrochloric acid solutions with concentrations (by volume) of 20%, 35%, and 40%, respectively. How many liters of each solution should she use if she wishes to obtain 200 liters of solution with a concentration of 34.25% and insists on using 30 liters more of the 35% solution than of the 20% solution?

37. Find the equation and radius of the circle that goes through $(0, 0)$, $(4, 0)$, and $(72/25, 96/25)$. (*Hint:* Writing the equation in the form $(x - h)^2 + (y - k)^2 = r^2$ is not the best way to start. Is there another way to write the equation of a circle?)

38. Determine a, b, and c so that

$$\frac{-12x + 6}{(x - 1)(x + 2)(x - 3)} = \frac{a}{x - 1} + \frac{b}{x + 2} + \frac{c}{x - 3}$$

Solve the nonlinear systems in Problems 39–46 by using algebraic methods. (Hint: You may use the three operations discussed in this section. Another useful algebraic maneuver is to solve one equation for one unknown in terms of the other and then substitute in the other equation.)

39. $\begin{aligned} x + 2y &= 10 \\ x^2 + y^2 - 10x &= 0 \end{aligned}$

40. $\begin{aligned} x + y &= 10 \\ x^2 + y^2 - 10x - 10y &= 0 \end{aligned}$

41. $\begin{aligned} x^2 + y^2 - 4x + 6y &= 12 \\ x^2 + y^2 + 10x + 4y &= 96 \end{aligned}$

42. $\begin{aligned} x^2 + y^2 - 16y &= 45 \\ x^2 + y^2 + 4x - 20y &= 65 \end{aligned}$

43. $\begin{aligned} y &= 4x^2 - 2 \\ y &= x^2 + 1 \end{aligned}$

44. $\begin{aligned} x &= 3y^2 - 5 \\ x &= y^2 + 3 \end{aligned}$

45. $\begin{aligned} x^2 + y^2 &= 4 \\ x + 2y &= 2\sqrt{5} \end{aligned}$

46. $\begin{aligned} x - \log y &= 1 \\ \log y^x &= 2 \end{aligned}$

47. Find the dimensions of a rectangle whose diagonal and perimeter measure 25 and 62 meters, respectively.

48. A certain rectangle has an area of 120 square inches. Increasing the width by 4 inches and decreasing the length by 3 inches increases the area by 24 square inches. Find the dimensions of the original rectangle.

As early as Section 1.4, we learned how to use a graphics calculator to solve a system of two equations in two unknowns geometrically. Use this method to solve the systems in Problems 47–54 (accurate to two decimal places). You will need to begin by solving each equation for y.

49. $\begin{aligned} 0.3x - 0.4y &= 1.4 \\ 0.5x + 0.9y &= 2.7 \end{aligned}$

50. $\begin{aligned} 1.5x + 3.6y &= 7.8 \\ 3.4x - 5.3y &= -1.4 \end{aligned}$

51. $\begin{aligned} x + 3y &= 11 \\ y &= -0.2x^2 + x + 8 \end{aligned}$

52. $\begin{aligned} y &= 0.4x^2 - 2x - 5 \\ y &= -0.5x^2 + 3x + 6 \end{aligned}$

53. $\begin{aligned} x + y &= 1 \\ x^4 + y^4 &= 500 \end{aligned}$

54. $\begin{aligned} 4x^2 + y^2 &= 36 \\ x^4 + y^4 &= 500 \end{aligned}$

55. Find the area of the triangle bounded by the y-axis and the lines $5x + 6y = 20$ and $3x - 6y = 23$.

56. **Challenge.** The ABC company reported the following statistics about its employees.
   (a) Average length of service for all employees: 15.9 years
   (b) Average length of service for male employees: 16.5 years
   (c) Average length of service for female employees: 14.1 years
   (d) Average hourly wage for all employees: $21.40
   (e) Average hourly wage for male employees: $22.50
   (f) Number of male employees: 300
   For reasons not stated, the company did not report the number of female employees or their average hourly wage, but you can figure them out. Do so.

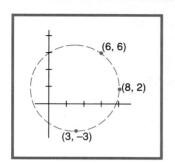

Figure 1

TEASER Find the center and radius of the circle that passes through the points $(3, -3)$, $(8, 2)$, and $(6, 6)$. (*Hint*: First determine the equation of this circle in the form $x^2 + y^2 + Ax + By + C = 0$. Then complete squares.)

6.2 SOLVING SYSTEMS USING MATRICES

In mathematics, we are always looking for shortcuts that will reduce the labor involved in solving a problem. Take the problem of solving a system of linear equations as an example. You have seen the labor required to solve a simple system of three equations in three unknowns. Imagine the labor involved in solving a system of 100 equations in 100 unknowns, a common problem in some business applications. Thinking about this problem led Arthur Cayley (1821–1895) to introduce the concept of a matrix. A **matrix** (the plural is matrices) is just a rectangular array of numbers; Figure 1 is a typical example.

Cayley looked at the process we use in solving a system of linear equations and noted that the operations involved are performed on the coefficients of the unknowns. Why not work only with those coefficients, he thought. Instead of working with the system of equations on the left in Figure 2, why not work with its matrix shown on the right?

$$
\begin{array}{l}
2x + 3y - z = 1 \\
x + 4y - z = 4 \\
3x + y + 2z = 5
\end{array}
\qquad
\begin{bmatrix}
2 & 3 & -1 & 1 \\
1 & 4 & -1 & 4 \\
3 & 1 & 2 & 5
\end{bmatrix}
$$

Figure 2

SOLVING A SYSTEM USING MATRICES

Watch what happens to the matrix as we solve the simple system of Figure 2 using the method of Section 6.1.

■ **Example 1.** Solve the system of Figure 2.

Solution.

$$
\begin{array}{r}
2x + 3y - z = 1 \\
x + 4y - z = 4 \\
3x + y + 2z = 5
\end{array}
\qquad
\begin{bmatrix}
2 & 3 & -1 & 1 \\
1 & 4 & -1 & 4 \\
3 & 1 & 2 & 5
\end{bmatrix}
$$

Interchange the first and second equation.

$$\begin{aligned} x + 4y - z &= 4 \\ 2x + 3y - z &= 1 \\ 3x + y + 2z &= 5 \end{aligned} \qquad \begin{bmatrix} 1 & 4 & -1 & 4 \\ 2 & 3 & -1 & 1 \\ 3 & 1 & 2 & 5 \end{bmatrix}$$

Add -2 times the first equation to the second; then add -3 times the first equation to the third equation.

$$\begin{aligned} x + 4y - z &= 4 \\ -5y + z &= -7 \\ -11y + 5z &= -7 \end{aligned} \qquad \begin{bmatrix} 1 & 4 & -1 & 4 \\ 0 & -5 & 1 & -7 \\ 0 & -11 & 5 & -7 \end{bmatrix}$$

Multiply the second equation by $-1/5$.

$$\begin{aligned} x + 4y - z &= 4 \\ y - \tfrac{1}{5}z &= \tfrac{7}{5} \\ -11y + 5z &= -7 \end{aligned} \qquad \begin{bmatrix} 1 & 4 & -1 & 4 \\ 0 & 1 & -\tfrac{1}{5} & \tfrac{7}{5} \\ 0 & -11 & 5 & -7 \end{bmatrix}$$

Add 11 times the second equation to the third equation.

$$\begin{aligned} x + 4y - z &= 4 \\ y - \tfrac{1}{5}z &= \tfrac{7}{5} \\ \tfrac{14}{5}z &= \tfrac{42}{5} \end{aligned} \qquad \begin{bmatrix} 1 & 4 & -1 & 4 \\ 0 & 1 & -\tfrac{1}{5} & \tfrac{7}{5} \\ 0 & 0 & \tfrac{14}{5} & \tfrac{42}{5} \end{bmatrix}$$

Now the system is in triangular form and can be solved by back substitution. The result is $z = 3$, $y = 2$, and $x = -1$; we report the solution as $(-1, 2, 3)$. ∎

EQUIVALENT MATRICES

Building on what we saw in Example 1, we say that matrix **A** is equivalent to matrix **B** (and write **A** → **B**) if **B** can be obtained from **A** by applying the operations below a finite number of times.

Operation 1 ($r_{i,j}$). *Interchanging rows i and j*
Operation 2 (kr_i). *Multiplying row i by a nonzero number k*
Operation 3 ($kr_i + r_j$). *Replacing row j by the sum of that row and k times row i*

Our notation should be clear, but observe how we label each step in the following example.

■ **Example 2.** Use the operations above to reduce the matrix of Figure 1 to triangular form, that is, so that all entries below the main northwest-southeast diagonal are 0.

TRIANGULAR FORM

The main diagonal goes through the upper left entry of the matrix. A matrix is in triangular form if all entries below this diagonal are 0.

Solution.

$$\begin{bmatrix} -2 & 4 & -2 & 7 \\ 1 & -3 & -4 & 0 \\ 5 & 2 & 5 & \frac{1}{2} \end{bmatrix} \xrightarrow{\ r_{1,2}\ } \begin{bmatrix} 1 & -3 & -4 & 0 \\ -2 & 4 & -2 & 7 \\ 5 & 2 & 5 & \frac{1}{2} \end{bmatrix}$$

$$\xrightarrow[-5r_1 + r_3]{2r_1 + r_2} \begin{bmatrix} 1 & -3 & -4 & 0 \\ 0 & -2 & -10 & 7 \\ 0 & 17 & 25 & \frac{1}{2} \end{bmatrix}$$

$$\xrightarrow{\ \frac{1}{2}r_2\ } \begin{bmatrix} 1 & -3 & -4 & 0 \\ 0 & -1 & -5 & \frac{7}{2} \\ 0 & 17 & 25 & \frac{1}{2} \end{bmatrix} \xrightarrow{\ 17r_2 + r_3\ } \begin{bmatrix} 1 & -3 & -4 & 0 \\ 0 & -1 & -5 & \frac{7}{2} \\ 0 & 0 & -60 & 60 \end{bmatrix} \quad \blacksquare$$

■ **Example 3.** Write down the system of equations that corresponds to the original matrix of Example 2. Then write down an equivalent triangular system and solve it.

Solution. The original system is

$$\begin{aligned} -2x + 4y - 2z &= 7 \\ x - 3y - 4z &= 0 \\ 5x + 2y + 5z &= \frac{1}{2} \end{aligned}$$

Referring to the final matrix in Example 2, we see that a corresponding triangular system is

$$\begin{aligned} x - 3y - 4z &= 0 \\ -y - 5z &= \frac{7}{2} \\ -60z &= 60 \end{aligned}$$

Solving this system yields $z = -1$, $y = 3/2$, and $x = 1/2$. In other words, the solution is $(1/2,\ 3/2,\ -1)$. ■

We remark that if $\mathbf{A} \to \mathbf{B}$, then $\mathbf{B} \to \mathbf{A}$. Also if $\mathbf{A} \to \mathbf{B}$ and $\mathbf{B} \to \mathbf{C}$, then $\mathbf{A} \to \mathbf{C}$, a fact that we have already used in our solutions to Examples 2 and 3.

FOUR EQUATIONS IN FOUR UNKNOWNS

So far, our examples have involved three equations in three unknowns. This limitation allowed us to keep the explanations simple. Our next example steps the level up one notch.

■ **Example 4.** Solve the system

$$w + 2x - y + 3z = 0$$
$$2w + 3x + 4y + 6z = 1$$
$$-3w - 4x + 2y + 2z = -2$$
$$-x + 3y = 3$$

Solution. We work with the corresponding matrices.

$$\begin{bmatrix} 1 & 2 & -1 & 3 & 0 \\ 2 & 3 & 4 & 6 & 1 \\ -3 & -4 & 2 & 2 & -2 \\ 0 & -1 & 3 & 0 & 3 \end{bmatrix} \xrightarrow[3r_1 + r_3]{-2r_1 + r_2} \begin{bmatrix} 1 & 2 & -1 & 3 & 0 \\ 0 & -1 & 6 & 0 & 1 \\ 0 & 2 & -1 & 11 & -2 \\ 0 & -1 & 3 & 0 & 3 \end{bmatrix}$$

$$\xrightarrow[-r_2 + r_4]{2r_2 + r_3} \begin{bmatrix} 1 & 2 & -1 & 3 & 0 \\ 0 & -1 & 6 & 0 & 1 \\ 0 & 0 & 11 & 11 & 0 \\ 0 & 0 & -3 & 0 & 2 \end{bmatrix} \xrightarrow{\frac{1}{11} r_3} \begin{bmatrix} 1 & 2 & -1 & 3 & 0 \\ 0 & -1 & 6 & 0 & 1 \\ 0 & 0 & 1 & 1 & 0 \\ 0 & 0 & -3 & 0 & 2 \end{bmatrix}$$

$$\xrightarrow{3r_3 + r_4} \begin{bmatrix} 1 & 2 & -1 & 3 & 0 \\ 0 & -1 & 6 & 0 & 1 \\ 0 & 0 & 1 & 1 & 0 \\ 0 & 0 & 0 & 3 & 2 \end{bmatrix}$$

Solving the triangular system corresponding to the last matrix gives $z = 2/3$, $y = -2/3$, $x = -5$, and $w = 22/3$. We report the solution as $(22/3, -5, -2/3, 2/3)$. ∎

SYSTEMS WITH NO OR INFINITELY MANY SOLUTIONS

The matrix method works perfectly well in situations that have no or infinitely many solutions.

■ **Example 5.** Solve the system

$$x + 2y - z = 5$$
$$2x - y + 2z = 4$$
$$3x + y + z = -2$$

Solution. We reduce the corresponding matrix to triangular form.

$$\begin{bmatrix} 1 & 2 & -1 & 5 \\ 2 & -1 & 2 & 4 \\ 3 & 1 & 1 & -2 \end{bmatrix} \xrightarrow[-3r_1 + r_3]{-2r_1 + r_2} \begin{bmatrix} 1 & 2 & -1 & 5 \\ 0 & -5 & 4 & -6 \\ 0 & -5 & 4 & -17 \end{bmatrix}$$

$$\xrightarrow{-r_2 + r_3} \begin{bmatrix} 1 & 2 & -1 & 5 \\ 0 & -5 & 4 & -6 \\ 0 & 0 & 0 & -11 \end{bmatrix}$$

It is obvious from the last row (which corresponds to $0z = -11$) that the system has no solution. ∎

■ Example 6. Solve the system

$$x - 2y + 3z = 12$$
$$3x - 4y - 2z = 9$$
$$5x - 8y + 4z = 33$$

Solution.

$$\begin{bmatrix} 1 & -2 & 3 & 12 \\ 3 & -4 & -2 & 9 \\ 5 & -8 & 4 & 33 \end{bmatrix} \xrightarrow[-5r_1 + r_3]{-3r_1 + r_2} \begin{bmatrix} 1 & -2 & 3 & 12 \\ 0 & 2 & -11 & -27 \\ 0 & 2 & -11 & -27 \end{bmatrix}$$

$$\xrightarrow{-r_2 + r_3} \begin{bmatrix} 1 & -2 & 3 & 12 \\ 0 & 2 & -11 & -27 \\ 0 & 0 & 0 & 0 \end{bmatrix}$$

The last row corresponds to $0z = 0$, which says that z can be any real number. In terms of z, the second row says that $2y - 11z = -27$; that is, $y = 11z/2 - 27/2$. When this expression is substituted in $x - 2y + 3z = 12$, we find that $x = 8z - 15$. Thus, we may describe the solution as the set of all triples of the form $(8z - 15, 11z/2 - 27/2, z)$. ■

SOLVING A SYSTEM ON A CALCULATOR

Many graphics calculators are programmed to handle matrices. Some allow one to perform the row operations described above mechanically and thereby to transform a matrix to triangular form by pressing the correct sequence of keys. You might wish to explore this possibility on your calculator.

Perhaps a more significant comment is this: Because the process of solving a system of equations can be described by an **algorithm** (an explicit series of steps), this process can be completely automated on a computer. In fact, computer software packages are available for solving systems of equations.

TEASER SOLUTION

Recall that we are to determine A, B, and C so that the points $(3, -3)$, $(8, 2)$, and $(6, 6)$ satisfy the equation $x^2 + y^2 + Ax + By + C = 0$. This means that we must solve for the unknowns A, B, and C in

$$9 + 9 + 3A - 3B + C = 0$$
$$64 + 4 + 8A + 2B + C = 0$$
$$36 + 36 + 6A + 6B + C = 0$$

or equivalently

$$C - 3B + 3A = -18$$
$$C + 2B + 8A = -68$$
$$C + 6B + 6A = -72$$

Turning to matrix form, we may write

$$\begin{bmatrix} 1 & -3 & 3 & -18 \\ 1 & 2 & 8 & -68 \\ 1 & 6 & 6 & -72 \end{bmatrix} \xrightarrow[\;-r_1 + r_3\;]{\;-r_1 + r_2\;} \begin{bmatrix} 1 & -3 & 3 & -18 \\ 0 & 5 & 5 & -50 \\ 0 & 9 & 3 & -54 \end{bmatrix}$$

$$\xrightarrow{\;\frac{1}{5}r_2\;} \begin{bmatrix} 1 & -3 & 3 & -18 \\ 0 & 1 & 1 & -10 \\ 0 & 9 & 3 & -54 \end{bmatrix} \xrightarrow{\;-9r_2 + r_3\;} \begin{bmatrix} 1 & -3 & 3 & -18 \\ 0 & 1 & 1 & -10 \\ 0 & 0 & -6 & 36 \end{bmatrix}$$

Back substitution gives the solution $A = -6$, $B = -4$, and $C = -12$.
We conclude that the circle through the given three points has equation

$$x^2 + y^2 - 6x - 4y - 12 = 0$$

which is equivalent (by completing the squares) to

$$(x - 3)^2 + (y - 2)^2 = 25$$

The required circle has center $(3, 2)$ and radius 5.

PROBLEM SET 6.2

A. Skills and Techniques

Write the matrix of each system in standard form.

1. $2x - y = 4$
   $x - 3y = -2$
2. $x + 2y = 13$
   $11x - y = 0$
3. $x - 2y + z = 3$
   $2x + y = 5$
   $x + y + 3z = -4$
4. $x + 4z = 10$
   $2y - z = 0$
   $3x - y = 20$
5. $2x = 3y - 4$
   $3x + 2 = -y$
6. $x = 4y + 3$
   $y = -2x + 5$
7. $x = 5$
   $2y + x - z = 4$
   $3x - y + 13 = 5z$
8. $z = 2$
   $2x - z = -4$
   $x + 2y + 4z = -8$

Regard each matrix in Problems 9–20 as the matrix of a linear system of equations. Without actually solving the system, determine whether it has a unique solution, infinitely many solutions, or no solution.

9. $\begin{bmatrix} 1 & -2 & 3 \\ 0 & 1 & -4 \end{bmatrix}$

10. $\begin{bmatrix} 2 & 5 & 0 \\ 0 & -3 & 5 \end{bmatrix}$

11. $\begin{bmatrix} 1 & -3 & 5 \\ 2 & -6 & -10 \end{bmatrix}$

12. $\begin{bmatrix} 2 & 1 & -4 \\ -6 & -3 & 12 \end{bmatrix}$

13. $\begin{bmatrix} 1 & -2 & 4 & -2 \\ 0 & 3 & 1 & 4 \\ 0 & 0 & 1 & -3 \end{bmatrix}$

14. $\begin{bmatrix} 5 & 4 & 0 & -11 \\ 0 & 1 & -4 & 0 \\ 0 & 0 & 2 & -4 \end{bmatrix}$

15. $\begin{bmatrix} 2 & 1 & 5 & 4 \\ 0 & 3 & -2 & 10 \\ 0 & 3 & -2 & 10 \end{bmatrix}$

16. $\begin{bmatrix} 4 & 1 & -3 & 5 \\ 0 & 0 & 1 & -4 \\ 0 & 0 & 1 & -4 \end{bmatrix}$

17. $\begin{bmatrix} 3 & 2 & -1 & 0 \\ 0 & 1 & 0 & -4 \\ 0 & 1 & 0 & 5 \end{bmatrix}$

18. $\begin{bmatrix} -1 & 5 & 6 & -3 \\ 0 & 0 & 0 & 0 \\ 0 & 0 & 0 & 4 \end{bmatrix}$

19. $\begin{bmatrix} 1 & 2 & 3 & 4 & 5 \\ 0 & 3 & 2 & 1 & 0 \\ 0 & 0 & 0 & 3 & -4 \\ 0 & 0 & 0 & -9 & 15 \end{bmatrix}$

20. $\begin{bmatrix} 0 & 0 & 0 & 2 & 3 \\ 0 & 0 & 3 & 4 & 5 \\ 0 & 4 & 5 & 6 & 7 \\ 5 & 6 & 7 & 8 & 9 \end{bmatrix}$

Use matrix methods to solve each system or show that it has no solution.

21. $x + 2y = 5$
$2x - 5y = -8$

22. $2x + 4y = 16$
$3x - y = 10$

23. $3x - 2y = 1$
$-6x + 4y = -2$

24. $x + 3y = 12$
$5x + 15y = 12$

25. $3x - 2y + 5z = -10$
$y - 4z = 8$
$2y + z = 7$

26. $4x + 5y + 2z = 25$
$y - 2z = 7$
$5y + z = 2$

27. $x + y - 3z = 10$
$2x + 5y + z = 18$
$5x + 8y - 8z = 48$

28. $x - 4y + z = 18$
$2x - 7y - 2z = 4$
$3x - 11y - z = 22$

29. $2x + 5y + 2z = 6$
$x + 2y - z = 3$
$3x - y + 2z = 9$

30. $x - 2y + 3z = -2$
$3x - 6y + 9z = -6$
$-2x + 4y - 6z = 0$

31. $x + 1.2y - 2.3z = 8.1$
$1.3x + 0.7y + 0.4z = 6.2$
$0.5x + 1.2y + 0.5z = 3.2$

32. $3x + 2y = 4$
$3x - 4y + 6z = 16$
$3x - y + z = 6$

33. $3x - 2y + 4z = 0$
$x - y + 3z = 1$
$4x + 2y - z = 3$

34. $-4x + y - z = 5$
$4x - y + z = -5$
$-24x + 6y - 6z = 10$

35. $2x + 4y - z = 8$
$4x + 9y + 3z = 42$
$8x + 17y + z = 58$

36. $2x + y + 2z + 3w = 2$
$y - 2z + 5w = 2$
$z + 3w = 4$
$2z + 7w = 4$

B. Applications and Extensions

37. Find a, b, and c so that the parabola $y = ax^2 + bx + c$ passes through the points $(-2, -32)$, $(1, 4)$, and $(3, -12)$.

38. Find the equation, center, and radius of a circle that passes through $(3, -3)$, $(8, 2)$, and $(6, 6)$.

39. Find the equation, center, and radius of the circle that passes through $(-8, 0)$, $(-1, -1)$, and $(0, -4)$.

40. Solve for α, β, and γ, assuming that these numbers are in the interval between 0 and π.

$$\begin{aligned} \cos \alpha + \cos \beta + 5 \cos \gamma &= 0 \\ 2 \cos \alpha + 4 \cos \beta + \cos \gamma &= -1 \\ 6 \cos \alpha + 2 \cos \beta - 3 \cos \gamma &= 2 \end{aligned}$$

41. Find angles α, β, γ, and δ in Figure 3 given that $\alpha - \beta + \gamma - \delta = 110°$.

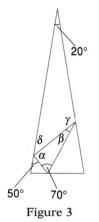

Figure 3

42. A chemist plans to mix three different nitric acid solutions with concentrations of 25%, 40%, and 50% to form 100 liters of a 32% solution. If he insists on using twice as much of the 25% solution as of the 40% solution, how many liters of each kind should he use?

43. Determine a, b, and c so that

$$\frac{2x^2 - 21x + 44}{(x - 2)^2(x + 3)} = \frac{a}{x - 2} + \frac{b}{(x - 2)^2} + \frac{c}{x + 3}$$

(*Hint:* First clear of fractions.)

44. Pablo plans to prepare a 100-pound mixture of peanuts, almonds, and cashews that is to sell at $2.88 per pound. Separately, these three kinds of nuts sell for $1.50, $3.50, and $4.50 per pound, respectively. How many pounds of each kind of nut should he use if he is required to use 6 pounds more of peanuts than of almonds?

45. The local garden store stocks three brands of phosphate-potash-nitrogen fertilizer with compositions indicated in the following table.

Brand	Phosphate	Potash	Nitrogen
A	10%	30%	60%
B	20%	40%	40%
C	20%	30%	50%

Soil analysis shows that Wanda Wiseankle needs fertilizer for her garden that is 19% phosphate, 34% potash, and 47% nitrogen. Can she obtain the right mixture by mixing the three brands? If so, how many pounds of each should she mix together to get 100 pounds of the desired blend?

46. Some graphing calculators perform matrix operations. The TI-81 and TI-82 will perform the three operations of this section though it requires some study of the instruction book to master the techniques for doing so. Use such a calculator to transform the matrix shown below to triangular form and then solve the corresponding system of equations.

$$\begin{bmatrix} 1 & 2 & 3 & 4 & 5 \\ 6 & 7 & 8 & 9 & 0 \\ 2 & 1 & 3 & 2 & -1 \\ 4 & 5 & -3 & 6 & 2 \end{bmatrix}$$

47. Use a calculator to transform the following matrix to triangular form and then solve (if possible) the corresponding system of equations.

$$\begin{bmatrix} 1 & 2 & 3 & 4 & 5 \\ 6 & 7 & 8 & 9 & 0 \\ 2 & 3 & 4 & 5 & 6 \\ 7 & 8 & 9 & 0 & 1 \end{bmatrix}$$

48. **Challenge.** Tom, Dick, and Harry are good friends but have very different work habits. Together, they contracted to paint three identical houses. Tom and Dick painted the first house in 72/5 hours; Tom and Harry painted the second house in 16 hours; Dick and Harry painted the third house in 144/7 hours. How long would it have taken each boy to paint a house alone?

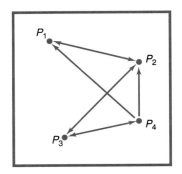

TEASER   A small airline serves the four cities P_1, P_2, P_3, and P_4 with one-way or round-trip service as indicated by the arrows in the accompanying diagram. The related matrix **A** (Figure 1) has 1 in the ij position (ith row, jth column) if there is a flight from P_i to P_j, and it has 0 in that position otherwise. Calculate $\mathbf{A}^2$ and $\mathbf{A}^3$ and give these matrices a physical interpretation.

$$A = \begin{bmatrix} 0 & 1 & 0 & 0 \\ 1 & 0 & 1 & 0 \\ 0 & 1 & 0 & 1 \\ 1 & 1 & 1 & 0 \end{bmatrix}$$

Figure 1

6.3   THE ALGEBRA OF MATRICES

When Arthur Cayley introduced matrices, he had more in mind than the application described in Section 6.2. There, matrices simplified the solving of systems of equations. Cayley realized that these number boxes could be studied independently of their application to solving equations. With appropriate definitions for addition and multiplication, he could create a new mathematical system with many interesting properties. Probably, he did not foresee very many uses for these new objects. If he could have looked ahead to our day, he would have seen matrices being applied in physics, chemistry, economics, and many other disciplines.

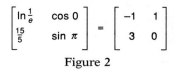

$$\begin{bmatrix} \ln\frac{1}{e} & \cos 0 \\ \frac{15}{5} & \sin \pi \end{bmatrix} = \begin{bmatrix} -1 & 1 \\ 3 & 0 \end{bmatrix}$$

Figure 2

To simplify the initial discussion, we restrict our attention to 2×2 matrices, that is, to matrices with two rows and two columns. Two such matrices are said to be **equal** if the entries in the corresponding positions have the same value (Figure 2).

ADDITION, SUBTRACTION, AND SCALAR MULTIPLICATION

Addition is straightforward; we add matrices by adding entries in corresponding positions.

$$\begin{bmatrix} -1 & 3 \\ 12 & 5 \end{bmatrix} + \begin{bmatrix} 5 & -4 \\ -9 & 3 \end{bmatrix} = \begin{bmatrix} -1+5 & 3-4 \\ 12-9 & 5+3 \end{bmatrix} = \begin{bmatrix} 4 & -1 \\ 3 & 8 \end{bmatrix}$$

It is easy to see that the commutative and associative properties hold for addition.

1. (Commutativity, $+$) $\mathbf{A} + \mathbf{B} = \mathbf{B} + \mathbf{A}$
2. (Associativity, $+$) $\mathbf{A} + (\mathbf{B} + \mathbf{C}) = (\mathbf{A} + \mathbf{B}) + \mathbf{C}$

The matrix

$$\mathbf{0} = \begin{bmatrix} 0 & 0 \\ 0 & 0 \end{bmatrix}$$

plays the role of the zero for matrices. And the additive inverse of

$$\mathbf{A} = \begin{bmatrix} a & b \\ c & d \end{bmatrix}$$

is

$$-\mathbf{A} = \begin{bmatrix} -a & -b \\ -c & -d \end{bmatrix}$$

Thus, we have the following.

3. (Additive identity) There is a matrix $\mathbf{0}$ satisfying $\mathbf{0} + \mathbf{A} = \mathbf{A} + \mathbf{0} = \mathbf{A}$.
4. (Additive inverses) For each matrix $\mathbf{A}$, there is a matrix $-\mathbf{A}$ satisfying $\mathbf{A} + (-\mathbf{A}) = -\mathbf{A} + \mathbf{A} = \mathbf{0}$.

We define subtraction of matrices by

$$\mathbf{A} - \mathbf{B} = \mathbf{A} + (-\mathbf{B})$$

and scalar multiplication by

$$k\begin{bmatrix} a & b \\ c & d \end{bmatrix} = \begin{bmatrix} ka & kb \\ kc & kd \end{bmatrix}$$

Scalar multiplication satisfies the expected properties,

5. $k(\mathbf{A} + \mathbf{B}) = k\mathbf{A} + k\mathbf{B}$

6. $(k + m)\mathbf{A} = k\mathbf{A} + m\mathbf{A}$

7. $(km)\mathbf{A} = k(m\mathbf{A}) = m(k\mathbf{A})$

■ **Example 1.** Given that

$$\mathbf{A} = \begin{bmatrix} -3 & 6 \\ 5 & 2 \end{bmatrix} \quad \text{and} \quad \mathbf{B} = \begin{bmatrix} 7 & -2 \\ -3 & -5 \end{bmatrix}$$

calculate (a) $3\mathbf{A}$, (b) $-2\mathbf{B}$, and (c) $3\mathbf{A} - 2\mathbf{B}$.

Solution.

(a) $3\mathbf{A} = \begin{bmatrix} -9 & 18 \\ 15 & 6 \end{bmatrix}$ (b) $-2\mathbf{B} = \begin{bmatrix} -14 & 4 \\ 6 & 10 \end{bmatrix}$

(c) $3\mathbf{A} - 2\mathbf{B} = 3\mathbf{A} + (-2\mathbf{B}) = \begin{bmatrix} -23 & 22 \\ 21 & 16 \end{bmatrix}$ ■

MULTIPLICATION

At first glance, our definition of multiplication may seem strange, but it is what is needed for applications that lie ahead. Stated in words, we multiply the rows of the left matrix by the columns of the right matrix in pairwise entry fashion and add the results.

Here it is in symbols.

$$\begin{bmatrix} a & b \\ c & d \end{bmatrix} \begin{bmatrix} A & B \\ C & D \end{bmatrix} = \begin{bmatrix} aA + bC & aB + bD \\ cA + dC & cB + dD \end{bmatrix}$$

■ **Example 2.** Calculate $\mathbf{AB}$ and $\mathbf{BA}$, given that

$$\mathbf{A} = \begin{bmatrix} 2 & -3 \\ -1 & 4 \end{bmatrix} \quad \text{and} \quad \mathbf{B} = \begin{bmatrix} -5 & 2 \\ 3 & 4 \end{bmatrix}$$

Solution.

$$\mathbf{AB} = \begin{bmatrix} 2 & -3 \\ -1 & 4 \end{bmatrix} \begin{bmatrix} -5 & 2 \\ 3 & 4 \end{bmatrix} = \begin{bmatrix} 2(-5) + (-3)(3) & 2(2) + (-3)(4) \\ -1(-5) + 4(3) & -1(2) + 4(4) \end{bmatrix}$$

$$= \begin{bmatrix} -19 & -8 \\ 17 & 14 \end{bmatrix}$$

$$\mathbf{BA} = \begin{bmatrix} -5 & 2 \\ 3 & 4 \end{bmatrix} \begin{bmatrix} 2 & -3 \\ -1 & 4 \end{bmatrix} = \begin{bmatrix} (-5)(2) + 2(-1) & (-5)(-3) + 2(4) \\ 3(2) + 4(-1) & 3(-3) + 4(4) \end{bmatrix}$$

$$= \begin{bmatrix} -12 & 23 \\ 2 & 7 \end{bmatrix}$$ ■

Note something very important that we discovered in Example 2. Multiplication of matrices is not commutative. But two other familiar properties do hold.

8. (Associativity, $\cdot$) $\mathbf{A} \cdot (\mathbf{B} \cdot \mathbf{C}) = (\mathbf{A} \cdot \mathbf{B}) \cdot \mathbf{C}$
9. (Distributivity, $\cdot$) $\mathbf{A} \cdot (\mathbf{B} + \mathbf{C}) = \mathbf{A} \cdot \mathbf{B} + \mathbf{A} \cdot \mathbf{C}$
$$(\mathbf{B} + \mathbf{C}) \cdot \mathbf{A} = \mathbf{B} \cdot \mathbf{A} + \mathbf{C} \cdot \mathbf{A}$$

LARGER MATRICES AND COMPATIBILITY

We have been illustrating with 2×2 matrices, but all that we have said is valid for larger size matrices provided there is **compatibility.** Two matrices are compatible for addition if they are the same size. Two matrices are compatible for multiplication if the left matrix has the same number of columns as the right matrix has rows. For example, we may multiply a 2×3 matrix by a 3×5 matrix; the result is a 2×5 matrix.

■ **Example 3.** Let

$$\mathbf{A} = \begin{bmatrix} 2 & -3 & 4 \\ -1 & 4 & 2 \end{bmatrix} \quad \mathbf{B} = \begin{bmatrix} 5 & 2 \\ -2 & 3 \end{bmatrix}$$

$$\mathbf{C} = \begin{bmatrix} 1 & 3 & 0 \\ 2 & -4 & 1 \\ 4 & -1 & -2 \end{bmatrix} \quad \mathbf{D} = \begin{bmatrix} 3 & -2 \\ 2 & -1 \\ -3 & 4 \end{bmatrix}$$

Calculate if possible: (a) **CD,** (b) **AC,** (c) **CA,** (d) **AD + B.**

Solution.

(a) $\mathbf{CD} = \begin{bmatrix} 1 & 3 & 0 \\ 2 & -4 & 1 \\ 4 & -1 & -2 \end{bmatrix} \begin{bmatrix} 3 & -2 \\ 2 & -1 \\ -3 & 4 \end{bmatrix}$

$$= \begin{bmatrix} 1(3) + 3(2) + 0(-3) & 1(-2) + 3(-1) + 0(4) \\ 2(3) - 4(2) + 1(-3) & 2(-2) - 4(-1) + 1(4) \\ 4(3) - 1(2) - 2(-3) & 4(-2) - 1(-1) - 2(4) \end{bmatrix}$$

$$= \begin{bmatrix} 9 & -5 \\ -5 & 4 \\ 16 & -15 \end{bmatrix}$$

(b) $\mathbf{AC} = \begin{bmatrix} 2 & -3 & 4 \\ -1 & 4 & 2 \end{bmatrix} \begin{bmatrix} 1 & 3 & 0 \\ 2 & -4 & 1 \\ 4 & -1 & -2 \end{bmatrix} = \begin{bmatrix} 12 & 14 & -11 \\ 15 & -21 & 0 \end{bmatrix}$

(c) **CA** is undefined; **C** and **A** are incompatible in this order.

(d) $\mathbf{AD} + \mathbf{B} = \begin{bmatrix} 2 & -3 & 4 \\ -1 & 4 & 2 \end{bmatrix} \begin{bmatrix} 3 & -2 \\ 2 & -1 \\ -3 & 4 \end{bmatrix} + \begin{bmatrix} 5 & 2 \\ -2 & 3 \end{bmatrix}$

$$= \begin{bmatrix} -12 & 15 \\ -1 & 6 \end{bmatrix} + \begin{bmatrix} 5 & 2 \\ -2 & 3 \end{bmatrix} = \begin{bmatrix} -7 & 17 \\ -3 & 9 \end{bmatrix}$$

■

USING A CALCULATOR

Many graphics calculators (including the TI-81 and TI-82) have the ability to manipulate matrices up to size 6×6 or even larger. We urge you to learn how your machine does this; you will be surprised at the ease with which complicated matrix calculations can be done.

■ **Example 4.** Use a calculator to calculate **AB** and then **AB** + 3**C** given that

$$A = \begin{bmatrix} 3 & -3 & 4 & 2 \\ -5 & 12 & 3 & -6 \\ 2 & -5 & 0 & 5 \\ 21 & 0 & 9 & -3 \end{bmatrix} \qquad B = \begin{bmatrix} 4 & -5 \\ 3 & 2 \\ -5 & 0 \\ 6 & -3 \end{bmatrix} \qquad C = \begin{bmatrix} -6 & -2 \\ 4 & 3 \\ 8 & -7 \\ -1 & 2 \end{bmatrix}$$

Solution. Inputting these three matrices is the time-consuming part of the job. Once they are entered, one only needs to press a few keys to get **AB** and then a few more to get **AB** + 3**C**. The corresponding screen outputs are shown in Figure 3. ■

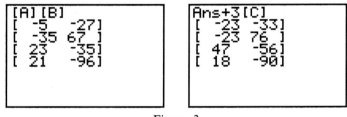

Figure 3

A BUSINESS APPLICATION

The DEF Company sells precut lumber for two types of summer cottages, standard and deluxe. The standard model requires 30,000 board-feet of lumber and 100 worker-hours of cutting; the deluxe model takes 40,000 board-feet of lumber and 110 worker-hours of cutting. This year, the DEF Company buys its lumber at $0.20 per board-foot and pays its laborers $9 per hour. Next year it expects these costs to be $0.25 and $10, respectively. This information can be displayed in matrix form as follows.

	Requirements **A**			Unit Cost **B**	
	Lumber	Labor		This year	Next year
Standard	30,000	100	Lumber	$0.20	$ 0.25
Deluxe	40,000	110	Labor	$9.00	$10.00

Now we ask whether the product matrix **AB** has economic significance. It does: It gives the total dollar cost of standard and deluxe cottages both for this year and next. You can see this from the following calculation.

$$\mathbf{AB} = \begin{bmatrix} (30{,}000)(0.20) + (100)(9) & (30{,}000)(0.25) + (100)(10) \\ (40{,}000)(0.20) + (110)(9) & (40{,}000)(0.25) + (110)(10) \end{bmatrix}$$

$$= \begin{array}{c} \text{This year} \quad \text{Next year} \\ \begin{bmatrix} \$6900 & \$\ 8500 \\ \$8990 & \$11{,}110 \end{bmatrix} \begin{array}{l} \text{Standard} \\ \text{Deluxe} \end{array} \end{array}$$

TEASER SOLUTION

We input **A** into a calculator, then simply press keys to obtain $\mathbf{A}^2 = \mathbf{A} \cdot \mathbf{A}$ and $\mathbf{A}^3 = \mathbf{A}^2 \cdot \mathbf{A}$. The screen outputs for these three matrices are shown in Figure 4.

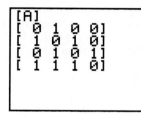
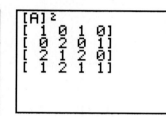

Figure 4

As for the interpretation of $\mathbf{A}^k$, we claim that the ij entry in this matrix gives the number of routes from P_i to P_j having k legs. Thus, the entry of 2 in the first column of $\mathbf{A}^2$ means there are exactly two two-legged routes from P_3 to P_1 (they are $P_3{\to}P_2{\to}P_1$ and $P_3{\to}P_4{\to}P_1$). Note that we cannot get from P_1 to P_4 in one leg—the entry in the (1, 4) position of **A** is 0—or in a two-legged trip—the entry in the (1, 4) position of $\mathbf{A}^2$ is 0—but we can make it in one three-legged trip—the entry in the (1, 4) position of $\mathbf{A}^3$ is 1.

PROBLEM SET 6.3

A. Skills and Techniques

In each case, calculate $A + B$, $A - B$, and $3A$ by hand.

1. $\mathbf{A} = \begin{bmatrix} 2 & -1 \\ 3 & 7 \end{bmatrix}$, $\mathbf{B} = \begin{bmatrix} 6 & 5 \\ -2 & 3 \end{bmatrix}$

2. $\mathbf{A} = \begin{bmatrix} -1 & 0 \\ 5 & 4 \end{bmatrix}$, $\mathbf{B} = \begin{bmatrix} 2 & -2 \\ 3 & 7 \end{bmatrix}$

3. $\mathbf{A} = \begin{bmatrix} 3 & -2 & 5 \\ 4 & 0 & -3 \end{bmatrix}$, $\mathbf{B} = \begin{bmatrix} 2 & 6 & -1 \\ 4 & 3 & -3 \end{bmatrix}$

4. $\mathbf{A} = \begin{bmatrix} 1 & 2 & 3 \\ 4 & 5 & 6 \\ 7 & 8 & 9 \end{bmatrix}$, $\mathbf{B} = \begin{bmatrix} -1 & -2 & -2 \\ -4 & -5 & -6 \\ -7 & -8 & -9 \end{bmatrix}$

In Problems 5–12, calculate AB and BA by hand. If a product is undefined, say so.

5. $\mathbf{A} = \begin{bmatrix} 2 & -1 \\ 3 & 7 \end{bmatrix}$, $\mathbf{B} = \begin{bmatrix} 6 & 5 \\ -2 & 3 \end{bmatrix}$

6. $\mathbf{A} = \begin{bmatrix} -1 & 0 \\ 5 & 4 \end{bmatrix}$, $\mathbf{B} = \begin{bmatrix} 2 & -2 \\ 3 & 7 \end{bmatrix}$

7. $\mathbf{A} = \begin{bmatrix} 1 & -1 & 2 \\ 3 & 4 & -4 \\ 2 & 1 & 3 \end{bmatrix}$, $\mathbf{B} = \begin{bmatrix} 0 & 2 & -3 \\ 1 & 2 & 3 \\ -1 & -2 & 4 \end{bmatrix}$

8. $\mathbf{A} = \begin{bmatrix} -2 & 5 & 1 \\ 0 & -2 & 3 \\ 1 & 2 & -1 \end{bmatrix}$, $\mathbf{B} = \begin{bmatrix} -3 & 4 & 1 \\ 2 & 5 & 1 \\ 1 & 2 & 3 \end{bmatrix}$

9. $A = \begin{bmatrix} 1 & -2 & 3 & 4 \\ 3 & 2 & -5 & 1 \end{bmatrix}$, $B = \begin{bmatrix} 1 & 2 \\ 3 & 4 \end{bmatrix}$

10. $A = \begin{bmatrix} -1 & 3 \\ 4 & 2 \\ 1 & 5 \end{bmatrix}$, $B = \begin{bmatrix} -1 & 2 & 3 & 4 \\ 0 & -3 & 2 & 1 \end{bmatrix}$

11. $A = \begin{bmatrix} 3 & 1 & -1 \\ 2 & 4 & 2 \\ -3 & 2 & -1 \end{bmatrix}$, $B = \begin{bmatrix} 1 \\ 2 \\ 3 \end{bmatrix}$

12. $A = \begin{bmatrix} 1 & 2 & -1 \end{bmatrix}$, $B = \begin{bmatrix} 4 & 3 \\ 0 & 2 \\ -1 & 4 \end{bmatrix}$

13. Calculate **AB** and **BA** for

$$A = \begin{bmatrix} 0 & 0 \\ 0 & 0 \end{bmatrix} \quad B = \begin{bmatrix} 2 & -1 \\ 3 & 4 \end{bmatrix}$$

14. State the general property illustrated by Problem 13.

15. Find **X** if

$$\begin{bmatrix} 2 & 1 & -3 \\ 1 & 5 & 0 \end{bmatrix} + X = 2 \begin{bmatrix} -1 & 4 & 3 \\ -2 & 0 & 4 \end{bmatrix}$$

16. Solve for **X**.

$$-3X + 2 \begin{bmatrix} 1 & -2 \\ 5 & 6 \end{bmatrix} = - \begin{bmatrix} 5 & -14 \\ 8 & 15 \end{bmatrix}$$

17. Calculate **A(B + C)** and **AB + AC** for

$$A = \begin{bmatrix} 2 & -1 \\ 3 & 4 \end{bmatrix} \quad B = \begin{bmatrix} 2 & 4 \\ 6 & 1 \end{bmatrix} \quad C = \begin{bmatrix} -1 & -2 \\ 3 & 6 \end{bmatrix}$$

What property is illustrated?

18. Calculate **(A + B)(A − B)** and $A^2 − B^2$ for

$$A = \begin{bmatrix} 3 & -2 \\ 1 & 4 \end{bmatrix} \quad B = \begin{bmatrix} 6 & -3 \\ 2 & 5 \end{bmatrix}$$

Why are your answers different?

19. Find the entry in the third row and second column of the product

$$\begin{bmatrix} 1.39 & 4.13 & -2.78 \\ 4.72 & -3.69 & 5.41 \\ 8.09 & -6.73 & 5.03 \end{bmatrix} \begin{bmatrix} 5.45 & 6.31 \\ 7.24 & -5.32 \\ 6.06 & 1.34 \end{bmatrix}$$

20. Find the entry in the second row and first column of the product in Problem 19.

*In Problems 21–26, use a calculator with matrix capabilities to calculate **D** given the following.*

$$A = \begin{bmatrix} 2 & -3 & 4 \\ 1 & 0 & 3 \\ -1 & 2 & 6 \end{bmatrix} \quad B = \begin{bmatrix} -2 & 5 & 0 \\ 3 & -2 & -5 \\ 5 & -3 & -6 \end{bmatrix}$$

21. $D = A^2 − B^2$
22. $D = (A + B)(A − B)$
23. $D = (A − B)^3$
24. $D = A^3 − 3A^2B + 3AB^2 − B^3$
25. $D = A^4 + 3B^4$
26. $D = A^3B^3 − (AB)^3$

B. Applications and Extensions

27. Let **A** and **B** be 3×4 matrices, **C** a 4×3 matrix, and **D** a 3×3 matrix. Which of the following do not make sense, that is, do not satisfy the compatibility conditions?
    (a) **AB**
    (b) **AC**
    (c) **AC − D**
    (d) **(A − B)C**
    (e) **(AC)D**
    (f) **A(CD)**
    (g) A^2
    (h) $(CA)^2$
    (i) **C(A + 2B)**

28. Calculate **AB** and **BA** for

$$A = \begin{bmatrix} 1 & 2 & 3 & 4 \end{bmatrix} \quad B = \begin{bmatrix} 2 \\ 1 \\ -1 \\ -2 \end{bmatrix}$$

29. Show that if **AB** and **BA** both make sense, then **AB** and **BA** are both square matrices.

30. If $(A + B)^2 = A^2 + 2AB + B^2$, what conclusions can you draw about **A** and **B**?

31. If the *i*th row of **A** consists of all zeros, what is true about the *i*th row of **AB** (assuming **AB** makes sense)?

32. Let

$$A = \begin{bmatrix} 0 & a \\ 0 & 0 \end{bmatrix} \quad B = \begin{bmatrix} 0 & a & b \\ 0 & 0 & c \\ 0 & 0 & 0 \end{bmatrix}$$

Calculate A^2 and B^3 and then make a conjecture.

33. A matrix of the form

$$A = \begin{bmatrix} 1 & a & b \\ 0 & 1 & c \\ 0 & 0 & 1 \end{bmatrix}$$

where *a*, *b*, and *c* are any real numbers is called a **Heisenberg matrix.** What is true about the product of two such matrices?

34. Let

$$\mathbf{A} = \begin{bmatrix} 1 & 2 & 0 \\ 0 & 1 & 0 \\ 0 & 0 & 1 \end{bmatrix}$$

Calculate $\mathbf{A}^2$, $\mathbf{A}^3$, $\mathbf{A}^4$, and conjecture the form of $\mathbf{A}^n$.

35. Let

$$\mathbf{B} = \begin{bmatrix} 1 & 0 & 3 \\ 0 & 1 & 0 \\ 0 & 0 & 1 \end{bmatrix}$$

Calculate $\mathbf{B}^2$, $\mathbf{B}^3$, $\mathbf{B}^4$, and conjecture the form of $\mathbf{B}^n$.

36. Let

$$\mathbf{A} = \begin{bmatrix} 3 & 0 & 0 \\ 0 & -4 & 0 \\ 0 & 0 & 5 \end{bmatrix}$$

If **B** is any 3×3 matrix, what does multiplication on the left by **A** do to **B**? Multiplication on the right by **A**?

37. Calculate $\mathbf{A}^2$ and $\mathbf{A}^3$ for the matrix **A** of Problem 36. State a general result about raising a diagonal matrix to a positive integral power.

38. Art, Bob, and Curt work for a company that makes Flukes, Gizmos, and Horks. They are paid for their labor on a piecework basis, receiving $1 for each Fluke, $2 for each Gizmo, and $3 for each Hork. Below are matrices **U** and **V,** representing their outputs on Monday and Tuesday. Matrix **X** is the wage per unit matrix.

	Monday's Output U				Tuesday's Output V				Wage per Unit X
	F	G	H		F	G	H		
Art	4	3	2	Art	3	6	1	F	1
Bob	5	1	2	Bob	4	2	2	G	2
Curt	3	4	1	Curt	5	1	3	H	3

Compute the following matrices and decide what they represent.
(a) **UX**   (b) **VX**   (c) **U** + **V**   (d) (**U** + **V**)**X**

39. Four friends, A, B, C, and D, have unlisted telephone numbers. Whether or not one person knows another's number is indicated by the matrix **U** below, where 1 indicates knowing and 0 indicates not knowing. For example, the 1 in row 3, column 1 means that C knows A's number.

$$\mathbf{U} = \begin{array}{c@{}c} & \begin{array}{cccc} A & B & C & D \end{array} \\ \begin{array}{c} A \\ B \\ C \\ D \end{array} & \begin{bmatrix} 0 & 0 & 1 & 0 \\ 0 & 0 & 1 & 0 \\ 1 & 0 & 0 & 1 \\ 0 & 1 & 0 & 0 \end{bmatrix} \end{array}$$

(a) Calculate $\mathbf{U}^2$.
(b) Interpret $\mathbf{U}^2$ in terms of the possibility of each person's being able to get a telephone message to another.
(c) Can D get a message to A via one other person?
(d) Interpret $\mathbf{U}^3$.
(e) Is it always possible to get a message from one person to another via at most two intermediates?

40. Use a calculator to calculate various powers of **A** and **B**. Then conjecture the general form of $\mathbf{A}^n$ and $\mathbf{B}^n$.

$$\mathbf{A} = \begin{bmatrix} 1 & 2 & 0 & 0 \\ 0 & 1 & 0 & 0 \\ 0 & 0 & 1 & 3 \\ 0 & 0 & 0 & 1 \end{bmatrix} \qquad \mathbf{B} = \begin{bmatrix} 1 & 0 & 0 & 3 \\ 0 & 1 & 0 & 0 \\ 0 & 0 & 1 & 0 \\ 0 & 0 & 0 & 1 \end{bmatrix}$$

41. Use a calculator to calculate higher and higher powers of each matrix (for example, it is easy to calculate $\mathbf{A}^2$, $\mathbf{A}^4$, $\mathbf{A}^8$, ... by using the squaring operation repeatedly). Determine the entries in the matrix that is being approached by $\mathbf{A}^n$ as n gets larger and larger.

(a) $\mathbf{A} = \begin{bmatrix} 0.1 & 0.9 \\ 0.5 & 0.5 \end{bmatrix}$ (b) $\mathbf{A} = \begin{bmatrix} 0.5 & 0.4 & 0.1 \\ 0.4 & 0.3 & 0.3 \\ 0.1 & 0.5 & 0.4 \end{bmatrix}$

(c) $\mathbf{A} = \begin{bmatrix} 0.1 & 0.2 & 0.3 & 0.4 \\ 0.2 & 0.2 & 0.4 & 0.2 \\ 0.3 & 0.3 & 0.3 & 0.1 \\ 0.1 & 0.1 & 0.4 & 0.4 \end{bmatrix}$

Note something about the form of these three matrices. Then make a conjecture about when $\mathbf{A}^n$ will converge to a limiting matrix as n gets larger and larger. Test your conjecture on other matrices.

42. **Challenge.** Find the four square roots of the matrix

$$\begin{bmatrix} 7 & 10 \\ 15 & 22 \end{bmatrix}$$

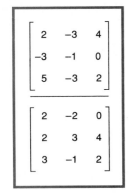

TEASER Very few math books would ask you to make sense out of a quotient of two matrices like that shown at the left. Nevertheless, we ask you to give a reasonable interpretation to this expression and then to calculate the answer, which should be a 3×3 matrix.

6.4 INVERSES OF MATRICES

Let's stay with 2×2 matrices for now. In the set of 2×2 matrices, there is a special matrix denoted by **0,** namely, the one with all four entries 0. This matrix is the additive identity; that is, $\mathbf{0} + \mathbf{A} = \mathbf{A} + \mathbf{0} = \mathbf{A}$ for every 2×2 matrix **A.** Is there a multiplicative identity in the system of 2×2 matrices that acts like the number 1 in the real number system? In other words, is there a 2×2 matrix **I** satisfying $\mathbf{I} \cdot \mathbf{A} = \mathbf{A} \cdot \mathbf{I} = \mathbf{A}$ for all 2×2 matrices **A?** The answer is yes; in fact,

$$\mathbf{I} = \begin{bmatrix} 1 & 0 \\ 0 & 1 \end{bmatrix}$$

Note that

$$\begin{bmatrix} 1 & 0 \\ 0 & 1 \end{bmatrix}\begin{bmatrix} a & b \\ c & d \end{bmatrix} = \begin{bmatrix} a & b \\ c & d \end{bmatrix} = \begin{bmatrix} a & b \\ c & d \end{bmatrix}\begin{bmatrix} 1 & 0 \\ 0 & 1 \end{bmatrix}$$

You can check that

$$\mathbf{I} = \begin{bmatrix} 1 & 0 & 0 \\ 0 & 1 & 0 \\ 0 & 0 & 1 \end{bmatrix}$$

is the multiplicative identity in the system of 3×3 matrices, and you should be able to guess at its form for the system of $n \times n$ matrices.

MULTIPLICATIVE INVERSES OF 2×2 MATRICES

Recall that 1/3 is the multiplicative inverse of 3 because

$$\frac{1}{3} \cdot 3 = 3 \cdot \frac{1}{3} = 1$$

You will also recall that 3^{-1} is another symbol for 1/3. We say that the matrix **B** is the multiplicative inverse of **A** if

$$\mathbf{B} \cdot \mathbf{A} = \mathbf{A} \cdot \mathbf{B} = \mathbf{I}$$

and, if such a matrix **B** exists, we denote it by $\mathbf{A}^{-1}$. Only one number, namely

0, fails to have a multiplicative inverse, whereas many matrices (including **0**) fail to have a multiplicative inverse, as we shall see. However, if you write down a 2 × 2 matrix at random, it will very likely have an inverse.

■ **Example 1.** Show that matrix **A** has an inverse where

$$\mathbf{A} = \begin{bmatrix} 6 & 7 \\ 1 & 2 \end{bmatrix}$$

Solution. We claim that

$$\mathbf{A}^{-1} = \begin{bmatrix} \dfrac{2}{5} & -\dfrac{7}{5} \\ -\dfrac{1}{5} & \dfrac{6}{5} \end{bmatrix}$$

To establish this, we simply calculate two matrix products to see that we get **I** in each case.

$$\begin{bmatrix} \dfrac{2}{5} & -\dfrac{7}{5} \\ -\dfrac{1}{5} & \dfrac{6}{5} \end{bmatrix}\begin{bmatrix} 6 & 7 \\ 1 & 2 \end{bmatrix} = \begin{bmatrix} 1 & 0 \\ 0 & 1 \end{bmatrix} \quad \text{and} \quad \begin{bmatrix} 6 & 7 \\ 1 & 2 \end{bmatrix}\begin{bmatrix} \dfrac{2}{5} & -\dfrac{7}{5} \\ -\dfrac{1}{5} & \dfrac{6}{5} \end{bmatrix} = \begin{bmatrix} 1 & 0 \\ 0 & 1 \end{bmatrix} \quad ■$$

You are probably asking two questions. How did we know that the matrix **A** of Example 1 had an inverse, and, knowing this, how did we find it? A beautiful theorem answers both questions.

THEOREM (MULTIPLICATIVE INVERSES)

The matrix

$$\mathbf{A} = \begin{bmatrix} a & b \\ c & d \end{bmatrix}$$

has a multiplicative inverse if and only if $D = ad - bc$ is nonzero. If $D \neq 0$, then

$$\mathbf{A}^{-1} = \begin{bmatrix} \dfrac{d}{D} & -\dfrac{b}{D} \\ -\dfrac{c}{D} & \dfrac{a}{D} \end{bmatrix}$$

Thus, the number D determines whether a matrix has a multiplicative inverse. We call D the **determinant** of the matrix. Every square matrix has associated with it a number called its determinant. We will study this concept in detail in Section 6.5.

■ **Example 2.** Determine whether each of the following matrices has a multiplicative inverse, and if so, find it.

(a) $\mathbf{A} = \begin{bmatrix} 2 & -5 \\ 5 & -9 \end{bmatrix}$ (b) $\mathbf{B} = \begin{bmatrix} 3 & -5 \\ -6 & 10 \end{bmatrix}$

Solution. (a) $D = (2)(-9) - (-5)(5) = 7$

$$A^{-1} = \begin{bmatrix} -\dfrac{9}{7} & \dfrac{5}{7} \\ -\dfrac{5}{7} & \dfrac{2}{7} \end{bmatrix}$$

(b) $D = (3)(10) - (-5)(-6) = 0$. Thus **B** does not have an inverse. ■

INVERSES FOR LARGER MATRICES (HAND CALCULATION)

There is a theorem like the one above for square matrices of any size, which Cayley found in 1858. The theorem says that a square matrix has an inverse if and only if its determinant is nonzero, and it gives a formula for the inverse when it exists. Rather than introducing and discussing this complicated formula, we are going to explain a simple process that always produces the inverse when it exists. Briefly described, it is this. Take any square matrix **A** and extend it by writing the corresponding identity matrix **I** next to it on the right. By using the row operations of Section 6.2, attempt to reduce **A** to the identity matrix while simultaneously performing the same operations on **I**. If you can reduce **A** to **I**, you will simultaneously turn **I** into A^{-1}. If you cannot reduce **A** to **I**, then **A** has no inverse.

■ **Example 3.** Use the above method to obtain the inverse of the matrix of Example 1.

Solution.

$$\begin{bmatrix} 6 & 7 & 1 & 0 \\ 1 & 2 & 0 & 1 \end{bmatrix} \xrightarrow{r_{1,2}} \begin{bmatrix} 1 & 2 & 0 & 1 \\ 6 & 7 & 1 & 0 \end{bmatrix}$$

$$\xrightarrow{-6r_1 + r_2} \begin{bmatrix} 1 & 2 & 0 & 1 \\ 0 & -5 & 1 & -6 \end{bmatrix} \xrightarrow{-\frac{1}{5}r_2} \begin{bmatrix} 1 & 2 & 0 & 1 \\ 0 & 1 & -\dfrac{1}{5} & \dfrac{6}{5} \end{bmatrix}$$

$$\xrightarrow{-2r_2 + r_1} \begin{bmatrix} 1 & 0 & \dfrac{2}{5} & -\dfrac{7}{5} \\ 0 & 1 & -\dfrac{1}{5} & \dfrac{6}{5} \end{bmatrix}$$

The desired inverse appears on the right in the last display; you will note that it agrees with the answer we gave in Example 1. ■

■ **Example 4.** Use the method in Example 3 to obtain the inverse of

$$A = \begin{bmatrix} 2 & 6 & 6 \\ 2 & 7 & 6 \\ 2 & 7 & 7 \end{bmatrix}$$

Solution.

$$\begin{bmatrix} 2 & 6 & 6 & 1 & 0 & 0 \\ 2 & 7 & 6 & 0 & 1 & 0 \\ 2 & 7 & 7 & 0 & 0 & 1 \end{bmatrix} \xrightarrow{\frac{1}{2}r_1} \begin{bmatrix} 1 & 3 & 3 & \frac{1}{2} & 0 & 0 \\ 2 & 7 & 6 & 0 & 1 & 0 \\ 2 & 7 & 7 & 0 & 0 & 1 \end{bmatrix}$$

$$\xrightarrow[\substack{-2r_1 + r_3}]{-2r_1 + r_2} \begin{bmatrix} 1 & 3 & 3 & \frac{1}{2} & 0 & 0 \\ 0 & 1 & 0 & -1 & 1 & 0 \\ 0 & 1 & 1 & -1 & 0 & 1 \end{bmatrix} \xrightarrow{-r_2 + r_3} \begin{bmatrix} 1 & 3 & 3 & \frac{1}{2} & 0 & 0 \\ 0 & 1 & 0 & -1 & 1 & 0 \\ 0 & 0 & 1 & 0 & -1 & 1 \end{bmatrix}$$

$$\xrightarrow{-3r_2 + r_1} \begin{bmatrix} 1 & 0 & 3 & \frac{7}{2} & -3 & 0 \\ 0 & 1 & 0 & -1 & 1 & 0 \\ 0 & 0 & 1 & 0 & -1 & 1 \end{bmatrix}$$

$$\xrightarrow{-3r_3 + r_1} \begin{bmatrix} 1 & 0 & 0 & \frac{7}{2} & 0 & -3 \\ 0 & 1 & 0 & -1 & 1 & 0 \\ 0 & 0 & 1 & 0 & -1 & 1 \end{bmatrix}$$

Thus, the desired inverse is

$$\mathbf{A}^{-1} = \begin{bmatrix} \frac{7}{2} & 0 & -3 \\ -1 & 1 & 0 \\ 0 & -1 & 1 \end{bmatrix}$$

■

MATRIX INVERSES ON A CALCULATOR

Many graphics calculators are programmed to compute the multiplicative inverse of a square matrix automatically.

■ **Example 5.** Use a calculator to find the inverse of the matrix of Example 4.

Solution. The left screen in Figure 1 shows matrix **A**; the right screen, obtained by using the ⌐x⌐ key, shows $\mathbf{A}^{-1}$.　■

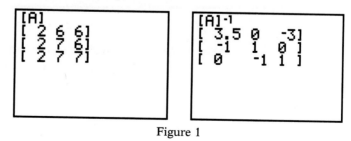

Figure 1

■ **Example 6.** Use a calculator to find the inverses of

$$\mathbf{A} = \begin{bmatrix} 1 & 2 & -1 & 0 \\ 2 & 3 & 7 & 0 \\ 3 & -1 & 2 & \dfrac{3}{2} \\ 6 & 3 & -3 & 2 \end{bmatrix} \quad \text{and} \quad \mathbf{B} = \begin{bmatrix} 1 & 2 & -1 \\ 2 & 3 & 7 \\ -3 & -5 & -6 \end{bmatrix}$$

Solution. Figure 2 shows $\mathbf{A}$ and $\mathbf{A}^{-1}$. *Note:* To display all of $\mathbf{A}^{-1}$ on the screen, you may have to scroll to the right using the arrow keys. Matrix $\mathbf{B}$ does not have an inverse although your calculator (because of approximation errors) may fail to recognize this fact and report an answer (the TI-82 gives an error code). If it does give an answer, the entries will be very large and should arouse your suspicions. To convince yourself that $\mathbf{B}$ does not have an inverse, try the hand-calculation method. ∎

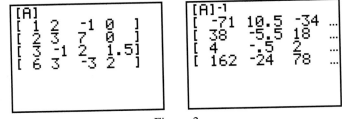

Figure 2

APPLICATIONS TO SYSTEMS OF EQUATIONS

A system of n equations in n unknowns can be written in matrix form $\mathbf{AX} = \mathbf{B}$ and then solved by the simple algebraic maneuver of multiplying both sides *on the left* by $\mathbf{A}^{-1}$. This procedure gives the solution $\mathbf{A}^{-1}\mathbf{AX} = \mathbf{X} = \mathbf{A}^{-1}\mathbf{B}$. It seems almost too easy, but it does, of course, involve the nontrivial task of finding $\mathbf{A}^{-1}$. Here is an illustration.

■ **Example 7.** Solve the following system of equations

$$\begin{aligned} 2x + 6y + 6z &= 2 \\ 2x + 7y + 6z &= -3 \\ 2x + 7y + 7z &= -5 \end{aligned}$$

Solution. First, we write this system in the form

$$\begin{bmatrix} 2 & 6 & 6 \\ 2 & 7 & 6 \\ 2 & 7 & 7 \end{bmatrix} \begin{bmatrix} x \\ y \\ z \end{bmatrix} = \begin{bmatrix} 2 \\ -3 \\ -5 \end{bmatrix}$$

Note that the coefficient matrix is the matrix $\mathbf{A}$ for which we found the inverse in Example 4. Multiplying both sides on the left by $\mathbf{A}^{-1}$ gives

$$\begin{bmatrix} x \\ y \\ z \end{bmatrix} = \begin{bmatrix} 3.5 & 0 & -3 \\ -1 & 1 & 0 \\ 0 & -1 & 1 \end{bmatrix} \begin{bmatrix} 2 \\ -3 \\ -5 \end{bmatrix} = \begin{bmatrix} 22 \\ -5 \\ -2 \end{bmatrix}$$

In principle, the inverse
matrix method of solving
a system of equations
works beautifully. If $\mathbf{A}^{-1}$
exists, the method gives
the unique solution. If
$\mathbf{A}^{-1}$ does not exist, then
either the system has no
solution or it has infi-
nitely many solutions.
But the method works
only for systems that
have the same number of
equations as of un-
knowns. Moreover, find-
ing the inverses of large
matrices, even with a
computer, is time-
consuming and prone to
errors caused by round-
ing off. Thus, computer
software packages tend
to use a version of the
method of Section 6.2 in
solving systems of equa-
tions.

You should check that $x = 22$, $y = -5$, and $z = -2$ is the solution to the given system. ∎

■ **Example 8.** Let $\mathbf{A}$ be the 4×4 matrix for which we found $\mathbf{A}^{-1}$ in Example 6 and let $\mathbf{X}$ be the 4×1 matrix with entries w, x, y, and z. Use a calculator to solve $\mathbf{AX} = \mathbf{C}$, where $\mathbf{C}$ is the 4×1 matrix with entries 1, 1, 1, and 2.

Solution. Symbolically, the solution is $\mathbf{X} = \mathbf{A}^{-1}\mathbf{C}$. For the product on the right, our calculator gives the 4×1 matrix with entries -43.5, 23.5, 2.5, and 100. ∎

TEASER SOLUTION

We are to interpret and calculate the quotient of the matrices $\mathbf{A}$ and $\mathbf{B}$ where

$$\mathbf{A} = \begin{bmatrix} 2 & -3 & 4 \\ -3 & -1 & 0 \\ 5 & -3 & 2 \end{bmatrix} \quad \text{and} \quad \mathbf{B} = \begin{bmatrix} 2 & -2 & 0 \\ 2 & 3 & 4 \\ 3 & -1 & 2 \end{bmatrix}$$

We interpret $\mathbf{A/B}$ to mean $\mathbf{AB}^{-1}$. A calculator makes this an easy product to calculate.

$$\mathbf{AB}^{-1} = \begin{bmatrix} 2 & -3 & 4 \\ -3 & -1 & 0 \\ 5 & -3 & 2 \end{bmatrix} \begin{bmatrix} 2.5 & 1 & -2 \\ 2 & 1 & -2 \\ -2.75 & -1 & 2.5 \end{bmatrix} = \begin{bmatrix} -12 & -5 & 12 \\ -9.5 & -4 & 8 \\ 1 & 0 & 1 \end{bmatrix}$$

PROBLEM SET 6.4

A. Skills and Techniques

Without use of a calculator, determine whether or not each of the following matrices has an inverse and, if it does, find the inverse.

1. $\begin{bmatrix} 2 & 3 \\ -1 & -1 \end{bmatrix}$

2. $\begin{bmatrix} 4 & 3 \\ 1 & 2 \end{bmatrix}$

3. $\begin{bmatrix} 4 & 2 \\ 6 & 3 \end{bmatrix}$

4. $\begin{bmatrix} 6 & -14 \\ 0 & 2 \end{bmatrix}$

5. $\begin{bmatrix} 0 & 3 \\ 2 & 4 \end{bmatrix}$

6. $\begin{bmatrix} 1 & 3 \\ 2 & 4 \end{bmatrix}$

7. $\begin{bmatrix} 1 & -2 & 1 \\ 3 & 0 & 2 \\ 1 & 2 & \frac{1}{2} \end{bmatrix}$

8. $\begin{bmatrix} 3 & 0 & 0 \\ 0 & 4 & 0 \\ 0 & 0 & 5 \end{bmatrix}$

9. $\begin{bmatrix} 1 & 3 & 4 \\ 2 & 1 & -1 \\ 4 & 7 & 7 \end{bmatrix}$

10. $\begin{bmatrix} -2 & 4 & 2 \\ 3 & 5 & 6 \\ 1 & 9 & 8 \end{bmatrix}$

11. $\begin{bmatrix} 1 & 1 & 1 \\ 1 & -1 & 2 \\ 3 & 2 & 0 \end{bmatrix}$

12. $\begin{bmatrix} 2 & 1 & 1 \\ 1 & 3 & 1 \\ -1 & 4 & 0 \end{bmatrix}$

13. $\begin{bmatrix} 3 & 1 & 2 \\ 4 & 1 & -6 \\ 1 & 0 & 1 \end{bmatrix}$

14. $\begin{bmatrix} 2 & 4 & 6 \\ 3 & 2 & -5 \\ 2 & 3 & 1 \end{bmatrix}$

15. $\begin{bmatrix} 2 & -2 & 4 \\ 5 & 3 & 2 \\ 3 & 5 & -2 \end{bmatrix}$

16. $\begin{bmatrix} 1 & 2 & 1 & 1 \\ 0 & 2 & 3 & 2 \\ 0 & 0 & 1 & 3 \\ 0 & 0 & 0 & 4 \end{bmatrix}$

17. $\begin{bmatrix} 1 & 1 & 1 & 1 \\ 1 & 1 & 1 & -1 \\ 1 & 1 & -1 & 1 \\ 1 & -1 & 1 & 1 \end{bmatrix}$

18. $\begin{bmatrix} 1 & 1 & 1 & 1 \\ 1 & 1 & 1 & 1 \\ 1 & 1 & -1 & 1 \\ 1 & -1 & 1 & 1 \end{bmatrix}$

Solve the following systems by making use of the inverses you found in Problems 11–14. Begin by writing the system in the matrix form $AX = B$.

19.
$$\begin{aligned} x + y + z &= 2 \\ x - y + 2z &= -1 \\ 3x + 2y &= 5 \end{aligned}$$

20.
$$\begin{aligned} 2x + y + z &= 4 \\ x + 3y + z &= 5 \\ -x + 4y &= 0 \end{aligned}$$

21.
$$\begin{aligned} 3x + y + 2z &= 3 \\ 4x + y - 6z &= 2 \\ x + z &= 6 \end{aligned}$$

22.
$$\begin{aligned} 2x + 4y + 6z &= 9 \\ 3x + 2y - 5z &= 2 \\ 2x + 3y + z &= 4 \end{aligned}$$

Using a calculator with matrix capabilities, invert each matrix in Problems 23–26. Then use the inverse to solve the associated system of equations.

23. $\begin{bmatrix} 3 & -1 & 2 \\ 4 & 5 & -3 \\ 7 & 2 & 3 \end{bmatrix}$ $\begin{aligned} 3x - y + 2z &= 2 \\ 4x + 5y - 3z &= -1 \\ 7x + 2y + 3z &= 4 \end{aligned}$

24. $\begin{bmatrix} 6 & 1 & -3 \\ 4 & 4 & 2 \\ 5 & -1 & -4 \end{bmatrix}$ $\begin{aligned} 6x + y - 3z &= -9 \\ 4x + 4y + 2z &= 2 \\ 5x - y - 4z &= -4 \end{aligned}$

25. $\begin{bmatrix} 2 & 2 & 3 & 4 \\ 4 & 3 & 2 & 1 \\ 5 & 2 & -1 & 3 \\ -3 & -2 & 5 & 1 \end{bmatrix}$ $\begin{aligned} 2x + 2y + 3z + 4w &= -2 \\ 4x + 3y + 2z + w &= -3 \\ 5x + 2y - z + 3w &= 5 \\ -3x - 2y + 5z + w &= 1 \end{aligned}$

26. $\begin{bmatrix} -7 & 2 & -3 & 5 \\ -5 & 4 & 2 & 3 \\ 6 & 1 & 5 & -3 \\ 5 & -2 & -3 & -4 \end{bmatrix}$ $\begin{aligned} -7x + 2y - 3z + 5w &= -5 \\ -5x + 4y + 2z + 3w &= -6 \\ 6x + y + 5z - 3w &= -7 \\ 5x - 2y - 3z - 4w &= -8 \end{aligned}$

27. Find **X**, with entries accurate to three decimal places, given that $AX = C$ and

$$A = \begin{bmatrix} 0.2 & 0.7 & 0.9 & -2 & 0.7 \\ 0.3 & 0.1 & -4 & 0.5 & 3 \\ 0.4 & 2 & 0.3 & -3 & 0.5 \\ 0.2 & 3 & 5 & -8 & 0.3 \\ 0.3 & 0.4 & 0.5 & 0.6 & 0.7 \end{bmatrix} \quad C = \begin{bmatrix} 0.3 & 0.5 \\ 0.2 & 0.6 \\ 0.4 & 0.2 \\ 0.3 & 0.3 \\ 0.1 & -1 \end{bmatrix}$$

28. Find $(A^{-1})^3$, with entries accurate to three decimal places, for the matrix **A** of Problem 27.

B. Applications and Extensions

29. Find the multiplicative inverse of

$$\begin{bmatrix} 2 & 0 & 0 \\ 0 & 3 & 0 \\ 0 & 0 & -4 \end{bmatrix}$$

30. Give a formula for U^{-1} if

$$U = \begin{bmatrix} a & 0 & 0 \\ 0 & b & 0 \\ 0 & 0 & c \end{bmatrix}$$

When does the matrix **U** fail to have an inverse?

31. Let **A** and **B** be 3×3 matrices with inverses A^{-1} and B^{-1}. Show that **AB** has an inverse given by $B^{-1}A^{-1}$. (*Hint:* The product in either order must be **I**.)

32. Show that

$$\begin{bmatrix} 1 & -1 \\ 3 & -3 \end{bmatrix} \begin{bmatrix} 2 & -4 \\ 2 & -4 \end{bmatrix} = \begin{bmatrix} 0 & 0 \\ 0 & 0 \end{bmatrix}$$

Thus $AB = 0$, but neither **A** nor **B** is **0**. This is another way that matrices differ from ordinary numbers.

33. Suppose that $AB = 0$ and **A** has a multiplicative inverse. Show that $B = 0$. See Problem 32.

34. Let **C** be a 3×3 matrix and **A** and **B** be as follows:

$$A = \begin{bmatrix} 1 & 0 & 0 \\ 0 & 1 & 2 \\ 0 & 0 & 1 \end{bmatrix} \quad B = \begin{bmatrix} 1 & 0 & 0 \\ 0 & 1 & -2 \\ 0 & 0 & 1 \end{bmatrix}$$

(a) What happens to **C** under the multiplication **AC**? Under the multiplication **BC**?
(b) What do you conclude about the matrices **A** and **B**?

35. Let

$$\mathbf{A} = \begin{bmatrix} 1 & 0 & 0 \\ 1 & 1 & 0 \\ 1 & 0 & 1 \end{bmatrix}$$

(a) Calculate $\mathbf{A}^{-1}$.
(b) Calculate $\mathbf{A}^2$, $\mathbf{A}^3$, and $\mathbf{A}^{-2} = (\mathbf{A}^{-1})^2$.
(c) Conjecture the form of $\mathbf{A}^n$ for any integer n.

36. Find the inverse of the matrix.

$$\begin{bmatrix} 1 & 1 & 1 & 1 \\ 1 & 2 & 2 & 2 \\ 1 & 2 & 1 & 1 \\ 1 & 2 & 1 & 2 \end{bmatrix}$$

37. Consider the matrices

$$\mathbf{A} = \begin{bmatrix} 0 & 1 & 0 \\ 0 & 0 & 1 \\ 1 & 0 & 0 \end{bmatrix} \quad \text{and} \quad \mathbf{B} = \begin{bmatrix} 0 & 1 & 0 & 0 \\ 0 & 0 & 1 & 0 \\ 0 & 0 & 0 & 1 \\ 1 & 0 & 0 & 0 \end{bmatrix}$$

(a) Show that $\mathbf{A}^3 = \mathbf{I}$ and $\mathbf{B}^4 = \mathbf{I}$.
(b) What does (a) allow you to conclude about $\mathbf{A}^{-1}$ and $\mathbf{B}^{-1}$?
(c) Conjecture a generalization of (a).

38. Show that the inverse of a Heisenberg matrix (see Problem 33 of Section 6.3) is a Heisenberg matrix.

39. Let

$$\mathbf{A} = \begin{bmatrix} 1 & 1 & 1 & 1 \\ 1 & 2 & 3 & 4 \\ 1 & 3 & 6 & 10 \\ 1 & 4 & 10 & 20 \end{bmatrix} \quad \mathbf{X} = \begin{bmatrix} x \\ y \\ z \\ w \end{bmatrix}$$

Solve the equation $\mathbf{AX} = \mathbf{B}$ for $\mathbf{B}$ equal to:

(a) $\begin{bmatrix} 4 \\ 3 \\ 2 \\ 1 \end{bmatrix}$ (b) $\begin{bmatrix} 1 \\ 0 \\ 1 \\ 0 \end{bmatrix}$ (c) $\begin{bmatrix} 1 \\ -2 \\ 3 \\ -4 \end{bmatrix}$

40. Let

$$\mathbf{A} = \begin{bmatrix} 1 & 1 & 1 & 1 & 1 \\ 1 & 2 & 3 & 4 & 5 \\ 1 & 3 & 6 & 10 & 15 \\ 1 & 4 & 10 & 20 & 35 \\ 1 & 5 & 15 & 35 & 70 \end{bmatrix}$$

$$\mathbf{X} = \begin{bmatrix} x \\ y \\ z \\ u \\ v \end{bmatrix} \quad \mathbf{B} = \begin{bmatrix} 2 \\ -1 \\ 3 \\ -4 \\ 5 \end{bmatrix}$$

Find $\mathbf{A}^{-1}$ and solve $\mathbf{AX} = \mathbf{B}$.

41. The matrices

$$\mathbf{A} = \begin{bmatrix} 1 & \frac{1}{2} \\ 1 & 1 \\ 2 & 3 \end{bmatrix} \quad \text{and} \quad \mathbf{B} = \begin{bmatrix} 1 & \frac{1}{2} & \frac{1}{3} \\ \frac{1}{2} & \frac{1}{3} & \frac{1}{4} \\ \frac{1}{3} & \frac{1}{4} & \frac{1}{5} \end{bmatrix}$$

and their $n \times n$ generalizations are called **Hilbert matrices;** they play an important role in numerical analysis.

(a) Find $\mathbf{A}^{-1}$ and $\mathbf{B}^{-1}$.
(b) Let $\mathbf{C}$ be the column matrix with entries (11/6, 13/12, 47/60). Solve the equation $\mathbf{BX} = \mathbf{C}$ for $\mathbf{X}$.

42. Challenge. Let $\mathbf{A}$ be the 4×4 version of the Hilbert matrix (see Problem 41). Find $\mathbf{A}^{-2} = (\mathbf{A}^{-1})^2$, assuming that all entries of $\mathbf{A}^{-1}$ are known to be integers.

TEASER   Let $\mathbf{A}$ be a square matrix. The equation $\det(\mathbf{A} - x\mathbf{I}) = 0$ is called the **eigen equation,** and its solutions are called the **eigenvalues** of the matrix. These values play a very important role both in the theoretical study of matrices and in their applications, as you will learn if you take a course called linear algebra. Find the solutions to the eigen equation shown at the left accurate to two decimal places.

$$\begin{vmatrix} 2-x & 5 & 3 \\ 0 & -2-x & 0 \\ 1 & 0 & 1-x \end{vmatrix} = 0$$

6.5 DETERMINANTS

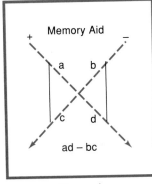

Figure 1

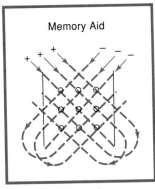

Figure 2

The history of determinants is long and illustrious, with roots going back to the seventeenth century. Associated with every square matrix **A** is a single number called its **determinant** and denoted by either det **A** or |**A**|. We met this concept for 2 × 2 matrices in Section 6.4. For a 2 × 2 matrix, we now introduce the determinant symbol

$$\begin{vmatrix} a & b \\ c & d \end{vmatrix}$$

to which we give the numerical value $ad - bc$. Note the use of the vertical bars in contrast to the brackets we use to denote a matrix. We will refer to both the symbol above and its value as a determinant. The memory device in Figure 1 may help you to remember the value of this determinant. Keep in mind what we learned in Section 6.4 for 2 × 2 matrices and their corresponding (second-order) determinants. A 2 × 2 matrix has a multiplicative inverse if and only if its determinant is nonzero.

THIRD-ORDER DETERMINANTS

We define a third-order determinant (determinant of a 3 × 3 matrix) and its value by

$$\begin{vmatrix} a_1 & b_1 & c_1 \\ a_2 & b_2 & c_2 \\ a_3 & b_3 & c_3 \end{vmatrix} = a_1 b_2 c_3 + a_3 b_1 c_2 + a_2 b_3 c_1 - a_3 b_2 c_1 - a_2 b_1 c_3 - a_1 b_3 c_2$$

Admittedly, the formula is complicated, and so we introduce a device (Figure 2) for remembering it. Note that each term is a product of three factors; three have plus signs and three have minus signs.

■ **Example 1.** Calculate the value of

$$\begin{vmatrix} -3 & 2 & 3 \\ 1 & 4 & -2 \\ 5 & -3 & 2 \end{vmatrix}$$

Solution. Following the pattern in Figure 2, we obtain

$$(-3)(4)(2) + (2)(-2)(5) + (3)(-3)(1)$$
$$- (3)(4)(5) - (2)(1)(2) - (-3)(-3)(-2) = -99 \quad ■$$

PROPERTIES OF DETERMINANTS

We want to know what happens to a determinant when we perform the row operations of Section 6.2.

1. *Interchanging two rows changes the sign of the determinant.*
2. *Multiplying a row by k multiplies the value of the determinant by k.*
3. *Replacing a row by the sum of that row and a multiple of another row does not change the value of the determinant.*

Though they are harder to prove in the third-order case, we emphasize that these three properties hold for both second- and third-order determinants (and for the higher-order determinants to be defined later). We offer only one proof, a proof of property 3 in the second-order case.

$$\begin{vmatrix} a & b \\ c + ka & d + kb \end{vmatrix} = a(d + kb) - b(c + ka)$$

$$= ad + akb - bc - bka = ad - bc = \begin{vmatrix} a & b \\ c & d \end{vmatrix}$$

■ **Example 2.** Use the properties above to simplify the evaluation of

$$\begin{vmatrix} 2 & -4 & 6 \\ 3 & -5 & -2 \\ 5 & -2 & 4 \end{vmatrix}$$

Solution. Our goal is to reduce to triangular form, using the three properties listed earlier (circled numbers above an equal sign refer to those properties).

$$\begin{vmatrix} 2 & -4 & 6 \\ 3 & -5 & -2 \\ 5 & -2 & 4 \end{vmatrix} \overset{②}{=} 2\begin{vmatrix} 1 & -2 & 3 \\ 3 & -5 & -2 \\ 5 & -2 & 4 \end{vmatrix} \overset{③}{=} 2\begin{vmatrix} 1 & -2 & 3 \\ 0 & 1 & -11 \\ 0 & 8 & -11 \end{vmatrix}$$

$$\overset{③}{=} 2\begin{vmatrix} 1 & -2 & 3 \\ 0 & 1 & -11 \\ 0 & 0 & 77 \end{vmatrix} = 2(77) = 154 \qquad ■$$

In evaluating the last determinant, we used the following important fact.

The determinant of a triangular matrix (all entries below the main diagonal are 0) is the product of the diagonal elements.

This statement is true for third-order determinants because, of the six terms, all except the term composed of the product of the main diagonal entries have a zero factor.

HIGHER-ORDER DETERMINANTS

First, we introduce the standard notation for an $n \times n$ matrix.

$$\begin{bmatrix} a_{11} & a_{12} & a_{13} & \cdots & a_{1n} \\ a_{21} & a_{22} & a_{23} & \cdots & a_{2n} \\ a_{31} & a_{32} & a_{33} & \cdots & a_{3n} \\ \vdots & \vdots & \vdots & & \vdots \\ a_{n1} & a_{n2} & a_{n3} & \cdots & a_{nn} \end{bmatrix}$$

Note the use of the double subscript on each entry: the first subscript gives the row in which a_{ij} is located and the second gives the column. For example, a_{32} is the entry in the third row and second column.

Associated with each entry a_{ij} in an $n \times n$ matrix is a determinant M_{ij} of order $n - 1$ called the **minor** of a_{ij}. It is obtained by taking the determinant

a_{ij} stands. For example, the minor M_{13} of a_{13} in the 4×4 matrix

$$\begin{bmatrix} a_{11} & a_{12} & a_{13} & a_{14} \\ a_{21} & a_{22} & a_{23} & a_{24} \\ a_{31} & a_{32} & a_{33} & a_{34} \\ a_{41} & a_{42} & a_{43} & a_{44} \end{bmatrix}$$

is the third-order determinant.

$$\begin{vmatrix} a_{21} & a_{22} & a_{24} \\ a_{31} & a_{32} & a_{34} \\ a_{41} & a_{42} & a_{44} \end{vmatrix}$$

Here is the definition to which we have been leading.

$$\begin{vmatrix} a_{11} & a_{12} & \cdots & a_{1n} \\ a_{21} & a_{22} & \cdots & a_{2n} \\ \vdots & \vdots & & \vdots \\ a_{n1} & a_{n2} & \cdots & a_{nn} \end{vmatrix} = a_{11}M_{11} - a_{12}M_{12} + a_{13}M_{13} \cdots + (-1)^{n+1}a_{1n}M_{1n}$$

There are two important questions to answer regarding this definition.

Does this definition really define? Only if the minors M_{ij} can be evaluated. They are themselves determinants, but here is the key point: They are of order $n - 1$, one less than the order of the determinant we started with. They can, in turn, be expressed in terms of determinants of order $n - 2$, and so on, using the same definition. Thus, for example, a fifth-order determinant can be expressed in terms of fourth-order determinants, and these fourth-order determinants can be expressed in terms of third-order determinants. But we know how to evaluate third-order determinants.

Is this definition consistent with the earlier definition when applied to third-order determinants? Yes, for if we apply it to a general third-order determinant, we get

$$\begin{vmatrix} a_1 & b_1 & c_1 \\ a_2 & b_2 & c_2 \\ a_3 & b_3 & c_3 \end{vmatrix} = a_1 \begin{vmatrix} b_2 & c_2 \\ b_3 & c_3 \end{vmatrix} - b_1 \begin{vmatrix} a_2 & c_2 \\ a_3 & c_3 \end{vmatrix} + c_1 \begin{vmatrix} a_2 & b_2 \\ a_3 & b_3 \end{vmatrix}$$

$$= a_1b_2c_3 - a_1c_2b_3 - b_1a_2c_3 + b_1c_2a_3 + c_1a_2b_3 - c_1b_2a_3$$

This is the same value we gave earlier.

Evaluating a fourth-order (or higher-order) determinant is a daunting task, but it is simplified by the properties enunciated earlier and a special property we consider next.

EXPANSION ACCORDING TO ANY ROW OR COLUMN

Our definition expressed the value of a determinant in terms of the entries and minors of the first row; we call it an expansion according to the first row. It is a remarkable fact that we can expand a determinant according to any row or column (and always get the same answer).

Before we can show what we mean, we must explain a sign convention. We associate a plus or minus sign with every position in a matrix. To the ij position, we assign a plus sign if $i + j$ is even and a minus sign otherwise. Thus for a 4×4 matrix, we have this pattern of signs.

$$\begin{bmatrix} + & - & + & - \\ - & + & - & + \\ + & - & + & - \\ - & + & - & + \end{bmatrix}$$

There is always a $+$ in the upper left position and then the signs alternate.

With this understanding about signs, we may expand according to any row or column. For example, to evaluate a fourth-order determinant, we can expand according to the second column if we wish. We multiply each entry in that column by its minor, prefixing each product with a plus or minus sign according to the pattern above. Then we add the results.

$$\begin{vmatrix} a_{11} & a_{12} & a_{13} & a_{14} \\ a_{21} & a_{22} & a_{23} & a_{24} \\ a_{31} & a_{32} & a_{33} & a_{34} \\ a_{41} & a_{42} & a_{43} & a_{44} \end{vmatrix} = -a_{12}M_{12} + a_{22}M_{22} - a_{32}M_{32} + a_{42}M_{42}$$

■ **Example 3.** Evaluate the following determinant

$$\begin{vmatrix} 6 & 0 & 4 & -1 \\ 2 & 0 & -1 & 4 \\ -2 & 4 & -2 & 3 \\ 4 & 0 & 5 & -4 \end{vmatrix}$$

Solution. It is obviously best to expand according to the second column, since three of the four resulting terms are zero. The single nonzero term is just $(-1)(4)$ times the minor M_{32}, namely,

$$-4 \begin{vmatrix} 6 & 4 & -1 \\ 2 & -1 & 4 \\ 4 & 5 & -4 \end{vmatrix}$$

We could now evaluate this third-order determinant using the device in Figure 2. But having seen the usefulness of zeros, let us take a different tack. It is easy to get two zeros in the first column. Simply add -3 times the second row to the first row and -2 times the second row to the third row. We get

$$-4 \begin{vmatrix} 0 & 7 & -13 \\ 2 & -1 & 4 \\ 0 & 7 & -12 \end{vmatrix}$$

Finally, expand according to the first column.

$$(-4)(-1)(2) \begin{vmatrix} 7 & -13 \\ 7 & -12 \end{vmatrix} = 8(-84 + 91) = 56$$

The reason for the factor of -1 is that the entry 2 is in a minus position in the 3×3 pattern of signs. ■

■ **Example 4.** Evaluate

$$\begin{vmatrix} 2 & -1 & 3 & 0 \\ 1 & 0 & 4 & -1 \\ 0 & 3 & 6 & 2 \\ 1 & 1 & -1 & -1 \end{vmatrix}$$

Solution. We expand according to the first column, after introducing two more zeros.

$$\begin{vmatrix} 2 & -1 & 3 & 0 \\ 1 & 0 & 4 & -1 \\ 0 & 3 & 6 & 2 \\ 1 & 1 & -1 & -1 \end{vmatrix} = \begin{vmatrix} 0 & -3 & 5 & 2 \\ 0 & -1 & 5 & 0 \\ 0 & 3 & 6 & 2 \\ 1 & 1 & -1 & -1 \end{vmatrix} = - \begin{vmatrix} -3 & 5 & 2 \\ -1 & 5 & 0 \\ 3 & 6 & 2 \end{vmatrix}$$

$$= - \begin{vmatrix} -6 & -1 & 0 \\ -1 & 5 & 0 \\ 3 & 6 & 2 \end{vmatrix} = -2 \begin{vmatrix} -6 & -1 \\ -1 & 5 \end{vmatrix} = 62 \quad ■$$

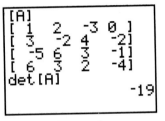

Figure 3

USING A CALCULATOR

No matter how you approach it, evaluation of determinants by hand is hard work and time-consuming. Fortunately, many calculators do this work automatically, once the data are entered. Figure 3 shows the screen for one such calculation. We suggest that you use your calculator to reevaluate each of the determinants in Examples 1–4. Then try the following example.

■ **Example 5.** Using a calculator, evaluate

$$\begin{vmatrix} 1 & 2 & -1 \\ 2 & 3 & 7 \\ -3 & -5 & -6 \end{vmatrix}$$

BEWARE

There are several different ways of evaluating determinants, including some that we have not mentioned. Depending on the method your calculator uses, it may sometimes give only a good approximation to the answer even if the entries are integers.

Solution. The value of this determinant is actually 0 (if you add the first and second rows to the third, you get a row of zeros). Yet our TI-81 calculator gives 9×10^{-12} as the answer (our TI-82 gives 0). We remind you again that calculators are not infallible; they have limitations that you must keep in mind (see the box titled Beware). Compare this example with Example 6 of Section 6.4. See also Problem 49. ■

AN APPLICATION

Our discussion of determinants has only hinted at their uses. Here is one that we think you are prepared to appreciate.

A square matrix has a multiplicative inverse if and only if its determinant is nonzero. Consequently, a system of n equations in n unknowns has a unique solution if and only if the determinant of coefficients is nonzero.

TEASER SOLUTION

The eigen equation

$$\begin{vmatrix} 2-x & 5 & 3 \\ 0 & -2-x & 0 \\ 1 & 0 & 1-x \end{vmatrix} = 0$$

is equivalent to the cubic equation $-x^3 + x^2 + 7x + 2 = 0$, as you can discover by a hand calculation. Using a graphics calculator to solve this equation leads to the three solutions (eigenvalues): -2.00, -0.30, and 3.30. Substitute -2 for x in the determinant, and you will see that it is an exact solution; the other two are two-decimal approximations.

The real power of some graphics calculators (for example, the TI-81 and TI-82) can be illustrated by doing this problem another way. To find the eigenvalues of a square matrix $\mathbf{A}$, enter $\mathbf{A}$ and $\mathbf{I}$. Then ask the calculator to graph $y = \det(\mathbf{A} - x\mathbf{I})$. Finally, determine the solutions of $\det(\mathbf{A} - x\mathbf{I}) = 0$ by finding where $y = 0$ using a root finding process.

PROBLEM SET 6.5

A. Skills and Techniques

Evaluate each of the following determinants by hand. Be sure to look for the easiest way in each case.

1. $\begin{vmatrix} 4 & 0 \\ 0 & -2 \end{vmatrix}$

2. $\begin{vmatrix} 8 & 0 \\ 5 & 0 \end{vmatrix}$

3. $\begin{vmatrix} 3 & 2 \\ 5 & 6 \end{vmatrix}$

4. $\begin{vmatrix} 5 & 3 \\ 5 & -3 \end{vmatrix}$

5. $\begin{vmatrix} -1 & -7 & 9 \\ 0 & 5 & 4 \\ 0 & 0 & 10 \end{vmatrix}$

6. $\begin{vmatrix} 3 & -2 & 1 \\ 0 & 0 & 0 \\ 1 & 5 & -8 \end{vmatrix}$

7. $\begin{vmatrix} 3 & 0 & 8 \\ 10 & 0 & 2 \\ -1 & 0 & -9 \end{vmatrix}$

8. $\begin{vmatrix} 9 & 0 & 0 \\ 0 & 0 & -2 \\ 0 & 4 & 0 \end{vmatrix}$

9. $\begin{vmatrix} 3 & 0 & 0 \\ -2 & 5 & 4 \\ 1 & 2 & -9 \end{vmatrix}$

10. $\begin{vmatrix} 4 & 8 & -2 \\ 1 & -2 & 0 \\ 2 & 4 & 0 \end{vmatrix}$

11. $\begin{vmatrix} 2 & -3 & 2 \\ 1 & 0 & -4 \\ -1 & 0 & 6 \end{vmatrix}$

12. $\begin{vmatrix} 3 & 1 & -5 \\ 2 & -2 & 7 \\ 1 & 0 & -1 \end{vmatrix}$

13. $\begin{vmatrix} 1.1 & 2.2 & 3.3 \\ 4.4 & 5.5 & 6.6 \\ 5.5 & 7.7 & 9.9 \end{vmatrix}$

14. $\begin{vmatrix} 1 & 2 & 3 \\ 0 & 2 & 3 \\ 0 & 0 & 3 \end{vmatrix}$

15. $\begin{vmatrix} 1 & 1 & 1 \\ 1 & 2 & 3 \\ 1 & 3 & 6 \end{vmatrix}$

16. $\begin{vmatrix} 1 & 1 & 1 \\ 1 & 2 & 4 \\ 1 & 3 & 9 \end{vmatrix}$

17. $\begin{vmatrix} 3 & 2 & -4 \\ 1 & 0 & 5 \\ 4 & -2 & 3 \end{vmatrix}$

18. $\begin{vmatrix} 2 & 4 & 1 \\ 1 & 3 & 6 \\ 2 & 3 & -1 \end{vmatrix}$

19. $\begin{vmatrix} 3 & 5 & -10 \\ 2 & 4 & 6 \\ -3 & -5 & 12 \end{vmatrix}$

20. $\begin{vmatrix} 2 & -1 & 2 \\ 4 & 3 & 4 \\ 7 & -5 & 10 \end{vmatrix}$

21. $\begin{vmatrix} 3 & 0 & 0 & 0 \\ -1 & 1 & 4 & 2 \\ 2 & 0 & 2 & -3 \\ -4 & 0 & 1 & 5 \end{vmatrix}$

22. $\begin{vmatrix} 0 & 5 & 0 & 0 \\ 1 & -3 & 0 & 2 \\ 4 & 1 & 2 & 8 \\ -3 & 2 & 0 & 5 \end{vmatrix}$

23. $\begin{vmatrix} 1 & 2 & -3 & 1 & 2 \\ -1 & 0 & 2 & 5 & -3 \\ 5 & 0 & 0 & -2 & 4 \\ 0 & 0 & 0 & 6 & 3 \\ 0 & 0 & 0 & 2 & -7 \end{vmatrix}$.

24. $\begin{vmatrix} 1 & 2 & 3 & 4 & 5 \\ 2 & 1 & 1 & 1 & 1 \\ 3 & 1 & 1 & 1 & 1 \\ 4 & 1 & 1 & 1 & 1 \\ 5 & 1 & 1 & 1 & 1 \end{vmatrix}$

Check your answers to several of the above problems using a calculator that has matrix capabilities. Then evaluate each of the following determinants.

25. $\begin{vmatrix} 5.1 & -3.2 & 2.6 \\ 1.3 & 4.5 & 2.3 \\ 3.4 & -2.2 & 1.9 \end{vmatrix}$

26. $\begin{vmatrix} 2.3 & 1.9 & -4.3 \\ 3.4 & -2.5 & 7.6 \\ 5.1 & -4.3 & 8.1 \end{vmatrix}$

27. $\begin{vmatrix} 2 & -3 & 4 & 5 \\ 2 & -3 & 4 & 7 \\ 1 & 6 & 4 & 5 \\ 2 & 6 & 4 & -8 \end{vmatrix}$

28. $\begin{vmatrix} 2 & 2 & 3 & 7 \\ 1 & 2 & 3 & -2 \\ 4 & -3 & 9 & 6 \\ 1 & 2 & 3 & -1 \end{vmatrix}$

29. $\begin{vmatrix} 1 & -2 & 1 & 4 \\ -2 & 5 & -3 & 1 \\ 0 & 7 & -4 & 2 \\ 3 & -2 & 2 & 6 \end{vmatrix}$

30. $\begin{vmatrix} 1 & -2 & 0 & -4 \\ 3 & -4 & 3 & -10 \\ 2 & 1 & -2 & 1 \\ 4 & -5 & 1 & 4 \end{vmatrix}$

31. $\begin{vmatrix} 2 & -1 & 3 & 4 & -7 \\ 4 & 2 & -3 & 1 & 5 \\ -3 & 6 & 2 & -1 & 6 \\ 4 & -2 & -9 & 5 & 3 \\ -2 & -3 & 4 & 6 & -2 \end{vmatrix}$

32. $\begin{vmatrix} 2.5 & 1 & -2 & 3 & 7 \\ 1.6 & -2 & -3 & -4 & 6 \\ -3.9 & 2 & -5 & 7 & 2 \\ 4.1 & -3 & 6 & -1 & 4 \\ 2.2 & 4 & -3 & 0 & 2.1 \end{vmatrix}$

Refer to the Teaser Solution (especially its final paragraph), and then find all the real eigenvalues of each of the following matrices, accurate to two decimal places.

33. $\begin{bmatrix} 1 & 2 & 3 \\ 0 & 2 & 3 \\ 1 & 3 & 4 \end{bmatrix}$

34. $\begin{bmatrix} 2 & -1 & -1 \\ 3 & 4 & 2 \\ 0 & -1 & -1 \end{bmatrix}$

35. $\begin{bmatrix} 1.2 & -2 & 1 & 3 \\ -2.1 & 4 & 2 & 1.1 \\ 2 & 1.2 & 0 & -2 \\ 2 & 2 & 2 & 2 \end{bmatrix}$

36. $\begin{bmatrix} 1 & 2 & 3 & 4 \\ -1 & -3 & 2 & -1 \\ 0 & -2 & 1 & 3 \\ 4 & 3 & 2 & 1 \end{bmatrix}$

B. Applications and Extensions

37. Let

$$\begin{vmatrix} a_1 & b_1 & c_1 \\ a_2 & b_2 & c_2 \\ a_3 & b_3 & c_3 \end{vmatrix} = 12$$

Use the properties of determinants to evaluate:

(a) $\begin{vmatrix} a_1 & a_2 & a_3 \\ b_1 & b_2 & b_3 \\ c_1 & c_2 & c_3 \end{vmatrix}$

(b) $\begin{vmatrix} a_3 & b_3 & c_3 \\ a_2 & b_2 & c_2 \\ a_1 & b_1 & c_1 \end{vmatrix}$

(c) $\begin{vmatrix} a_1 & b_1 & c_1 \\ a_2 & b_2 & c_2 \\ 3a_3 & 3b_3 & 3c_3 \end{vmatrix}$

(d) $\begin{vmatrix} a_1 + 3a_3 & b_1 + 3b_3 & c_1 + 3c_3 \\ a_2 & b_2 & c_2 \\ a_3 & b_3 & c_3 \end{vmatrix}$

38. Show that the value of a third-order determinant is 0 if either of the following is true.
(a) Two rows are proportional.
(b) The sum of two rows is the third row.

39. Solve for x.

$$\begin{vmatrix} x^2 & x & 1 \\ 1 & 2 & 3 \\ 4 & 2 & 1 \end{vmatrix} = 0$$

40. Show that for all x,

$$\begin{vmatrix} x^3 & x^2 & x \\ 3x^2 & 4x & 5 \\ 2x^2 & 3x & 4 \end{vmatrix} = 0$$

41. Evaluate the determinant below. (*Hint:* Subtract the first row from each of the second and third rows.)

$$\begin{vmatrix} n+1 & n+2 & n+3 \\ n+4 & n+5 & n+6 \\ n+7 & n+8 & n+9 \end{vmatrix}$$

42. If $C = AB$, where A and B are square matrices of the same size, then their determinants satisfy $|C| = |A||B|$. Prove this fact for 2×2 matrices.

43. Suppose that A and B are 3×3 matrices with $|A| = -2$. Use Problem 42 to evaluate (a) $|A^5|$; (b) $|A^{-1}|$; (c) $|B^{-1}AB|$; (d) $|3A^3|$.

44. Show that

$$\begin{vmatrix} a_1 & b_1 & 0 & 0 \\ a_2 & b_2 & 0 & 0 \\ 0 & 0 & c_1 & d_1 \\ 0 & 0 & c_2 & d_2 \end{vmatrix} = \begin{vmatrix} a_1 & b_1 \\ a_2 & b_2 \end{vmatrix} \begin{vmatrix} c_1 & d_1 \\ c_2 & d_2 \end{vmatrix}$$

45. Evaluate the following determinant the easy way.

$$\begin{vmatrix} 1 & 2 & 0 & 0 \\ 2 & 3 & 0 & 0 \\ 0 & 0 & 4 & -2 \\ 0 & 0 & 3 & 2 \end{vmatrix}$$

46. Find the eigenvalues of the following matrix without using a calculator.

$$\begin{bmatrix} 2.12 & 3.14 & -1.61 & 1.72 \\ 0 & -2.36 & 5.91 & 7.82 \\ 0 & 0 & 1.46 & 3.34 \\ 0 & 0 & 0 & 3.31 \end{bmatrix}$$

47. For what values of k does each system fail to have a unique solution?

(a) $\begin{aligned} kx + y &= 4 \\ x + ky &= 2 \end{aligned}$

(b) $\begin{aligned} kx - 4y + 2z &= 5 \\ 2ky + 3z &= -1 \\ 2x - 4y + 2z &= 7 \end{aligned}$

48. Consider the following determinant equation.

$$\begin{vmatrix} x & y & 1 \\ a & b & 1 \\ c & d & 1 \end{vmatrix} = 0 \qquad \text{where } (a, b) \neq (c, d)$$

(a) Show that the above equation is the equation of a line in the xy-plane.
(b) How can you tell immediately that the points (a, b) and (c, d) are on this line?
(c) Write a determinant equation for the line that passes through the points $(5, -1)$ and $(4, 11)$.

49. Explain why a determinant with all integer entries must be an integer.

50. Explain why a determinant with all rational entries must be a rational number.

51. Evaluate the determinants of each of the following matrices, whose entries come from Pascal's triangle. Note that each matrix has a border of 1s at the left and on top; every other entry is the sum of the neighbor above and the one to the left (see Section 8.5).

$$A_2 = \begin{bmatrix} 1 & 1 \\ 1 & 2 \end{bmatrix} \quad A_3 = \begin{bmatrix} 1 & 1 & 1 \\ 1 & 2 & 3 \\ 1 & 3 & 6 \end{bmatrix}$$

$$A_4 = \begin{bmatrix} 1 & 1 & 1 & 1 \\ 1 & 2 & 3 & 4 \\ 1 & 3 & 6 & 10 \\ 1 & 4 & 10 & 20 \end{bmatrix} \quad A_5 = \begin{bmatrix} 1 & 1 & 1 & 1 & 1 \\ 1 & 2 & 3 & 4 & 5 \\ 1 & 3 & 6 & 10 & 15 \\ 1 & 4 & 10 & 20 & 35 \\ 1 & 5 & 15 & 35 & 70 \end{bmatrix}$$

52. On the basis of the results in Problem 51, conjecture the value of $D_n = \det A_n$. Then support your conjecture by describing a systematic way of evaluating D_n.

53. Find the inverse of each matrix in Problem 51 and note that all entries are integers.

54. It is shown in linear algebra that if a matrix A has integer entries and $D = \det A \neq 0$, then the entries of A^{-1} are rational numbers of the form m/D, where m and D are integers. Thus, in this case, $B = DA^{-1}$ has integer entries. This fact suggests a way of using a calculator to find the inverse of a matrix that avoids the long decimals that we got in Section 6.4. Here are the steps:
(i) Enter A.
(ii) Calculate $B = DA^{-1}$.
(iii) Report the answer by giving the number D and the matrix B and stating that $A^{-1} = (1/D)B$.

Use your calculator to find D and $\mathbf{B}$ and also to check that $\mathbf{A} \cdot (1/D)\mathbf{B} = \mathbf{I}$ for each of the following.

(a) $\mathbf{A} = \begin{bmatrix} 2 & -1 & 3 \\ 7 & 6 & 4 \\ 1 & 2 & -5 \end{bmatrix}$

(b) $\mathbf{A} = \begin{bmatrix} 2 & -3 & 4 & 6 \\ 3 & -2 & 5 & -4 \\ 1 & 5 & -7 & 8 \\ -2 & 3 & 4 & -9 \end{bmatrix}$

55. Follow the instructions of Problem 54 for the following matrices.

(a) $\mathbf{A} = \begin{bmatrix} -4 & 5 & 3 \\ 2 & -3 & 6 \\ 5 & -4 & 9 \end{bmatrix}$

(b) $\mathbf{A} = \begin{bmatrix} 10 & -1 & 2 & 3 \\ -4 & -2 & 5 & 2 \\ 3 & 2 & 1 & 6 \\ 4 & -4 & 2 & -3 \end{bmatrix}$

56. Let $\mathbf{A}_3, \mathbf{A}_4, \mathbf{A}_5$ be the 3×3, 4×4, and 5×5 Hilbert matrices (see Problem 41 of Section 6.4). Calculate:
(a) $|\mathbf{A}_3|$ (b) $|\mathbf{A}_4|$ (c) $|\mathbf{A}_5|$
Conjecture what happens to $|\mathbf{A}_n|$ as n grows larger and larger.

57. Refer to Problem 56 and let

$$\mathbf{A} = \mathbf{A}_5, \quad \mathbf{B} = \begin{bmatrix} 1 \\ 1 \\ 1 \\ 1 \\ 1 \\ 1 \end{bmatrix}, \quad \text{and} \quad \mathbf{C} = \begin{bmatrix} 1 \\ 1 \\ 1.01 \\ 1 \\ 1 \end{bmatrix}$$

Calculate:
(a) $\mathbf{A}^{-1}$ (b) $\mathbf{A}^{-1}\mathbf{B}$ (c) $\mathbf{A}^{-1}\mathbf{C}$
What difficulty do you see in trying to solve a system of linear equations in which the coefficient matrix is a Hilbert matrix?

58. Challenge. Evaluate the determinants

$$E_1 = \begin{vmatrix} 2 & 1 \\ 1 & 2 \end{vmatrix}$$

$$E_2 = \begin{vmatrix} 2 & 1 & 0 \\ 1 & 2 & 1 \\ 0 & 1 & 2 \end{vmatrix}$$

$$E_3 = \begin{vmatrix} 2 & 1 & 0 & 0 \\ 1 & 2 & 1 & 0 \\ 0 & 1 & 2 & 1 \\ 0 & 0 & 1 & 2 \end{vmatrix}$$

Now generalize by conjecturing the value of the nth-order determinant E_n that has 2s on the main diagonal, 1s adjacent to this diagonal on either side, and 0s elsewhere. Then prove your conjecture. (*Hint:* Expand according to the first row to show that $E_n = 2E_{n-1} - E_{n-2}$, and from this expansion argue that your conjecture must be correct.)

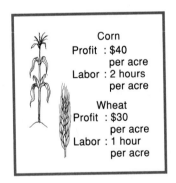

TEASER Susan Brown has 480 acres of land on which she can grow either corn or wheat. She figures that she has 800 hours of labor available during the crucial summer season. Given the profit margins and labor requirements shown at the left, how many acres of each should she plant to maximize her profit? What is this maximum profit?

Corn
Profit : $40 per acre
Labor : 2 hours per acre

Wheat
Profit : $30 per acre
Labor : 1 hour per acre

6.6 SYSTEMS OF INEQUALITIES

From systems of equations, which we have emphasized so far, we move to systems of inequalities. Such systems can involve dozens of variables and be very complicated, but we will restrict our attention to inequalities involving just two variables. A typical system of the kind we have in mind is the following.

$$3x - 6y \leq 12$$
$$4x + 6y \leq 30$$
$$6x - 5y \geq 3$$

The set of points (x, y) that satisfy all of these inequalities simultaneously will be called the **solution set;** some authors call it the feasible set. Our first task is to determine this solution set, which we will do by graphing it. We begin by analyzing the case of one linear inequality in two variables.

THE GRAPH OF $AX + BY \leq C$

The graph of $ax + by = c$ is a line; the graph of $ax + by \leq c$ is a half-plane. To see why this is so, consider the inequality $3x - 6y \leq 12$. We may rewrite this, after solving for y, as

$$y \geq \frac{1}{2}x - 2$$

The graph of those points that satisfy $y = x/2 - 2$ is the line shown in Figure 1; the graph of those points that satisfy $y > x/2 - 2$ consists of all points above that line and is shown shaded in Figure 1.

In general, the graph of $ax + by \leq c$ consists of all points on the line $ax + by = c$ together with all points on one side of it. You can tell which side by taking a test point, say $(0, 0)$, and checking whether it satisfies the inequality. If it does, all points on the same side as the test point are in the graph.

■ **Example 1.** Sketch the graph of $4x + 6y \leq 30$.

Solution. First we graph the line $4x + 6y = 30$. Since $(0, 0)$ satisfies the inequality, all points on its side of the line are in the graph and so are shaded. The result is shown in Figure 2. ■

THE GRAPH OF A SYSTEM OF INEQUALITIES

Consider again the system of inequalities mentioned earlier.

$$3x - 6y \leq 12$$
$$4x + 6y \leq 30$$
$$6x - 5y \geq 3$$

Figure 3 shows how its solution set is gradually restricted as we add one

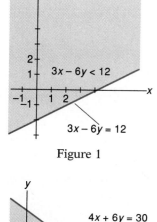

Figure 1

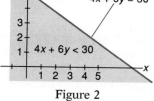

Figure 2

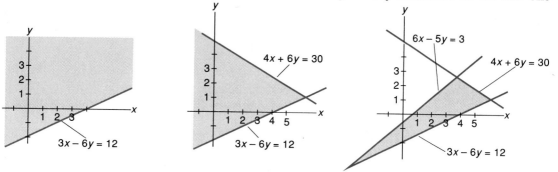

Figure 3

inequality after another. The required set is the polygonal set in the right-hand picture.

■ **Example 2.** Sketch the solution set for the following system of inequalities.

$$2x - y \geq -5$$
$$x + y \leq 11$$
$$3x - y \leq 13$$
$$x \geq 0, \quad y \geq 0$$

Solution. Figure 4 shows the solution set. As a check on its correctness, note that (1, 1) satisfies all 5 inequalities. ■

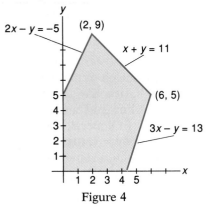

Figure 4

Each of the solution sets we have met is convex. A set is **convex** if it contains the line segment connecting any two of its points. This means that a convex set has no holes or dents. Figure 5 shows a nonconvex set. The solution set to a system of linear inequalities, being the intersection of half-planes, is always convex.

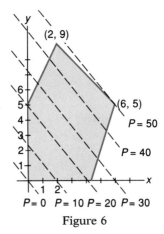

Not convex

Figure 5

LINEAR PROGRAMMING PROBLEMS

The type of problem we take up next occurs often in business and industry. It deals with maximizing (or minimizing) a linear function subject to linear inequalities, called **constraints.** We call this a **linear programming problem.** The Teaser problem that opens this section is such a problem and will be solved later. We begin with something simpler.

■ **Example 3.** Maximize $P = 5x + 4y$ subject to the constraints of Example 2.

Solution. Figure 6 shows the solution set found in Example 2 with the addition of several dashed lines. These dashed lines are the graphs of $5x + 4y = P$ for various values of P. Imagine the leftmost of these dashed lines to move across the solution set in a northeasterly direction, maintaining a constant slope. As it does so, P steadily increases, reaching its maximum value of 50 at the corner point (6, 5). ■

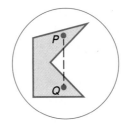

Figure 6

Reasoning as in Example 3 leads us to a remarkable conclusion.

312 • Systems of Equations and Inequalities

If a linear function, subject to linear constraints, has a maximum value (or minimum value), this value is always attained at a corner point of the solution set determined by the constraints.

This means that to solve a linear programming problem, all that is necessary is to find the corner points of the solution set, evaluate the function to be maximized at these points, and determine which of this finite set of values is largest. What might seem like a hopeless task (maximizing a function on an *infinite* set) turns out to be a matter of evaluating a function at a few points.

AN APPLICATION

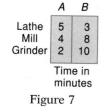

	A	B
Lathe	5	3
Mill	4	8
Grinder	2	10

Time in minutes

Figure 7

Here is a simplified manufacturing problem that is easily solved by our method.

■ **Example 4.** A small parts company finds that its regular production schedule leaves its lathe, mill, and grinder idle for 61, 88, and 95 minutes, respectively, each working day. It is considering using this idle time to manufacture two new items, A and B, which can be sold for $8 and $20 apiece, respectively. Figure 7 shows the number of minutes required on each machine to manufacture one unit of A and one unit of B. How many items of each type should the company make each day to maximize the additional daily revenue and what is this maximum additional revenue?

Solution. If you guessed from the selling prices that the company should use all its idle time to make item B, you are wrong. Let x and y denote the respective number of items of A and B to be manufactured. Obviously, x and y must be nonnegative. This fact, together with the time constraints, leads to the inequalities

$$5x + 3y \leq 61$$
$$4x + 8y \leq 88$$
$$2x + 10y \leq 95$$
$$x \geq 0, \quad y \geq 0$$

> **INTEGER ANSWERS**
>
> Linear programming in which the answers must be integers is called **integer programming.** Special techniques for achieving this have been developed. In Example 4, we simply evaluated R at integer points of the solution set near (5, 8.5). This led us to choose the point (6, 8).

The revenue produced is $R = 8x + 20y$ in dollars. Thus, we must solve the linear programming problem: Maximize R subject to the above constraints.

Figure 8 shows the graph of the solution set together with the coordinates

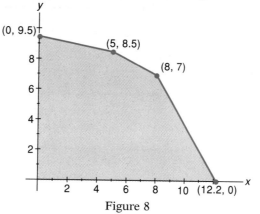

Figure 8

of the corner points (obtained by solving two equations simultaneously). Figure 9 shows the value of R at each of these points. We conclude that the company should make 5 items of A and 8.5 items of B each day, with a resulting revenue of $210. Someone may insist that only finished items should be made each day. In that case, a good choice would be 6 items of A and 8 items of B with a resulting revenue of $208. ∎

Corner	$R = 8x + 20y$
(0, 0)	0
(0, 9.5)	190
(5, 8.5)	210
(8, 7)	204
(12.2, 0)	97.6

Figure 9

NONLINEAR SYSTEMS OF INEQUALITIES

Of course, many problems of the real world turn out to be nonlinear. This nonlinearity greatly complicates the theory, but sometimes a kind of reasoning similar to that used above can help us solve such problems too.

∎ **Example 5.** Graph the solution set for the following system of inequalities. Then maximize $P = 3x + 2y$ on this set.

$$y \geq 2x^2$$
$$y \leq 2x + 4$$

Solution. Here we use a graphics calculator to draw the corresponding equations and to shade the region between them (Figure 10). The line $3x + 2y = -4$ is also shown at the lower left. As the line $3x + 2y = P$ moves upward and to the right, maintaining its slope of $-3/2$, P increases until it reaches its maximum at the upper corner of the solution set. This point, obtained by solving $y = 2x^2$ and $y = 2x + 4$ simultaneously, has coordinates (2, 8). Thus, P has a maximum value of $3(2) + 2(8) = 22$. ∎

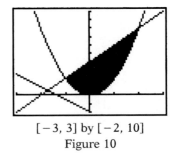

[−3, 3] by [−2, 10]
Figure 10

TEASER SOLUTION

Suppose Susan Brown plants x acres of corn and y acres of wheat. Here are the constraints.

Land constraint:	$x + y \leq 480$
Labor constraint:	$2x + y \leq 800$
Nonnegativity constraints:	$x \geq 0, \quad y \geq 0$

Her task is to maximize $P = 40x + 30y$ subject to these constraints. The solution set corresponding to these constraints is shown in Figure 11, with the corners labeled. Figure 12 shows the value of P at each of the corner points. Susan Brown should plant 320 acres of corn and 160 acres of wheat, giving her a maximum profit of $17,600.

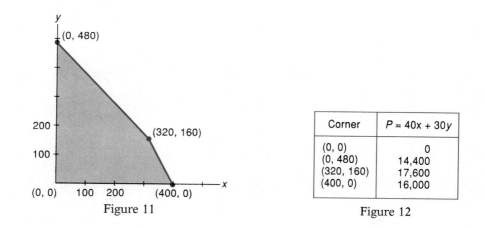

Figure 11

Corner	$P = 40x + 30y$
(0, 0)	0
(0, 480)	14,400
(320, 160)	17,600
(400, 0)	16,000

Figure 12

Suppose that the price of corn goes up so that Susan Brown can expect a profit of $80 per acre on corn but still only $30 per acre on wheat. What should she do then? Figure 13 shows that in this case she should plant 400 acres of corn and no wheat even though to do so she must leave 80 acres of land idle.

Corner	$P = 80x + 30y$
(0, 0)	0
(0, 480)	14,400
(320, 160)	30,400
(400, 0)	32,000

Figure 13

A. Skills and Techniques

Graph the solution set of each inequality.

1. $4x + y \leq 8$
2. $2x + 5y \leq 20$
3. $x \leq 3$
4. $y \leq -2$
5. $4x - y \geq 8$
6. $2x - 5y \geq -20$

Graph the solution set of the given system of inequalities. Label the coordinates of each corner point.

7. $4x + y \leq 8$
   $2x + 3y \leq 14$
   $x \geq 0 \quad y \geq 0$
8. $2x + 5y \leq 20$
   $4x + y \leq 22$
   $x \geq 0 \quad y \geq 0$
9. $4x + y \leq 8$
   $x - y \leq -2$
   $x \geq 0$
10. $2x + 5y \leq 20$
    $x - 2y \geq 1$
    $y \geq 0$

Find the maximum and minimum values of the given linear function subject to the indicated inequalities.

11. $P = 2x + y$; the inequalities of Problem 7
12. $P = 3x + 2y$; the inequalities of Problem 8
13. $P = 2x - y$; the inequalities of Problem 9
14. $P = 3x - 2y$; the inequalities of Problem 10

Solve each linear programming problem.

15. Minimize $5x + 2y$ subject to
    $x + y \geq 4$
    $x \geq 2$
    $y \geq 0$
16. Minimize $2x - y$ subject to
    $x - 2y \geq 2$
    $y \geq 2$
17. Minimize $2x + y$ subject to
    $4x + y \geq 7$
    $2x + 3y \geq 6$
    $x \geq 1$
    $y \geq 0$
18. Minimize $3x + 2y$ subject to
    $x - 2y \leq 2$
    $x - 2y \geq -2$
    $3x - 2y \geq 10$

19. Maximize $2x - y$ subject to
    $4x + y \leq 8$
    $x - y \geq -2$
    $x \geq 0 \quad y \geq 0$
20. Maximize $3x - 2y$ subject to
    $2x + 5y \leq 20$
    $x - 2y \leq 1$
    $x \geq 0$

Use a graphics calculator to display the solution set for the given inequalities. Then maximize P subject to these inequalities.

21. $y \leq 4x - x^2$
    $y \leq x$
    $x \geq 0 \quad y \geq 0$
    $P = x + 2y$
22. $y \geq 2^x$
    $y \leq 8$
    $x \geq 0$
    $P = 5x + y$
23. $y \leq \ln x$
    $x \leq 8$
    $y \geq 0$
    $P = 10x + 3y$
24. $x^2 + y^2 \geq 9$
    $0 \leq x \leq 3$
    $0 \leq y \leq 3$
    $P = x + 4y$
25. $y \leq -x^2 - 3x + 8$
    $y \geq x^3 - x - 2$
    $x \geq -1$
    $P = x + 3y$
26. $y \geq (0.4)^x$
    $y \leq -0.5(x + 12)(x - 3)$
    $x \leq 2$
    $P = x + 10y$

B. Applications and Extensions

27. Find the maximum and minimum values of $x + y$ (if they exist) subject to the following inequalities.

    $$x - y \geq -1$$
    $$x - 2y \leq 1$$
    $$3x + y \leq 10$$
    $$x \geq 0 \quad y \geq 0$$

28. Find the maximum and minimum values of $x - 2y$ (if they exist) subject to the inequalities of Problem 27.

29. Sketch the graph of the solution set for

$$x^2 + y^2 \leq 25$$
$$x^2 + (y - 7)^2 \geq 32$$

Then find the maximum and minimum values of $P = 2x + 2y$ on this set. (*Hint:* What happens to P as the line $2x + 2y = P$ moves across the set from left to right? Where does this moving line first touch and last leave the set?)

30. Rachel has received a lump sum of $100,000 on her retirement, which she will invest in safe *AA* bonds paying 12% annual interest and risky *BBB* bonds paying 15% annual interest. She plans to put at least $10,000 more in the safe bonds than the risky ones but wants at least two-thirds as much in *BBB* bonds as in *AA* bonds. How much should she invest in each type of bond to maximize her annual income while maintaining the constraints? (*Suggestion:* To keep the numbers small, measure money in thousands of dollars.)

31. A company makes a single product on two production lines, *A* and *B*. A labor force of 900 hours per week is available, and weekly running costs shall not exceed $1500. It takes 4 hours to produce one item on production line *A* and 3 hours on production line *B*. The cost per item is $5 on line *A* and $6 on line *B*. Find the largest number of items that can be produced in one week.

32. An oil refinery has a maximum production of 2000 barrels of oil per day. It produces two types of oil; type *A*, which is used for gasoline, and type *B*, which is used for heating oil. There is a requirement that a least 300 barrels of type *B* be produced each day. If the profit is $3 a barrel for type *A* and $2 a barrel for type *B*, find the maximum profit per day.

33. A manufacturer of trailers wishes to determine how many camper units and how many house trailers she should produce in order to make the best possible use of her resources. She has 42 units of wood, 56 worker-weeks of labor, and 16 units of aluminum. (Assume that all other needed resources are available and have no effect on her decision.) The amount of each resource needed to produce each camper and each trailer is given below.

	Wood	Worker-Weeks	Aluminum
Per camper	3	7	3
Per trailer	6	7	1

If the manufacturer realizes a profit of $600 on a camper and $800 on a trailer, what should be her production in order to maximize her profit?

34. A shoemaker has a supply of 100 square feet of type *A* leather, which is used for soles, and 600 square feet of type *B* leather, used for the rest of the shoe. The average shoe uses 1/4 square feet of type *A* leather and 1 square foot of type *B* leather. The average boot uses 1/4 square feet and 3 square feet of types *A* and *B* leather, respectively. If shoes sell at $40 a pair and boots at $60 a pair, find the maximum income.

35. Suppose that the minimum monthly requirements for one person are 60 units of carbohydrates, 40 units of protein, and 35 units of fat. Two foods *A* and *B* contain the following numbers of units of the three diet components per pound.

	Carbohydrates	Protein	Fat
A	5	3	5
B	2	2	1

If food *A* costs $3 a pound and food *B* costs $1.40 a pound, how many pounds of each should a person purchase per month to minimize the cost?

36. A grain farmer has 100 acres available for sowing oats and wheat. The seed oats cost $5 per acre and the seed wheat costs $8 per acre. The labor costs are $20 per acre for oats and $12 per acre for wheat. The farmer expects an income from oats of $220 per acre and from wheat of $250 per acre. How many acres of each crop should he sow to maximize his profit, if he does not wish to spend more than $620 for seed and $1800 for labor?

37. Sketch the polygon with vertices $(0, 3)$, $(4, 7)$, $(3, 0)$, and $(2, 4)$ taken in cyclic order. Then, find the maximum value of $|y - 2x| + y + x$ on this polygon.

38. Draw the graph of the set $D = \{(x, y): x^4 + y^4 \leq 6561\}$. Then think of a way to use your calculator to maximize $P = 2x + y$ on this set. Find this maximum value accurate to two decimal places. (*Hint:* First, convince yourself that the maximum value of P occurs at a boundary point of D in the first quadrant.)

39. Maximize $P = 0.2x^3 + y^2$ on the set D of Problem 38. Again, you should be able to convince yourself that the maximum value of P occurs at a boundary point of D in the first quadrant.

40. **Challenge.** If $P = (a, b)$ is a point in the plane, then $tP = (ta, tb)$. Thus, if $0 \leq t \leq 1$, the set of points of the form $tP + (1 - t)Q$ is just the line segment PQ (see Section 1.2 for material on averages). It follows that a set A is convex if whenever P and Q are in A, all points of the form $tP + (1 - t)Q$, $0 \leq t \leq 1$, are also in A.

(a) Let P, Q, and R be three fixed points in the plane and consider the set H of all points of the form $t_1P + t_2Q + t_3R$, where the ts are nonnegative and $t_1 + t_2 + t_3 = 1$. Show that H is convex.

(b) Let P, Q, R, and S be four fixed points in the plane and consider the set K of all points of the form $t_1P + t_2Q + t_3R + t_4S$, where the ts are nonnegative and $t_1 + t_2 + t_3 + t_4 = 1$. Show that K is convex.

(c) Describe the sets H and K geometrically.

CHAPTER 6 REVIEW PROBLEM SET

In Problems 1–10, write True or False in the blank. If false, tell why.

_____ 1. It is possible for a system of three equations in three unknowns to have no solution.

_____ 2. It is possible for a system of three equations in three unknowns to have exactly three solutions.

_____ 3. Matrix multiplication of compatible matrices is associative.

_____ 4. Multiplying each row of a 2×2 matrix by 3 multiplies its determinant by 3.

_____ 5. If two rows of a 3×3 matrix are multiples of each other, then the determinant of the matrix is 0.

_____ 6. A square triangular matrix in which the main diagonal elements are nonzero has a nonzero determinant.

_____ 7. Adding a multiple of one column to another column does not change the value of the determinant of a square matrix.

_____ 8. Interchanging two rows of a square matrix does not change the value of its determinant.

_____ 9. If $\mathbf{A}$ and $\mathbf{B}$ are 2×2 matrices, then $\mathbf{AB} = \mathbf{BA}$.

_____ 10. Every nonzero square matrix has an inverse.

In Problems 11–20, solve the given system or show that it has no solution.

11. $3x + 2y = 7$
    $2x - y = 7$

12. $y = 3x - 2$
    $6x - 2y = 5$

13. $\dfrac{3}{x} + \dfrac{1}{y} = 9$
    $\dfrac{2}{x} - \dfrac{3}{y} = -5$

14. $x^2 + y^2 = 13$
    $2x^2 - 3y^2 = -19$

15. $2x + y - 4z = 3$
    $3y + z = 7$
    $z = -2$

16. $2x - 3y + z = 4$
    $3y - 5z = 6$
    $-9y + 15z = 18$

17. $x - 2y + 4z = 16$
    $2x - 3y - z = 4$
    $x + 3y + 2z = 5$

18. $x - 3y + z = 6$
    $2x + y - 2z = 5$
    $-4x - 2y + 4z = -10$

19. $x^2 + 2y^2 = 13$
    $2x^2 - 3y^2 = -44$

20. $5 \cdot 2^x - 3^y = 1$
    $2^{x+2} + 3^{y+2} = 40$

In Problems 21–28, perform the indicated operations by hand and express as a single matrix. You may wish to confirm your results using your calculator.

21. $\begin{bmatrix} 2 & -1 & 4 \\ 0 & 1 & 3 \end{bmatrix} + 2\begin{bmatrix} 1 & 0 & 3 \\ 4 & -1 & 5 \end{bmatrix}$

22. $\begin{bmatrix} 5 & -4 \\ 3 & 1 \end{bmatrix}\begin{bmatrix} 2 & 5 \\ 3 & -1 \end{bmatrix}$

23. $4\begin{bmatrix} 2 \\ -1 \\ 0 \end{bmatrix} - 3\begin{bmatrix} 8 \\ -5 \\ 9 \end{bmatrix}$

24. $\begin{bmatrix} 2 & -3 \\ 4 & 1 \\ 0 & 5 \end{bmatrix}\begin{bmatrix} 4 & 0 & 1 \\ 2 & -3 & 6 \end{bmatrix}$

25. $\begin{bmatrix} 1 & 0 \\ 0 & 1 \end{bmatrix}\begin{bmatrix} 5 & 2 \\ -3 & 4 \end{bmatrix}\begin{bmatrix} b & 0 \\ 0 & b \end{bmatrix}$

26. $\begin{bmatrix} 6 & 11 \\ 0 & 0 \end{bmatrix}\begin{bmatrix} 11 & 0 \\ -6 & 0 \end{bmatrix}$

27. $\begin{bmatrix} 4 & 3 \\ 2 & 3 \end{bmatrix}\left\{\begin{bmatrix} 4 & 3 \\ 2 & 3 \end{bmatrix}^{-1} + \begin{bmatrix} 1 & 0 \\ 0 & 1 \end{bmatrix}\right\}$

28.

$$\begin{bmatrix} 1 & -2 & 3 \\ 0 & 2 & 1 \\ 1 & -2 & 4 \end{bmatrix} \begin{bmatrix} 1 & 0 & 0 \\ 0 & \frac{1}{2} & 0 \\ 0 & 0 & \frac{1}{3} \end{bmatrix}^{-1}$$

29. Find the inverse of the left matrix in Problem 28.

30. Write the following system in matrix form and then use the result of Problem 29 to solve it.

$$\begin{aligned} x - 2y + 3z &= 2 \\ 2y + z &= -4 \\ x - 2y + 4z &= 0 \end{aligned}$$

Evaluate each determinant by hand.

31. $\begin{vmatrix} 3 & -4 \\ -5 & 6 \end{vmatrix}$

32. $\begin{vmatrix} \begin{bmatrix} 5 & -2 \\ -10 & 4 \end{bmatrix}^2 \end{vmatrix}$

33. $\begin{vmatrix} 3 & 0 & -2 \\ -3 & 1 & 0 \\ 1 & 5 & 4 \end{vmatrix}$

34. $\begin{vmatrix} 5 & 1 & 19 \\ 0 & -\dfrac{1}{2} & 1 \\ 0 & 0 & 4 \end{vmatrix}$

35. $\begin{vmatrix} 2 & -1 & 4 \\ 1 & 3 & -2 \\ 4 & 5 & 0 \end{vmatrix}$

36. $\begin{vmatrix} 2 & -1 & 4 \\ 1 & 3 & -2 \\ 4 & 5 & 8 \end{vmatrix}$

37. $\begin{vmatrix} 3 & 1 & 0 & 0 \\ 5 & 2 & 0 & 0 \\ 0 & 0 & 4 & -1 \\ 0 & 0 & -6 & 2 \end{vmatrix}$

38. $\begin{vmatrix} 1 & 0 & -2 & 2 \\ 0 & 0 & -3 & 1 \\ -3 & 2 & 5 & -7 \\ -1 & 0 & 3 & 4 \end{vmatrix}$

Use a calculator to evaluate each expression involving matrices A and B below.

$$A = \begin{vmatrix} 1 & 2 & 3 & 4 \\ 5 & 6 & 7 & 8 \\ 9 & 10 & 11 & 12 \\ 13 & 14 & 15 & 16 \end{vmatrix} \quad B = \begin{vmatrix} 1 & 2 & 3 & 4 \\ 2 & 3 & 5 & 7 \\ 3 & 5 & 8 & 12 \\ 4 & 8 & 13 & 20 \end{vmatrix}$$

39. $A - 3B$

40. AB^2

41. $\det A$

42. $\det B$

43. B^{-1}

44. A^{-1}

Use a calculator to solve (if possible) each system of equations. Note that Problems 45 and 46 are related to B above.

45.
$$\begin{aligned} x + 2y + 3z + 4w &= 5 \\ 2x + 3y + 5z + 7w &= -1 \\ 3x + 5y + 8z + 12w &= -2 \\ 4x + 8y + 13z + 20w &= 3 \end{aligned}$$

46.
$$\begin{aligned} x + 2y + 3z + 4w &= 5 \\ 2x + 3y + 5z + 7w &= -10 \\ 3x + 5y + 8z + 12w &= -2 \\ 4x + 8y + 13z + 20w &= 31 \end{aligned}$$

47.
$$\begin{aligned} 0.6x + 0.2y + 0.1z &= 0.6 \\ 4.3x + 2.9y - 9.2z &= 153.4 \\ -7.4x + 7.3y + 2.8z &= 4.5 \end{aligned}$$

48.
$$\begin{aligned} 800a - 400b + 300c - 550d &= 535 \\ -375a + 225b - 100c - 100d &= -65 \\ 975a + 800b + 750c - 500d &= 1005 \\ -500a - 650b + 750c + 925d &= -1172.5 \end{aligned}$$

In Problems 49–52, graph the solution set for the given system of inequalities and label the coordinates of the vertices (corner points).

49.
$$\begin{aligned} x + y &\leq 7 \\ -3x + 4y &\geq 0 \\ x \geq 0 \quad y &\geq 0 \end{aligned}$$

50.
$$\begin{aligned} 2x + y &\leq 8 \\ x + y &\leq 6 \\ 3x + 2y &\geq 12 \end{aligned}$$

51.
$$\begin{aligned} 3x + y &\leq 15 \\ x + y &\leq 7 \\ x \geq 0 \quad y &\geq 0 \end{aligned}$$

52.
$$\begin{aligned} x - 2y + 4 &\geq 0 \\ x + y - 11 &\geq 0 \\ x \geq 0 \quad y &\geq 0 \end{aligned}$$

53. Find the maximum value of $P = 2x + y$ subject to the constraints of Problem 49.

54. Find the maximum value of $P = x + 2y$ subject to the constraints of Problem 50.

55. Find (if possible) the maximum and the minimum values of $P = x + 2y$ subject to the constraints of Problem 51.

56. Find (if possible) the maximum and the minimum values of $P = 2x + 3y$ subject to the constraints of Problem 52.

57. A metallurgist has three alloys containing 30%, 40%, and 60% nickel, respectively. How much of each should be melted and combined to obtain 100 grams of an alloy containing 45% nickel, provided that 10 grams more of the 60% alloy is used than the 30% alloy?

58. A manufacturer of personal computers makes $100 on model A and $120 on model B. Daily production for model A can vary between 50 and 110 and for model B between 75 and 100, but total daily production cannot exceed 180 units. How many units of each model should be produced each day to maximize profit?

59. All Squares College uses a nonstandard grading system. Use a calculator with matrix capabilities to determine the point value the college assigns to each letter grade in calculating students' grade point averages (GPAs), given the results shown in Figure 1 for five of its seniors. Assume that each course grade carries 1 credit.

60. Using matrix multiplication on your calculator, determine what the GPAs of the five students in Problem 59 would be if a standard grading system were used, with $A = 4$, $B = 3$, and so on.

Student	Grades					GPA
K. Green	7A	4B	6C	7D	1F	7.16
A. Gonzales	12A	12B	1C	0D	0F	12.16
V. Field	1A	16B	6C	1D	1F	7.40
Y. Li	5A	6B	9C	4D	1F	6.96
F. Johnson	3A	4B	12C	2D	4F	5.36

Figure 1

7

ANALYTIC GEOMETRY

Johannes Kepler studied mathematics and astronomy at the University of Tübingen but did his most important work at the observatory in Prague. Convinced that God had designed the world in an aesthetically pleasing manner, he was attracted to the beauty and harmony of Copernicus's heliocentric system, which puts the sun rather than the earth at the center of the solar system. Studying data on the motions of the planets led him to conjecture his famous three laws: (1) Planets move in ellipses (Section 7.2) with the sun at a focus, (2) the line from the sun to a planet sweeps out equal areas in equal times, and (3) the square of the period is proportional to the cube of the mean distance from the sun.

Then came Isaac Newton, the greatest of English mathematicians. Born on Christmas day of 1642, the young Newton showed little academic promise

but liked to build kites, clocks, and other gadgets. Fortunately his mother sent him off to Cambridge University. There Newton studied mathematics and pondered the outstanding problems of physics. In a short period of 18 months, while the university was in recess because of the bubonic plague, Newton discovered the general binomial theorem, invented the elements of calculus, proposed the theory of colors, and stated the universal law of gravitation (the inverse square law of attraction). On the basis of this latter law, Newton derived Kepler's three laws as simple mathematical consequences. The eminent Joseph-Louis Lagrange remarked that Newton was the greatest genius that ever lived and the most fortunate, since only once can the system of the universe be established. Newton is buried at Westminster Abbey with England's famous heroes.

Johannes Kepler (1571–1630)

Isaac Newton (1642–1727)

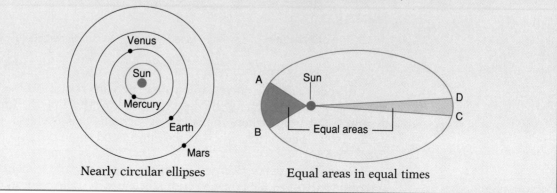

Nearly circular ellipses

Equal areas in equal times

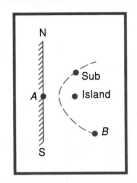

7.1 PARABOLAS

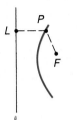

Figure 1

The question raised in the Teaser is a special case of a very general class of problems that we want to consider. Let ℓ be a fixed line (the **directrix**) and let *F* be a fixed point not on the line (the **focus**), as shown in Figure 1. Let the point *P* move in such a way that the ratio of the distance $d(P, F)$ from the focus to the distance $d(P, L)$ from the line ℓ is a positive constant *e* (the **eccentricity**); that is, so that

$$d(P, F) = ed(P, L)$$

The point *P* traces a curve that is called an **ellipse** if $0 < e < 1$, a **parabola** if $e = 1$, and a **hyperbola** if $e > 1$. In Figure 2, we show these three curves for the cases $e = 1/2$, $e = 1$, and $e = 2$.

Several things can be observed from Figure 2. The parabola is an open-armed curve with the arms opening wider and wider. The condition $e < 1$ forces the ellipse to fold its arms, forming a closed curve. Most surprising, the hyperbola, while opening its arms wide, creates another pair of arms behind its back, so to speak, that exactly mimics the pair in front.

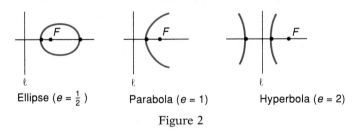

Ellipse ($e = \frac{1}{2}$)   Parabola ($e = 1$)   Hyperbola ($e = 2$)

Figure 2

In each case, the curves are symmetric with respect to the line through the focus perpendicular to the directrix. We call this line the (major) **axis of symmetry.** A point where the curve crosses this axis is called a **vertex.** The parabola has one vertex, whereas the ellipse and the hyperbola have two vertices.

These three curves were of immense interest to the ancient Greeks, who referred to them as the **conic sections** because of their relation to sections of a cone (Figure 3). If we pass a plane through a cone of two nappes, we obtain

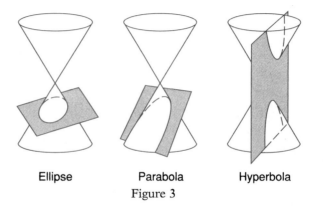

Ellipse   Parabola   Hyperbola

Figure 3

as the intersection an ellipse, parabola, or hyperbola, depending on the angle the plane makes with the axis of the cone. (We may also obtain a circle, a point, a line, or two intersecting lines, but that is a matter for a later discussion.)

Our interest is in deriving equations for these three curves and in using those equations to answer questions about the curves. The parabola is in some ways the simplest of the three curves, and that is where we begin our investigation. You will recall that we studied parabolas in Section 2.3 but from a quite different perspective.

DERIVING THE EQUATION OF A PARABOLA

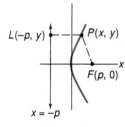

Figure 4

Place the parabola in the coordinate system so that its axis is the x-axis and its vertex is at the origin (Figure 4). Let the focus F be to the right of the origin at $(p, 0)$; then the directrix is the line $x = -p$. If $P(x, y)$ is any point on the parabola, it must satisfy

$$\boxed{d(P, F) = d(P, L)}$$

which, because of the distance formula, assumes the form

$$\sqrt{(x - p)^2 + (y - 0)^2} = \sqrt{(x + p)^2 + (y - y)^2}$$

This equation is equivalent to the result obtained by squaring both sides, that is,

$$x^2 - 2px + p^2 + y^2 = x^2 + 2px + p^2$$

which, in turn, simplifies to

$$\boxed{y^2 = 4px}$$

We call this final equation the **standard equation of the parabola.** Because replacing y by $-y$ results in an equivalent equation, its graph is symmetric with respect to the x-axis. The number p is the distance from the vertex to the focus. Thus, $y^2 = 8x$ is the equation of a horizontal parabola, opening to the right, with vertex at the origin and focus at $(2, 0)$.

The equation just derived has three variants. If we interchange the roles of x and y (obtaining $x^2 = 4py$), we have the equation of a parabola that opens up with the y-axis as its axis of symmetry. The corresponding parabolas which open to the left and down have equations $y^2 = -4px$ and $x^2 = -4py$, respectively. All of these variants are conveniently summarized in Figure 5.

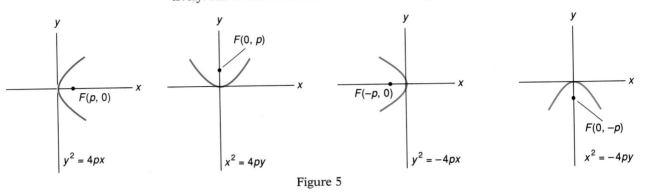

Figure 5

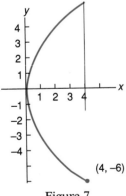

Figure 6

■ **Example 1.** Determine the equation of the parabola with vertex at the origin and directrix $x = 1/2$ (Figure 6).

Solution. This parabola opens left and has $p = 1/2$, so its equation is $y^2 = -4(1/2)x$; that is, $y^2 = -2x$. ■

■ **Example 2.** Determine the focus and directrix of the parabola with equation $y = -x^2/10$.

Solution. First, we rewrite this equation in standard form, that is, as $x^2 = -10y = -4(5/2)y$. This is the equation of a parabola opening downward with $p = 5/2$. Thus, the focus is the point $(0, -5/2)$, and the directrix is the line $y = 5/2$. ■

■ **Example 3.** Find the equation of the parabola (Figure 7) with vertex at the origin that opens right and passes through the point $(4, -6)$. Then determine the coordinates of the focus and also of the two points on the parabola that are $29/4$ units from the focus.

Solution. This parabola has an equation of the form $y^2 = 4px$, so $(-6)^2 = 4p(4)$, from which we conclude that $p = 36/16 = 9/4$. Thus, the equation of the parabola is $y^2 = 9x$, and its focus is at $(9/4, 0)$. A point $29/4$ units from the focus will also be $29/4$ units from the directrix $x = -9/4$ and therefore will have x-coordinate $-9/4 + 29/4 = 5$. Substituting this value for x in the equation $y^2 = 9x$ gives $y = \pm\sqrt{45} = \pm3\sqrt{5}$. Thus, the required points have coordinates $(5, \pm3\sqrt{5})$. ■

Figure 7

TRANSLATIONS

We raise now the question of what happens to the equation of a parabola when it is translated (shifted without changing orientation) so that its vertex is at (h, k) rather than at $(0, 0)$. A similar question has already been addressed in Sections 1.3 and 2.7. The result is that we must replace x by $x - h$ and y by

$y - k$ in the standard form (or one of its variants). Thus, if we translate the parabola (Figure 8) with equation $y^2 = 4px$ so that its vertex is at (h, k), the resulting parabola has equation

$$(y - k)^2 = 4p(x - h)$$

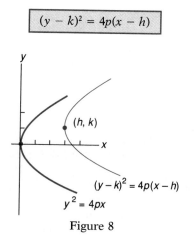

Figure 8

■ **Example 4.** Determine the equation of the horizontal parabola with vertex (3, 4) and focus (5, 4).

Solution. We note first that p, the distance from the vertex to the focus, is 2 and that the parabola opens right. We conclude that the equation of the parabola is

$$(y - 4)^2 = 4 \cdot 2(x - 3) = 8(x - 3)$$

or after simplifying

$$y^2 - 8y + 40 = 8x$$ ■

Note that the final equation of Example 4 is quadratic in y and linear in x. This example and earlier ones suggest an important observation. An equation in x and y that is second-degree in one variable and first-degree in the other is the equation of a parabola that has its axis of symmetry parallel to one of the coordinate axes. A procedure known as **completing the square** shows why this is true. Our next example illustrates what we mean.

■ **Example 5.** Determine the vertex, focus, and directrix of the parabola whose equation is

$$x^2 - 6x = 3y + 6$$

Solution. To complete the square for $x^2 + 2bx$, we must add b^2. Thus to complete the square for $x^2 - 6x$, we must add 9. Of course, we cannot just add 9 to an expression and pretend we haven't changed the result. If we add 9 to both sides of an equation, however, we obtain an equivalent equation.

Consequently, we may write the given equation as

$$x^2 - 6x + 9 = 3y + 15$$

or

$$(x - 3)^2 = 3(y + 5)$$

This is the equation of a parabola that turns up and has $p = 3/4$. Its vertex is $(3, -5)$, its focus is $(3, -17/4)$, and its directrix is $y = -5 - 3/4 = -23/4$. ∎

RELATION TO EARLIER WORK

We claimed in Section 2.3 that the graph of $y = ax^2 + bx + c$ is always a parabola. Note that this equation may be written as

$$y = a\left(x^2 + \frac{b}{a}x + \frac{b^2}{4a^2}\right) + \frac{4ac - b^2}{4a}$$

or

$$\frac{1}{a}\left(y - \frac{4ac - b^2}{4a}\right) = \left(x + \frac{b}{2a}\right)^2$$

This is an equation of a vertical parabola with vertex at $(-b/2a, (4ac - b^2)/4a)$.

∎ **Example 6.** Describe the parabola with equation $y = -2x^2 + 5x - 3$ and determine its vertex and focus.

Solution. Rather than apply the result above, we return to the basic technique of completing the square.

$$y = -2x^2 + 5x - 3$$
$$y + 3 = -2\left(x^2 - \frac{5}{2}x\right)$$
$$y + 3 - \frac{25}{8} = -2\left(x^2 - \frac{5}{2}x + \frac{25}{16}\right)$$
$$-\frac{1}{2}\left(y - \frac{1}{8}\right) = \left(x - \frac{5}{4}\right)^2$$

We conclude that the parabola turns down, with vertex at $(5/4, 1/8)$ and focus at $(5/4, 0)$. ∎

Cross section of a parabolic mirror with light source at focus

THE OPTICAL PROPERTY

Consider a circular cup-shaped mirror with a parabolic cross section. If a light source is placed at the focus, all rays will be reflected from the mirror along lines parallel to the axis of the mirror (Figure 9). This property of parabolas is used in the design of some flashlight reflectors, car head lamps, satellite dishes, and many large telescopes. For more information, see Problems 56 and 57.

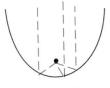

Figure 9

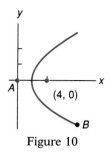

Figure 10

TEASER SOLUTION

The information needed to solve the Teaser is shown in Figure 10. Clearly, the path of the submarine is a parabola with vertex at (2, 0), focus at (4, 0), and directrix the line $x = 0$. Thus the equation of the path is $8(x - 2) = y^2$. Since the point B is 9 miles from the focus, it is also 9 miles from the directrix, so its x-coordinate is 9. Solving the equation $8(9 - 2) = y^2$ for y gives $y = \pm\sqrt{56}$. The point B is 9 miles east and $\sqrt{56} \approx 7.48$ miles south of A; therefore, it is $\sqrt{81 + 56} \approx 11.70$ miles from A.

PROBLEM SET 7.1

A. Skills and Techniques

Determine the equation of the parabola with vertex at the origin and satisfying the given conditions.

1. The focus is at $(-6, 0)$.
2. The focus is at $(0, 6)$.
3. The directrix is $y = -5$.
4. The directrix is $x = 3$.
5. The directrix is $y = 4$.
6. The focus is $(0, -1.5)$.
7. The parabola opens up and goes through $(-2, 8)$.
8. The parabola opens left and goes through $(-2, 2)$.
9. The parabola goes through $(6, \pm2)$.
10. The parabola goes through $(\pm3, -3)$.

Determine the focus and directrix of the parabola with the given equation.

11. $y = \dfrac{1}{2}x^2$

12. $y = -3x^2$
13. $x = -2y^2$

14. $x = \dfrac{1}{4}y^2$

15. $y = -\dfrac{2}{9}x^2$

16. $y = \dfrac{3}{8}x^2$

Write the equation of the parabola satisfying the given conditions.

17. Vertex $(4, -3)$, focus $(6, -3)$
18. Vertex $(-2, 5)$, focus $(-2, 8)$
19. Focus $(2, -2)$, directrix $y = 1$
20. Focus $(-1, 3)$, directrix $x = 3$
21. Vertex $(-1, 1)$, opens up, goes through $(1, 3)$
22. Vertex $(2, -3)$, opens down, goes through $(0, -11)$

In Problems 23–32, determine the vertex and focus of the parabola with the given equation.

23. $4(x + 2) = (y - 3)^2$
24. $-2(y + 1) = (x - 2)^2$
25. $x^2 + 4x + 2y = 2$
26. $-x^2 + 6x + 3y = 1$
27. $y^2 - 8y - 4x = 0$
28. $y^2 + 10y - 2x = 0$
29. $2x^2 + 4x + 2y = 10$
30. $-2x^2 - 8x + 4y = -6$
31. $y = -4x^2 - 4x$
32. $y = 2x^2 + 12x + 1$
33. Find the two points where the parabola $y = 2x^2 - 3x + 5$ and the line $y = 4$ intersect.
34. Find the two points where the parabola $4(y - 1) = (x + 2)^2$ and the line $y = 2$ intersect.
35. Draw the graphs of the parabola $y = 2x^2 - 3x + 1$ and the line $y = 2x$. Then determine their points of intersection accurate to two decimal places.
36. Follow the instructions of Problem 35 for the parabola $y = -3x^2 + 4x - 1$ and the line $y = 2x - 5$.
37. How wide is the parabola $y = 2x^2 - 4x$ at a level 4 units above the vertex?
38. Determine the equation of the parabola that has vertex at $(4, 0)$ and is 8 units wide at a level 4 units above the vertex.

B. Applications and Extensions

39. A paraboloidal reflector is made by revolving the part of the parabola $16y = x^2$ below $y = 2$ about its axis of symmetry. Where should a light source be placed to produce a beam with parallel rays?
40. A paraboloidal mirror, 4 feet in diameter and 0.5 feet deep, uses the sun's rays to heat objects. Where should an object be placed for most effective heating?

41. The cables for the central span of a suspension bridge are parabolic, as in Figure 11. If the towers are 1000 feet apart and rise 500 feet above the bridge floor and if the cables drop to the bridge floor at the center, how long is the supporting strut 120 feet from the center?

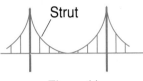

Figure 11

42. If the towers for a suspension bridge are 800 meters apart and the cables are attached to them at points 400 meters above the bridge floor, how long is the vertical strut that is 100 meters from the tower? Assume that the vertex of the parabola is on the floor of the bridge.

43. A door in the shape of a parabolic arch (Figure 12) is 12 feet high at the center and 5 feet wide at the base. A rectangular box 9 feet tall is to be slid through the door. What is the widest the box can be?

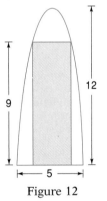

Figure 12

44. A parabolic door opening is 10 feet high at the center and is 2 feet wide at a height of 8 feet. How wide is it at the base?

45. The focus of a paraboloidal mirror is 10 centimeters above its vertex. A light ray leaving a source at the focus is reflected from the mirror to an absorbing element 10 centimeters to the right of and 30 centimeters above the focus. How far did this ray travel?

46. The chord of a parabola through the focus and perpendicular to the axis is called the **latus rectum** of the parabola. Find the length of the latus rectum for the parabola $4px = y^2$.

47. The chord of a parabola that is perpendicular to the axis and 1 unit from the vertex has length 3 units. How long is its latus rectum?

48. The path of a projectile fired from ground level is a parabola opening down. If the greatest height reached by the projectile is 100 meters and if its range (horizontal reach) is 800 meters, what is the horizontal distance from the point of firing to the point where the projectile first reaches a height of 64 meters?

49. A projectile shot from the ground follows a parabolic path and reaches its maximum height of 100 meters after traveling a horizontal distance of M meters. After 40 more meters of horizontal travel, it is at height 80 meters. How far from its starting point will it land?

50. Suppose that a submarine has been ordered to follow a path that keeps it equidistant from a circular island of radius r and a straight shoreline that is $2p$ units from the edge of the island. Derive the equation of the submarine's path, assuming that the shoreline has equation $x = -p$ and that the center of the island is on the x-axis.

51. Find the distance between the two points of the parabola $y = -x^2 + 4x + 9$ that have x-coordinates -2 and 3.

52. Find the area of the triangle whose three vertices lie on the parabola $y = 2x^2 - 3x$ at points with x-coordinates -2, 0, and 3.

53. Use a calculator to draw the graphs of $y = x^2$, $y = 10x - 21$, and $y = -0.1x + 9.3$, using the same axes. Note that all three graphs go through $A = (3, 9)$. Determine the other points B and C where the lines intersect the parabola and then calculate the area of the triangle ABC.

54. By experimenting with a calculator, determine k so that the line $y = (5/3)x + k$ is tangent to the parabola $y = (2/7)x^2$. Also find the coordinates (a, b) of the point of tangency accurate to one decimal place.

55. An equilateral triangle is inscribed in the parabola $y^2 = 4px$ with one vertex at the origin. Find the length of a side of the triangle.

56. Consider a line ℓ, two fixed points P and Q on the same side of ℓ, and a (variable) point R on ℓ. Use Figure 13 to show that the distance $\overline{PR} + \overline{RQ}$ is minimized precisely when $\alpha = \beta$. (*Note:* A light ray is known to be reflected from a mirror ℓ so that the angle of incidence equals the angle of reflection; thus a light ray from P to ℓ to Q picks the shortest path.)

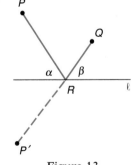

Figure 13

57. (Optical property of the parabola) Imagine the parabola $y^2 = 4px$ of Figure 14 to be a mirror with points F, R, G, and H as indicated and with RG parallel to the x-axis.
(a) Show that $\overline{FR} + \overline{RG} = 2p$.
(b) Show that $\overline{FH} + \overline{HG} > 2p$.
Conclude from Problem 56 that a light ray from the focus to a parabolic mirror is reflected parallel to the axis of the parabola.

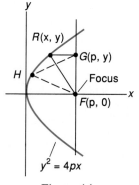

Figure 14

58. Challenge. Consider the parabola $y = x^2$ (Figure 15). Let T_1 be the triangle with vertices on this parabola at a, c, and b with c midway between a

and b. Let T_2 be the union of the two triangles, with vertices on the parabola at a, d, c and c, e, b, respectively, with d midway between a and c and e midway between c and b. In a similar manner, let T_3 be the union of four triangles with vertices on the parabola, and so on.
(a) Show that the area of T_1 is given by $A(T_1) = (b - a)^3/8$.
(b) Show that $A(T_2) = A(T_1)/4$.
(c) Find the area of the curved parabolic segment below the line PQ. (*Hint:* $1 + 1/4 + 1/16 + \cdots = 4/3$.)

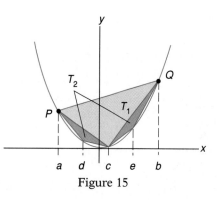

Figure 15

TEASER A door has the shape of an elliptical arch (a semi-ellipse) that is 10 feet wide and 4 feet high at the center. A box 2 feet high is to be pushed through the door. How wide can it be?

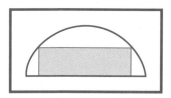

7.2 ELLIPSES

Recall that we defined an **ellipse** to be the set of points P satisfying

$$d(P, F) = e\,d(P, L)$$

where F is a fixed point called the focus, L is the nearest point on a fixed line called the directrix, and e is a number between 0 and 1. The line through the focus perpendicular to the directrix is an axis of symmetry, and the ellipse intersects this axis in two points A and A', which are called vertices (Figure 1).

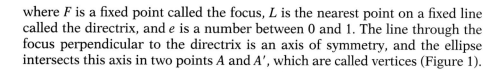

Figure 1

DERIVING THE EQUATION OF AN ELLIPSE

Place the ellipse in the coordinate system so that its axis of symmetry is the x-axis and its two vertices are $A(a, 0)$ and $A'(-a, 0)$. Let $F(c, 0)$ be the focus and $x = k$ the directrix (Figure 2). Then the defining equation, applied first with $P = A$ and then with $P = A'$ gives

$$a - c = e(k - a) = ek - ea$$

and

$$a + c = e(k + a) = ek + ea$$

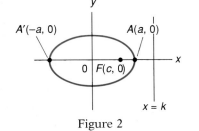

Figure 2

When these two equations are solved for c and k, we obtain $c = ae$ and $k = a/e$. The results are shown in Figure 3.

Now let $P(x, y)$ be any point on the ellipse, and let $L(a/e, y)$ be its projection on the directrix $x = a/e$. Then the distance formula turns the defining equation into

$$\sqrt{(x - ae)^2 + (y - 0)^2} = e\sqrt{(x - a/e)^2 + (y - y)^2}$$

or after squaring both sides,

$$x^2 - 2aex + a^2e^2 + y^2 = e^2\left(x^2 - \frac{2ax}{e} + \frac{a^2}{e^2}\right)$$

This equation simplifies to

$$(1 - e^2)x^2 + y^2 = a^2(1 - e^2)$$

and after division by $a^2(1 - e^2)$ to

$$\frac{x^2}{a^2} + \frac{y^2}{a^2(1 - e^2)} = 1$$

Because this equation contains x and y only to even powers, it is clear that the ellipse is symmetric with respect to both axes and the origin (the **center** of the ellipse). The axis through the two vertices is the **major axis;** the axis through the center and perpendicular to the major axis is the **minor axis.** Note that because of the symmetry, there must be a second focus at $(-ae, 0)$ and a second directrix at $x = -a/e$.

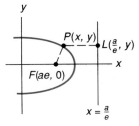

Figure 3

THE STANDARD EQUATION OF THE ELLIPSE

Because $0 < e < 1$, the number $a^2(1 - e^2)$ is positive. Let $b = a\sqrt{1 - e^2}$ so that the equation derived above takes the form

$$\frac{x^2}{a^2} + \frac{y^2}{b^2} = 1$$

This is called the **standard equation of the ellipse.** The number $2a$ is the **major diameter,** and $2b$ is the **minor diameter.** Moreover, if $c = ae$ is the x-coordinate of the right focus, then a, b, and c satisfy the Pythagorean relationship $a^2 = b^2 + c^2$. All these relationships are summarized in Figure 4; note the significance of the shaded triangle, which we call the **fundamental triangle** of the ellipse.

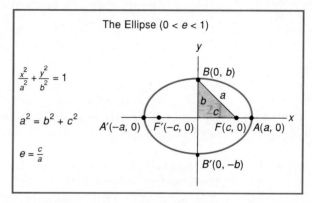

Figure 4

■ **Example 1.** Determine the major diameter, minor diameter, and the coordinates of the two foci of the ellipse. Then sketch its graph.

$$\frac{x^2}{25} + \frac{y^2}{9} = 1$$

Solution. Since $a = 5$ and $b = 3$, the major diameter is 10 and the minor diameter is 6. The relationship $a^2 = b^2 + c^2$ implies that $c = 4$ and that the coordinates of the foci are $(\pm 4, 0)$. The graph is shown in Figure 5. ■

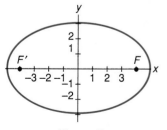

Figure 5

■ **Example 2.** Determine the equation of the ellipse with vertices $(\pm 8, 0)$ that goes through $(4, 2)$.

Solution. Since $a = 8$, the equation has the form

$$\frac{x^2}{64} + \frac{y^2}{b^2} = 1$$

and so

$$\frac{16}{64} + \frac{4}{b^2} = 1$$

Solving this equation for b^2 gives $b^2 = 16/3$. We conclude that the equation of the ellipse is

$$\frac{x^2}{64} + \frac{y^2}{16/3} = 1$$

■

The ellipses studied so far are called **horizontal ellipses** because their major axes are horizontal. If we interchange the roles of x and y in the standard equation, we obtain

$$\frac{x^2}{b^2} + \frac{y^2}{a^2} = 1$$

which is the equation of a **vertical ellipse.** For example, the graph of

$$\frac{x^2}{4} + \frac{y^2}{9} = 1$$

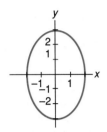

Figure 6

is the vertical ellipse shown in Figure 6. You can determine from the equation whether an ellipse is horizontal or vertical by noting whether the larger denominator is associated with the x or the y term. Thus, $x^2/18 + y^2/14 = 1$ is the equation of a horizontal ellipse, whereas $x^2/4 + y^2/5 = 1$ is the equation of a vertical ellipse.

■ **Example 3.** Write the equations and determine the eccentricities for the three vertical ellipses with vertices at $(0, \pm 8)$ and foci at (a) $(0, \pm 7)$, (b) $(0, \pm 4)$, (c) $(0, \pm 1)$. Then sketch their graphs together.

Solution. We use $b^2 = a^2 - c^2$ and $e = c/a$ together with the given information to obtain the results.

(a) $\dfrac{x^2}{15} + \dfrac{y^2}{64} = 1, e = 7/8$

(b) $\dfrac{x^2}{48} + \dfrac{y^2}{64} = 1, e = 4/8 = 1/2$

(c) $\dfrac{x^2}{63} + \dfrac{y^2}{64} = 1, e = 1/8$

The three graphs are shown in Figure 7. ■

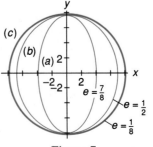

Figure 7

Be sure to note the significance of the size of the number e in Figure 7. The closer e is to 1, the more eccentric (long and narrow) is the ellipse; the closer e is to 0, the more nearly circular is the ellipse. Thus if we imagine c to shrink to 0, the two foci coalesce and we obtain a circle (the circle $x^2/a^2 + y^2/a^2 = 1$). For this reason, a circle is sometimes called an ellipse of eccentricity 0, but see the box titled Is a Circle an Ellipse?

TRANSLATIONS

If in the standard form $x^2/a^2 + y^2/b^2 = 1$ we replace x by $x - h$ and y by $y - k$, we obtain

$$\frac{(x - h)^2}{a^2} + \frac{(y - k)^2}{b^2} = 1$$

This is the equation of an ellipse that has been translated (shifted without changing its orientation) so that the center is at (h, k) rather than at the origin.

■ **Example 4.** Determine the equation of the horizontal ellipse with center at $(1, -2)$, major diameter 8, and minor diameter 6. Sketch its graph.

Solution. The given information implies that $a = 4$ and $b = 3$; therefore, the equation of the ellipse is

$$\frac{(x - 1)^2}{16} + \frac{(y + 2)^2}{9} = 1$$

The graph is shown in Figure 8. ■

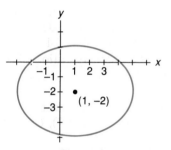

Figure 8

■ **Example 5.** Determine the equation of the ellipse with vertices at $(3, -1)$ and $(3, 9)$ and with a focus at $(3, 6)$.

Solution. This information implies that the ellipse is vertical, with center at $(3, 4)$, $a = 5$, and $c = 2$. It follows that $b^2 = a^2 - c^2 = 25 - 4 = 21$. The required equation is

$$\frac{(x - 3)^2}{21} + \frac{(y - 4)^2}{25} = 1$$

■

All the equations for ellipses that we have considered are quadratic in x and in y. Do all equations that are quadratic in x and y correspond to ellipses? The answer is no. By completing the squares, however, we can always identify those that do.

■ **Example 6.** Determine whether $x^2 + 6x + 2y^2 - 8y = -1$ is the equation of an ellipse, and, if it is, give its center and major and minor diameters.

Solution. We attempt to transform this equation to the standard form for an ellipse.

$$x^2 + 6x + 2y^2 - 8y = -1$$

$$x^2 + 6x + 9 + 2(y^2 - 4y + 4) = -1 + 9 + 8$$

$$(x + 3)^2 + 2(y - 2)^2 = 16$$

$$\frac{(x + 3)^2}{16} + \frac{(y - 2)^2}{8} = 1$$

This is the equation of a horizontal ellipse with center at $(-3, 2)$, major diameter 8, and minor diameter $2\sqrt{8}$.

If the original coefficient of y^2 had been -2 instead of 2, the equation would not have corresponded to an ellipse (but rather to a hyperbola, the subject of Section 7.3). ∎

THE STRING PROPERTY

We have mentioned that because of symmetry, the ellipse has two directrices and two foci (see Figure 9). Let $P(x, y)$ be any point on the ellipse. By applying the definition of an ellipse to P, first using the left focus and directrix and then the right focus and directrix, we obtain

$$d(P, F') = e\left(x + \frac{a}{e}\right) = ex + a \qquad d(P, F) = e\left(\frac{a}{e} - x\right) = a - ex$$

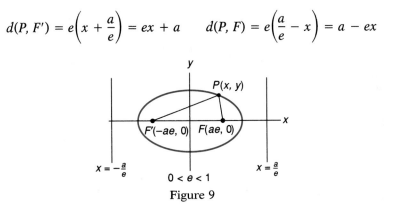

Figure 9

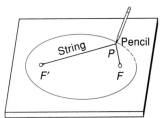

Figure 10

Thus,

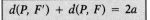

$$d(P, F') + d(P, F) = 2a$$

a result which we call the **string property** of the ellipse. Let a string of length $2a$ be tacked down at its ends F' and F. With a pencil P pulling the string taut (Figure 10), we may draw the ellipse. This property is often taken as the definition of an ellipse.

THE OPTICAL PROPERTY

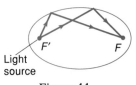

Figure 11

Imagine a curved mirror with elliptical cross sections to have a light source at one focus (Figure 11). It can be shown (see Problem 46) that each light ray is reflected back through the other focus. This fact is used in the design of various optical instruments, including some telescopes, and a medical device called a lithotripter, in which shock waves are used to dissolve kidney stones without surgery. It is also the basis of the "whispering gallery" effect that can be observed in the U.S. Capitol and the Mormon Tabernacle in Salt Lake City, Utah. A speaker standing at one focus can be heard whispering by a listener standing many feet away at the other focus.

TEASER SOLUTION

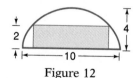

Figure 12

Figure 12 displays the information we need to find the maximum width for the box. For the indicated ellipse, $a = 5$ and $b = 4$, so we may take its equation to be

$$\frac{x^2}{25} + \frac{y^2}{16} = 1$$

Our task is to determine x when $y = 2$, that is, to solve $x^2/25 + 2^2/16 = 1$. This gives $x = 5\sqrt{3}/2$ and so the maximum width for the box is $5\sqrt{3} \approx 8.66$ feet.

PROBLEM SET 7.2

A. Skills and Techniques

Each equation represents an ellipse. Write it in standard form and then determine the coordinates of the vertices and the foci.

1. $100x^2 + 36y^2 = 3600$
2. $25x^2 + 169y^2 = (25)(169)$
3. $9x^2 + 16y^2 = 144$
4. $9x^2 + 4y^2 = 36$
5. $20x^2 + 36y^2 = 720$
6. $8x^2 + 9y^2 = 72$
7. $\frac{9}{16}x^2 + \frac{1}{16}y^2 = 1$
8. $9x^2 + 25y^2 = 1$
9. $4k^2x^2 + k^2y^2 = k^4$
10. $k^2x^2 + (k^2 + 1)y^2 = k^2$

Write the equation (in standard form) for the ellipse that satisfies the following conditions.

11. Vertices $(\pm 5, 0)$, foci $(\pm 4, 0)$
12. Vertices $(0, \pm 13)$, foci $(0, \pm 12)$
13. Foci $(0, \pm 6)$, minor diameter 8
14. Center $(0, 0)$, focus $(6, 0)$, major diameter 20
15. Horizontal, center $(0, 0)$, major diameter 14, minor diameter 4
16. Vertices $(\pm 10, 0)$, curve goes through $(0, \pm 6)$
17. Vertices $(\pm 9, 0)$, curve goes through $(3, \sqrt{8})$
18. Ends of minor diameter at $(\pm 4, 0)$, curve goes through $(\sqrt{2}, 4\sqrt{3})$
19. Vertices $(\pm 6, 0)$, eccentricity 1/3

20. Center $(0, 0)$, endpoint of minor axis $(6, 0)$, eccentricity 4/5
21. Vertices $(0, 0)$ and $(7, 0)$, minor diameter 4
22. Vertices $(2, -2)$ and $(2, 8)$, one focus at $(2, 6)$
23. Center $(2, -2)$, one focus at $(2, 3)$, eccentricity 2/3
24. Center $(-3, -1)$, one vertex at $(-3, 2)$, minor diameter 4

In Problems 25–30, determine the center and vertices of ellipses with the following equations.

25. $\dfrac{(x + 3)^2}{16} + \dfrac{(y - 4)^2}{36} = 1$
26. $\dfrac{(x - 5)^2}{49} + \dfrac{(y + 2)^2}{25} = 1$
27. $2x^2 + 4x + y^2 - 6y = 39$
28. $3x^2 - 18x + 2y^2 + 8y + 23 = 0$
29. $x^2 + 10x + 4y^2 + 8y = 7$
30. $-2x^2 + 12x - 4y^2 - 16y = 10$
31. An elliptical fireplace arch (a semi-ellipse) has a base 5 feet long and height at the center of 3 feet. How high is the arch 1 foot from the center?
32. An elliptical pan (obtained by revolving the bottom half of a horizontal ellipse about its minor axis) has diameter 12 inches and depth 2 inches. How deep is it 3 inches from the center?

B. Applications and Extensions

33. The earth's elliptical orbit has eccentricity $e = 0.0167$ and major diameter 304,000,000 kilometers.

Find the earth's greatest distance from the sun. Assume that the earth and sun are point masses, with the sun at a focus (Figure 13).

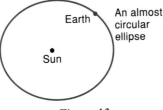

Figure 13

34. Halley's comet has an elliptical orbit with the sun at a focus. Its major and minor diameters are 36.18 and 9.12 astronomical units (AU), respectively. Determine its closest approach to the sun in AU.

35. How long is the diameter of the ellipse $36x^2 + 100y^2 = 3600$ that passes through the point $(8, -3.6)$?

36. Find the length of the diameter of $16x^2 + 9y^2 = 144$ that has slope 2.

37. Find the intersection points of the ellipse $9x^2 + 16y^2 = 144$ and the parabola $y = x^2$.

38. How long is the latus rectum (chord through the focus perpendicular to the major axis) for the ellipse $x^2/a^2 + y^2/b^2 = 1$?

39. Assume that the center of the earth (a sphere of radius 4000 miles) is at one focus of the elliptical path of a satellite. If the satellite's nearest approach to the surface of the earth is 2000 miles and its farthest distance away is 10,000 miles, what are the major and minor diameters of the elliptical path?

40. A and B are the foci of an ellipse and C is on the ellipse. Determine the major and minor diameters of the ellipse given that $\overline{AB} = 8$, $\overline{BC} = 6$, and $\overline{AC} = 10$.

41. The **area of the ellipse** $x^2/a^2 + y^2/b^2 = 1$ is πab. Find the area of the ellipse $11x^2 + 7y^2 = 77$.

42. A square with sides parallel to the coordinate axes is inscribed in the ellipse $b^2x^2 + a^2y^2 = a^2b^2$. Determine the area of the square.

43. A dog's collar is attached by a ring to a loop of rope 42 feet long. The loop of rope is thrown over two stakes 12 feet apart (Figure 14).

Figure 14

(a) How much area can the dog cover?
(b) If the dog manages to nudge the rope over the top of one of the stakes, how much would this increase the area it can cover?

44. Graph the ellipse $25x^2 + 169y^2 = 4225$ and the circle $x^2 + y^2 = 169$. Then find the area of the region outside the ellipse and inside the circle.

45. Let P be a point on a 16-foot ladder 7 feet from the top end. As the ladder slides with its top end against a wall (the y-axis) and its bottom end along the ground (the x-axis), P traces a curve. Find the equation of this curve.

46. (Optical property of the ellipse) In Figure 15, let P and Q be the foci of an ellipse, R be a point on the ellipse, ℓ be the tangent line at R, and R' be any other point on ℓ.

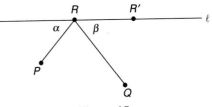

Figure 15

(a) Show that $\overline{PR'} + \overline{R'Q} > \overline{PR} + \overline{RQ}$.
(b) Show that $\alpha = \beta$. (See Problem 56 of Section 7.1)
We can conclude that a light ray from one focus P of an elliptical mirror is reflected back through the other focus.

47. Use a graphics calculator to determine the intersection points of $2x + 3y = 8$ and $x^2/64 + y^2/49 = 1$ accurate to two decimal places. (*Hint:* You will need to replace the second equation by the two equations that you get when you solve for y.)

48. Superimpose the graphs of $3x + y = 10$ and $3x + y = 20$ on the ellipse of Problem 47. Then figure out a way to maximize $P = 3x + y$ on the set consisting of the boundary and the interior of the ellipse.

49. Use a calculator to graph the parabola $y = x^2 + 4x$ and the ellipse $x^2/64 + y^2/49 = 1$ in the same plane, and find their points of intersection, accurate to two decimal places.

50. Use an algebraic method to find the points of intersection of the circle $x^2 + y^2 = 60$ and the ellipse $x^2/100 + y^2/36 = 1$, accurate to two decimal places. Then check your answer by doing the same problem graphically, using a calculator.

51. On the elliptical orbit of a planet, the point farthest from the sun is called the **aphelion** and the point nearest the sun is called the **perihelion.** If the ratio of Jupiter's distance from the sun at aphelion to its distance at perihelion is 21:19, determine the eccentricity of the orbit. Assume that Jupiter and the sun are point masses, with the sun at a focus of the ellipse.

52. **Challenge.** The cushion of a pool table has the shape of the ellipse $x^2/25 + y^2/16 = 1$ (Figure 16) with its only pocket at the focus $(3, 0)$. Toward what exact point P on the far cushion should a player aim a ball lying at $(-3.5, -1.2)$ in order for it to rebound into the pocket?

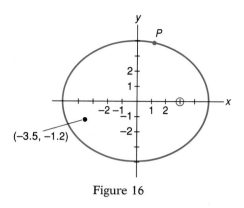

Figure 16

TEASER   The right cushion of the billiard table shown in the diagram has the shape of a branch of the hyperbola $x^2 - y^2/24 = 1$. A ball struck at $(-5, 0)$ hits this cushion at a point with y-coordinate 1 and rebounds as shown in the diagram. Where does the ball hit the left cushion whose equation is $x = -6$? (*Hint:* The point $(-5, 0)$ is one of the foci of the hyperbola.)

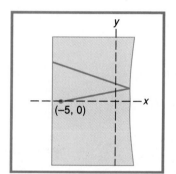

7.3   HYPERBOLAS

A hyperbola is the set of points P satisfying

$$d(P, F) = ed(P, L)$$

where $e > 1$, F is a fixed point (the focus), and L is the nearest point on a fixed line (the directrix). The line through the focus perpendicular to the directrix is an axis of symmetry, called the **major axis.** The hyperbola intersects this axis in two points, A' and A, which are called **vertices** (Figure 1).

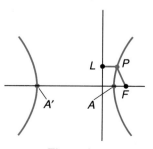

Figure 1

We place the hyperbola in the coordinate system so that the points A' and A are at $(-a, 0)$ and $(a, 0)$. Following the same procedure used for the ellipse in Section 7.2, we find that the focus is at $(ae, 0)$ and that the directrix has equation $x = a/e$ (Figure 2). As with the ellipse, this procedure leads to the equation

$$\frac{x^2}{a^2} + \frac{y^2}{a^2(1 - e^2)} = 1$$

but, because $e > 1$, $1 - e^2$ is now a negative number.

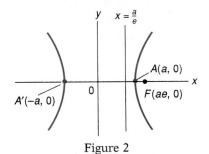

Figure 2

THE STANDARD EQUATION OF THE HYPERBOLA

Let $b = a\sqrt{e^2 - 1}$. Then the equation above takes the form

$$\frac{x^2}{a^2} - \frac{y^2}{b^2} = 1$$

which we call the **standard equation of the hyperbola.** As with the ellipse, the number $2a$ is the distance between the two vertices, and the foci are at $(\pm c, 0)$, where $c = ae$. But the new Pythagorean relationship is $a^2 + b^2 = c^2$, which differs from that for an ellipse. Figure 3 summarizes this information.

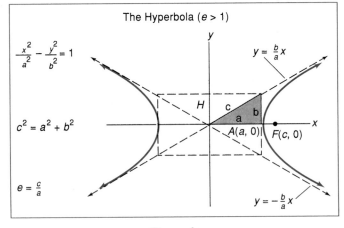

Figure 3

Once again, a **fundamental triangle** plays an important role; it determines the central rectangle in the figure. We claim that the diagonals of this rectangle (which have equations $y = \pm bx/a$) are slant asymptotes for the hyperbola. To show that they are, we solve the standard equation for y, obtaining

$$y = \pm\frac{b}{a}\sqrt{x^2 - a^2}$$

And note that, for large x, this equation behaves like $y = \pm bx/a$.

■ **Example 1.** Use a calculator to show the graphs of $y = \pm2\sqrt{x^2 - 9}/3$ and $y = \pm2x/3$, thereby lending support to the assertion about asymptotes made above.

Solution. The four graphs are shown in Figure 4. ■

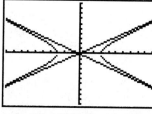

$[-10, 10]$ by $[-10, 10]$

Figure 4

■ **Example 2.** Determine the foci and the asymptotes of the hyperbola with equation $3x^2 - 9y^2 = 27$.

Solution. In standard form, the equation is

$$\frac{x^2}{9} - \frac{y^2}{3} = 1$$

so $a = 3$, $b = \sqrt{3}$, and $c = \sqrt{a^2 + b^2} = \sqrt{12} = 2\sqrt{3}$. Thus the foci are at $(\pm2\sqrt{3}, 0)$, and the asymptotes are $y = \pm\sqrt{3}x/3$. ■

If we interchange the roles of x and y in the standard equation, we obtain

$$\frac{y^2}{a^2} - \frac{x^2}{b^2} = 1$$

This is the equation of a vertical hyperbola. For example, $y^2/16 - x^2/19 = 1$ is the equation of a vertical hyperbola with vertices at $(0, \pm4)$. Note that whether a hyperbola is horizontal or vertical is not related to the size of the denominators but rather to which term receives the minus sign.

■ **Example 3.** Determine the equation of the hyperbola with foci at $(0, \pm6)$ and $e = 3/2$. Then sketch the graph, showing the asymptotes.

Solution. Since $c = 6$ and $c = ae$, we conclude that $a = (2/3)(6) = 4$ and $b^2 = c^2 - a^2 = 36 - 16 = 20$. Thus, the equation of the hyperbola is

$$\frac{y^2}{16} - \frac{x^2}{20} = 1$$

and its graph appears in Figure 5. ■

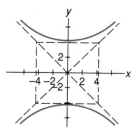

Figure 5

TRANSLATIONS

If in the standard form $x^2/a^2 - y^2/b^2 = 1$ we replace x by $x - h$ and y by $y - k$, we obtain

$$\frac{(x - h)^2}{a^2} - \frac{(y - k)^2}{b^2} = 1$$

the equation of a horizontal hyperbola with center at (h, k).

339 • 7.3 Hyperbolas

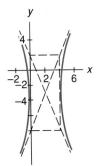

Figure 6

Example 4. Determine the vertices and foci of the hyperbola with equation

$$\frac{(x-2)^2}{4} - \frac{(y+3)^2}{25} = 1$$

Then sketch the graph.

Solution. The hyperbola is horizontal with $a = 2$, $b = 5$, $c = \sqrt{4 + 25} = \sqrt{29}$, and center at $(2, -3)$. Thus, the vertices are at $(0, -3)$ and $(4, -3)$; the foci are at $(2 \pm \sqrt{29}, -3)$. The graph is sketched in Figure 6. ∎

Example 5. Determine whether $-2x^2 - 12x + y^2 + 8y = 18$ is the equation of a hyperbola, and if it is find its vertices.

Solution. We use the familiar completing the squares process.

$$-2x^2 - 12x + y^2 + 8y = 18$$
$$-2(x^2 + 6x + 9) + y^2 + 8y + 16 = 18 - 18 + 16$$
$$-2(x + 3)^2 + (y + 4)^2 = 16$$
$$\frac{(y+4)^2}{16} - \frac{(x+3)^2}{8} = 1$$

This is the equation of a vertical hyperbola with center at $(-3, -4)$ and $a = 4$. Thus, the vertices are at $(-3, -8)$ and $(-3, 0)$. ∎

THE STRING PROPERTY

Because of the symmetry of the hyperbola, it has two foci and two directrices (Figure 7). If the point P is on its right branch,

$$d(P, F') = e\left(x + \frac{a}{e}\right) = ex + a, \qquad d(P, F) = e\left(x - \frac{a}{e}\right) = ex - a$$

and so $d(P, F') - d(P, F) = 2a$. A similar calculation for P on the left branch yields $d(P, F') - d(P, F) = -2a$. Thus, in either case,

$$\boxed{|d(P,F') - d(P,F)| = 2a}$$

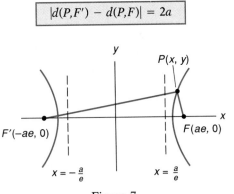

Figure 7

This fact, which we refer to as the **string property** of the hyperbola, is often taken as the definition of a hyperbola.

Tie a knot in a piece of string at K so that the difference in the lengths of the two segments is $2a$. Tack the ends down at F' and F, insert a pencil as shown in Figure 8, and pull on the knot K. You will draw one branch of a hyperbola with foci at F' and F.

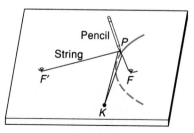

Figure 8

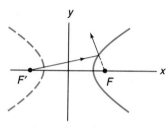

Figure 9

APPLICATIONS

The hyperbola, too, has an optical property, as illustrated in Figure 9. If we imagine the right branch of the hyperbola to be a mirror, then a light ray from the left focus F', upon hitting the mirror, will be reflected away along the line that passes through the right focus F. The optical properties of the parabola and the hyperbola are combined in one design for a reflecting telescope (Figure 10).

Another important application of the hyperbola occurs in the design of LORAN, a system for long-range navigation. A ship at sea can determine the difference $2a$ in its distance from two fixed transmitters by measuring the difference in reception times of synchronized radio signals. This measurement puts its position on a hyperbola, with the two transmitters F' and F as foci. If another pair of transmitters G' and G are used, the ship must be at the intersection of the two corresponding hyperbolas (Figure 11), and its position is fixed. A related application is described in Problem 50.

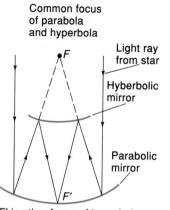

Common focus of parabola and hyperbola

Light ray from star

Hyberbolic mirror

Parabolic mirror

F' is other focus of hyperbola.
Put eyepiece here.

Figure 10

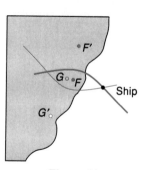

Ship

Figure 11

Figure 12 shows the information we need to solve the Teaser. The path of the ball begins at the focus $(-5, 0)$, so the ball will rebound from the hyperbolic cushion on a line through the other focus at $(5, 0)$. Noting that it hits this cushion at a point with y-coordinate 1, we determine the x-coordinate of the impact point to be $x = \sqrt{25/24}$. Thus, the line of rebound has the equation (point-slope form)

$$y - 0 = \frac{1 - 0}{\sqrt{25/24} - 5}(x - 5)$$

To find the point where it hits the left cushion is to substitute $x = -6$ in this equation and determine y. We get $y \approx 2.76$; the ball will hit the left cushion at a point 2.76 units above the line through the foci.

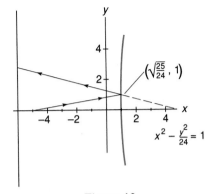

Figure 12

PROBLEM SET 7.3

A. Skills and Techniques

Use a graphics calculator to draw the graph of each hyperbola and its asymptotes. Note that you will first have to solve for y in the equation of the hyperbola.

1. $25x^2 - 16y^2 = 400$
2. $9x^2 - 4y^2 = 36$
3. $4y^2 - x^2 = 16$
4. $25y^2 - 16x^2 = 400$

Rewrite the equation of the hyperbola in standard form. Then determine its vertices, foci, and asymptotes. Finally, sketch its graph.

5. $9x^2 - 16y^2 = 144$
6. $25x^2 - 144y^2 = 3600$

7. $9x^2 - 16y^2 = -144$
8. $25x^2 - 144y^2 = -3600$
9. $4x^2 - y^2 = 16$
10. $9x^2 - 4y^2 = 144$
11. $4y^2 - 16x^2 = 16$
12. $4y^2 - 3x^2 = 12$

Find the equation of the hyperbola that satisfies the indicated conditions.

13. Vertices $(\pm 1, 0)$, foci $(\pm 4, 0)$
14. Vertices $(\pm 4, 0)$, foci $(\pm 5, 0)$
15. Vertices $(\pm 5, 0)$, asymptotes $y = \pm 2x$
16. Vertices $(\pm 3, 0)$, asymptotes $y = \pm x$
17. Vertices $(0, \pm 3)$, goes through $(2, 5)$
18. Vertices $(\pm 3, 0)$, goes through $(2\sqrt{3}, 9)$
19. Foci $(\pm 6, 0)$, eccentricity 2

20. Vertices $(0, \pm 6)$, eccentricity 4/3
21. Vertices $(1, 0)$ and $(1, 8)$, a focus at $(1, 9)$
22. Foci $(1, 2)$ and $(11, 2)$, a vertex at $(2, 2)$
23. Foci $(3, 4)$ and $(9, 4)$, eccentricity 3
24. Vertices $(-1, -1)$ and $(-1, 7)$, eccentricity 2

Determine the foci and vertices of each hyperbola.

25. $16(x + 2)^2 - 9(y - 3)^2 = 144$
26. $25(x - 3)^2 - 144(y + 1)^2 = 3600$
27. $4y^2 - (x + 3)^2 = 16$
28. $(y - 2)^2 - (x - 3)^2 = 4$

Use the completing the squares process to write each equation in standard form. Determine the character of its graph (horizontal or vertical, hyperbola or ellipse). Then give the center and the distance between the vertices.

29. $2x^2 - 4x - y^2 - 6y = 0$
30. $2x^2 - 4x - 4y^2 + 16y = 4$
31. $2x^2 - 4x + y^2 - 6y = 0$
32. $2x^2 + 8x + 4y^2 + 16y = 4$
33. $-2x^2 - 4x + y^2 - 6y = -11$
34. $-4x^2 + 8x + 4y^2 + 16y = 4$

B. Applications and Extensions

35. How long is the focal chord (chord through a focus perpendicular to the major axis) of the hyperbola $16x^2 - 9y^2 = 144$?
36. How long is the focal chord of the hyperbola $b^2x^2 - a^2y^2 = a^2b^2$?
37. Assume that the parabolic mirror in Figure 10 has equation $x^2 = 20y$ and the hyperbolic mirror has eccentricity 5/4. Find the equation of the hyperbola.
38. Alternatively, suppose that the hyperbolic mirror in Figure 10 has equation $25y^2 - 144x^2 = 3600$. Determine the equation of the parabola.
39. Draw the graph of the hyperbola $3x^2 - y^2 = 3$. Find the exact distance between the two points on the right branch of the hyperbola $3x^2 - y^2 = 3$ that have y-coordinates 3 and 12.
40. Draw the graph of $25y^2 - 144x^2 = 3600$ and superimpose the graph of $y = 3.5x - 2$. Find the distance between the two intersection points of these graphs accurate to two decimal places.
41. Determine the exact coordinates of the two points where the graph of the line $2y - x = 0$ intersects the hyperbola $3x^2 - y^2 = 33$.
42. Find the equation of the line through $(3, 1)$ that intersects $3x^2 - y^2 = 3$ in only one point.
43. Find the equation of the horizontal hyperbola with center at $(0, 0)$ if this hyperbola and the line $y = 3(x - 3)$ intersect in the point $(3, 0)$ and only there.
44. Find the eccentricity of the horizontal hyperbola with asymptotes $y = \pm x$.

45. A certain comet is traveling a hyperbolic path with the sun at a focus, the earth at the center, and eccentricity 5 (Figure 13). How close will the comet come to the earth? (*Hint:* Pretend that the sun and the earth are point masses 93 million miles apart.)

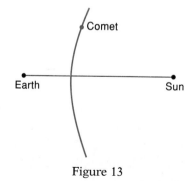

Figure 13

46. A light ray from the left focus of the hyperbola $16(x - 5)^2 - 9y^2 = 144$ hits the silvered right branch at $(33/4, 5/3)$ and is reflected to be absorbed at the point $(a, 10)$, as shown in Figure 14. Determine a.

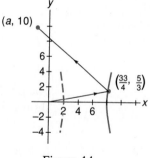

Figure 14

47. A ball shot from $(-5, 0)$ hit the right branch of the hyperbolic bangboard $x^2/16 - y^2/9 = 1$ at the point $(8, 3\sqrt{3})$. What was the ball's y-coordinate when its x-coordinate was 10?
48. The rectangle $PQRS$ with sides parallel to the coordinate axes is inscribed in the hyperbola $x^2/4 - y^2/9 = 1$ as shown in Figure 15. Find the exact coordinates of P if the area of the rectangle is $6\sqrt{5}$.

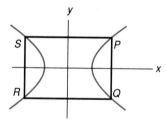

Figure 15

49. Lu, located at $(0, -2200)$, fired a rifle. The sound echoed off a cliff at $(0, 2200)$ to Brian, located at the point (x, y). Brian heard this echo 6 seconds after he heard the original shot. Find the xy-equation of the curve on which Brian is located. Assume that distances are in feet and that sound travels 1100 feet per second.

50. Challenge. Anton, Bernie, and Cindy, located at $(-8, 0)$, $(8, 0)$, and $(8, 10)$, respectively, recorded the exact times when they heard an explosion. On comparing notes, they discovered that Bernie and Cindy heard the explosion at the same time but that Anton heard it 12 seconds later. Assuming that distances are in kilometers and that sound travels 1/3 kilometer per second, determine the point of the explosion.

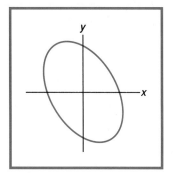

TEASER The graph of the equation $x^2 + xy + y^2 = 12$ is the ellipse shown in the diagram. Find its major and minor diameters.

7.4 ROTATIONS

The process of completing the squares, illustrated several times in this chapter, shows that the graph of the second-degree equation

$$Ax^2 + Cy^2 + Dx + Ey + F = 0$$

is either a **principal conic** (parabola, ellipse, hyperbola) or one of its limiting forms (parallel lines, single line, circle, point, empty set, intersecting lines). That each of these nine graphs is a possibility is demonstrated in Figure 1.

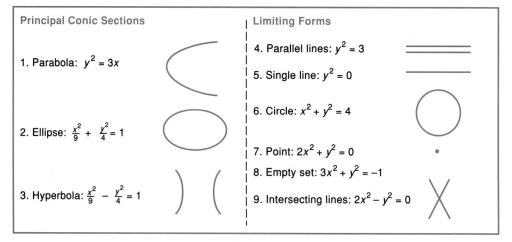

Principal Conic Sections

1. Parabola: $y^2 = 3x$

2. Ellipse: $\frac{x^2}{9} + \frac{y^2}{4} = 1$

3. Hyperbola: $\frac{x^2}{9} - \frac{y^2}{4} = 1$

Limiting Forms

4. Parallel lines: $y^2 = 3$

5. Single line: $y^2 = 0$

6. Circle: $x^2 + y^2 = 4$

7. Point: $2x^2 + y^2 = 0$

8. Empty set: $3x^2 + y^2 = -1$

9. Intersecting lines: $2x^2 - y^2 = 0$

Figure 1

Example 1. Identify the graph of each of the following equations.
(a) $x^2 + 4y^2 - 6x - 8y + 12 = 0$
(b) $x^2 + 4y^2 - 6x - 8y + 13 = 0$
(c) $x^2 + 4y^2 - 6x - 8y + 14 = 0$
(d) $x^2 - 4y^2 - 6x + 8y + 5 = 0$

Solution. We use the completing the squares process.

(a)
$$x^2 - 6x + 4(y^2 - 2y) = -12$$
$$x^2 - 6x + 9 + 4(y^2 - 2y + 1) = -12 + 13$$
$$(x - 3)^2 + 4(y - 1)^2 = 1$$

We recognize this as the equation of a horizontal ellipse.
(b) Exactly the same process as in (a) transforms this equation to

$$(x - 3)^2 + 4(y - 1)^2 = 0$$

which is the equation of the point (3, 1).
(c) This equation transforms to

$$(x - 3)^2 + 4(y - 1)^2 = -1$$

whose graph is the empty set.
(d) Completing the squares changes this equation to

$$(x - 3)^2 - 4(y - 1)^2 = 0$$

Because the left side is a difference of squares, this can be rewritten as

$$((x - 3) - 2(y - 1))((x - 3) + 2(y - 1)) = 0$$

This equation in turn is equivalent to

$$x - 2y - 1 = 0 \quad \text{or} \quad x + 2y - 5 = 0$$

which we recognize as equations of intersecting lines. ∎

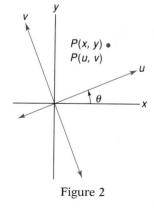

Figure 2

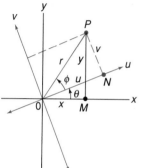

Figure 3

THE ROTATION FORMULAS

The equation

$$x^2 + xy + y^2 - 12 = 0$$

introduced in the Teaser is of second degree, but doesn't quite fit the pattern of previous examples because of the appearance of the xy-term. Yet we claim that its graph is an ellipse. Let's see why it is.

Introduce a new pair of axes, called the u- and v-axes, into the xy-plane (Figure 2). These axes have the same origin as the old x- and y-axes, but are rotated so that the positive u-axis makes an angle θ with the positive x-axis. A point P then has two sets of coordinates: (x, y) and (u, v). How are they related?

Draw a line segment from the origin O to P, let r denote the length of OP, and let ϕ denote the angle from the u-axis to OP. Then x, y, u, and v will have the geometric interpretations indicated in Figure 3.

Looking at the right triangle *OPM,* we see that

$$\cos(\phi + \theta) = \frac{x}{r}$$

so

$$x = r\cos(\phi + \theta) = r(\cos\phi\cos\theta - \sin\phi\sin\theta)$$
$$= (r\cos\phi)\cos\theta - (r\sin\phi)\sin\theta$$

Next, the right triangle *OPN* tells us that $u = r\cos\phi$ and $v = r\sin\phi$. Thus

$$\boxed{x = u\cos\theta - v\sin\theta}$$

Similarly

$$y = r\sin(\phi + \theta) = r(\sin\phi\cos\theta + \cos\phi\sin\theta)$$
$$= (r\sin\phi)\cos\theta + (r\cos\phi)\sin\theta$$

$$\boxed{y = u\sin\theta + v\cos\theta}$$

We call the boxed results **rotation formulas.**

To show how we use these formulas to identify graphs of second-degree equations with *xy*-terms, we consider an even simpler equation than the one in the Teaser.

■ **Example 2.**   Use a rotation of 45° to identify the graph of $xy = 1$.

Solution. The rotation formulas take the form

$$x = u\cos 45° - v\sin 45° = \frac{\sqrt{2}}{2}(u - v)$$

$$y = u\sin 45° + v\cos 45° = \frac{\sqrt{2}}{2}(u + v)$$

When we make these substitutions in $xy = 1$, we obtain

$$\frac{\sqrt{2}}{2}(u - v)\frac{\sqrt{2}}{2}(u + v) = 1$$

which simplifies to

$$\frac{u^2}{2} - \frac{v^2}{2} = 1$$

Note that the curve has not changed; we have simply represented it by equations in two different coordinate systems, one rotated with respect to the other. In the new *uv*-system, we recognize the curve as a hyperbola; it was the same hyperbola in the old *xy*-system. Figure 4 illustrates our point.   ■

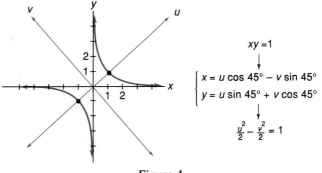

Figure 4

THE GENERAL SECOND-DEGREE EQUATION

How do we know what rotation to make? Consider the most general second-degree equation in x and y:

$$Ax^2 + Bxy + Cy^2 + Dx + Ey + F = 0$$

If we make the substitutions

$$x = u \cos \theta - v \sin \theta$$

$$y = u \sin \theta + v \cos \theta$$

this equation takes the form

$$au^2 + buv + cv^2 + du + ev + f = 0$$

where a, b, c, d, e, and f are numbers that depend on θ. We could find values for all of them, but we really care only about b. When we do the necessary algebra, we find

$$b = B(\cos^2 \theta - \sin^2 \theta) - 2(A - C) \sin \theta \cos \theta$$
$$= B \cos 2\theta - (A - C) \sin 2\theta$$

We would like to have $b = 0$; that is,

$$B \cos 2\theta = (A - C) \sin 2\theta$$

This will occur if

$$\boxed{\cot 2\theta = \frac{A - C}{B}}$$

This formula is the answer to our question: to eliminate the cross-product (xy) term, choose θ so that it satisfies this formula. In the example, $xy = 1$, we have $A = 0$, $B = 1$, and $C = 0$, so we choose θ to satisfy

$$\cot 2\theta = \frac{0 - 0}{1} = 0$$

One angle that works is $\theta = 45°$. We could also use $\theta = 135°$ or $\theta = -225°$, but it is customary to choose a first-quadrant angle.

■ **Example 3.** By making a rotation, remove the cross-product term from $4x^2 + 2\sqrt{3}xy + 2y^2 + 10\sqrt{3}x + 10y = 5$ and thereby identify its graph. Sketch the graph in the xy-plane.

Solution. The angle of rotation θ must satisfy

$$\cot 2\theta = \frac{A - C}{B} = \frac{4 - 2}{2\sqrt{3}} = \frac{1}{\sqrt{3}}$$

This means that $2\theta = 60°$ and so $\theta = 30°$. When we use this value of θ in the rotation formulas, we obtain

$$x = u \cdot \frac{\sqrt{3}}{2} - v \cdot \frac{1}{2} = \frac{\sqrt{3}u - v}{2}$$

$$y = u \cdot \frac{1}{2} + v \cdot \frac{\sqrt{3}}{2} = \frac{u + \sqrt{3}v}{2}$$

Substituting these in the original equation gives

$$4\frac{(\sqrt{3}u - v)^2}{4} + 2\sqrt{3}\frac{(\sqrt{3}u - v)(u + \sqrt{3}v)}{4}$$
$$+ 2\frac{(u + \sqrt{3}v)^2}{4} + 10\sqrt{3}\frac{\sqrt{3}u - v}{2} + 10\frac{u + \sqrt{3}v}{2} = 5$$

After collecting terms and simplifying, we have

$$5u^2 + v^2 + 20u = 5$$

Next we complete the squares.

$$5(u^2 + 4u + 4) + v^2 = 5 + 20$$

$$\frac{(u + 2)^2}{5} + \frac{v^2}{25} = 1$$

In the uv-system, this is the equation of a vertical ellipse with center at $(-2, 0)$. Its graph is shown in Figure 5. ■

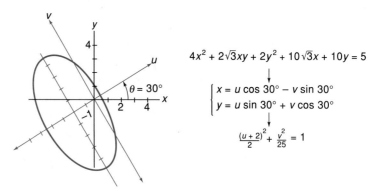

$$4x^2 + 2\sqrt{3}xy + 2y^2 + 10\sqrt{3}x + 10y = 5$$
$$\downarrow$$
$$\begin{cases} x = u \cos 30° - v \sin 30° \\ y = u \sin 30° + v \cos 30° \end{cases}$$
$$\downarrow$$
$$\frac{(u + 2)^2}{2} + \frac{v^2}{25} = 1$$

Figure 5

The rotation angle in Example 3 turned out to be a special angle, namely $\theta = 30°$. Our next example illustrates a procedure for handling a nonspecial angle.

■ **Example 4.** Use a rotation to eliminate the cross-product term in $x^2 + 24xy + 8y^2 = 136$. Identify the corresponding curve and sketch its graph in the xy-plane.

Solution. The rotation angle θ satisfies

$$\cot 2\theta = \frac{A - C}{B} = \frac{1 - 8}{24} = -\frac{7}{24}$$

but this does not determine a special angle. However, we can place 2θ in standard position with $(-7, 24)$ on its terminal side, and thereby conclude that $\cos 2\theta = -7/25$. Applying the half-angle identities (Section 5.4) gives

$$\sin \theta = \sqrt{\frac{1 + 7/25}{2}} = \frac{4}{5} \qquad \cos \theta = \sqrt{\frac{1 - 7/25}{2}} = \frac{3}{5}$$

Thus, the rotation formulas take the form

$$x = \frac{3u - 4v}{5} \qquad y = \frac{4u + 3v}{5}$$

CALCULATOR HINT

Can we use a graphics calculator to draw the graphs of general quadratic equations? The answer is yes. But in order to do so, we must first use the quadratic formula to solve the equation for y in terms of x. This gives two solutions: $y = f(x)$ and $y = g(x)$, which can be graphed in the standard way. See Problems 37–42.

All of this work was preliminary; our main task is to substitute these expressions for x and y in the original equation and simplify.

$$\left(\frac{3u - 4v}{5}\right)^2 + 24\left(\frac{3u - 4v}{5}\right)\left(\frac{4u + 3v}{5}\right) + 8\left(\frac{4u + 3v}{5}\right)^2 = 136$$

After multiplying by 25 and collecting terms, we have

$$425u^2 - 200v^2 = 136 \cdot 25$$

or

$$\frac{u^2}{8} - \frac{v^2}{17} = 1$$

This is the equation of a hyperbola. Its graph is shown in Figure 6. ■

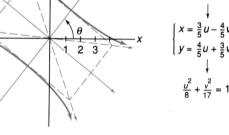

$$x^2 + 24xy + 8y^2 = 136$$
$$\downarrow$$
$$\begin{cases} x = \frac{3}{5}u - \frac{4}{5}v \\ y = \frac{4}{5}u + \frac{3}{5}v \end{cases}$$
$$\downarrow$$
$$\frac{u^2}{8} + \frac{v^2}{17} = 1$$

Figure 6

The condition $\cot 2\theta = (A - C)/B = (1 - 1)/1 = 0$ determines the rotation angle $\theta = 45°$ and the rotation formulas

$$x = \frac{\sqrt{2}}{2}(u - v) \qquad y = \frac{\sqrt{2}}{2}(u + v)$$

When these expressions are substituted in

$$x^2 + xy + y^2 = 12$$

we obtain

$$\frac{3}{2}u^2 + \frac{1}{2}v^2 = 12$$

or equivalently,

$$\frac{u^2}{8} + \frac{v^2}{24} = 1$$

This is the equation of an ellipse with major diameter $2\sqrt{24} \approx 9.80$ and minor diameter $2\sqrt{8} \approx 5.66$.

PROBLEM SET 7.4

A. Skills and Techniques

Without actually graphing, identify the graph (line, vertical parabola, horizontal ellipse, etc.) of each equation. In many cases, you will need to begin by completing the squares.

1. $x^2 + y^2 + 12x - 2y + 33 = 0$
2. $x^2 + y^2 + 12x - 2y + 40 = 0$
3. $x^2 + y^2 + 12x - 2y + 37 = 0$
4. $x^2 + 4y^2 + 12x - 8y + 36 = 0$
5. $x^2 - 4y^2 + 12x - 8y + 28 = 0$
6. $x^2 + 4y^2 + 12x - 8y + 40 = 0$
7. $4x^2 - 16x + y - 8 = 0$
8. $4x^2 - 16x + 12 = 0$
9. $4x^2 + 16x + 4y^2 - 8y = 0$
10. $x^2 + 2x + 4y^2 - 8y = 0$
11. $4x^2 - 16x + y^2 - 8y + 6 = 0$
12. $4x^2 - 16x - y^2 - 8y - 2 = 0$
13. $4x^2 - 16x + y^2 - 8y + 32 = 0$
14. $4x^2 - 16x + y^2 - 8y + 40 = 0$
15. $4x^2 - 16x - 9y^2 + 18y + 7 = 0$

16. $4x^2 - 16x - 9y^2 + 18y + 8 = 0$
17. $4x^2 - 16x + 9y + 18 = 0$
18. $4x^2 - 16x + 15 = 0$

Use the rotation formulas $x = u \cos \theta - v \sin \theta$ and $y = u \sin \theta + v \cos \theta$ to transform the given xy-equation to a uv-equation.

19. $x^2 + 4y^2 = 16; \theta = 90°$
20. $4y^2 - x^2 = 4; \theta = 90°$
21. $y^2 = 4\sqrt{2}x; \theta = 45°$
22. $x^2 = -\sqrt{2}y + 3; \theta = 45°$
23. $x^2 - xy + y^2 = 4; \theta = 45°$
24. $x^2 + 3xy + y^2 = 10; \theta = 45°$
25. $6x^2 - 24xy - y^2 = 30; \theta = \cos^{-1}\left(\frac{3}{5}\right)$
26. $3x^2 - \sqrt{3}xy + 2y^2 = 39; \theta = 60°$

Use a suitable rotation to eliminate the cross-product term. Then, if necessary, complete the squares to put the equation in standard form. Finally, sketch the graph of the equation in the xy-plane.

27. $3x^2 + 10xy + 3y^2 + 8 = 0$
28. $2x^2 + xy + 2y^2 = 90$
29. $4x^2 - 3xy = 18$
30. $4xy - 3y^2 = 64$
31. $x^2 - 2\sqrt{3}xy + 3y^2 - 12\sqrt{3}x - 12y = 0$
32. $x^2 + 2\sqrt{3}xy + 3y^2 + 8\sqrt{3}x - 8y = 0$
33. $13x^2 + 6\sqrt{3}xy + 7y^2 - 32 = 0$
34. $17x^2 + 12xy + 8y^2 + 17 = 0$
35. $9x^2 - 24xy + 16y^2 - 60x + 80y + 75 = 0$
36. $16x^2 + 24xy + 9y^2 - 20x - 15y - 150 = 0$

Use a calculator to draw and identify the graph of each of the following quadratic equations. Hint: First use the quadratic formula to solve for y in terms of x. For example,

$$9x^2 - 6xy + y^2 + 4x - 10 = 0$$

is equivalent to

$$y = \frac{6x \pm \sqrt{36x^2 - 36x^2 - 16x + 40}}{2}$$

which simplifies to

$$y = 3x \pm \sqrt{-4x + 10}$$

Graphing these two equations produces a parabola, as you should check.

37. $3x^2 + 2xy + y^2 = 10$
38. $2x^2 - 4xy - y^2 = 8$
39. $x^2 - 2xy - y^2 + 6x = 4$
40. $4x^2 - 4xy + y^2 - 5x - 8 = 0$
41. $16x^2 - 8xy + y^2 - 9x - 5 = 0$
42. $4x^2 - 3xy + 2y^2 - 6x = 0$

B. Applications and Extensions

43. Name the conic with equation $y^2 + ax^2 = x$ for the various values of a.

44. A curve C goes through the three points $(-1, 2)$, $(0, 0)$, and $(3, 6)$. Write the equation for C if C is
(a) a vertical parabola
(b) a horizontal parabola
(c) a circle

45. Without any algebra, determine the uv-equation corresponding to the equation $x^2/16 + y^2/9 = 1$ when the axes are rotated through $90°$. Then do the algebra to corroborate your answer.

46. Without any algebra, determine the uv-equation corresponding to $(x - 2\sqrt{2})^2 + (y - 2\sqrt{2})^2 = 16$ when the axes are rotated through $45°$.

47. Draw the graph of the equation $4x^2 - 2xy + y^2 = 32$, which is equivalent to $y = x \pm \sqrt{32 - 3x^2}$. The graph is an ellipse and can therefore be circumscribed by a rectangle with sides parallel to the x- and y-axes. Find the vertices of this rectangle. (*Suggestion:* Use the ⟨TRACE⟩ feature of your calculator and watch the coordinates of the cursor.)

48. Consider the ellipse of Problem 47. Superimpose the line $y = k(x + 6)$ and experiment with various values of k until you find the line through $(-6, 0)$ that is tangent to the lower half of the ellipse. Also find the point of tangency.

49. Transform the equation $(y^2 - x^2)(y + x) = 8\sqrt{2}$ to a uv-equation by rotating the axes through $45°$. Sketch the graph, showing both sets of axes.

50. The graph of $x \cos \alpha + y \sin \alpha = d$ is a line. Show that the perpendicular distance from the origin to this line is $|d|$ by making a rotation of axes through the angle α.

51. Use Problem 50 to show that the perpendicular distance from the origin to the line $ax + by = c$ is $|c|/\sqrt{a^2 + b^2}$.

52. Use the result of Problem 51 to find the perpendicular distance from the origin to the line $5x + 12y = 39$.

53. When $Ax^2 + Bxy + Cy^2 = K$ is transformed to $au^2 + buv + cv^2 = K$ by a rotation of axes, it turns out that $A + C = a + c$ and $B^2 - 4AC = b^2 - 4ac$. (Ambitious students will find showing this to be a straightforward but somewhat lengthy algebraic exercise.) Use these results to transform $x^2 - 8xy + 7y^2 = 9$ to $au^2 + cv^2 = 9$ without actually carrying out the rotation.

54. Recall that the area of an ellipse with major diameter $2a$ and minor diameter $2b$ is πab. Use the first sentence of Problem 53 to show that if $A + C$ and $4AC - B^2$ are both positive, then $Ax^2 + Bxy + Cy^2 = 1$ is the equation of an ellipse with area $2\pi/\sqrt{4AC - B^2}$.

55. Find the exact distance between the two vertices of the conic with equation $x^2 + 24xy + 8y^2 = 136$.

56. **Challenge.** Find exactly that point in the first quadrant on the graph of $x^2 + 14xy + 49y^2 = 100$ that is closest to the origin.

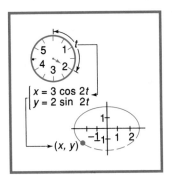
7.5 PARAMETRIC EQUATIONS

The idea of defining a curve by means of parametric equations was introduced very briefly in Section 2.8. Here, we want to explore this concept much more thoroughly. We will discover that it has many applications, including the study of motion problems that arise naturally in physics.

Imagine that the xy-coordinates of a point on a curve are specified, not by giving a relationship between x and y, but rather by telling how x and y are related to a third variable. For example, you may imagine that as time t advances from $t = a$ to $t = b$, the point $P(x, y)$ traces out a curve in the plane. Then both x and y are functions of t. That is,

$$x = f(t) \quad \text{and} \quad y = g(t) \quad a \leq t \leq b$$

We call the boxed equations **parametric equations** of a curve, with t as parameter. A **parameter** is simply an auxiliary variable on which other variables depend.

GRAPHING PARAMETRIC EQUATIONS

We can certainly graph parametric equations by hand. Simply assign values to the parameter t, calculate x and y, plot the corresponding points (x, y), and connect the points with a smooth curve. We can do this faster and more accurately with graphics calculators, which are programmed to do this job automatically.

■ **Example 1.**   Use a calculator to draw the graphs of
(a) $x = 2t - 1, y = (t + 2)^2 - 8, -4 \leq t \leq 2$
(b) $x = 8 \cos t, y = 8 \sin t, 0 \leq t \leq \pi$

Solution. After putting our calculator in parametric mode and setting the indicated range values for t, we obtain the curves shown in Figure 1. (You may obtain slightly different-looking curves, depending on how you set the

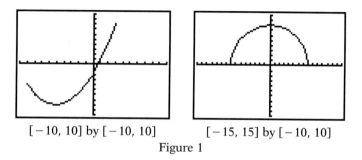

[−10, 10] by [−10, 10]   [−15, 15] by [−10, 10]

Figure 1

range values for x and y.) Our curves look suspiciously like a piece of a parabola and a semicircle. We will verify that this is the case shortly. ∎

ELIMINATING THE PARAMETER

Sometimes we can eliminate the parameter; sometimes we cannot. If $x = t - 1$ and $y = t^3$ are parametric equations for a curve, then $t = x + 1$ and so $y = (x + 1)^3$; we have eliminated the parameter t. But we would have a hard time eliminating the parameter in the equations $x = t^3 - 3t + 2$ and $y = t^5 - 3t^3 + 4$.

∎ **Example 2.** Eliminate the parameter t from the parametric equations of Example 1.

Solution.

(a) We solve for t in the first equation ($t = (x + 1)/2$), and substitute in the second equation. Then we simplify.

$$y = \left(\frac{x + 1}{2} + 2\right)^2 - 8$$

$$y = \left(\frac{x + 5}{2}\right)^2 - 8$$

$$4(y + 8) = (x + 5)^2$$

This is the equation of a parabola; the condition $-4 \le t \le 2$ in the parametric form has the effect of restricting this parabola to the part shown in Figure 1.

(b) Note that $x^2 + y^2 = 64 \cos^2 t + 64 \sin^2 t = 64$, which is the equation of a circle. The condition $0 \le t \le \pi$ restricts the graph to the top half of this circle. ∎

PARAMETRIC EQUATIONS FOR A LINE

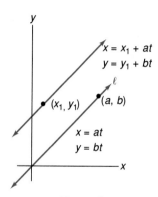

Figure 2

Consider a line ℓ that passes through the points $(0, 0)$ and (a, b), as shown in Figure 2. We claim that $x = at$ and $y = bt$ are a pair of parametric equations for this line. To see why, note that $t = 0$ and $t = 1$ yield the given points. Moreover, if we eliminate the parameter t (solve for t in the first equation, substitute in the second), we get $y = (b/a)x$, which we recognize as the equation of a line. This exercise shows, incidentally, that the parametric equations for a curve are not unique, because we have many choices for a and b.

Next we translate the above line, replacing x by $x - x_1$ and y by $y - y_1$. We get

$$x = x_1 + at \quad \text{and} \quad y = y_1 + bt$$

These are parametric equations for a line through (x_1, y_1) parallel to the line with which we started.

■ **Example 3.** Find parametric equations for the line through the points $(3, -1)$ and $(7, 5)$.

Solution. This line is parallel to the line through the origin and $(7 - 3, 5 + 1) = (4, 6)$. Thus, the parametric equations of the given line may be written as

$$x = 3 + 4t \quad \text{and} \quad y = -1 + 6t \qquad\blacksquare$$

PARAMETRIC EQUATIONS FOR THE CONICS

There are straightforward ways to represent each of the conic sections in parametric form. Here we will consider only vertical or horizontal conics; the more general situation will be discussed later in the section.

■ **Example 4 (Parabolas).** Write parametric equations for the parabola $4(y - 2) = (x + 1)^2$ and use your calculator (in parametric mode) to graph the result.

Solution. Solving for y gives

$$y = 2 + \frac{1}{4}(x + 1)^2$$

A simple parametric form of this is

$$x = t \quad \text{and} \quad y = 2 + \frac{1}{4}(t + 1)^2$$

The graph is shown in Figure 3. $\qquad\blacksquare$

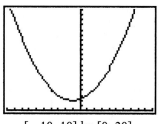

$[-10, 10]$ by $[0, 20]$,
$-10 \le t \le 8$

Figure 3

■ **Example 5 (Ellipses).** Show that

$$x = a \cos t, \quad y = b \sin t, \quad 0 \le t \le 2\pi$$

are parametric equations for an ellipse (or a circle if $a = b$). Then draw the graph of

$$x = 3 \cos t, \quad y = 5 \sin t, \quad 0 \le t \le 2\pi$$

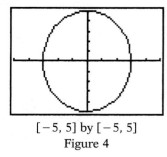

[−5, 5] by [−5, 5]

Figure 4

Solution.

$$\left(\frac{x}{a}\right)^2 + \left(\frac{y}{b}\right)^2 = \cos^2 t + \sin^2 t = 1$$

which is the equation of an ellipse (or a circle if $a = b$). The graph of the case $a = 3, b = 5$ is shown in Figure 4. ∎

■ **Example 6 (Hyperbolas).** Show that

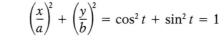

are parametric equations for a hyperbola. Then draw the graph for the case $a = 3$ and $b = 5$.

Solution.

$$\left(\frac{x}{a}\right)^2 - \left(\frac{y}{b}\right)^2 = \sec^2 t - \tan^2 t = 1$$

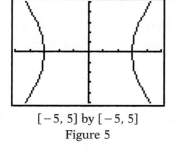

[−5, 5] by [−5, 5]

Figure 5

is the equation of a horizontal hyperbola. The required graph is shown in Figure 5. ∎

THE CYCLOID

For the conic sections, parametric representation is an optional and often useful device. For some important curves, however, parametric representation is almost essential. One such curve is the **cycloid,** which is the path of a point on the rim of a wheel rolling along a straight road.

Consider a wheel of radius a that is free to roll along the x-axis, and suppose a point P on its rim is initially at the origin. Let C denote the center of the wheel, T the wheel's point of tangency with the x-axis, and t the radian measure of angle TCP (Figure 6). Then the arc PT and the segment OT have

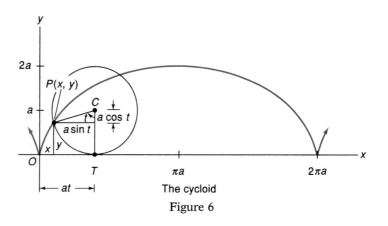

The cycloid

Figure 6

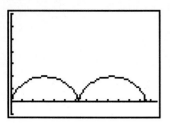

[0, 68] by [−5, 35]

Figure 7

the same length; thus, the center C of the rolling circle is at (at, a). A little trigonometry shows that

$$x = at - a \sin t = a(t - \sin t)$$
$$y = a - a \cos t = a(1 - \cos t)$$

These are the parametric equations for a cycloid.

■ **Example 7.** Draw the graph of the cycloid with $a = 5$ for $0 \le t \le 4\pi$.

Solution. A calculator-drawn graph is shown in Figure 7. ■

■ **Example 8.** If a circle of radius b rolls inside a circle of radius $a = 4b$ (Figure 8), it will generate a **hypocycloid** with parametric equations

$$x = a \cos^3 t, \qquad y = a \sin^3 t, \qquad 0 \le t \le 2\pi$$

Draw the graph for the case $a = 2$, using a calculator.

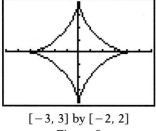

[−3, 3] by [−2, 2]

Figure 9

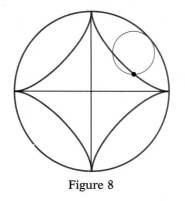

Figure 8

Solution. Figure 9 shows the required graph. ■

Rotated Conics

Recall the rotation formulas of Section 7.4, namely,

$$x = u \cos \theta - v \sin \theta$$

$$y = u \sin \theta + v \cos \theta$$

If we now substitute the parametric equations $u = f(t)$ and $v = g(t)$ for a curve in the above equations and graph, we will have that same curve in the xy-system rotated through the angle θ.

■ **Example 9.** Obtain parametric equations for the result of rotating the ellipse

$$u = 8 \cos t, \qquad v = 6 \sin t$$

through an angle $\theta = 32°$. Then use a calculator to draw the graph of this rotated ellipse.

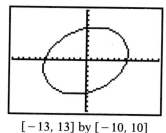

Figure 10

formulas, we obtain

$$x = (8 \cos t)\cos 32° − (6 \sin t)\sin 32°$$

$$y = (8 \cos t)\sin 32° + (6 \sin t)\cos 32°$$

The graph of the curve determined by these parametric equations is shown in Figure 10. ∎

AN APPLICATION TO PROJECTILE MOTION

It can be shown that the path of a projectile fired at a speed of v_0 feet per second from the origin at an angle θ with respect to the horizontal (Figure 11) has parametric equations

$$x = (v_0 \cos \theta)t \quad \text{and} \quad y = −16t^2 + (v_0 \sin \theta)t$$

in feet after t seconds. We are assuming that air resistance is negligible.

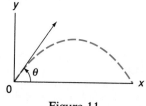

Figure 11

MOVING OBJECTS

You can actually watch objects move along their paths in real time on some calculators, including the TI-81 and TI-82. Use parametric representation with time as the parameter. If more than one object is involved, use simultaneous mode. You will be surprised what a realistic picture this method gives. These ideas are explored in Problems 57–62.

∎ **Example 10.** A projectile is shot from the origin at ground level with a speed of 64 feet per second and at an angle of 60° with the horizontal. Find each of the following.
(a) the xy-equation of the path
(b) the total time of flight
(c) the horizontal distance traveled
(d) the maximum height of the projectile

Solution. $x = 32t$ and $y = −16t^2 + 32\sqrt{3}t$
(a) Since $t = x/32$,

$$y = −16\left(\frac{x}{32}\right)^2 + 32\sqrt{3}\left(\frac{x}{32}\right)$$

$$= −\frac{1}{64}x^2 + \sqrt{3}x$$

which is the equation of a parabola.
(b) We find the values of t for which $y = 0$.

$$y = −16t^2 + 32\sqrt{3}t = −16t(t − 2\sqrt{3})$$

Thus $y = 0$ when $t = 0$ (time of firing) and when $t = 2\sqrt{3}$ (time of landing). The time of flight is $2\sqrt{3}$ seconds.
(c) The range is $32(2\sqrt{3}) = 64\sqrt{3}$ feet.
(d) The maximum height occurs when $t = (1/2)(2\sqrt{3}) = \sqrt{3}$. At that time,

$$y = −16(\sqrt{3})^2 + 32\sqrt{3}(\sqrt{3}) = 48 \text{ feet}$$

TEASER SOLUTION

The parametric equations

$$x = 3 \cos 2t \quad \text{and} \quad y = 2 \sin 2t$$

put the particle at time $t = 7\pi/12$ at the point

$$\left(3 \cos \frac{7\pi}{6}, 2 \sin \frac{7\pi}{6}\right) = \left(-\frac{3\sqrt{3}}{2}, -1\right) \approx (-2.6, -1)$$

Furthermore,

$$\left(\frac{x}{3}\right)^2 + \left(\frac{y}{2}\right)^2 = \cos^2 2t + \sin^2 2t = 1$$

so the path of the particle is a horizontal ellipse with major diameter 6 and minor diameter 4.

PROBLEM SET 7.5

A. Skills and Techniques

Use a graphics calculator to draw the graph of each of the following parametric equations. If you can, identify the graph by name.

1. $x = 1 + 3t, y = -2 + 2t; -3 \le t \le 2$
2. $x = -4 + 2t, y = 6 - 3t; -1 \le t \le 4$
3. $x = 2t - 1, y = t^2 - 2; -2 \le t \le 3$
4. $x = 0.5t^2, y = 3 - 2t; -2 \le t \le 4$
5. $x = t^3, y = 2t^2; -2 \le t \le 2$
6. $x = 2^t, y = 3t; -3 \le t \le 3$
7. $x = \dfrac{4 - t^2}{1 + t^2}, y = \dfrac{8t}{1 + t^2}; -10 \le t \le 10$
8. $x = \dfrac{4t^2}{1 + t^2}, y = \dfrac{t^3}{1 + t^2}; -10 \le t \le 10$
9. $x = 6 \cos t, y = 6 \sin t; 0 \le t \le 2\pi$
10. $x = 2 + 5 \cos t, y = -3 + 5 \sin t; 0 \le t \le 2\pi$
11. $x = 6 \cos t, y = 8 \sin t; 0 \le t \le 2\pi$
12. $x = 9 \cos t, y = 4 \sin t; 0 \le t \le 2\pi$
13. $x = -1 + 8 \cos t, y = 2 + 3 \sin t; 0 \le t \le \pi$
14. $x = 1 + 4 \cos t, y = -2 + 7 \sin t; 0 \le t \le \pi$
15. $x = 4 \sec t, y = 6 \tan t; 0 \le t \le 2\pi$
16. $x = 8 \tan t, y = 4 \sec t; 0 \le t \le 2\pi$
17. $x = 4t - 2 \sin t, y = 4 - 2 \cos t; -4\pi \le t \le 4\pi$
18. $x = t - 8 \sin t, y = 1 - 8 \cos t; 0 \le t \le 4\pi$

Eliminate the parameter in each of the following parametric equations, thereby obtaining the corresponding xy-equation. Name the graph of this equation.

19. $x = 3t - 1, y = 2t + 3$
20. $x = 3t - 1, y = 2t^2$

21. $x = 2s, y = s^2 - s + 2$
22. $x = s^2 + 3s - 4, y = 2s - 1$
23. $x = 2 \cos t, y = 4 \sin t$
24. $x = 3 \sin t, y = 2 \cos t$
25. $x = 2 \sec t, y = 3 \tan t$
26. $x = \dfrac{1 - t^2}{1 + t^2}, y = \dfrac{2t}{1 + t^2}$

Write parametric equations for the curves described in Problems 27–40.

27. The line through $(-1, 2)$ and $(7, 6)$
28. The line through $(2, -3)$ and $(-2, 0)$
29. The line through $(1, 2)$ with slope 3
30. The line through $(-2, 4)$ with slope -2
31. The parabola $y = x^2 - 3x + 1$
32. The parabola with focus $(2, -1)$ and directrix $y = 5$
33. The horizontal ellipse with center at the origin, major diameter 10, and minor diameter 6
34. The ellipse with center at the origin, a focus at $(0, 12)$, and a vertex at $(0, 13)$
35. The circle of radius 4 centered at $(2, 3)$
36. The ellipse of Problem 33 but centered at $(2, 3)$
37. The hyperbola $x^2 - (y^2/4) = 1$
38. The hyperbola with center at the origin, a vertex at $(4, 0)$, and a focus at $(5, 0)$
39. The path of a point on the rim of a wheel of radius 15 rolling along the x-axis, assuming the point is initially at the origin
40. The path of the point, initially at $(8, 0)$, on the rim of a circle of radius 2, which rolls inside a central circle of radius 8

Determine parametric equations for each curve and use a calculator to draw its graph in the xy-plane.

41. The ellipse $x^2/25 + y^2/16 = 1$ rotated through an angle of 45°
42. The ellipse $x^2/16 + y^2/36 = 1$ rotated through an angle of 60°
43. The hyperbola $x^2/16 - y^2/16 = 1$ rotated through an angle of 42°
44. The hyperbola $x^2/36 - y^2/16 = 1$ rotated through an angle of 58°

Determine the maximum height and horizontal distance traveled for each projectile, assuming it is shot from ground level at the given angle with the ground and with the indicated velocity.

45. $\theta = 45°$, $v_0 = 96$ feet per second
46. $\theta = 60°$, $v_0 = 2000$ feet per second

B. Applications and Extensions

47. Write parametric equations for each curve.
    (a) The line through $(-1, 3)$ that is parallel to the line with parametric equations $x = 2 - 3t$ and $y = 1 + 4t$
    (b) The circle with center $(-3, 4)$ that goes through the origin
    (c) The ellipse with center $(0, 0)$ and eccentricity $5/13$ that has a focus at $(0, 5)$
48. Draw the graph of $x = \sqrt{2t + 1}$, $y = \sqrt{8t}$; $t \geq 0$. Then identify this curve.
49. Show that all of the following parameterizations represent the same curve. Name this curve.
    (a) $x = 2 \cos t$, $y = 2 \sin t$; $0 \leq t \leq \pi/2$
    (b) $x = \sqrt{t}$, $y = \sqrt{4 - t}$; $0 \leq t \leq 4$
    (c) $x = t + 1$, $y = \sqrt{3 - 2t - t^2}$; $-1 \leq t \leq 1$
    (d) $x = (2 - 2t)/(1 + t)$, $y = 4\sqrt{t}/(1 + t)$; $0 \leq t \leq 1$
50. Draw the graph of $x = 5 \sin t$, $y = 5 \tan t$; $-\pi/2 < t < \pi/2$. Then determine the corresponding xy-equation.
51. Draw the graph of the parametric equations $x = 6t^{-1/2} \cos t$, $y = 6t^{-1/2} \sin t$; $t > 0$. Give an appropriate name to this graph.
52. Draw the graph of the parametric equations $x = 5 \sin^2 t - 4 \cos^2 t$, $y = 4 \cos^2 t + 5 \sin^2 t$; $0 \leq t \leq \pi/2$. Make a conjecture as to what curve this is. Prove your conjecture. What curve do you get if you change the domain for t to $\pi/2 \leq t \leq \pi$?
53. Modify the text discussion of the cycloid (and its accompanying diagram) to handle the case where the point P is $b > a$ units from the center of the wheel (a flanged wheel, as on a train locomotive). The path of P is called a **prolate cycloid.** Determine the resulting parametric equations and then graph the case where $a = 2$, $b = 3$, and $0 \leq t \leq 15$.

54. Define two (important) functions called the **hyperbolic sine** and **hyperbolic cosine** by
$$\sinh t = \frac{e^t - e^{-t}}{2} \qquad \cosh t = \frac{e^t + e^{-t}}{2}$$
    (a) Show that $\cosh^2 t - \sinh^2 t = 1$.
    (b) Show that $x = a \cosh t$, $y = a \sinh t$ give a parameterization for one branch of a hyperbola.
55. Draw graphs of the following parametric equations. Refer to the discussion of the cycloid and note how small changes in its parametric equations lead to some very different graphs. In each case, allow t to range over the interval $-2\pi \leq t \leq 2\pi$.
    (a) $x = 4(t - \sin t)$, $y = 4(1 - \cos t)$
    (b) $x = 4(t - \sin 2t)$, $y = 4(1 - \cos t)$
    (c) $x = 4(t - \sin 3t)$, $y = 4(1 - \cos t)$
    (d) $x = 4(t - \sin 3t)$, $y = 4(1 - \cos 2t)$
56. Draw the graphs of the following parametric equations, all related in some way to the hypocycloid (the first is actually a hypocycloid). In each case, allow t to range over the interval $0 \leq t \leq 2\pi$.
    (a) $x = 6 \cos t + 2 \cos 3t$, $y = 6 \sin t - 2 \sin 3t$
    (b) $x = 8 \cos t + \cos 8t$, $y = 8 \sin t - \sin 8t$
    (c) $x = 9 \cos t - 3 \cos 3t$, $y = 9 \sin t - 3 \sin 3t$
    (d) $x = 8 \cos t - \cos 8t$, $y = 8 \sin t - \sin 8t$
57. A batted ball leaves a point 3 feet above home plate at an angle of 25° from the horizontal with a velocity of 130 feet per second (Figure 12). Assuming negligible air resistance, we can write the parametric equations for its position at $t \geq 0$ as
$$x = (130 \cos 25°)t, \; y = 3 - 16t^2 + (130 \sin 25°)t$$

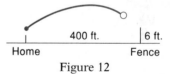

| | 400 ft. | 6 ft. |
Home | | Fence

Figure 12

Is it a home run? That is, will the ball clear a 6-foot fence 400 feet away? To analyze this question, do the following.
    (a) Draw the path of the ball using a graphics calculator with Tmin = Xmin = 0, Tmax = 6, Xmax = 550, Ymin = -10, Ymax = 150.
    (b) Determine t when $x = 400$.
    (c) Determine the height of the ball when $x = 400$.
58. Refer to Problem 57. By experiment, determine the angle among the multiples of 5° that gives the longest home run, that is, the angle that allows the ball to land on the ground farthest from home plate.
59. Use algebra to find the initial angle θ that produces the maximum horizontal range for a projectile shot from ground level. Then write a formula for this range in terms of θ and the initial velocity v_0. Does your result confirm what you found in Problem 58?

60. Note (Figure 13) that the circle $x^2 + y^2 = 5$ and the ray $y = (3/4)x$, $x \geq 0$ intersect in the point $(4, 3)$. Suppose particles A and B move so that at time t their positions are given by the following parametric equations

A: $x = 5 \cos t$, $y = 5 \sin t$; $0 \leq t \leq 2\pi$
B: $x = 4t$, $y = 3t$; $t \geq 0$

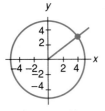

Figure 13

(a) Convince yourself that the paths of the two particles are the curves shown in Figure 13.
(b) Convince yourself that the moving particles do not collide even though their paths intersect. On some graphics calculators, including the TI-81 and TI-82, you may choose simultaneous mode and actually watch the two particles move on the screen.
(c) Demonstrate that the particles do not collide by finding the minimum distance between them, that is, by minimizing

$$D(t) = \sqrt{(5 \cos t - 4t)^2 + (5 \sin t - 3t)^2}$$

61. Refer to Problem 60. Two particles A and B are moving in the plane so that at time t their positions are as follows.

A: $x = 10t$, $y = -5t^2 + 25t$
B: $x = 57 + 10 \cos t$, $y = 10 + \sin t$

Draw their paths using a graphics calculator (in simultaneous mode) with Xmin $= Y$min $= 0$, Xmax $= 80$, and Ymax $= 40$.

(a) Estimate to two-decimal-place accuracy the shortest distance between the particles.
(b) Estimate to two-decimal-place accuracy the shortest distance between their paths.

62. Challenge. Juana is riding a Ferris wheel of radius 20 feet that turns counterclockwise at the rate of 0.8 radians per second. The center of the Ferris wheel is at $(190, 20)$, and Juana is at $(210, 20)$ at $t = 0$. Also at $t = 0$, Dorothy threw a ball from the origin at an angle 1.10 radians from the horizontal with a velocity of 80 feet per second (see Figure 14). Convince yourself that the appropriate parametric equations for Juana and the ball are as follows.

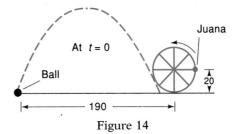

Figure 14

Juana: $x = 190 + 20 \cos 0.8t$, $y = 20 + 20 \sin 0.8t$
Ball: $x = (80 \cos 1.10)t$, $y = -16t^2 + (80 \sin 1.10)t$

(a) Use a graphics calculator (in simultaneous mode) to decide if Juana can catch the ball. Assume that she is a point with arms 1-foot long.
(b) Determine how close the ball will get to Juana accurate to one decimal place.
(c) Experiment by changing the angle at which the ball is thrown, or by changing the initial speed, or by changing both until you find conditions that will allow Juana to catch the ball.

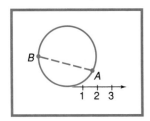

TEASER The diagram shows the graph of the polar equation $r = 4 \sin \theta$ together with the points A and B on this graph corresponding to $\theta = \pi/6$ and $\theta = 3\pi/4$. Find the distance between A and B; that is, find the length of the dotted line segment.

The interplay between equations and their graphs has played a dominant role in this book. The idea goes back to two French mathematicians, Pierre Fermat and René Descartes, who noted that, once perpendicular x- and y-axes are introduced in the plane, points can be specified by giving their x- and y-coordinates. The resulting system, the **Cartesian coordinate system,** is the backdrop for every graph that we have drawn.

Now it is time to say that other kinds of coordinate systems play useful roles in mathematics; in particular, this is true of the **polar coordinate system.**

POLAR COORDINATES

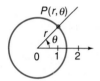

Figure 1

In place of two perpendicular axes, as in Cartesian coordinates, we introduce in the plane a single horizontal ray, called the **polar axis,** emanating from a fixed point O, called the **pole.** We assume that the polar axis has a positive number scale with 0 at the pole. Any point P, other than the pole, is the intersection of a unique circle with center O and a unique ray emanating from O (Figure 1). If r is the radius of the circle and θ is the angle (actually one of the angles) the ray makes with the polar axis, then (r, θ) are **polar coordinates** for P.

Points specified by polar coordinates are easiest to plot if we use polar graph paper. The grid on this paper consists of concentric circles and rays emanating from their common center. Figure 2 shows such a grid and a few points with their polar coordinates.

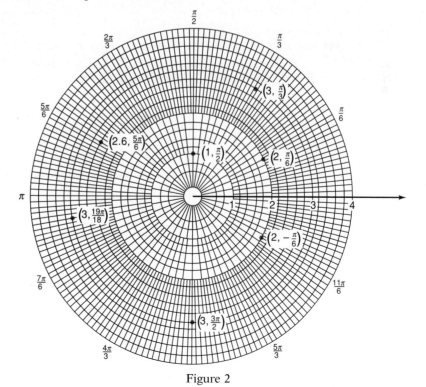

Figure 2

Of course, we can measure an angle θ in degrees or in radians. More significantly, notice that although a pair of polar coordinates (r, θ) determines a unique point, each point has many different pairs of polar coordinates. For example,

$$\left(2, \frac{3\pi}{2}\right) \quad \left(2, -\frac{\pi}{2}\right) \quad \left(2, \frac{7\pi}{2}\right) \quad (2, -450°)$$

are all coordinates for the same point.

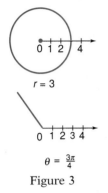

$r = 3$

$\theta = \frac{3\pi}{4}$

Figure 3

POLAR GRAPHS

The simplest polar equations are $r = k$ and $\theta = k$, where k is a constant. The graph of the first is a circle; the graph of the second is a ray emanating from the pole. Examples are shown in Figure 3. Polar equations such as

$$r = 4 \sin^2 \theta \quad \text{and} \quad r = 1 + \cos 2\theta$$

are more complicated. To graph such equations (by hand), we suggest making a table of (r, θ) values, plotting the corresponding points, and then connecting those points with a smooth curve.

■ **Example 1.** Sketch the graph of the polar equation

$$r = \frac{1}{1 - \cos \theta}$$

Solution. In Figure 4, we have constructed a table of values and drawn the corresponding graph. It looks suspiciously like a parabola, and in Section 7.7, we will verify that this suspicion is correct. ■

r	θ
—	0
3.4	$\pi/4$
1	$\pi/2$
.6	$3\pi/4$
.5	π
.6	$5\pi/4$
1	$3\pi/2$
3.4	$7\pi/4$
—	2π

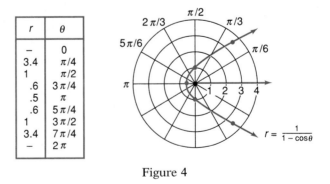

Figure 4

■ **Example 2.** Sketch the graph of $r = 2(1 + \cos \theta)$.

Solution. A table of values and the graph are shown in Figure 5. ■

362 • Analytic Geometry

r	θ
4	0
3.73	π/6
3	π/3
2	π/2
1	2π/3
0	π
1	4π/3
2	3π/2
3	5π/3
4	2π

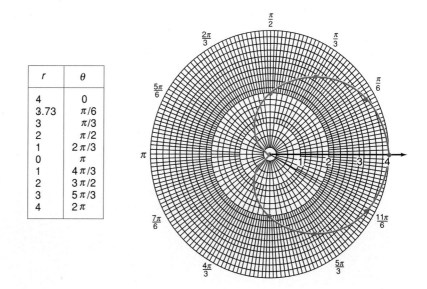

$$r = 2(1 + \cos \theta)$$

Figure 5

The equation in Example 2 is a special case of the following general forms

$$r = a \pm b \cos \theta \quad \text{and} \quad r = a \pm b \sin \theta$$

with a and b positive. Their graphs are called **limaçons,** with the special case $a = b$ giving a curve called a **cardioid** (a heart-shaped curve as in Example 2). Limaçons have the general shapes shown in Figure 6. (In the case $a < b$, we allow r to be negative, a matter we consider next.)

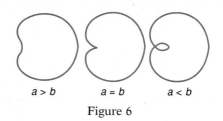

Figure 6

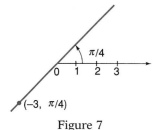

Figure 7

ALLOWING R TO BE NEGATIVE

It is often useful to allow r to be negative. By the point $(-3, \pi/4)$, we shall mean the point 3 units from the pole on the ray in the opposite direction from the ray $\theta = \pi/4$ (Figure 7).

■ **Example 3.** Allowing r to be negative, sketch the graph of $r = 2 \sin 2\theta$.

Solution. We first construct an extended table of values; then we sketch the graph (Figure 8). Note that the four leaves of the graph correspond to the four parts (a), (b), (c), and (d) of the table of values. For example, leaf (b) results from those values of θ between $\pi/2$ and π where r is negative. The complete graph is referred to as a four-leaved rose. ∎

θ	0	$\frac{\pi}{12}$	$\frac{\pi}{6}$	$\frac{\pi}{4}$	$\frac{\pi}{3}$	$\frac{5\pi}{12}$	$\frac{\pi}{2}$	$\frac{7\pi}{12}$	$\frac{3\pi}{4}$	$\frac{11\pi}{12}$	π	$\frac{5\pi}{4}$	$\frac{3\pi}{2}$	$\frac{7\pi}{4}$	2π
2θ	0	$\frac{\pi}{6}$	$\frac{\pi}{3}$	$\frac{\pi}{2}$	$\frac{2\pi}{3}$	$\frac{5\pi}{6}$	π	$\frac{7\pi}{6}$	$\frac{3\pi}{2}$	$\frac{11\pi}{6}$	2π	$\frac{5\pi}{2}$	3π	$\frac{7\pi}{2}$	4π
r	0	1	$\sqrt{3}$	2	$\sqrt{3}$	1	0	-1	-2	-1	0	2	0	-2	0
			a						b			c		d	

$r = 2 \sin 2\theta$

Figure 8

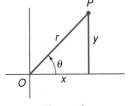

Figure 9

RELATION TO CARTESIAN COORDINATES

Let the positive x-axis of the Cartesian coordinate system serve as the polar axis of a polar coordinate system, the origin coinciding with the pole (Figure 9). Then the Cartesian coordinates (x, y) and polar coordinates (r, θ) of a point P are related by the following equations

$$x = r \cos \theta \qquad y = r \sin \theta$$
$$x^2 + y^2 = r^2 \qquad \tan \theta = \frac{y}{x}$$

■ **Example 4.** (a) Find the Cartesian coordinates of the point with polar coordinates $(4, \pi/6)$. (b) Find the polar coordinates of the point with Cartesian coordinates $(-3, \sqrt{3})$. (c) Change the Cartesian equation $(x^2 + y^2)^2 = x^2 - y^2$ to a polar equation.

Solution.

(a) $x = 4 \cos \dfrac{\pi}{6} = 4 \dfrac{\sqrt{3}}{2} = 2\sqrt{3}$ and $y = 4 \sin \dfrac{\pi}{6} = 4 \cdot \dfrac{1}{2} = 2$. So the Cartesian coordinates are $(2\sqrt{3}, 2)$.

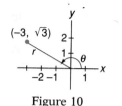

Figure 10

(b) Refer to Figure 10 and note that $r = \sqrt{(-3)^2 + (\sqrt{3})^2} = \sqrt{12} = 2\sqrt{3}$ and $\tan\theta = -\sqrt{3}/3$. Since the point is in the second quadrant, we choose $5\pi/6$ as an appropriate value for θ. One choice of polar coordinates is $(2\sqrt{3}, 5\pi/6)$.

(c) Replacing $x^2 + y^2$ by r^2, x by $r\cos\theta$, and y by $r\sin\theta$, we get

$$(r^2)^2 = r^2\cos^2\theta - r^2\sin^2\theta$$

$$r^4 = r^2(\cos^2\theta - \sin^2\theta)$$

$$r^2 = \cos 2\theta \quad \blacksquare$$

POLAR GRAPHS WITH A GRAPHICS CALCULATOR

Some graphics calculators, including the TI-82, have a polar graphing mode; with those that do, it is a simple matter to draw the polar graph of $r = f(\theta)$. Other calculators, including the TI-81, can draw the polar graph of $r = f(\theta)$ only after we rewrite it in the parametric form

$$x = f(\theta)\cos\theta \quad \text{and} \quad y = f(\theta)\sin\theta$$

■ **Example 5.** Use a graphics calculator to draw the graphs of Examples 1, 2, and 3, namely, the graphs of

(a) $r = \dfrac{1}{1 - \cos\theta}$ (b) $r = 2(1 + \cos\theta)$ (c) $r = 2\sin 2\theta$

Solution. The TI-82, in polar mode, makes this a trivial exercise. On the TI-81, you must first rewrite the equations in parametric form, as indicated above, and replace θ by T. The graphs are shown in Figures 11, 12, and 13.

$\blacksquare$

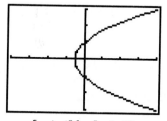

[−4, 4] by [−3, 3]
Figure 11

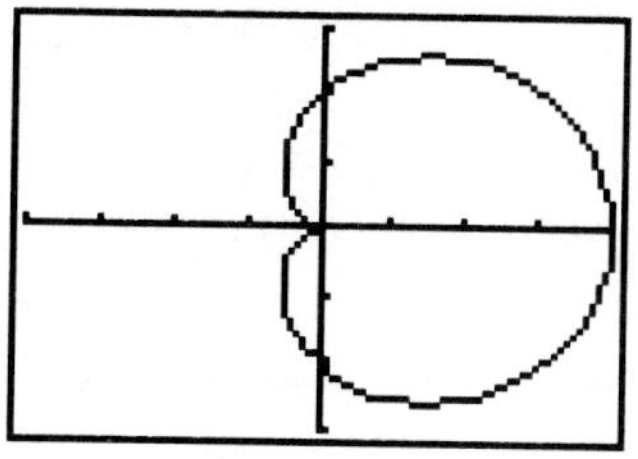

[−4, 4] by [−3, 3]
Figure 12

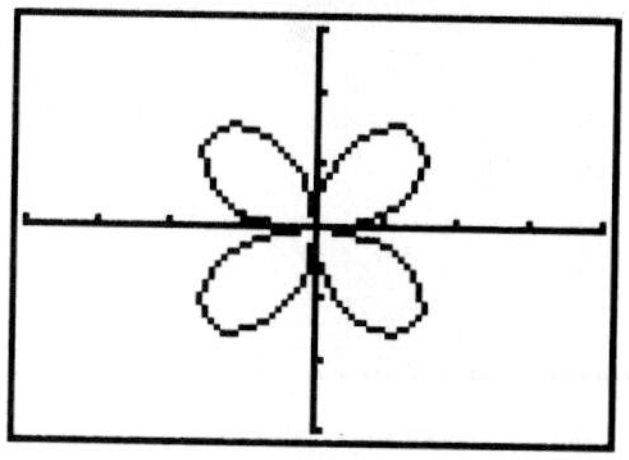

[−4, 4] by [−3, 3]
Figure 13

■ **Example 6.** Change $r = 2\sin 3\theta$ and $r = 1 + 2\cos\theta$ to parametric form and then use a calculator to draw their graphs.

Solution. We can rewrite these equations as

$$x = 2\sin 3\theta\cos\theta, \qquad y = 2\sin 3\theta\sin\theta$$

and

$$x = (1 + 2\cos\theta)\cos\theta, \qquad y = (1 + 2\cos\theta)\sin\theta$$

After replacing θ by T and using parametric mode, we obtained the graphs shown in Figures 14 and 15.

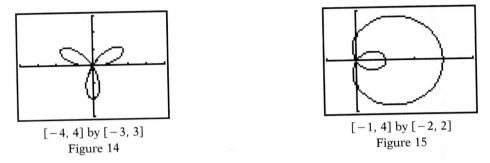

$[-4, 4]$ by $[-3, 3]$
Figure 14

$[-1, 4]$ by $[-2, 2]$
Figure 15

TEASER SOLUTION

Since $r = 4 \sin \theta$, the polar coordinates of the points A and B corresponding to $\theta = \pi/6$ and $\theta = 3\pi/4$ are $(2, \pi/6)$ and $(2\sqrt{2}, 3\pi/4)$, respectively. Using the relations $x = r \cos \theta$ and $y = r \sin \theta$, we find the Cartesian coordinates of the two points to be $(\sqrt{3}, 1)$ and $(-2, 2)$. From the distance formula, we conclude that the distance between these points is

$$\sqrt{(\sqrt{3} + 2)^2 + (1 - 2)^2} = \sqrt{3 + 4\sqrt{3} + 4 + 1}$$
$$= \sqrt{8 + 4\sqrt{3}} \approx 3.86$$

PROBLEM SET 7.6

A. Skills and Techniques

Plot the points with the given polar coordinates. Polar graph paper makes this easy; without it, you will have to estimate the angles.

1. $(3, \pi/4)$
2. $(2, \pi/3)$
3. $(4, 3\pi/2)$
4. $(3, 3)$
5. $(3, -5\pi/6)$
6. $(4, -5\pi/4)$
7. $(-2, \pi/3)$
8. $(-3, 3)$

Use the process of making a table of values, plotting the corresponding points, and connecting these points with a smooth curve to graph each of the following polar equations.

9. $r = 5 \cos \theta$
10. $r = 6 \sin \theta$
11. $r = 3(1 + \sin \theta)$
12. $r = 2(1 - \cos \theta)$

13. $r = 4 + 2 \cos \theta$
14. $r = 5 + 3 \sin \theta$
15. $r = 2 + 4 \cos \theta$
16. $r = 3 + 5 \sin \theta$
17. $r = 6 \cos 3\theta$
18. $r = 6 \cos 2\theta$

Confirm your answers to Problems 9–18 by using a graphics calculator. Then use your calculator to draw the graphs of the polar equations in Problems 19–34.

19. $r = \dfrac{\theta}{3}$

20. $r = \dfrac{2}{\theta}$

21. $r = 4 \csc \theta$
22. $r = 4 \sec \theta$
23. $r = 4 \cot \theta$
24. $r = 4 \tan \theta$

25. $r = \dfrac{6}{1 + \sin \theta}$

26. $r = \dfrac{6}{1 + 0.5 \sin \theta}$

27. $r = \dfrac{12}{3 - \cos \theta}$

28. $r = \dfrac{12}{1 - 3 \cos \theta}$

29. $r = 4 - 4 \cos \theta$

30. $r = 4 - 3 \cos \theta$

31. $r = 4 \cos \theta + 7 \sin \theta$

32. $r = 3 \cos \theta + 8 \sin \theta$

33. $r = 6 \cos 4\theta$

34. $r = 6 \cos 5\theta$

35. Write the Cartesian coordinates for the point whose polar coordinates are as shown. Do the first few mentally, but feel free to use a calculator on the later ones (some calculators, including the TI-82, are programmed to make these conversions).
(a) $(4, \pi/4)$
(b) $(6, \pi/6)$
(c) $(3, \pi)$
(d) $(2, 3\pi/2)$
(e) $(8, 11\pi/6)$
(f) $(-2, \pi/3)$
(g) $(2, 3.5)$
(h) $(4, 7.64)$
(i) $(-4, -2.3)$

36. Follow the instructions of Problem 35, but now assume that you are to change from Cartesian to polar coordinates. Use a positive r and the smallest nonnegative angle (in radians).
(a) $(4, 0)$
(b) $(0, 3)$
(c) $(-3, 0)$
(d) $(2, -2)$
(e) $(-\sqrt{3}, 1)$
(f) $(-\sqrt{3}, -3)$
(g) $(4, 3)$
(h) $(-3.4, 6.2)$
(i) $(\pi, -\sqrt{2})$

Transform each Cartesian equation to a polar equation.

37. $x^2 + y^2 = 25$

38. $y = x^2$

39. $x^2 + y^2 = 2x$

40. $y = 3$

41. $x^2 + (y - 1)^2 = 1$

42. $(x - 2)^2 + y^2 = 4$

Transform each polar equation to a Cartesian equation. In some cases, it may help to use an algebraic trick such as first multiplying both sides by r.

43. $r = 8$

44. $r = \dfrac{6}{\cos \theta}$

45. $r = 3 \cos \theta$

46. $r = 5 \sin \theta$

47. $r = \cos 2\theta$

48. $r = \sin 2\theta$

49. $r^2 = \cos^2 \theta$

50. $r^2 = \tan \theta$

51. $r^2 = \dfrac{3}{1 + \cos^2 \theta}$

52. $r \sin\left(\theta + \dfrac{\pi}{3}\right) = 4$

B. Applications and Extensions

53. Determine (in polar coordinates) the points of intersection of the graphs of each pair of polar equations.
(a) $r = 4 \cos \theta, r \cos \theta = 1$
(b) $r = 2\sqrt{3} \sin \theta, r = 2(1 + \cos \theta)$

54. Find the polar coordinates of the midpoint of the line segment joining the points with polar coordinates $(4, 2\pi/3)$ and $(8, \pi/6)$.

55. Show that the distance d between the points with polar coordinates (r_1, θ_1) and (r_2, θ_2) is given by
$$d = \sqrt{r_1^2 + r_2^2 - 2r_1 r_2 \cos(\theta_2 - \theta_1)}$$
and use this to check the answer given to the Teaser.

56. Draw the graph of $r = 8 \cos(\theta - \pi/3)$. Then show that the circle of radius a and center (a, α) has polar equation $r = 2a \cos(\theta - \alpha)$. Refer to Figure 16.

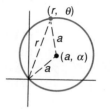

Figure 16

57. Draw the polar rectangle $0 < a < r < b$, $\alpha < \theta < \beta$, $\beta - \alpha < \pi$. Then obtain a formula for its area.

58. Show that the graph of $r = a \cos \theta + b \sin \theta$ is a circle and give its center (Cartesian coordinates) and its radius.

59. To see the potential beauty and complexity of polar graphs, use your calculator to draw the graph of $r = 6(\cos^4 4\theta + \sin 3\theta)$ for $0 \le \theta \le 2\pi$. To get a good graph, you will want to use a small θ-step, say 0.01.

60. Follow the instructions of Problem 59 with $\sin 3\theta$ replaced by $\sin 4\theta$.

61. Use your calculator to draw the graph of $r = 2/(1 + b \sin \theta)$ for $b = 0.5, 0.8, 1.0, 1.2,$ and 1.5. Conjecture the name of the curve in each case.

62. Draw the graph of $r = 4 + b \sin \theta$ for $b = 2, 3, 4,$ 5, and 6, observing how the shape of the graph changes.

63. Draw the graph of $r = 6 \sin n\theta$ for $n = 2, 3, 4,$ and 5 for $0 \le \theta \le 2\pi$. Make a conjecture about the number of petals for the graph of $r = 6 \sin n\theta$.

64. Draw the graph of $r = 6 \sin 2.9\theta$ for $0 \le \theta \le 30$.

65. Draw the graph of $r = 6 \sin 2.5\theta$ for the interval $0 \le \theta \le 2\pi$. Determine the smallest interval $0 \le \theta \le M$ that makes the graph start to repeat.

66. **Challenge.** Show that the graph of $r = 6 \sin q\theta$ will ultimately repeat if q is a rational number.

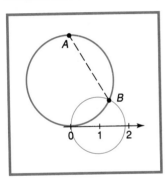

TEASER   Find the distance between the highest point of the circle $r = 3 \sin \theta$ and the first-quadrant point where this circle intersects the circle $r = \sqrt{3} \cos \theta$. That is, find the distance between points A and B in the diagram.

7.7   POLAR EQUATIONS OF CONICS

The principal conics (ellipse, parabola, hyperbola) serve to describe the motions of the moon, planets, comets, space probes, and even of thrown baseballs. In this context, it is the polar equations of the conics, rather than their Cartesian equations, that prove most useful. For example, the derivation of Kepler's laws of motion, found in many calculus books, uses the polar equations of the conics.

Our discussion begins with the polar equations of lines and circles and moves on to the principal conics.

THE POLAR EQUATION OF A LINE

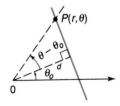

Figure 1

If a line passes through the pole, it has the exceedingly simple equation $\theta = \theta_0$, with θ_0 a constant. More generally, consider a line that does not go through the pole, but is some distance d from it. Let θ_0 be the angle from the polar axis to the perpendicular drawn from the pole to the given line (Figure 1). Then if $P(r, \theta)$ is a point on the line,

$$\cos(\theta - \theta_0) = \frac{d}{r}$$

or

$$r = \frac{d}{\cos(\theta - \theta_0)}$$

■ **Example 1.**   Write the polar equation of the line shown in Figure 2.

Solution. Note that $d = 2.7$ and $\theta_0 = \pi/3$. Thus

$$r = \frac{2.7}{\cos\left(\theta - \frac{\pi}{3}\right)}$$

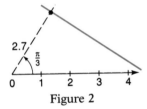

Figure 2

■

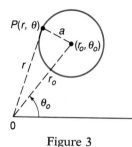

Figure 3

THE POLAR EQUATION OF A CIRCLE

If the circle is centered at the pole, its polar equation is simply $r = a$, where a is the radius of the circle. If the center is at (r_0, θ_0), then by the law of cosines (Figure 3)

$$a^2 = r^2 + r_0^2 - 2rr_0 \cos(\theta - \theta_0)$$

which is too complicated to be of much use. If the circle passes through the pole, however, so that $r_0 = a$, this equation simplifies to

$$r^2 = 2ra \cos(\theta - \theta_0)$$

or, after dividing by r,

$$\boxed{r = 2a \cos(\theta - \theta_0)}$$

The cases $\theta_0 = 0$ and $\theta_0 = \pi/2$ are particularly nice. The first gives $r = 2a \cos \theta$, and the second gives $r = 2a \cos(\theta - \pi/2)$, which is equivalent to $r = 2a \sin \theta$. These two cases are illustrated in Figure 4.

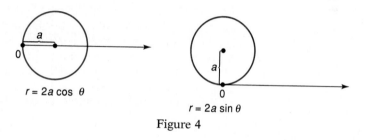

$r = 2a \cos \theta$

$r = 2a \sin \theta$

Figure 4

■ **Example 2.** Use a calculator to draw the graph of the circle

$$r = 6 \cos\left(\theta - \frac{\pi}{3}\right)$$

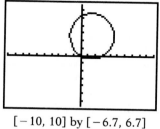

[−10, 10] by [−6.7, 6.7]

Figure 5

Solution. Some calculators allow direct graphing of polar equations. Others require first changing $r = f(\theta)$ to the parametric form $x = f(\theta) \cos \theta$, $y = f(\theta) \sin \theta$. Thus in our example, we can write

$$x = 6 \cos\left(\theta - \frac{\pi}{3}\right) \cos \theta, \qquad y = 6 \cos\left(\theta - \frac{\pi}{3}\right) \sin \theta$$

The required graph is shown in Figure 5. ■

■ **Example 3.** Write the polar equation of a circle of radius 8 and center at $(8, \pi/2)$.

Solution. Note that this circle goes through the pole. Its polar equation is $r = 16 \sin \theta$. ■

POLAR EQUATIONS FOR THE PRINCIPAL CONICS

In many applications, it is natural to put a focus of the conic at the pole. Since doing so also simplifies the polar equation, we assume that the focus of our conic is at the pole and that the directrix is d units away, as in Figure 6. The equation $d(P, F) = e\, d(P, L)$ that defines a principal conic then takes the form

$$r = e(d - r\cos(\theta - \theta_0))$$

which is equivalent to

$$r = \frac{ed}{1 + e\cos(\theta - \theta_0)}$$

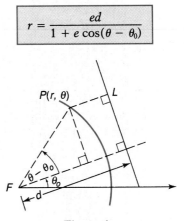

$P(r, \theta)$ L

F

Figure 6

As an example, consider a case where $\theta_0 = 0$, namely,

$$r = \frac{2}{1 + \dfrac{1}{2}\cos\theta} = \frac{\dfrac{1}{2}\cdot 4}{1 + \dfrac{1}{2}\cos\theta}$$

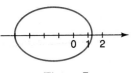

Figure 7

Since $e = 1/2$, the graph is an ellipse, the one shown in Figure 7. On the other hand,

$$r = \frac{12}{3 + 4\cos\theta} = \frac{4}{1 + \dfrac{4}{3}\cos\theta} = \frac{\dfrac{4}{3}\cdot 3}{1 + \dfrac{4}{3}\cos\theta}$$

is a hyperbola with $e = 4/3$ and $d = 3$. Examples 4 and 5 give a complete discussion of the cases $\theta_0 = 0,\ \pi/2,\ \pi,$ and $3\pi/2$.

■ **Example 4.**   Refer to Figure 8. If the directrix is perpendicular to the polar axis, then $\theta_0 = 0$ or $\theta_0 = \pi$, and the polar equation of the conic takes one of the two forms

$$\theta_0 = 0 \qquad\qquad \theta_0 = \pi$$

$$r = \frac{ed}{1 + e\cos\theta} \qquad r = \frac{ed}{1 - e\cos\theta}$$

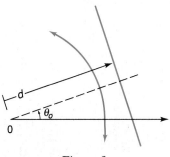

Figure 8

In the first case, the directrix is to the right of the focus; in the second, it is to the left.

Identify each of the following conics by name, find its eccentricity, and write the *xy*-equation of its directrix.

(a) $r = \dfrac{4}{3 + 3 \cos \theta}$ (b) $r = \dfrac{5}{2 - 3 \cos \theta}$

Solution.

(a) Divide numerator and denominator by 3, obtaining

$$r = \frac{4/3}{1 + \cos \theta}$$

The conic is a parabola, since the eccentricity $e = 1$. The equation of the directrix is $x = 4/3$.

(b) The equation can be rewritten as

$$r = \frac{\dfrac{5}{2}}{1 - \dfrac{3}{2} \cos \theta} = \frac{\dfrac{3}{2} \cdot \dfrac{5}{3}}{1 - \dfrac{3}{2} \cos \theta}$$

The conic is a hyperbola with $e = 3/2$ and directrix $x = -5/3$.   ∎

■ **Example 5.** Refer again to Figure 8. If the directrix is parallel to the polar axis and above it, then $\theta_0 = \pi/2$; if it is parallel to and below the polar axis, then $\theta_0 = 3\pi/2$. The corresponding equations can be simplified to

$$\theta_0 = \frac{\pi}{2} \qquad\qquad \theta_0 = \frac{3\pi}{2}$$

$$r = \frac{ed}{1 + e \sin \theta} \qquad\qquad r = \frac{ed}{1 - e \sin \theta}$$

(a) Derive the first of these equations.

(b) Identify the conic $r = 5/(2 - \sin \theta)$ by name, give its eccentricity, and write the *xy*-equation of its directrix.

Solution.

(a) The equation of the conic with $\theta_0 = \pi/2$ is

$$r = \frac{ed}{1 + e\cos(\theta - \pi/2)}$$

Since

$$\cos\left(\theta - \frac{\pi}{2}\right) = \cos\theta\cos\frac{\pi}{2} + \sin\theta\sin\frac{\pi}{2} = \sin\theta$$

we get

$$r = \frac{ed}{1 + e\sin\theta}$$

(b) Dividing numerator and denominator by 2, we obtain

$$r = \frac{\dfrac{5}{2}}{1 - \dfrac{1}{2}\sin\theta} = \frac{\dfrac{1}{2}\cdot 5}{1 - \dfrac{1}{2}\sin\theta}$$

The conic is an ellipse with eccentricity 1/2. The directrix is below the polar axis and has xy-equation $y = -5$. ∎

■ **Example 6.** By hand, sketch the graph of the polar equation $r = 6/(1 + 2\cos\theta)$. Then confirm your answer by using a calculator to draw the graph.

Solution. We recognize this as the polar equation of a hyperbola ($e = 2$) with major axis along the polar axis. Next we make the small table of values shown in Figure 9 and plot the corresponding points (marked with dots). The points marked with a cross are obtained by symmetry ($\cos(-\theta) = \cos\theta$).

r	θ
2	0°
3	60°
6	90°
−8.2	150°
−6	180°

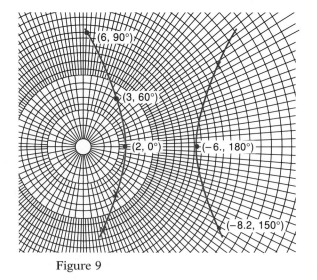

Figure 9

The corresponding calculator graph with asymptotes is shown in Figure 10.

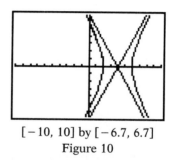

$[-10, 10]$ by $[-6.7, 6.7]$
Figure 10

SUMMARY

We summarize the most important results about polar equations for conics in Figure 11.

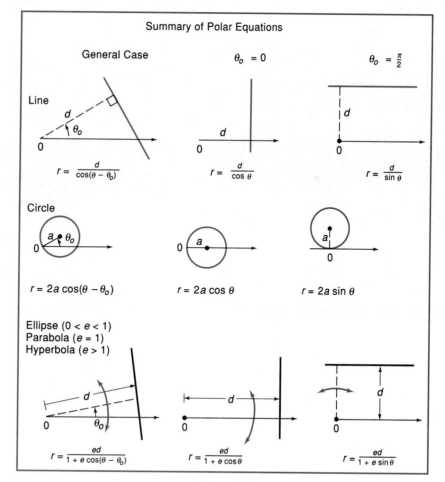

Summary of Polar Equations

General Case $\theta_0 = 0$ $\theta_0 = \frac{\pi}{2}$

Line

$r = \dfrac{d}{\cos(\theta - \theta_0)}$ $r = \dfrac{d}{\cos \theta}$ $r = \dfrac{d}{\sin \theta}$

Circle

$r = 2a \cos(\theta - \theta_0)$ $r = 2a \cos \theta$ $r = 2a \sin \theta$

Ellipse $(0 < e < 1)$
Parabola $(e = 1)$
Hyperbola $(e > 1)$

$r = \dfrac{ed}{1 + e \cos(\theta - \theta_0)}$ $r = \dfrac{ed}{1 + e \cos\theta}$ $r = \dfrac{ed}{1 + e \sin\theta}$

Figure 11

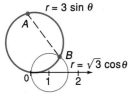

r = 3 sin θ

r = √3 cos θ

Figure 12

We find the first-quadrant intersection of the two circles (Figure 12) by solving $r = 3 \sin \theta$ and $r = \sqrt{3} \cos \theta$ simultaneously. We get

$$3 \sin \theta = \sqrt{3} \cos \theta$$

$$\tan \theta = \frac{\sqrt{3}}{3}$$

$$\theta = \frac{\pi}{6}$$

Thus, the polar coordinates of the point B are $(3/2, \pi/6)$. Using the equations $x = r \cos \theta$ and $y = r \sin \theta$ to convert to Cartesian coordinates, we find that B has Cartesian coordinates $(3\sqrt{3}/4, 3/4)$. Clearly A has Cartesian coordinates $(0, 3)$. Thus, the distance between A and B is

$$\sqrt{\left(\frac{3\sqrt{3}}{4}\right)^2 + \left(\frac{3}{4} - 3\right)^2} = \frac{\sqrt{27 + 81}}{4} = \frac{\sqrt{108}}{4} \approx 2.598$$

PROBLEM SET 7.7

A. Skills and Techniques

Write a polar equation for the indicated line.

1. The line with Cartesian equation $x = -2$
2. The line with Cartesian equation $y = 3$
3. The line 4 units from the pole whose perpendicular makes an angle of $\pi/6$ radians with the positive x-axis
4. The line with Cartesian equation $x - y = 6$
5. The line shown in Figure 13

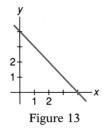

Figure 13

6. The line with slope $-\sqrt{3}$ that is 5 units from the pole and goes through the first quadrant.

Draw the graph of each line. Then write its Cartesian equation in the form $Ax + By = C$.

7. $r = \dfrac{6}{\cos \theta}$

8. $r = \dfrac{-5}{\sin \theta}$

9. $r = \dfrac{4}{\cos(\theta - \pi/6)}$

10. $r = \dfrac{10}{\cos(\theta - \pi/3)}$

11. $r = \dfrac{6}{\sin(\theta + \pi/3)}$

12. $r = \dfrac{3}{\cos(\theta + \pi/4)}$

Write a polar equation for the indicated circle.

13. Center at the pole and radius 3
14. Cartesian equation $x^2 + y^2 = 36$
15. Center $(5, \pi/2)$ and goes through the pole
16. Center $(4, \pi)$ and goes through the pole
17. Center $(5, \pi/3)$ and goes through the pole
18. Center $(4, 2\pi/3)$ and goes through the pole

Without actually graphing, identify the graph of each of the following polar equations by name.

19. $r = \dfrac{4}{1 + 0.6 \cos \theta}$

20. $r = \dfrac{6}{1 + 1.5 \cos \theta}$

21. $r = \dfrac{5}{2 + 4 \cos \theta}$

22. $r = \dfrac{6}{2 + 2 \cos \theta}$

23. $r = \dfrac{6}{\sin \theta + 2 \cos \theta}$

24. $r = \dfrac{6}{2 \cos \theta}$

25. $r = \dfrac{9}{3 + 3 \sin \theta}$

26. $r = \dfrac{9}{2 + 3 \sin \theta}$

Each of the following is the polar equation of a conic with focus at the pole. Use a calculator to draw its graph. Then determine the xy-equation of its directrix.

27. $r = \dfrac{4}{1 + 0.5 \cos \theta}$

28. $r = \dfrac{4}{1 + 1.5 \cos \theta}$

29. $r = \dfrac{12}{2 + 2 \sin \theta}$

30. $r = \dfrac{12}{3 + 2 \sin \theta}$

31. $r = \dfrac{12}{2 + 4 \cos(\theta - \pi/3)}$

32. $r = \dfrac{12}{4 + 2 \cos(\theta - \pi/6)}$

B. Applications and Extensions

33. Write a simple polar equation corresponding to each Cartesian equation.

(a) $(x + 9)^2 + y^2 = 81$
(b) $x + \sqrt{3}y = 4$
(c) $4(x + 1) = y^2$
(d) $16x = y^2$

34. Write a Cartesian equation for the curve whose polar equation is

$$\cos^2 \theta + \cos \theta \sin \theta - 6 \sin^2 \theta = 0$$

Then identify this curve.

35. Find the intersection points (in polar coordinates) of the circles with polar equations $r = 3 \sin \theta$ and $r = \sqrt{3} \cos \theta$. Then find the distance between these points.

36. Find the major and minor diameters of the ellipse with polar equation $r = 8/(2 + \cos \theta)$.

37. Generalize Problem 36 by expressing the major and minor diameters of the ellipse $r = ed/(1 + e \cos \theta)$, $0 < e < 1$, in terms of e and d.

38. Express the length of the latus rectum (chord through a focus perpendicular to the major axis) of the conic $r = ed/(1 + e \cos \theta)$ in terms of e and d.

39. Draw the graph of $r = 4/(1 + e \cos(\theta - 2))$ for $e = 0.5, 0.8, 1.0, 1.2,$ and 1.5, noting how the curve changes character.

40. Determine the major and minor diameters of the ellipse with polar equation $r = 6/(1 + 0.8 \cos(\theta - 2))$.

41. A comet follows the path $r = 6/(1 + 1.4 \sin(\theta - 1))$ with distance in astronomical units (AU). Determine its closest approach to the sun, assuming the sun is at a focus.

42. Suppose that meteors A and B are traveling in the same plane so that at time t their paths are given by the parametric equations

$$\text{A:} \quad x = \frac{5 \cos t}{1 + 0.2 \cos t} \qquad y = \frac{5 \sin t}{1 + 0.2 \cos t}$$

$$\text{B:} \quad x = \frac{5 \cos t}{1 - \cos t} \qquad y = \frac{5 \sin t}{1 - \cos t}$$

(a) Name the paths of the two meteors.
(b) Draw the two paths in simultaneous mode and guess whether the two meteors will collide.
(c) Justify your guess rigorously.

43. A comet follows an elliptical orbit of eccentricity 0.5 with the sun at a focus. If the closest approach to the sun is 2 astronomical units, determine L so that the polar equation of its path has the form $r = L/(1 + 0.5 \cos \theta)$ and use your calculator to draw this path. Then determine its maximum distance M from the sun.

44. Find, accurate to two decimal places, the (r, θ)-coordinates of the points of intersection of the polar curves $r = 4/(1 + 0.8 \cos(\theta - 2))$ and $r = 8 \cos \theta$.

45. Find the shortest distance (accurate to two decimal places) from the point with polar coordinates (10, 1) to the ellipse with polar equation $r = 8/(2 + \cos \theta)$. (*Hint:* See Problem 55 of Section 7.6.)

46. **Challenge.** Draw the graph of the limaçon $r = 6 + 3 \cos \theta$ and then find the lengths of its chords $\theta = 0$, $\theta = \pi/6$, $\theta = \pi/3$, and $\theta = \pi/2$. Guess at a result for this particular limaçon. Then prove a very general result about any limaçon of the form $r = b + a \cos \theta$, $b \geq a$.

CHAPTER 7 REVIEW PROBLEM SET

In Problems 1–10, write True or False in the blank. If false, tell why.

_____ **1.** It is 6 units from the focus to the directrix of the parabola $x^2 = 6y$.

_____ **2.** The parabola $(y - 2)^2 = 12(x + 3)$ opens to the right.

_____ **3.** The ellipse $(x - 4)^2/9 + y^2/16 = 1$ has vertices at $(4, \pm 4)$.

_____ **4.** The hyperbola $x^2/9 - y^2/25 = 1$ has eccentricity $\sqrt{34}/5$.

_____ **5.** The hyperbolas $4x^2 - 9y^2 = 36$ and $-4x^2 + 9y^2 = 36$ have the same asymptotes.

_____ **6.** The graph of the equation $x^2 + y^2 + 2x - 4y + 5 = 0$ is a single point.

_____ **7.** If $B \neq 0$, the graph of $x^2 + Bxy + y^2 = 1$ is one of these three: parabola, ellipse, hyperbola.

_____ **8.** The point with polar coordinates $(6\sqrt{2}, -135°)$ has Cartesian coordinates $(-6, -6)$.

_____ **9.** The line with parametric equations $x = 2 + 3t$, $y = -1 - 6t$ has slope 2.

_____ **10.** The Cartesian equation of the curve with parametric equations $x = \cos t$, $y = \cos 2t$ is $y = 2x^2 - 1$.

11. Determine the number of the description that best fits the given equation and write it in the blank.

_____ (a) $3x^2 + 3y^2 = 10$
_____ (b) $(x + 2)^2 + 4(y - 1)^2 = 12$
_____ (c) $(x + 2)^2 - 4(y - 1)^2 = -4$
_____ (d) $(x + 2)^2 - 4(y - 1)^2 = 4$
_____ (e) $4(x - 1)^2 + y^2 + 20 = 0$
_____ (f) $(y - 2)^2 = 4$
_____ (g) $(y - 2)^2 = 4x^2$
_____ (h) $y = 2x^2 - x - 6$
_____ (i) $(y - 2)^2 = 36 - (x + 4)^2$
_____ (j) $2x - 3y^2 = 6y - 4$
_____ (k) $(3x - 5y)^2 = 0$

(i)   a single point
(ii)   a vertical parabola
(iii) a circle
(iv)   a horizontal hyperbola
(v)   a horizontal ellipse
(vi)   parallel lines
(vii) the empty set
(viii) a vertical hyperbola
(ix)   a horizontal parabola
(x)   a single line
(xi)   intersecting lines

12. Find the vertex and focus of the parabola

$$(x + 1)^2 = -12(y - 2)$$

In Problems 13–15, find the equation of the parabola that satisfies the given conditions.

13. Vertex (0, 0), focus (5, 0)
14. Vertex (0, 0), opens down, passes through $(-2, -4)$
15. Vertex (3, 5), directrix $y = 1$
16. Light rays from the sun hit a paraboloidal reflector of diameter 6 feet and depth 1 foot. Where are the rays focused?
17. Determine the vertices and foci of the ellipse with equation $25(x + 2)^2 + 9(y - 1)^2 = 225$.

In Problems 18–20, find the equation of the ellipse that satisfies the given conditions.

18. Vertices (± 4, 0); the major diameter is twice the minor diameter.
19. Vertices (0, ± 6), eccentricity 1/2

20. Foci $(\pm 2\sqrt{5}, 0)$; $a = b + 2$

21. A semi-elliptical arch is 12 feet wide at its base and 8 feet high at the center. How wide is the arch at a height of 6 feet?

22. Determine the vertices and asymptotes of the hyperbola $25x^2 - 4y^2 = 100$.

Find the equation of the hyperbola that satisfies the given conditions.

23. Vertices $(0, \pm 6)$, eccentricity 3/2

24. Vertices $(\pm 4, 0)$; the point $(5, 3/2)$ is on the hyperbola.

25. Vertices $(\pm 5, 0)$; one asymptote $5y = 2x$.

Sketch the graph of the given equation.

26. $(x + 1)^2 = 12(y - 2)$
27. $4(x - 4)^2 - (y + 1)^2 = 16$
28. $(x + 3)^2 + 4y^2 = 16$

In Problems 29–31, use completing the squares to identify the conic with the given equation.

29. $x^2 + y^2 - 14x + 2y + 25 = 0$
30. $9x^2 + 4y^2 - 90x + 16y + 205 = 0$
31. $9x^2 - 4y^2 - 54x + 16y + 65 = 0$
32. Transform the equation $7x^2 - 4\sqrt{3}xy + 3y^2 = 36$ by rotating the axes through 60°. Then identify the corresponding curve.
33. By rotation of axes, eliminate the cross-product term from $3x^2 + 12xy + 8y^2 = 12$. Then find the distance between the foci of this conic.

Use a graphics calculator to find the required information accurate to two decimal places.

34. The points of intersection of the parabolas $y = 1.1x^2 + 2.2x + 3.3$ and $y = -4.4x^2 + 5.5x + 6.6$.
35. The points of intersection of the parabola $y = x^2 - x$ and the ellipse $4(x - 2)^2 + 5y^2 = 240$.
36. The distance between the two points of intersection of the line $y = 2.1x - 2.7$ with the parabola $y = 2x^2 - 2.5x - 6.1$.

Use a calculator to graph each pair of parametric equations and to guess at the name of the graph. Then eliminate the parameter t to confirm this guess.

37. $x = 4t - 5, y = -3t + 2$
38. $x = 4 \cos t, y = 3 \sin t, 0 \le t \le 2\pi$
39. $x = 2t, y = 12t^2 + 2t - 4$
40. $x = 9 \sin t, y = 3 \cos t, -\pi/2 \le t \le \pi/2$

41. $x = -5 \sec t, y = \tan t, -\pi/2 < t < \pi/2$
42. $x = 4 \sin t \cos t, y = 4 \sin t \sin t, 0 \le t \le \pi$

In Problems 43–48, name the curve determined by the given polar equation. Then draw its graph either by hand or with a graphics calculator.

43. $r = 4$
44. $r \cos \theta = 4$
45. $r = 4 \sin \theta$
46. $r = 4 \cos 3\theta$
47. $r = \dfrac{2}{1 + \sin \theta}$
48. $r = 2(1 + \sin \theta)$
49. Transform the equation $x^2 + y^2 + 4x - 2y = 0$ to a polar equation of the form $r = f(\theta)$.
50. Transform the polar equation
$$r = \frac{2 \sin 2\theta}{\cos^3 \theta - \sin^3 \theta}$$
to a Cartesian equation. (*Hint:* Clear of fractions and multiply both sides by r^2.)

51. Determine the number of the description that best fits the given polar equation and write it in the blank.

_____ (a) $r \cos \theta = -4$
_____ (b) $r = -4 \cos \theta$
_____ (c) $r = 3/(1 + \cos \theta)$
_____ (d) $r = 2/(1 + 4 \cos \theta)$
_____ (e) $r = 5/\cos(\theta - \pi/4)$
_____ (f) $r = 3/(2 + 3 \cos \theta)$
_____ (g) $r = 4/(4 - \cos \theta)$
_____ (h) $r = 2 \cos \theta - 4 \sin \theta$

(i)   a parabola
(ii)   a hyperbola with $e = 4$
(iii) an ellipse
(iv)   a circle
(v)   a hyperbola with $e = 3/2$
(vi)   a nonvertical line
(vii) a vertical line
(viii) none of the above

52. In each case, find the minimum distance from the pole to the curve with the given polar equation.

(a) $r = \dfrac{1}{1 - \cos \theta}$ (b) $r = 3 - 2 \sin \theta$

53. Determine the maximum distance from the origin to the curve with parametric equations $x = 8 \sin(t - 1), y = 6 \cos t, 0 \le t \le 2\pi$, accurate to two decimal places.

54. Determine the minimum distance from the origin to the curve with polar equation $r = 1 + 0.5\,\theta + \cos(\theta + 1)$, $0 \le \theta \le 2\pi$, accurate to two decimal places.

8

SEQUENCES, COUNTING, AND PROBABILITY

An action, like tossing a coin, which can result in only two possible outcomes—success or failure—is called a Bernoulli trial and the number of successes in independent Bernoulli trials is said to have the binomial distribution (Section 8.9). James Bernoulli was the first in a line of prominent Swiss mathematicians with the name Bernoulli. He worked in many fields but is perhaps most famous as a founder of the theory of probability. His book Ars Conjectandi contains the first general treatment of permutations and combinations (see Sections 8.6 and 8.7) and of the binomial distribution. James was quick to become proficient in the powerful new methods of calculus and his brother John actually wrote the first textbook on calculus (ca. 1692) although it was not published until years later.

Early workers in probability dealt largely with games of chance; this gave the subject a somewhat disreputable aura. It is only in the twentieth century that probability theory has achieved real respect-

ability. This is due to a large number of talented mathematicians among whom is David Blackwell, the only black member of the National Academy of Sciences. Few would have predicted that Blackwell, born in the backwaters of southern Illinois to parents with little education, would attain such distinction. That he did is a tribute to his ability and determination and to the encouragement of some enlightened teachers. Before retirement, he had become chair of the prestigious statistics department at the University of California at Berkeley and had made contributions to all branches of mathematics related to probability theory. Because we wanted to keep our discussion of graphics calculators as simple as possible, we have refrained from describing their programming features until now. The study of sequences provides a natural setting for introducing this powerful tool. Beginning in Section 8.3, we develop programs to generate sequences and use them thereafter in applications.

David Blackwell (1919–   )

James Bernoulli (1654–1705)

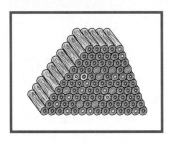

TEASER A pile of logs like that in the diagram has 5 logs in the top row and 34 logs in the bottom row. How many logs are in the pile?

8.1 ARITHMETIC SEQUENCES AND SUMS

The word *sequence* is used in ordinary language. For example, here is a sequence of steps that you might take in purchasing a car.

S_1: Save money for the down payment
S_2: Discuss the idea with knowledgeable people
S_3: Test drive several models
S_4: Make a choice
S_5: Arrange for financing

What characterizes this sequence is the notion that one step follows another in a definite order. There is a first step, a second step, a third step, and so on. It is this idea we want to discuss but in the context of numbers.

NUMBER SEQUENCES

A **number sequence** is an arrangement of numbers in which there is a first number, a second number, a third number, and so on. A typical example is the sequence

$$2, 4, 6, 8, 10, \ldots$$

of even numbers. The three dots indicate that the sequence continues indefinitely, following the pattern suggested by the initial terms. Because this sequence continues on and on without end, we may call it an *infinite* sequence; most of the sequences we consider will be infinite sequences.

To denote a general number sequence, we use the notation

$$a_1, a_2, a_3, a_4, a_5, \ldots$$

or similar expressions with other letters. Thus, for the even number sequence referred to above,

$$a_1 = 2$$

$$a_2 = 4$$

$$a_3 = 6$$

$$\vdots$$

Note that a_1 stands for the first term, a_2 for the second term, a_3 for the third term, and so on. The subscript indicates the position of the term in the sequence. For the general term, we use the symbol a_n, and occasionally we denote a whole sequence by the symbol $\{a_n\}$.

There is another more formal way to describe a sequence. A **number sequence** is a function whose domain is the set of positive integers. This means that it is a rule that associates with each positive integer n a number a_n. Conformity with Chapter 2 would suggest that we should use the notation $a(n)$, but tradition dictates that we use a_n instead. Recall that we usually specify functions by giving (explicit) formulas; this is true of sequences also.

Consider the sequence b_1, b_2, b_3, . . . where b_n is given by the formula

$$b_n = \frac{n + 1}{n}$$

Then

$$b_1 = \frac{1 + 1}{1} = 2$$

$$b_2 = \frac{2 + 1}{2} = \frac{3}{2}$$

$$b_3 = \frac{3 + 1}{3} = \frac{4}{3}$$

$$b_4 = \frac{4 + 1}{4} = \frac{5}{4}$$

■ **Example 1.** A sequence $\{p_n\}$ is determined by the formula $p_n = n^2 - n + 1$. Evaluate p_3 and p_{10}.

Solution. $p_3 = 3^2 - 3 + 1 = 7$, and $p_{10} = 10^2 - 10 + 1 = 91$. ■

Sequences are often specified by giving enough of the initial terms so that a pattern is apparent; a challenge is then to give an explicit formula for the nth term. Consider the sequence of even numbers 2, 4, 6, 8, . . . mentioned above and denoted there by a_1, a_2, a_3, a_4, ... Note that the value of a term is just twice its subscript, that is,

$$a_n = 2n$$

Knowing this, we can easily give the value of any term in the sequence. For example, $a_{39} = 2(39) = 78$, and $a_{710} = 2(710) = 1420$.

■ **Example 2.** Give an explicit formula for the sequence $\{a_n\}$ with initial terms 0, 1, 4, 9, 16, 25, 36, . . . and use it to determine a_{13}.

Solution. We note that the given terms are perfect squares. It is tempting to write $a_n = n^2$, but that solution would make $a_1 = 1^2 = 1$, whereas we want $a_1 = 0$. A little thought suggests $a_n = (n - 1)^2$, and therefore $a_{13} = (13 - 1)^2 = 144$. ■

ARITHMETIC SEQUENCES

Consider the following number sequences. When you see a pattern, fill in the boxes.

(a)   5, 9, 13, 17, $\square$, $\square$, . . .
(b)   2, 2.5, 3, 3.5, $\square$, $\square$, . . .
(c)   8, 5, 2, -1, $\square$, $\square$, . . .

What is it that these three sequences have in common? Simply this: In each case, you can get a term by adding a fixed amount to the preceding term. In (a), you add 4 each time, in (b) you add 0.5, and in (c) you add -3. Such a sequence is called an **arithmetic sequence** (AS for short). If we denote such a sequence by $a_1, a_2, a_3, \ldots$, then its nth term a_n is related to its $(n-1)$th term a_{n-1} as follows.

AS Recursion Formula: $a_n = a_{n-1} + d$

Here d, called the **common difference,** is the amount to be added each time ($d = 4$, $d = 0.5$, and $d = -3$ in our three examples). The boxed formula is called a **recursion formula** (or recurrence formula) because it relates the value of a term to a previous term. For our three examples, we may write the recursion formulas

$$a_n = a_{n-1} + 4$$

$$b_n = b_{n-1} + 0.5$$

$$c_n = c_{n-1} + (-3) = c_{n-1} - 3$$

■ **Example 3.**   Given that $e_1 = 4$ and the recursion formula $e_n = e_{n-1} + 0.3$, determine the first five terms of the sequence $\{e_n\}$.

Solution. $e_1 = 4$, $e_2 = 4 + 0.3 = 4.3$, $e_3 = 4.3 + 0.3 = 4.6$, $e_4 = 4.6 + 0.3 = 4.9$, and $e_5 = 4.9 + 0.3 = 5.2$.   ■

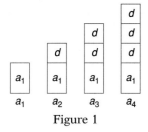

Figure 1

If we ask for the value of e_{291} in Example 3, the disadvantage of a recursion formula becomes clear. In order to determine this value, we must know the value of e_{290}, which requires knowledge of e_{289} and in effect means we must know the value of all previous terms. An **explicit formula,** that is, a formula that relates the value of a term directly to its subscript, would make it easy to answer questions like this. But can we find such a formula?

Figure 1 shows us how to obtain the explicit formula for any arithmetic sequence $\{a_n\}$. Notice that the number of ds to be added to a_1 is 1 less than the subscript of the term, that is,

AS Explicit Formula: $a_n = a_1 + (n-1)d$

■ **Example 4.** Determine e_{291} for the arithmetic sequence with $e_1 = 4$ and common difference $d = 0.3$ (see Example 3).

Solution. $e_{291} = 4 + 290(0.3) = 91.$ ■

SUMS OF ARITHMETIC SEQUENCES

There is an oft-told story about the famous mathematician Carl Gauss that aptly illustrates the next idea. We are not sure the story is true, but if it isn't, it should be.

When Gauss was about 10 years old, he was admitted to an arithmetic class. To keep the class busy, the teacher often assigned long addition problems. One day he asked his students to add the numbers from 1 to 100. He had hardly finished giving the assignment when young Gauss put his slate on the teacher's desk with the answer 5050 written on it.

Here is how Gauss thought about the problem (Figure 2). Each of the indicated pairs has 101 as its sum, and there are 50 such pairs. Thus the answer is $50(101) = 5050$. For a 10-year-old boy, that is good thinking.

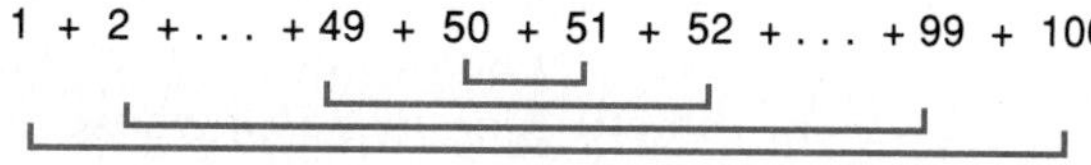

$$1 + 2 + \ldots + 49 + 50 + 51 + 52 + \ldots + 99 + 100$$

Figure 2

Gauss's trick works perfectly well on any arithmetic sequence where we want to add an even number of terms. And there is a slight modification that works whether the number of terms to be added is even or odd.

Suppose $a_1, a_2, a_3, \ldots$ is an arithmetic sequence and let

$$S_n = a_1 + a_2 + a_3 + \cdots + a_{n-1} + a_n$$

Write this sum twice, once forward and once backward, and then add.

$$
\begin{array}{rcccccccc}
S_n & = & a_1 & + & a_2 & + \cdots + & a_{n-1} & + & a_n \\
S_n & = & a_n & + & a_{n-1} & + \cdots + & a_2 & + & a_1 \\
\hline
2S_n & = & (a_1 + a_n) & + & (a_2 + a_{n-1}) & + \cdots + & (a_2 + a_{n-1}) & + & (a_1 + a_n)
\end{array}
$$

Each group on the right has the same sum, namely, $a_1 + a_n$. For example,

$$a_2 + a_{n-1} = a_1 + d + a_n - d = a_1 + a_n$$

There are n such groups and so

$$2S_n = n(a_1 + a_n)$$

> **AS Sum Formula:** $S_n = n\dfrac{a_1 + a_n}{2}$

You can remember this formula by noting that it says that the sum of n terms of an arithmetic sequence is n times the average of the first and last terms to be added.

In the case of Gauss's problem, the sum formula gives

$$S_{100} = 100 \frac{1 + 100}{2} = 100(50.5) = 5050$$

■ **Example 5.** Determine the sum of the first 200 terms of the arithmetic sequence $\{a_n\}$ with $a_1 = -5$ and $d = 3$, that is, for the sequence, $-5, -2, 1, 4, 7, \ldots$.

Solution. First we find the value of a_{200} from the AS explicit formula.

$$a_{200} = -5 + 199(3) = 592$$

Then the AS sum formula gives

$$S_{200} = 200 \frac{-5 + 592}{2} = 100(587) = 58{,}700 \qquad ■$$

ARITHMETIC SEQUENCES AND LINEAR FUNCTIONS

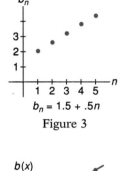

$b_n = 1.5 + .5n$

Figure 3

$b(x) = 1.5 + .5x$

Figure 4

We have said that a sequence is a function whose domain is the set of positive integers. Functions are best visualized by drawing their graphs. Consider the arithmetic sequence $\{b_n\}$ whose explicit formula is given by

$$b_n = 2 + (n - 1)(0.5) = 1.5 + 0.5n$$

Its graph is shown in Figure 3.

Even a cursory look at this graph suggests that the points lie along a straight line. Now consider the function

$$b(x) = 1.5 + 0.5x$$

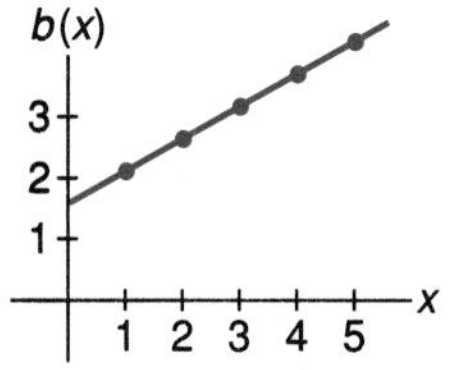

where x is allowed to be any real number. This is a linear function, being of the form $mx + q$ (see Section 2.2). Its graph is the line in Figure 4, and its values at $x = 1, 2, 3, \ldots$ are $b_1, b_2, b_3, \ldots$.

The relationship just illustrated between an arithmetic sequence and a linear function holds in general. An arithmetic sequence is just a linear function whose domain has been restricted to the positive integers.

TEASER SOLUTION

The numbers of logs in each row of the log pile, starting from the top, form the sequence $5, 6, 7, 8, \ldots$. To find the number of logs in the pile is to find the sum of a certain number of terms in an arithmetic sequence. How many terms? You should convince yourself that there are 30 rows, because the bottom row has 34 logs. According to the AS sum formula,

$$S_{30} = 30 \frac{5 + 34}{2} = 15(39) = 585$$

PROBLEM SET 8.1

A. Skills and Techniques

In each case, find the specified term of the sequence, given the indicated information.

1. a_{20}; $a_n = 2n - 5$
2. a_{50}; $a_n = \dfrac{3n + 25}{n}$
3. b_6; $b_n = (2n - 1)^2$
4. b_{30}; $b_n = 4 - 0.5n$
5. a_5; $a_n = a_{n-1} + 3$ and $a_1 = 2$
6. a_5; $a_n = 3a_{n-1}$ and $a_1 = 2$
7. b_6; $b_n = 2b_{n-1}$ and $b_1 = -1$
8. c_5; $c_n = c_{n-1} + 2$ and $c_1 = -1$
9. c_5; $c_n = c_{n-1} + 8(n - 1)$ and $c_1 = 1$
10. b_5; $b_n = b_{n-1} + 3n^2 - 3n$ and $b_1 = 1$

Discover a pattern and then give an explicit formula of the form $a_n = \ldots$ for each of the following sequences. Then determine a_{10}.

11. 1, 8, 27, 64, 125, . . .
12. 2, 4, 8, 16, 32, . . .
13. $\dfrac{1}{2}, \dfrac{2}{3}, \dfrac{3}{4}, \dfrac{4}{5}, \dfrac{5}{6}, \cdots$
14. $-1, 1, -1, 1, -1, \ldots$
15. 1, 4, 7, 10, 13, . . .
16. 5, 9, 13, 17, 21, . . .
17. 12, 7, 2, -3, -8, . . .
18. -24, -19, -14, -9, -4, . . .
19. 0, 3, 8, 15, 24, . . .
20. 2, -4, 8, -16, 32, . . .

Find the sum S_{30} of the first 30 terms of each of the following arithmetic sequences.

21. $a_1 = 6$ and $d = 4$
22. $a_1 = -5$ and $d = 7$
23. 2, 5, 8, 11, . . .
24. 5, 11, 17, 23, . . .
25. 7, 5, 3, 1, . . .
26. 3, -1, -5, -9, . . .

Calculate each of the sums in Problems 27–32.

27. $2 + 4 + 6 + \cdots + 200$
28. $1 + 3 + 5 + \cdots + 299$
29. $3 + 6 + 9 + \cdots + 198$
30. $4 + 8 + 12 + \cdots + 396$
31. $10 + 15 + 20 + \cdots + 200$
32. $9 + 12 + 15 + \cdots + 300$
33. The bottom rung of a tapered ladder (Figure 5) is 30 centimeters long and the top rung is 15 centimeters long. If there are 17 rungs, how many centimeters of rung material are needed to make the ladder, assuming no waste?

Figure 5

34. A clock strikes once at 1:00, twice at 2:00, and so on. How many times does it strike between 10:30 A.M. on Monday and 10:30 P.M. on Tuesday?
35. If 3, a, b, c, d, 7, . . . is an arithmetic sequence, find a, b, c, and d.
36. If 8, a, b, c, 5 is an arithmetic sequence, find a, b, and c.
37. How many multiples of 9 are there between 200 and 300? Find their sum.
38. If Ronnie is paid $10 on January 1, $20 on January 2, $30 on January 3, and so on, how much does he earn during January?

B. Applications and Extensions

39. Let a_n be the nth digit in the decimal expansion of $1/7 = 0.1428 \ldots$. Thus, $a_1 = 1$, $a_2 = 4$, $a_3 = 2$, and so on. Find a pattern and use it to determine a_8, a_{27}, and a_{53}.
40. Suppose that January 1 occurs on a Wednesday. Let a_n be the day of the week corresponding to the nth day of the year. Thus, $a_1 =$ Wednesday, $a_2 =$ Thursday, and so on. Determine a_{39}, a_{57}, and a_{84}.
41. If 15, a, b, c, d, 24 are the first six terms of the arithmetic sequence $\{a_n\}$, find a_{51}.
42. Find the sum of all multiples of 7 between 300 and 450.
43. At a club meeting with 300 people present, everyone shook hands with every other person exactly once. How many handshakes were there? (*Hint:* Person A shook hands with how many people, person B shook hands with how many people not already counted, and so on.)

44. Mary learned 20 new French words on January 1, 24 new French words on January 2, 28 new French words on January 3, and so on, through January 31. By how much did she increase her French vocabulary during January?

45. A pile of logs has 70 logs in the bottom layer, 69 logs in the second layer, and so on to the top layer with 10 logs. How many logs are in the pile?

46. Calculate the following sum.

$$-1^2 + 2^2 - 3^2 + 4^2 - 5^2 + 6^2 - \cdots - 99^2 + 100^2$$

(*Hint:* Group in a clever way.)

47. José will invest $1000 today at 9.5% (annual) simple interest. How much will this investment be worth 10 years from now? (*Note:* Under simple interest, only the original principal of $1000 draws interest.)

48. Roberto plans to invest $1000 today and at the beginning of each of the succeeding 9 years. How much will his total investment be worth 10 years from now if his investments draw 9.5% simple interest?

49. To pay off a loan of $6000, Ikeda agreed to pay at the end of each month interest corresponding to 1% of the unpaid balance of the principal and then $200 to reduce the principal. What was the total of all his payments?

50. Calculate $\ln 2 + \ln 2^2 + \ln 2^3 + \cdots + \ln 2^{100}$.

51. Calculate

$$\frac{1}{2} + \left(\frac{1}{3} + \frac{2}{3}\right) + \left(\frac{1}{4} + \frac{2}{4} + \frac{3}{4}\right) + \cdots$$
$$+ \left(\frac{1}{100} + \frac{2}{100} + \frac{3}{100} + \cdots + \frac{99}{100}\right)$$

52. Show that the sum of n consecutive positive integers plus n^2 is equal to the sum of the next n consecutive integers.

53. Approximately how long is the playing groove in a 33 1/3-rpm record that takes 18 minutes to play if the groove starts 6 inches from the center and ends 3 inches from the center? To approximate, assume that each revolution produces a groove that is circular.

54. The Greeks were enchanted with sequences that arose in a geometric way (Figures 6 and 7).

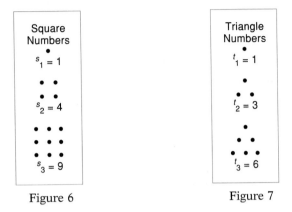

Figure 6   Figure 7

(a) Write explicit formulas for s_n and t_n.
(b) Write recursion formulas for s_n and t_n.

55. The numbers $1, 5, 12, 22, \ldots$ are called **pentagonal numbers.** See if you can figure out why and then guess at an explicit formula for p_n, the nth pentagonal number. Use diagrams.

56. **Challenge.** Find the sum of all the digits in the integers from 1 to 999,999.

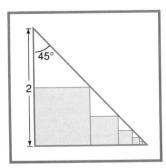

TEASER   Determine the total area of the infinite sequence of colored squares shown in the figure at the left.

Consider the following number sequences. When you see a pattern, use it to fill in the boxes.

(a)   3, 6, 12, 24, $\square$, $\square$, . . .
(b)   12, 4, 4/3, 4/9, $\square$, $\square$, . . .
(c)   0.2, 0.6, 1.8, 5.4, $\square$, $\square$, . . .

What is it that these three sequences have in common? In each case, multiplication by a fixed constant leads from one term to the next. In sequence (a), you multiply by 2, in (b) by 1/3, and in (c) by 3. We call such a sequence a **geometric sequence** (GS for short). Thus, if $a_1, a_2, a_3, \ldots$ is a geometric sequence, then its nth term a_n is related to its $(n - 1)$th term a_{n-1} as follows.

> **GS Recursion Formula:** $a_n = ra_{n-1}$

Here r is a fixed number called the **common ratio.**

■ **Example 1.**   Find the sixth term of the geometric sequence $\{a_n\}$, given that $a_1 = 27$ and $a_n = (2/3)a_{n-1}$.

Solution. $a_1 = 27, a_2 = (2/3)(27) = 18, a_3 = (2/3)(18) = 12, a_4 = (2/3)(12) = 8, a_5 = (2/3)(8) = 16/3, a_6 = (2/3)(16/3) = 32/9$. ■

Clearly, it would be advantageous to have an explicit formula for the nth term, and it is easy to find. Note that

$$a_2 = ra_1$$
$$a_3 = ra_2 = r(ra_1) = r^2a_1$$
$$a_4 = ra_3 = r(r^2a_1) = r^3a_1$$

In each case, the exponent on r is one less than the subscript on a. Our conclusion is as follows.

> **GS Explicit Formula:** $a_n = r^{n-1}a_1$

Thus, an explicit formula for the sequence of Example 1 is

$$a_n = \left(\frac{2}{3}\right)^{n-1}(27) = \frac{2^{n-1}}{3^{n-1}}3^3 = \frac{2^{n-1}}{3^{n-4}}$$

Note that for $n = 6$, this formula gives

$$a_6 = \frac{2^5}{3^2} = \frac{32}{9}$$

as expected.

A CONNECTION

The connecting link between geometric and arithmetic sequences is the logarithm function. Let

$$a_n = a_1r^{n-1}$$

determine a geometric sequence with a_1 and r positive. Then

$$b_n = \ln a_n$$
$$= \ln a_1 + (n - 1)\ln r$$

determines an arithmetic sequence with common difference

$$d = \ln r$$

■ **Example 2.** Find the 1000th term in the geometric sequence $\{b_n\}$ with first term 10 and common ratio 1.01.

Solution. $b_{1000} = (1.01)^{999}(10) \approx 207,516.3925.$ ■

■ **Example 3.** If Shaka invests $500 today at 6% interest compounded monthly, how much will his investment be worth at the end of 10 years (that is, at the end of 120 months)?

Solution. We studied this type of problem in Section 3.5. An annual rate of 6% corresponds to a rate of $1/2\% = 0.005$ per month. Thus, the values of the investment at the ends of months 1, 2, 3, ... form the geometric sequence $500(1.005)$, $500(1.005)^2$, $500(1.005)^3$, ... At the end of 120 months, the value of the investment will be $500(1.005)^{120} = \$909.70.$ ■

GEOMETRIC SEQUENCES AND EXPONENTIAL FUNCTIONS

Consider the geometric sequence $\{b_n\}$ with explicit formula

$$b_n = 36\left(\frac{1}{3}\right)^n$$

and the exponential function

$$b(x) = 36\left(\frac{1}{3}\right)^x$$

Their graphs are shown in Figures 1 and 2. It should be clear that the values of the exponential function $b(x)$ at the integers 1, 2, 3, are $b_1, b_2, b_3, \ldots$.

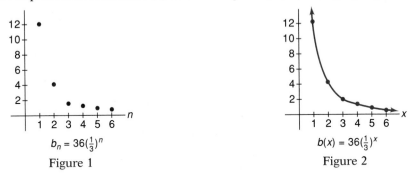

$$b_n = 36\left(\tfrac{1}{3}\right)^n$$

Figure 1

$$b(x) = 36\left(\tfrac{1}{3}\right)^x$$

Figure 2

What is true in this example is true in general. A geometric sequence is just an exponential function with its domain restricted to the positive integers.

SUMS OF GEOMETRIC SEQUENCES

There is an old legend about geometric sequences and chessboards. When the king of Persia learned to play chess, he was so enchanted with the game that he decided to reward the inventor, a man named Sessa. Calling Sessa to the palace, the king promised to fulfill any request Sessa might make. With an air of modesty, the wily Sessa asked for one grain of wheat for the first square of the chessboard, two for the second, four for the third, and so on (Figure 3).

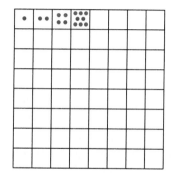

Figure 3

The king was amused at such an odd request; nevertheless, he called a servant and told him to get a bag of wheat and start counting. To the king's surprise, it soon became apparent that Sessa's request could never be fulfilled. The world's total production of wheat for a whole century would not be sufficient.

Sessa was asking for

$$1 + 2 + 2^2 + 2^3 + \cdots + 2^{63}$$

grains of wheat, the sum of the first 64 terms of the geometric sequence 1, 2, 4, 8, ... We are going to develop a formula for this sum and all others that arise from adding the terms of a geometric sequence.

Let $a_1, a_2, a_3, \ldots$ be a geometric sequence with ratio $r \neq 1$. As usual, let

$$S_n = a_1 + a_2 + a_3 + \cdots + a_n$$

which can be written

$$S_n = a_1 + a_1 r + a_1 r^2 + \cdots + a_1 r^{n-1}$$

Now multiply S_n by r, subtract the result from S_n, and use a little algebra to solve for S_n. We obtain

$$
\begin{aligned}
S_n &= a_1 + a_1 r + a_1 r^2 + \cdots + a_1 r^{n-1} \\
r S_n &= \quad\quad a_1 r + a_1 r^2 + \cdots + a_1 r^{n-1} + a_1 r^n \\
\hline
S_n - r S_n &= a_1 + 0 + 0 + \cdots + 0 - a_1 r^n \\
S_n(1 - r) &= a_1(1 - r^n)
\end{aligned}
$$

This derivation implies the following sum formula.

> **GS Sum Formula:** $\quad S_n = \dfrac{a_1(1 - r^n)}{1 - r} = \dfrac{a_1(r^n - 1)}{r - 1}, \quad r \neq 1$

Applying this formula to Sessa's problem (with $n = 64$, $a_1 = 1$, and $r = 2$) gives

$$S_{64} = \frac{1(2^{64} - 1)}{2 - 1} = 2^{64} - 1 \approx 1.84 \times 10^{19}$$

Thus, if a bushel of wheat contains 1 million grains, fulfilling Sessa's request would require more than 1.8×10^{13}, or 18 trillion, bushels of wheat.

■ **Example 4.** Find the sum of the first 20 terms of the geometric sequence 36, 12, 4, 4/3, ...

Solution. We apply the GS sum formula with $a_1 = 36$, $r = 1/3$, and $n = 20$.

$$S_{20} = \frac{36(1 - (1/3)^{20})}{1 - 1/3} = 54\left(1 - \left(\frac{1}{3}\right)^{20}\right) \approx 53.99999998 \quad\blacksquare$$

■ **Example 5.** Suppose Shaka of Example 3 were able to make payments of $500 at the end of each month for the next 120 months with interest at 6% compounded monthly. What would the total value of these payments be at the end of 120 months?

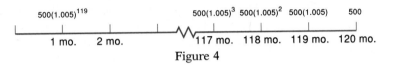

Figure 4

Solution. Refer to the time line in Figure 4. The last payment earns no interest; it will be worth $500. The next to the last payment earns interest for 1 month; it will be worth $500(1.005). The second to the last payment earns interest for 2 months; it will be worth $500(1.005)^2$. The very first payment (made at the end of 1 month) earns interest for 119 months and will be worth $500(1.005)^{119}$. To obtain the required answer, we must add the values of these 120 payments; that is, we must find the sum of a geometric sequence with $a_1 = 500$, $n = 120$, and $r = 1.005$. Our sum formula gives

$$S_{120} = \frac{500(1.005^{120} - 1)}{1.005 - 1} = \$81,939.67 \qquad \blacksquare$$

A series of equal payments made at equal intervals is called an **annuity.** Thus, Example 5 asked us to find the terminal value of an annuity of 120 monthly payments of $500 (with the first payment 1 month from now), assuming interest is at a rate of 6% compounded monthly.

THE SUM OF THE WHOLE SEQUENCE

We have developed formulas for adding up a finite number of terms of an arithmetic or geometric sequence. Can we make sense of the idea of adding up *infinitely* many terms of a sequence? For example, does the symbol

$$1 + 2 + 3 + 4 + \cdots$$

make any sense? (Here the three dots indicate that we are to continue adding indefinitely.) The answer is no, for the sum just keeps on growing without bound as we add more and more terms. But what about the symbol

$$\frac{1}{2} + \frac{1}{4} + \frac{1}{8} + \frac{1}{16} + \cdots$$

We will show that this symbol makes perfectly good sense.

To help us think about this second infinite addition, consider a string of length 1 meter. We may imagine cutting it into infinitely many pieces, as indicated in Figure 5. The pieces together make a string of length 1, so it seems natural to say that the infinite sum is 1.

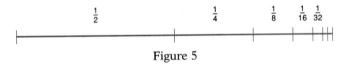

Figure 5

Let us look at the problem another way. The sum of the first n terms of the geometric sequence 1/2, 1/4, 1/8, . . . is

$$S_n = \frac{1/2(1 - (1/2)^n)}{1 - 1/2} = 1 - \left(\frac{1}{2}\right)^n$$

As n gets larger and larger (tends to infinity), $(1/2)^n$ gets smaller and smaller (approaches 0). Thus, S_n tends to 1 as n tends to infinity. We therefore say that 1 is the sum of all the terms of this sequence.

In the same manner, consider any geometric sequence with ratio r satisfying $|r| < 1$. We claim that as n grows larger and larger, r^n approaches 0. (As evidence, try calculating $(0.99)^{100}$, $(0.99)^{1000}$, and $(0.99)^{10000}$ on your calculator.) Thus, as n tends to infinity,

$$S_n = \frac{a_1(1 - r^n)}{1 - r}$$

approaches the value $a_1/(1 - r)$. We write

> **Geometric Series Formula:** $S_\infty = \dfrac{a_1}{1 - r}, \quad |r| < 1$

We emphasize that this formula is valid only for geometric sequences with $|r| < 1$. The subject of sums of infinite sequences goes under the name **infinite series** and is a major topic in calculus.

■ **Example 6.** Find the sum S_∞ of the infinite geometric series (see Example 4)

$$36 + 12 + 4 + \frac{4}{3} + \frac{4}{9} + \cdots$$

Solution. Since $r = 1/3$ and $a_1 = 36$, we obtain

$$S_\infty = \frac{36}{1 - 1/3} = 54$$

■

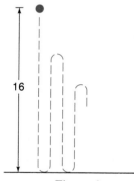

16

Figure 6

■ **Example 7.** A ball is dropped from a height of 16 feet. At each bounce it rises to a height of three-fourths the previous height (Figure 6). How far will it have traveled altogether (up and down) by the time it comes to rest?

Solution. The total distance L traveled is the sum of the "down" distances $(16 + 12 + 9 + \ldots)$ and the "up" distances $(12 + 9 + 27/4 + \ldots)$. Thus,

$$L = \frac{16}{1 - 3/4} + \frac{12}{1 - 3/4} = 64 + 48 = 112 \text{ feet}$$

■

The largest square has sides of length 1 giving it area 1; the next square has sides of length 1/2 giving it area 1/4; and so on. The sum of the areas of all the squares is

$$S_\infty = 1 + \frac{1}{4} + \frac{1}{16} + \frac{1}{64} + \cdots = \frac{1}{1 - 1/4} = \frac{4}{3}$$

If you are geometrically adept, you may notice that we can solve this problem without infinite series. Each colored square covers two-thirds of the vertical strip in which it is located, so it follows that the colored squares cover two-thirds of the area of the whole triangle; that is, they cover an area of $(2/3)(2) = 4/3$.

PROBLEM SET 8.2

A. Skills and Techniques

In each case, find the specified term of the given geometric sequence.

1. $a_{10}; a_n = 5 \cdot 2^n$

2. $b_6; b_n = 9\left(\frac{2}{3}\right)^n$

3. $c_8; c_n = \frac{2}{3}c_{n-1}$ and $c_1 = 27$

4. $a_5; a_n = 1.5a_{n-1}$ and $a_1 = 5$

5. $a_{20}; 1.02, 1.02^2, 1.02^3, \ldots$

6. $a_8; 64, 48, 36, \ldots$

Evaluate the sum S_{20} of the first 20 terms of each geometric sequence $\{a_n\}$.

7. $27, 18, 12, \ldots$

8. $64, 16, 4, \ldots$

9. $a_n = 3a_{n-1}$ and $a_1 = 2$

10. $a_n = \frac{2}{3}a_{n-1}$ and $a_1 = 6$

11. $a_n = e^n$

12. $a_n = 2\left(\frac{3}{4}\right)^n$

13. $a_n = 50(1.01)^{12n}$

14. $a_n = 200(1.005)^{12n}$

Evaluate the sum of each geometric series.

15. $3 + 2 + \frac{4}{3} + \frac{8}{9} + \cdots$

16. $16 + 12 + 9 + \frac{27}{4} + \cdots$

17. $100 + 80 + 64 + \frac{256}{5} + \cdots$

18. $16 - 8 + 4 - 2 + \cdots$

19. $200 + 200(1.01)^{-1} + 200(1.01)^{-2} + \cdots$

20. $0.4 + 0.04 + 0.004 + 0.0004 + \cdots$

21. $0.23 + 0.0023 + 0.000023 + \cdots$

22. $6.12 + 0.0012 + 0.000012 + \cdots$

A bar over a group of digits in a decimal indicates that this group of digits repeats indefinitely. For example, $0.\overline{32} = 0.32323232\ldots = 0.32 + 0.0032 + \ldots$. Use the formula for finding the sum of geometric series to write the decimals in Problems 23–28 as ratios of two integers.

23. $0.\overline{32}$

24. $0.\overline{54}$

25. $0.\overline{123}$

26. $2.\overline{88}$

27. $3.1\overline{45}$

28. $4.\overline{9}$

29. A certain culture of bacteria doubles every week. If there are 100 bacteria now, how many will there be after 10 full weeks?

30. A water lily grows so rapidly that each day it covers twice the area it covered the day before. At the end of 20 days, it completely covers a certain pond. If we start with two lilies, how long will it take to cover that same pond?

31. Johnny is paid $1 on January 1, $2 on January 2, $4 on January 3, and so on. Approximately how much will he earn during January?

32. If you were offered 1¢ today, 2¢ tomorrow, 4¢ the third day, and so on for 20 days or a lump sum of $10,000, which would you choose? Show why.

33. If $1 is put in the bank at 8% interest compounded annually, it will be worth $(1.08)^n$ dollars after n years. How much will $100 be worth after 10 years? When will the amount first exceed $250?

34. If $1 is put in the bank at 8% interest compounded quarterly, it will be worth $(1.02)^n$ dollars after n quarters. How much will $100 be worth after 10 years (40 quarters)? When will the amount first exceed $250?

35. Suppose Karen puts $100 in the bank today and $100 at the beginning of each of the following 9 years. If this money earns interest at 8% compounded annually, what will it be worth at the end of 10 years?

36. Ayanna makes 40 deposits of $25 each in a bank at intervals of three months, making the first deposit today. If money earns interest at 8% compounded quarterly, what will it all be worth at the end of 10 years (40 quarters)?

37. A ball is dropped from a height of 10 feet. At each bounce, it rises to a height of 1/2 the previous height. How far will it travel altogether (up and down) by the time it comes to rest?

38. Do Problem 37 assuming the ball rises to 2/3 its previous height at each bounce.

B. Applications and Extensions

39. Which of the following sequences are geometric, which are arithmetic, and which are neither?
(a) 130, 65, 32.5, 16.25, . . .
(b) $1, \dfrac{1}{2}, \dfrac{1}{3}, \dfrac{1}{4}, \ldots$
(c) $100(1.05), 100(1.07), 100(1.09), 100(1.11), \ldots$
(d) $100(1.05), 100(1.05)^2, 100(1.05)^3, \ldots$
(e) 1, 3, 6, 10, . . .
(f) $3, -6, 12, -24, \ldots$

40. Write an explicit formula for each sequence in Problem 39.

41. Suppose that the government pumps an extra $1 billion into the economy. Assume that each business and individual saves 25% of its income and spends the rest, so that of the initial $1 billion, 75% is respent by individuals and businesses. Of that amount, 75% is spent, and so on. What is the total increase in spending due to the government's action? (This is called the **multiplier effect** in economics.)

42. Given an arbitrary triangle of perimeter 10, a second triangle is formed by joining the midpoints of the first, a third triangle is formed by joining the midpoints of the second, and so on forever. Find the total length of all line segments in the resulting configuration.

43. Find the area of the painted region in Figure 7, which consists of an infinite sequence of 30°-60°-90° triangles.

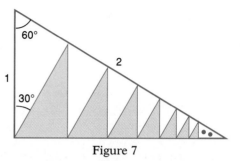

Figure 7

44. If the pattern in Figure 8 is continued indefinitely, what fraction of the area of the original square will be painted?

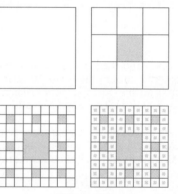

Figure 8

45. Each triangle in the descending chain (Figure 9) has its vertices at the midpoints of the sides of the next larger one. If the indicated pattern of painting is continued indefinitely, what fraction of the area of the original triangle will be painted? Does the original triangle need to be equilateral for this to be true?

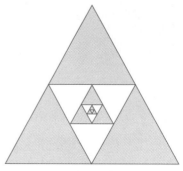

Figure 9

393 • 8.2 Geometric Sequences and Sums

46. Circles are inscribed in the triangles of Problem 45 as indicated in Figure 10. If the original triangle is equilateral, what fraction of the area is painted?

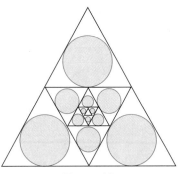

Figure 10

47. By considering $S - rS$, find S.
(a) $S = r + 2r^2 + 3r^3 + 4r^4 + \cdots, |r| < 1.$

(b) $S = \dfrac{1}{3} + 2\left(\dfrac{1}{3}\right)^2 + 3\left(\dfrac{1}{3}\right)^3 + 4\left(\dfrac{1}{3}\right)^4 + \cdots$

48. In a geometric sequence, the sum of the first two terms is 5 and the sum of the first six terms is 65. What is the sum of the first four terms? (*Hint:* Call the first term a and the ratio r. Then $a + ar = 5$, so $a = 5/(1 + r)$. This approach allows you to write 65 in terms of r alone. Solve for r.)

49. In one version of Zeno's paradox, Achilles can run 10 times as fast as the tortoise, but the tortoise has a 100-yard head start. Achilles cannot catch the tortoise, says Zeno, because when he runs 100 yards, the tortoise will have moved 10 yards ahead; when Achilles runs another 10 yards, the tortoise will have moved 1 yard ahead, and so on. Convince Zeno that Achilles will catch the tortoise and tell him exactly how many yards Achilles will have to run to do so.

50. Imagine a huge maze with infinitely many adjoining cells, each having a square base 1 meter by 1 meter. The first cell has walls 1 meter high, the second cell has walls 1/2 meter high, the third cell has walls 1/4 meter high, and so on.
(a) How much paint would it take to fill the maze with paint?
(b) How much paint would it take to paint the floors of all the cells?
(c) Explain this apparent contradiction.

51. Starting 100 miles apart, Tom and Joel ride toward each other on their bicycles, Tom going at 8 miles per hour and Joel at 12 miles per hour. Tom's dog, Corky, starts with Tom running toward Joel at 25 miles per hour. When Corky meets Joel, he immediately turns tail and heads back to Tom. Reaching Tom, Corky again turns tail and heads toward Joel, and so on. How far did Corky run by the time Tom and Joel met? This can be answered using geometric sequences, but if you are clever, you will find a better way.

52. **Challenge.** Sally walked 4 miles north, then 2 miles east, 1 mile south, 1/2 mile west, 1/4 mile north, and so on. If she continued this pattern indefinitely, how far from her initial point did she end?

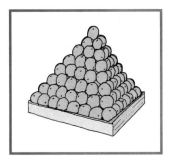

TEASER Oranges are often displayed in fruit markets in large pyramidal piles like the one shown at the left. Determine the number of oranges in a pile that has 40 tiers. (*Hint:* The top tier has one orange, the second tier has four oranges, and the third tier has nine oranges.)

8.3 GENERAL SEQUENCES AND PROGRAMMING

So far in this chapter, our attention has focused on two very special kinds of number sequences: arithmetic sequences and geometric sequences. Here we expand the discussion to arbitrary sequences. Consider the following sequences. When you see a pattern, fill in the boxes.

(a) $\dfrac{1}{3}, \dfrac{2}{4}, \dfrac{3}{5}, \dfrac{4}{6}, \dfrac{5}{7}, \Box, \Box, \ldots$

(b) $1, 3, 6, 10, 15, \Box, \Box, \ldots$

Clearly, these sequences are neither arithmetic nor geometric. Yet many students will immediately see that sequence (a) can be specified by the explicit formula $a_n = n/(n + 2)$. Similarly, sequence (b) is determined by giving $b_1 = 1$ together with the recursion formula $b_n = b_{n-1} + n$. Someone may also notice that sequence (b) can be specified by the explicit formula $b_n = n(n + 1)/2$.

A famous sequence, first discussed by the mathematician Leonardo Fibonacci (ca. 1170–1250) as a model for rabbit reproduction, is the subject of our next example.

■ **Example 1.** Find the pattern in the sequence 1, 1, 2, 3, 5, 8, 13, . . . and then give a formula that determines this sequence.

Solution. You will have a hard time giving an explicit formula for this sequence, although such a formula does exist (see Example 2 and Problem 44, Section 8.4). You may notice, however, that any term after the second is the sum of the previous two terms. Thus the **Fibonacci sequence** $\{f_n\}$ can be specified by giving the first two terms $f_1 = f_2 = 1$ together with the two-term recursion formula $f_n = f_{n-1} + f_{n-2}$. ■

Consider another famous sequence, the sequence of prime numbers

$$2, 3, 5, 7, 11, 13, 17, 19, \ldots$$

Although everyone sees the pattern in this sequence, no one knows how to give either an explicit or a recursion formula for it.

■ **Example 2.** Determine the first six terms of each of the following sequences.

(a) $a_n = \dfrac{1}{\sqrt{5}}\left(\left(\dfrac{1 + \sqrt{5}}{2}\right)^n - \left(\dfrac{1 - \sqrt{5}}{2}\right)^n\right)$

(b) $b_1 = 0, b_2 = 1$, and $b_n = \dfrac{(b_{n-2} + b_{n-1})}{2}, n > 2$

Solution. (a) Use your calculator on this sequence; you will discover that $a_1 = a_2 = 1$, $a_3 = 2$, $a_4 = 3$, $a_5 = 5$, and $a_6 = 8$. If you have been paying attention, you may note something interesting about these numbers. More about that later.

(b) $b_1 = 0, b_2 = 1, b_3 = \dfrac{0 + 1}{2} = \dfrac{1}{2}, b_4 = \dfrac{1 + 1/2}{2} = \dfrac{3}{4}, b_5 = \dfrac{1/2 + 3/4}{2} = \dfrac{5}{8}$, and

$b_6 = \dfrac{3/4 + 5/8}{2} = \dfrac{11}{16}$ ■

SIGMA NOTATION

Corresponding to the sequence $a_1, a_2, a_3, \ldots$ is the partial sum

$$S_n = a_1 + a_2 + a_3 + \cdots + a_n$$

We gave formulas for S_n in the case of arithmetic and geometric sequences, but it is not usually possible to do so for other sequences. The concept of adding up the terms of a sequence, however, is so important that mathematicians have introduced a special symbol for it. Using the Greek letter Σ (upper case *sigma*), we make the following definition.

$$\sum_{k=m}^{n} a_k = a_m + a_{m+1} + a_{m+2} + \cdots + a_n$$

Thus,

$$\sum_{k=3}^{6} a_k = a_3 + a_4 + a_5 + a_6$$

$$\sum_{k=1}^{100} k^2 = 1^2 + 2^2 + 3^2 + \cdots + 100^2$$

and

$$\sum_{k=1}^{\infty} \left(\frac{1}{3}\right)^k = \frac{1}{3} + \frac{1}{9} + \frac{1}{27} + \frac{1}{81} + \cdots = \frac{1/3}{1 - 1/3} = \frac{1}{2}$$

We used the GS series formula to evaluate the last sum.

■ **Example 3.** Determine the meaning of each of the following, and, if possible, calculate their values.

(a) $\displaystyle\sum_{k=1}^{31} \cos k\pi$

(b) $\displaystyle\sum_{k=1}^{50} (2k - 1)$

(c) $\displaystyle\sum_{k=1}^{8} 3^k$

(d) $\displaystyle\sum_{k=1}^{100} \left(\frac{1}{k}\right)$

Solution.

(a) $\displaystyle\sum_{k=1}^{31} \cos k\pi = \cos \pi + \cos 2\pi + \cos 3\pi + \cdots + \cos 31\pi$

$\qquad\qquad = -1 + 1 - 1 + \cdots - 1 = -1$

(b) $\displaystyle\sum_{k=1}^{50} (2k - 1) = 1 + 3 + 5 + \cdots + 99 = 50\left(\frac{1 + 99}{2}\right) = 2500$

(c) $\displaystyle\sum_{k=1}^{8} 3^k = 3 + 3^2 + 3^3 + \cdots + 3^8 = \frac{3(3^8 - 1)}{3 - 1} = 9840$

(d) $\displaystyle\sum_{k=1}^{100} \left(\frac{1}{k}\right) = 1 + \frac{1}{2} + \frac{1}{3} + \frac{1}{4} + \cdots + \frac{1}{100}$

We used the sum formulas for arithmetic and geometric sequences in (b) and (c). We know of no formula for the sum in (d), though if we had enough patience we could determine the sum of these 100 terms by hand or with a calculator. ■

Consider first the case where a sequence $\{a_n\}$ is given by an explicit formula. As in Example 2, suppose that

$$a_n = \frac{1}{\sqrt{5}}\left(\left(\frac{1+\sqrt{5}}{2}\right)^n - \left(\frac{1-\sqrt{5}}{2}\right)^n\right)$$

If we need just one term in this sequence, say a_{10}, we simply calculate

$$a_{10} = \frac{1}{\sqrt{5}}\left(\left(\frac{1+\sqrt{5}}{2}\right)^{10} - \left(\frac{1-\sqrt{5}}{2}\right)^{10}\right) = 55$$

by pressing a few keys on a calculator. If we anticipate needing many values of this sequence, however, it makes sense to program a calculator (or computer) so that it will produce any desired value of a_n when we simply enter the value of n. Here is such a program for the TI-81 or TI-82, to which we have given the name "AN." The comments explain or document the program.

Prgm AN	The program title
:Disp "N"	
:Input N	Enter desired n and store in N
:(1/√5)(((1+√5)/	Evaluate a_n and store the
2)∧N−((1−√5)/2)∧	result in A
N)→A	
:Disp "A(N)="	
:Disp A	Display a_n

■ **Example 4.**   Use the program above (or a similar one for your calculator) to obtain the values of a_5, a_{10}, a_{20}, and a_{30}.

Solution. We obtain 5, 55, 6765, and 832,040 for these four values.   ■

■ **Example 5.**   Modify the program above, by changing just one line, so that it will calculate the value of any term in the sequence $\{a_n\}$ where

$$a_n = \left(1 + \frac{1}{n}\right)^n$$

Then use this program to calculate a_{10}, a_{100}, a_{1000}, a_{10000}, a_{100000}, and $a_{1000000}$.

Solution. We change the third line to read

$$:(1+1/N)\wedge N \to A$$

The six required values are

$$2.59374246,\ 2.704813829,\ 2.716923932$$
$$2.718145927,\ 2.718268237,\ 2.718280469$$

If you conjecture that these values are converging toward the value of the constant e, you are right.   ■

What if a sequence $\{a_n\}$ is given by a recursion formula instead of an explicit formula? Is it still possible to program a calculator to give us the value of any term a_n by simply entering the value of n? Yes. Consider the sequence $\{a_n\}$, where $a_1 = 6$ and $a_n = (1/2)a_{n-1} + n$ for $n > 1$. Here is a program for the TI-81 or TI-82 that will do the job.

Prgm RECUR1	The program title
:Disp "A(1)"	
:Input A	Enter a_1 and store in A
:Disp "N"	
:Input N	Enter desired n and store in N
:1→K	Store 1 in K
:Lbl 1	
:K+1→K	Add 1 to K
:0.5A+K→A	Evaluate a_k and store in A
:If (K<N)	If $K < N$ go back to Lbl 1; otherwise continue
:Goto 1	
:Disp "A(N) ="	
:Disp A	Display a_n

Note that the program calculates a_k for every value of $k \le n$, just as you would in doing a problem by hand. You should check that this program gives the correct values $a_{10} = 18.01171875$ and $a_{20} = 38.00001144$.

■ **Example 6.** Modify the program above so that it will calculate any term in the Fibonacci sequence $\{a_n\}$, defined by $a_1 = a_2 = 1$ and $a_n = a_{n-1} + a_{n-2}$ for $n > 2$. Use this program to calculate a_5, a_{10}, a_{20}, and a_{30}.

Solution. This sequence has a two-term recursion, so we will have to change a lot more than one line. Here is the complete program.

Prgm RECUR2	
:Disp "A(1)"	
:Input C	Enter a_1 and store in C
:Disp "A(2)"	
:Input B	Enter a_2 and store in B
:Disp "N"	
:Input N	Enter desired n and store in N
:2→K	
:Lbl 1	
:K+1→K	
:B+C→A	Evaluate a_k and store in A
:B→C	Store new a_{k-2} in C
:A→B	Store new a_{k-1} in B
:If (K<N)	
:Goto 1	
:Disp "A(N)="	
:Disp A	Display a_n

Using RECUR2, we find $a_5 = 5$, $a_{10} = 55$, $a_{20} = 6765$, and $a_{30} = 832{,}040$. Note that these are the same values we got in Example 4. ∎

GENERATING SUMS ON A PROGRAMMABLE CALCULATOR

Consider next the problem of calculating $S_n = \sum_{k=1}^{n} a_k$ for any given value of n, where $\{a_k\}$ is a given sequence. Here is a program that will make this calculation if a_k is given by an explicit formula, such as $a_k = k^3 - 2k$.

Prgm SUMAN
:Disp "N"
:Input N   Enter desired n and store in N
:0→S
:0→K
:Lbl 1
:K+1→K
:K∧3−2K→A   Evaluate a_k and store in A
:S+A→S   Evaluate new sum, store in S
:If (K<N)
:Goto 1
:Disp "S(N) ="
:Disp S   Display S_n

Check that this program gives the value 215,295 when $n = 30$.

TI-82 HINT

On the TI-82, you can automatically sum up to 99 terms of a sequence if the sequence is given by an explicit formula. Open LIST, select **MATH** and then **sum.** Open LIST again and select **seq(.** Then to sum x^4 from $x = 3$ to $x = 51$, enter

$$x^4, x, 3, 51, 1$$

which should return the value 72431849.

■ **Example 7.** Modify the above program so it will calculate the value of

$$S_n = \sum_{k=1}^{n} \frac{1}{k} = 1 + \frac{1}{2} + \frac{1}{3} + \frac{1}{4} + \cdots + \frac{1}{n}$$

for any desired n. Then evaluate this sum when $n = 100$ (see Example 3).

Solution. We need to change only the line that evaluates a_k to read

$$:1/K→A$$

Our calculator gives $S_{100} = 5.187377518$. ∎

Suppose next that we want to calculate S_n but that the sequence $\{a_k\}$ is given by a two-term recursion formula. In particular, suppose $\{a_k\}$ is the Fibonacci sequence determined by $a_1 = a_2 = 1$ and $a_k = a_{k-1} + a_{k-2}$ for $k > 2$. The following program will calculate $S_n = \sum_{k=1}^{n} a_k$. It is a slight modification of the program in Example 6.

Prgm SUMREC2
:Disp "A(1)"
:Input C
:Disp "A(2)"
:Input B
:Disp "N"

```
:Input N
:B+C→S                    Store $a_1 + a_2$ in S
:2→K
:Lbl 1
:K+1→K
:B+C→A
:B→C
:A→B
:S+A→S                    Evaluate new sum, store in S
:If (K<N)
:Goto 1
:Disp "S(N) ="
:Disp S                   Display $S_n$
```

■ **Example 8.** Use the above program to calculate the sum of the first 40 Fibonacci numbers.

Solution. The answer is 267,914,295.   ■

TEASER SOLUTION

The problem of the pyramid of oranges is to find the sum

$$1^2 + 2^2 + 3^2 + 4^2 + \cdots + 40^2$$

That is, it is to evaluate $\sum_{k=1}^{n} k^2$ when $n = 40$. We simply modify the program described before Example 7 by writing $a_k = k^2$ and ask for the value when $n = 40$. We get 22,140.

PROBLEM SET 8.3

A. Skills and Techniques

For each sequence, determine a formula (explicit or recursion) based on the pattern in the indicated terms.

1. 0, 2, 2, 4, 6, 10, 16, . . .
2. 1, 3, 4, 7, 11, 18, 29, . . .
3. 0, 1, 3, 6, 10, 15, 21, . . .
4. 1, 1, 4, 10, 28, 76, . . .
5. 0, 7, 26, 63, 124, . . .
6. 2, 5, 10, 17, 26, . . .
7. 1, 1, 1, 3, 5, 9, 17, . . .
8. 1, 2, −1, 3, −4, 7, −11, . . .

Determine a_6 in each case.

9. $a_n = \dfrac{n-1}{n+1}$

10. $a_n = n^3 - 6n^2 + 11n - 6$

11. $a_n = (n-1)a_{n-1}$ given $a_1 = 2$

12. $a_n = \dfrac{a_{n-2}}{a_{n-1}}$ given $a_1 = 1$ and $a_2 = 2$

13. $a_n = \dfrac{2a_{n-1}}{a_{n-2}}$ given $a_1 = 1$ and $a_2 = 2$

14. $a_n = \sin\left(\dfrac{n\pi}{2}\right)$

15. $a_n = 2^{n-1}a_{n-1}$ given $a_1 = 1$

16. $a_n = (n-2)a_{n-2} + (n-1)a_{n-1}$ given $a_1 = a_2 = 1$

Calculate each sum without using a program.

17. $\displaystyle\sum_{k=1}^{5} 2^k$

18. $\displaystyle\sum_{k=1}^{100} (3+k)$

19. $\displaystyle\sum_{k=1}^{100} (2-5k)$

20. $\displaystyle\sum_{k=1}^{20} e^k$

21. $\displaystyle\sum_{k=1}^{21} \sin\left(\frac{k\pi}{2}\right)$

22. $\displaystyle\sum_{k=1}^{500} \cos\left(\left(2k + \frac{1}{4}\right)\pi\right)$

23. $\displaystyle\sum_{k=1}^{15} (-1)^k$

24. $\displaystyle\sum_{k=1}^{8} (\sqrt{3})^{k-1}$

Write a program (or modify an existing one) to calculate any term of the following sequence. Check its correctness by calculating a_1 and a_2 by hand and with your program. Then calculate the indicated terms.

25. $a_n = n^3 - 5n^2$; a_{75} and a_{100}

26. $a_n = \sqrt{n^2 - n + 1}$; a_{20} and a_{50}

27. $a_n = \dfrac{1}{\sqrt{3}}\left(\left(\dfrac{1 + \sqrt{3}}{2}\right)^n - \left(\dfrac{1 - \sqrt{3}}{2}\right)^n\right)$; a_8 and a_{16}

28. $a_n = \dfrac{1}{\sqrt{11}}\left(\left(\dfrac{1 + \sqrt{11}}{2}\right)^n - \left(\dfrac{1 - \sqrt{11}}{2}\right)^n\right)$; a_8 and a_{16}

29. $a_n = \dfrac{n}{n + 1/n}$; a_{100} and a_{1000}

30. $a_n = \dfrac{2^n}{n^2}$; a_{16} and a_{32}

Each of the following sequences is determined by a recursion formula. Write a program (or modify an existing one) to calculate a_n. After checking it with a couple of hand calculations, use it to calculate a_{20} and a_{100}.

31. $a_n = 2a_{n-1}$ and $a_1 = 1$

32. $a_n = \dfrac{n}{a_{n-1}}$ and $a_1 = 2$

33. $a_n = \dfrac{a_{n-1}}{\ln(a_{n-1})}$ and $a_1 = 100{,}000$

34. $a_n = 2 - 2a_{n-1}$ and $a_1 = 0$
35. $a_n = 3a_{n-1} - 2a_{n-2}$ with $a_1 = 2$ and $a_2 = 3$
36. $a_n = 3a_{n-1} - 2a_{n-2}$ with $a_1 = -1$ and $a_2 = -2$

37. $a_n = \dfrac{(a_{n-1} + a_{n-2})}{2}$ with $a_1 = 1$ and $a_2 = 5$

38. $a_n = \dfrac{(a_{n-1} + a_{n-2})}{1.9}$ with $a_1 = 1$ and $a_2 = 5$

Write a program (or modify an existing one) to calculate each sum for the indicated values of n. Begin by checking it on small values of n.

39. $\displaystyle\sum_{k=1}^{n} \frac{(k^2 + k)}{4}$; $n = 10$ and $n = 100$

40. $\displaystyle\sum_{k=1}^{n} k(k + 1)(k + 2)$; $n = 10$ and $n = 25$

41. $\displaystyle\sum_{k=1}^{n} \sqrt{k\sqrt{k}}$; $n = 18$ and $n = 55$

42. $\displaystyle\sum_{k=1}^{n} \sin\left(\frac{k\pi}{10}\right)$; $n = 15$ and $n = 20$

43. $\displaystyle\sum_{k=1}^{n} a_k$ where $a_k = \dfrac{k}{a_{k-1}}$ and $a_1 = 1$; $n = 25$ and $n = 100$

44. $\displaystyle\sum_{k=1}^{n} a_k$ where $a_k = \left(\dfrac{k}{k + 1}\right)a_{k-1}$ and $a_1 = 10$; $n = 100$ and $n = 1000$

45. $\displaystyle\sum_{k=1}^{n} a_k$ where $a_k = \dfrac{a_{k-1}}{k - 1} + \dfrac{a_{k-2}}{k - 2}$ with $a_1 = 11$ and $a_2 = 10$; $n = 10$ and $n = 50$

46. $\displaystyle\sum_{k=1}^{n} a_k$ where $a_k = a_{k-1} + a_{k-2} - k$ with $a_1 = 4$ and $a_2 = 5$; $n = 50$ and $n = 500$

B. Applications and Extensions

Use a program to calculate each sum. (Hint: Write as a difference of two sums.)

47. $\displaystyle\sum_{k=101}^{200} \frac{k^3 - 5k - 2}{k^2}$

48. $\displaystyle\sum_{k=21}^{40} \frac{2^k}{k^4}$

49. $\displaystyle\sum_{k=1001}^{3000} \ln(k^2 + 2k + 5)$

50. $\displaystyle\sum_{k=51}^{150} \sin k$

Let c be a constant. Here are three important properties of Σ.

(a) $\displaystyle\sum_{k=1}^{n} ca_k = c\sum_{k=1}^{n} a_k$

(b) $\displaystyle\sum_{k=1}^{n} (a_k + b_k) = \sum_{k=1}^{n} a_k + \sum_{k=1}^{n} b_k$

(c) $\displaystyle\sum_{k=1}^{n} c = nc$

Given that $\sum_{k=1}^{10} a_k = 300$ and $\sum_{k=1}^{10} b_k = -10$, calculate each of the sums in Problems 51–54.

51. $\displaystyle\sum_{k=1}^{10} (3a_k - 2b_k)$

52. $\displaystyle\sum_{k=1}^{10} (0.5a_k + 10b_k)$

53. $\displaystyle\sum_{k=1}^{10} (2a_k - 3)$

54. $\displaystyle\sum_{k=1}^{10} (2a_k - b_k - 2)$

55. It is known that $\sum_{k=1}^{\infty} (1/k^2) = \pi^2/M$, where M is an integer. Determine M.

56. Determine the smallest n so that $\sum_{k=1}^{n} (1/k)$ exceeds 5.

57. The note under the rock said that the treasure could be found by walking 1 foot in the direction N1°E, then 2 feet in the direction N2°E, then 3 feet in the direction N3°E, and so on for 90 segments. How far and in what direction from the rock is the treasure?

58. The ABC company is growing so that each year profits are $1.1x + 1.05y$, where x and y represent profits the previous year and the year before that, respectively. Predict the company's profits in the year 2010 if its profits in years 1991 and 1992 were $20,000 and $30,000, respectively.

59. Oranges are piled in a pyramidal shape so that each tier is rectangular and the top tier has a row of 10 oranges.

(a) Write a formula in sigma notation for the number of oranges in a pile with n tiers.

(b) Calculate the number of oranges in a pile with 30 tiers.

60. Challenge. In the song "The Twelve Days of Christmas," my true love gave me 1 present on the first day, $1 + 2$ presents on the second day, $1 + 2 + 3$ presents on the third day, and so on.

(a) Write a nice formula in sigma notation for the total number of presents T that I will receive in a Christmas with n days.

(b) Calculate T when $n = 12$ and when $n = 24$.

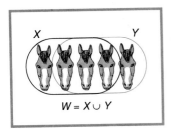

X   Y

$W = X \cup Y$

TEASER   What is wrong with the following argument?

Theorem. All horses in the world are of the same color.

Proof. Let P_n be the statement: All the horses in any set of n horses are identically colored. Certainly P_1 is true. Also the truth of P_k implies the truth of P_{k+1}. For suppose that P_k is true; that is, suppose that all horses in any set of k horses are identically colored. Then if W is any set of $k + 1$ horses, it is the union of two overlapping sets X and Y, each with k horses. (The situation for $k = 4$ is shown in the figure.) By supposition, the horses in X are identically colored and the horses in Y are identically colored. Since X and Y overlap, all horses in $W = X \cup Y$ are identically colored; that is, P_{k+1} is true. We conclude by *mathematical induction* that P_n is true for every positive integer n. The number of horses in the world is such an integer, so all the horses in the world are of the same color.

8.4   MATHEMATICAL INDUCTION

Some mathematics students are confused by the topic of mathematical induction, perhaps because it has little to do with the word *induction* as used in other disciplines. In science, for example, induction is a method of reasoning whereby on the basis of a number of specific instances one *conjectures* a general law about a whole population. In contrast, mathematical induction is a *method for proving* that all the statements in an infinite sequence of statements are true. The sequence $\{P_n\}$ in our opening Teaser is such a sequence of statements. That we claim to have used mathematical induction to prove something you know is false should make you wary. Pay careful attention to what follows.

MATHEMATICAL INDUCTION DESCRIBED

We begin with a formal characterization of **mathematical induction.**

> **THE PRINCIPLE OF MATHEMATICAL INDUCTION**
>
> Let $P_1, P_2, P_3, \ldots$ be a sequence of statements with the following two properties.
>
> 1. P_1 is true.
> 2. The truth of P_k implies the truth of P_{k+1} (abbreviated $P_k \Rightarrow P_{k+1}$).
>
> Then, P_n is true for every positive integer n.

Here is an analogy that may be helpful. Imagine an infinite column of standing dominoes (Figure 1). Suppose that someone pushes over the first domino in the column. Suppose further that the dominoes are spaced so that, when any domino falls, it topples the next one in the column. Don't you agree that ultimately every domino will fall?

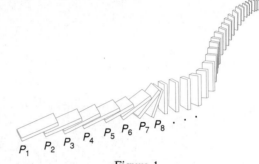

Figure 1

Our goal is to use mathematical induction to prove whole sequences of mathematical statements to be true. Here are four examples of sequences of statements that might interest us.

$$P_n: \quad \frac{1}{1 \cdot 2} + \frac{1}{2 \cdot 3} + \frac{1}{3 \cdot 4} + \cdots + \frac{1}{n(n+1)} = \frac{n}{n+1}$$

$Q_n: \quad n^2 - n + 41$ is a prime number

$R_n: \quad (a+b)^n = a^n + b^n$

$$S_n: \quad 1 + 2 + 3 + \cdots + n = \frac{n^2 + n - 6}{2}$$

To be sure that we understand the notation, let us write each of these statements for the case $n = 3$.

$$P_3: \quad \frac{1}{1 \cdot 2} + \frac{1}{2 \cdot 3} + \frac{1}{3 \cdot 4} = \frac{3}{4}$$

$Q_3: \quad 3^2 - 3 + 41$ is a prime number

$R_3: \quad (a+b)^3 = a^3 + b^3$

$$S_3: \quad 1 + 2 + 3 = \frac{3^2 + 3 - 6}{2}$$

n	$n^2 - n + 41$
1	41
2	43
3	47
4	53
5	61
6	71
7	83
.	.
.	.
.	.
40	1601
41	$1681 = 41^2$

Figure 2

Of these, P_3 and Q_3 are true, while R_3 and S_3 are false; you should verify these facts. A careful study of these four sequences will indicate the wide range of behavior that sequences of statements can display.

While it certainly is not obvious, we claim that P_n is true for every positive integer n; we are going to prove this claim soon. Q_n is a famous statement. It was thought by some to be true for all n and, in fact, it is true for $n = 1, 2, 3, \ldots, 40$ (see Figure 2). But it fails for $n = 41$, a fact that allows us to make an important point. Establishing the truth of Q_n for a finite number of cases, no matter how many, does not prove its truth for all n. Statements R_n and S_n are rather hopeless cases; R_n is true only for $n = 1$ and S_n is never true.

TWO ILLUSTRATIONS OF LEGITIMATE PROOFS

Our first example has already been mentioned.

■ **Example 1.** Prove that the statement

$$P_n: \quad \frac{1}{1 \cdot 2} + \frac{1}{2 \cdot 3} + \frac{1}{3 \cdot 4} + \cdots + \frac{1}{(n-1)n} + \frac{1}{n(n+1)} = \frac{n}{n+1}$$

is true for every positive integer n.

Solution. The first step is easy. P_1 is the statement

$$\frac{1}{1 \cdot 2} = \frac{1}{1+1}$$

which is clearly true.

To handle the second step ($P_k \Rightarrow P_{k+1}$), it is a good idea to write down the statements corresponding to P_k and P_{k+1} (at least on scratch paper). We get them by substituting k and $k + 1$ for n in the statement for P_n.

$$P_k: \quad \frac{1}{1 \cdot 2} + \frac{1}{2 \cdot 3} + \cdots + \frac{1}{(k-1)k} + \frac{1}{k(k+1)} = \frac{k}{k+1}$$

$$P_{k+1}: \quad \frac{1}{1 \cdot 2} + \frac{1}{2 \cdot 3} + \cdots + \frac{1}{k(k+1)} + \frac{1}{(k+1)(k+2)} = \frac{k+1}{k+2}$$

Notice that the left side of P_{k+1} is the same as that of P_k except for the addition of one more term, $1/((k+1)(k+2))$.

Suppose for the moment that P_k is true, and consider how this assumption allows us to simplify the left side of P_{k+1}.

$$\left[\frac{1}{1 \cdot 2} + \frac{1}{2 \cdot 3} + \cdots + \frac{1}{k(k+1)} \right] + \frac{1}{(k+1)(k+2)} = \frac{k}{k+1} + \frac{1}{(k+1)(k+2)}$$
$$= \frac{k(k+2) + 1}{(k+1)(k+2)}$$
$$= \frac{(k+1)^2}{(k+1)(k+2)}$$
$$= \frac{k+1}{k+2}$$

If you read this chain of equalities from top to bottom, you will see that we have established the truth of P_{k+1}, but under the *assumption that P_k is true*. That is, we have established that the truth of P_k implies the truth of P_{k+1}.

It follows from the principle of mathematical induction that P_n is true for every positive integer n. ∎

■ **Example 2.** Prove that the statement

$$P_n: \quad 1^2 + 2^2 + 3^2 + \cdots + n^2 = \frac{n(n+1)(2n+1)}{6}$$

is true for every positive integer n.

Solution. Again the first step is easy; P_1 is the true statement

$$1^2 = \frac{1(2)(3)}{6}$$

To show that $P_k \Rightarrow P_{k+1}$, we first write down these two statements

$$P_k: \quad 1^2 + 2^2 + 3^2 + \cdots + k^2 = \frac{k(k+1)(2k+1)}{6}$$

$$P_{k+1}: \quad 1^2 + 2^2 + 3^2 + \cdots + k^2 + (k+1)^2 = \frac{(k+1)(k+2)(2k+3)}{6}$$

Assuming that P_k is true, we can write the left side of P_{k+1} as shown in the following chain of equalities.

$$
\begin{aligned}
1^2 + 2^2 + 3^2 + \cdots + k^2 + (k+1)^2 &= \frac{k(k+1)(2k+1)}{6} + (k+1)^2 \\
&= \frac{(k+1)(k(2k+1) + 6(k+1))}{6} \\
&= \frac{(k+1)(2k^2 + 7k + 6)}{6} \\
&= \frac{(k+1)(k+2)(2k+3)}{6}
\end{aligned}
$$

Thus the truth of P_k does imply the truth of P_{k+1}. We conclude by mathematical induction that P_n is true for every positive integer n. Incidentally, the result just proved will be used in calculus. ∎

SOME COMMENTS ABOUT MATHEMATICAL INDUCTION

Students seldom have trouble with the verification step (showing that P_1 is true). The inductive step (showing that $P_k \Rightarrow P_{k+1}$) is harder and more subtle. In that step, we do not *prove* that either P_k or P_{k+1} is true, but rather that the truth of P_k implies the truth of P_{k+1}. For a vivid illustration of the difference, we point out that in the example

$$S_n: \quad 1 + 2 + 3 + \cdots + n = \frac{n^2 + n - 6}{2}$$

mentioned earlier, it is true that $S_k \Rightarrow S_{k+1}$ (as you will check in Problem 30), and yet not a single statement in the sequence $\{S_n\}$ is true. To put it another way, what $S_k \Rightarrow S_{k+1}$ means is that *if* S_k were true, *then* S_{k+1} would have to be true too. It is a little like saying that if spinach were ice cream, then children would want two servings of spinach at every meal.

Figure 3 and the analogy with a column of dominoes may help you think about the four examples we introduced early in our discussion.

Why They Fall and Why They Don't

$P_n: \dfrac{1}{1 \cdot 2} + \dfrac{1}{1 \cdot 2} + \cdots + \dfrac{1}{n(n+1)} = \dfrac{n}{n+1}$ P_1 is true. $P_k \Rightarrow P_{k+1}$	 $P_1 P_2 P_3 P_4 P_5 P_6 \ldots$ First domino is pushed over. Each falling domino pushes over the next one.
$Q_n : n^2 - n + 41$ is prime. $Q_1, Q_2, \ldots, Q_{40}$ are true. $Q_k \nRightarrow Q_{k+1}$	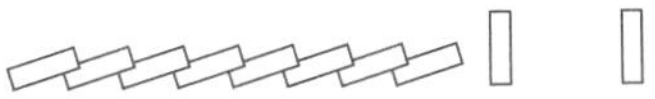 $Q_{35} Q_{36} Q_{37} Q_{38} Q_{39} Q_{40} \quad Q_{41} \qquad Q_{42}$ First 40 dominoes are pushed over. 41st domino remains standing.
$R_n : (a + b)^n = a^n + b^n$ R_1 is true. $R_k \nRightarrow R_{k+1}$	 $R_1 \quad R_2 \quad R_3 \quad R_4 \quad R_5 \quad R_6 \quad R_7$ First domino is pushed over but dominoes are spaced too far apart to push each other over.
$S_n : 1 + 2 + 3 + \ldots + n = \dfrac{n^2 + n - 6}{2}$ S_1 is false. $S_k \Rightarrow S_{k+1}$	 $S_1 S_2 S_3 S_4 S_5$ Spacing is just right but no one can push over the first domino.

Figure 3

TWO MORE EXAMPLES

Here is an example of a new type.

■ **Example 3.** Prove that the statement

$$P_n: x - y \text{ is a factor of } x^n - y^n$$

is true for every positive integer n.

Solution. Trivially, $x - y$ is a factor of $x - y$ because $x - y = 1(x - y)$; so P_1 is true.

Suppose that $x - y$ is a factor of $x^k - y^k$. This means that there is a polynomial $Q(x, y)$ such that

$$x^k - y^k = Q(x, y)(x - y)$$

Using this assumption, we may write

$$
\begin{aligned}
x^{k+1} - y^{k+1} &= x^{k+1} - x^k y + x^k y - y^{k+1} \\
&= x^k(x - y) + y(x^k - y^k) \\
&= x^k(x - y) + yQ(x, y)(x - y) \\
&= (x^k + yQ(x, y))(x - y)
\end{aligned}
$$

Thus $x - y$ is a factor of $x^{k+1} - y^{k+1}$. We have shown that $P_k \Rightarrow P_{k+1}$ and that P_1 is true; we therefore conclude that P_n is true for all n. ■

Our final example introduces a twist: a sequence of statements $\{P_n\}$ that is not true for every positive integer n but rather is true for every $n \geq 4$. We can demonstrate that this is so with a slight change in the principle of mathematical induction. We simply make the first step to be: show that P_4 is true.

■ **Example 4.** Show that the inequality

$$P_n: \quad 3^n > 2^n + 20$$

is true for every integer $n \geq 4$.

Solution. You might check that P_1, P_2, and P_3 are false. However,

$$P_4: \quad 3^4 > 2^4 + 20$$

is clearly true ($81 > 36$).

To show that $P_k \Rightarrow P_{k+1}$, we assume for the moment that P_k is true; that is, we assume that $3^k > 2^k + 20$. Then

$$3^{k+1} = 3(3^k) > 3(2^k + 20) > 2(2^k + 20) = 2^{k+1} + 40 > 2^{k+1} + 20$$

Therefore, P_{k+1} is true, provided P_k is true. We conclude that P_n is true for every integer $n \geq 4$. ■

TEASER SOLUTION

Now is the time to emphasize something we have glossed over. In proofs by induction, it is necessary that we show the implication $P_k \Rightarrow P_{k+1}$ is valid for every positive integer k (or in a case like Example 4, for ks that appear in the sequence of statements claimed to be true). In the pseudoproof in the Teaser, this implication is valid for every positive integer k with the exception of $k = 1$. When $k = 1$, the set W consisting of two horses cannot be thought of as the union of two overlapping sets X and Y. This one exception spoils the whole proof. Using the domino analogy, we can knock down the first domino, but it does not topple the second one.

PROBLEM SET 8.4

A. Skills and Techniques

In Problems 1–8, prove by mathematical induction that P_n is true for every positive integer n.

1. P_n: $1 + 2 + 3 + \cdots + n = \dfrac{n(n + 1)}{2}$

2. P_n: $1 + 3 + 5 + \cdots + (2n - 1) = n^2$

3. P_n: $3 + 7 + 11 + \cdots + (4n - 1) = n(2n + 1)$

4. P_n: $2 + 9 + 16 + \cdots + (7n - 5) = \dfrac{n(7n - 3)}{2}$

5. P_n: $1 \cdot 2 + 2 \cdot 3 + 3 \cdot 4 + \cdots + n(n + 1)$
$$= \frac{1}{3} n(n + 1)(n + 2)$$

6. P_n: $\dfrac{1}{1 \cdot 3} + \dfrac{1}{3 \cdot 5} + \dfrac{1}{5 \cdot 7} + \cdots + \dfrac{1}{(2n - 1)(2n + 1)}$
$$= \frac{n}{2n + 1}$$

7. P_n: $2 + 2^2 + 2^3 + \cdots + 2^n = 2(2^n - 1)$

8. P_n: $1^2 + 3^2 + 5^2 + \cdots + (2^n - 1)^2$
$$= \frac{n(2n - 1)(2n + 1)}{3}$$

In Problems 9–18, tell what you can conclude from the information given about the sequence of statements. For example, if you are given that P_4 is true and that $P_k \Rightarrow P_{k+1}$ for any k, then you can conclude that P_n is true for every integer $n \geq 4$.

9. P_8 is true and $P_k \Rightarrow P_{k+1}$.

10. P_8 is not true and $P_k \Rightarrow P_{k+1}$.

11. P_1 is true, but P_k does not imply P_{k+1}.

12. $P_1, P_2, \ldots, P_{1000}$ are all true.

13. P_1 is true and $P_k \Rightarrow P_{k+2}$.

14. P_{40} is true and $P_k \Rightarrow P_{k-1}$.

15. P_1 and P_2 are true; P_k and P_{k+1} together imply P_{k+2}.

16. P_1 and P_2 are true and $P_k \Rightarrow P_{k+2}$.

17. P_1 is true and $P_k \Rightarrow P_{4k}$.

18. P_1 is true, $P_k \Rightarrow P_{4k}$, and $P_k \Rightarrow P_{k-1}$.

Use mathematical induction to prove that each of the following is true for every positive integer n.

19. $x + y$ is a factor of $x^{2n} - y^{2n}$. (*Hint:* $x^{2k+2} - y^{2k+2} = x^{2k+2} - x^{2k}y^2 + x^{2k}y^2 - y^{2k+2}$.)

20. $x + y$ is a factor of $x^{2n-1} + y^{2n-1}$.

21. $n^2 - n$ is even (that is, has 2 as a factor).

22. $n^3 - n$ is divisible by 6.

In Problems 23–28, find the smallest positive integer N for which the given statement is true for $n \geq N$. Then prove that the statement is true for all integers greater than or equal to N.

23. $n + 5 < 2^n$

24. $3n \leq 3^n$

25. $\log_{10} n < n$ (*Hint:* $k + 1 < 10k$.)

26. $n^2 \leq 2^n$ (*Hint:* $k^2 + 2k + 1 = k(k + 2 + 1/k) < k(k + k)$.)

27. $(1 + x)^n \geq 1 + nx$, where $x \geq -1$

28. $|\sin nx| \leq |\sin x| \cdot n$ for all x

B. Applications and Extensions

29. These four formulas can all be proved by mathematical induction. We proved (b) in the text; you prove the others.

(a) $1 + 2 + 3 + \cdots + n = \dfrac{1}{2} n(n + 1)$

(b) $1^2 + 2^2 + 3^2 + \cdots + n^2 = \dfrac{1}{6} n(n + 1)(2n + 1)$

(c) $1^3 + 2^3 + 3^3 + \cdots + n^3 = \dfrac{1}{4} n^2(n + 1)^2$

(d) $1^4 + 2^4 + 3^4 + \cdots + n^4$
$$= \frac{1}{30} n(n + 1)(6n^3 + 9n^2 + n - 1)$$

From (a) and (c), another interesting formula follows, namely,

$1^3 + 2^3 + 3^3 + \cdots + n^3$
$$= (1 + 2 + 3 + \cdots + n)^2$$

30. Consider the statement

$$S_n: \quad 1 + 2 + 3 + \cdots + n = \frac{n^2 + n - 6}{2}$$

Show that
(a) $S_k \Rightarrow S_{k+1}$ for $k \geq 1$.
(b) S_n is not true for any positive integer n.

31. Use the results of Problem 29 to evaluate each of the following.

(a) $\displaystyle\sum_{k=1}^{100} (3k + 1)$

(b) $\displaystyle\sum_{k=1}^{10} (k^2 - 3k)$

(c) $\displaystyle\sum_{k=1}^{10} (k^3 + 3k^2 + 3k + 1)$

(d) $\displaystyle\sum_{k=1}^{n} (6k^2 + 2k)$

32. In the popular song titled "The Twelve Days of Christmas," my true love gave me 1 gift on the first day, $(2 + 1)$ gifts on the second day, $(3 + 2 + 1)$ gifts on the third day, and so on.
    (a) How many gifts did I get altogether during the 12 days?
    (b) How many gifts would I get altogether in a Christmas that had n days? Give a simple explicit formula.

33. Prove that for $n \geq 2$,

$$\left(1 - \frac{1}{4}\right)\left(1 - \frac{1}{9}\right)\left(1 - \frac{1}{16}\right) \cdots \left(1 - \frac{1}{n^2}\right) = \frac{n + 1}{2n}$$

34. Prove that for $n \geq 1$,

$$\frac{1}{\sqrt{1}} + \frac{1}{\sqrt{2}} + \frac{1}{\sqrt{3}} + \cdots + \frac{1}{\sqrt{n}} < 2\sqrt{n}$$

35. Prove that for $n \geq 3$,

$$\frac{1}{n + 1} + \frac{1}{n + 2} + \frac{1}{n + 3} + \cdots + \frac{1}{2n} > \frac{3}{5}$$

36. Prove that the number of diagonals in an n-sided convex polygon is $n(n - 3)/2$ for $n \geq 3$. The diagrams in Figure 4 show the situation for $n = 4$ and $n = 5$.

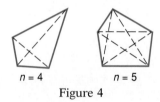

$n = 4$ $n = 5$

Figure 4

37. Prove that the sum of the measures of the interior angles in an n-sided polygon (without holes or self-intersections) is $(n - 2)180°$. What is the sum of the measures of the exterior angles of such a polygon?

38. Consider n lines in general position (no two parallel, no three meeting in the same point). Prove that I_n, the number of intersection points, is given by $I_n = (n^2 - n)/2$. (*Hint:* $I_{n+1} = I_n + n$.)

39. Refer to Problem 38. Prove that R_n, the number of regions, is given by $R_n = (n^2 + n + 2)/2$. (*Hint:* First convince yourself that $R_{n+1} = R_n + n + 1$.)

40. Let $f_1 = 1$, $f_2 = 1$, and $f_{n+2} = f_{n+1} + f_n$ for $n \geq 1$. Call f_n the Fibonacci sequence (see Section 8.3) and let $S_n = f_1 + f_2 + f_3 + \cdots + f_n$. Prove by mathematical induction that $S_n = f_{n+2} - 1$ for all n.

41. By considering adjoining geometric squares as in Figure 5, obtain a nice formula for $f_1^2 + f_2^2 + f_3^2 + \cdots + f_n^2$ in terms of f_n and f_{n+1}.

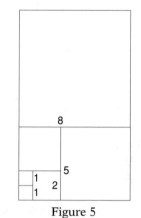

Figure 5

42. Use mathematical induction to prove the result discovered in Problem 41.

43. Let $a_0 = 0$, $a_1 = 1$, and $a_{n+2} = (a_{n+1} + a_n)/2$ for $n \geq 0$. Prove that for $n \geq 0$,

$$a_n = \frac{2}{3}\left(1 - \left(-\frac{1}{2}\right)^n\right)$$

(*Hint:* In the inductive step, show that P_k and P_{k+1} together imply P_{k+2}.)

44. **Challenge.** Let f_n be the Fibonacci sequence of Problem 40. Use mathematical induction (as in the hint to Problem 43) to prove that

$$f_n = \frac{1}{\sqrt{5}}\left(\left(\frac{1 + \sqrt{5}}{2}\right)^n - \left(\frac{1 - \sqrt{5}}{2}\right)^n\right)$$

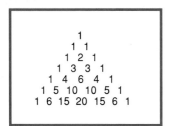

TEASER In the adjoining figure, we show row 0 through row 6 of a famous array of numbers that goes by the name Pascal's triangle. Determine the sum of the numbers in each row and use your answers to make a conjecture about the sum of the numbers in the *n*th row of the triangle. Prove your conjecture.

8.5 THE BINOMIAL FORMULA

The triangular array of numbers in our opening display has intrigued professional and amateur mathematicians for centuries. It is named for the gifted mathematician-philosopher Blaise Pascal (1623–1662), who wrote a treatise on the subject. We call immediate attention to the way the triangle is generated. The triangle has 1s down its slant edges, and any entry in the body of the triangle is the sum of its two neighbors in the row above. For example, the entry 15 is the sum of 5 and 10.

Of the hundreds of interesting facts about this array of numbers, certainly it is its relation to powers of a binomial that is most important. Consider the following results.

$$
\begin{aligned}
(x + y)^0 &= & 1 \\
(x + y)^1 &= & x + y \\
(x + y)^2 &= & x^2 + 2xy + y^2 \\
(x + y)^3 &= & x^3 + 3x^2y + 3xy^2 + y^3 \\
(x + y)^4 &= & x^4 + 4x^3y + 6x^2y^2 + 4xy^3 + y^4 \\
(x + y)^5 &= x^5 + 5x^4y + 10x^3y^2 + 10x^2y^3 + 5xy^4 + y^5
\end{aligned}
$$

The coefficients are the numbers in Pascal's triangle.

Suppose that you wanted to know the coefficient of $x^{18}y^{32}$ in the expansion of $(x + y)^{50}$. One way to proceed would be to generate Pascal's triangle, one row at a time, until you got to row 50 and then pick the 33rd number in that row. It would certainly be better to have an explicit formula for each coefficient, a matter to which we now turn.

THE BINOMIAL COEFFICIENT $\binom{N}{R}$

We define the symbol $\binom{n}{r}$, to be called a **binomial coefficient,** by

$$
\binom{n}{r} = \frac{n(n - 1)(n - 2) \cdots (n - r + 2)(n - r + 1)}{r(r - 1)(r - 2) \cdots 3 \cdot 2 \cdot 1}
$$

Here n and r are positive integers with $1 \le r \le n$. For example,

$$
\binom{5}{3} = \frac{5 \cdot 4 \cdot 3}{3 \cdot 2 \cdot 1} = 10 \qquad \binom{5}{4} = \frac{5 \cdot 4 \cdot 3 \cdot 2}{4 \cdot 3 \cdot 2 \cdot 1} = 5 \qquad \binom{5}{5} = \frac{5 \cdot 4 \cdot 3 \cdot 2 \cdot 1}{5 \cdot 4 \cdot 3 \cdot 2 \cdot 1} = 1
$$

are the last three coefficients in $(x + y)^5$. You can remember the definition of

$\binom{n}{r}$ by noting that both numerator and denominator are products of r integers starting at n and r, respectively, with the factors steadily decreasing by 1. Thus

$$\binom{12}{9} = \frac{12 \cdot 11 \cdot 10 \cdot 9 \cdot 8 \cdot 7 \cdot 6 \cdot 5 \cdot 4}{9 \cdot 8 \cdot 7 \cdot 6 \cdot 5 \cdot 4 \cdot 3 \cdot 2 \cdot 1} = \frac{\overset{2}{\cancel{12}} \cdot 11 \cdot 10}{3 \cdot 2 \cdot 1} = 220$$

In evaluating one of these symbols, do not fail to cancel all common factors in the denominator and numerator first. Cancellation is always possible, and the result is always an integer.

■ **Example 1.** Calculate $\binom{20}{5}$.

Solution.

$$\binom{20}{5} = \frac{\cancel{20} \cdot 19 \cdot \overset{3}{\cancel{18}} \cdot 17 \cdot 16}{\cancel{5} \cdot \cancel{4} \cdot \cancel{3} \cdot \cancel{2} \cdot 1} = 15{,}504 \qquad ■$$

The product in the denominator of $\binom{n}{r}$ occurs often enough in mathematics to be given a special name, **r factorial,** and a special symbol, $r!$.

$$r! = r(r-1)(r-2)\ldots 3 \cdot 2 \cdot 1$$

For example, $3! = 3 \cdot 2 \cdot 1 = 6$, $4! = 4 \cdot 3 \cdot 2 \cdot 1 = 24$, and $50!$ is so large that you need a calculator to evaluate it (some calculators have a special factorial key).

Consider again the definition of $\binom{n}{r}$. If we multiply both numerator and denominator by $(n-r)!$ we obtain

$$\binom{n}{r} = \frac{n!}{r!(n-r)!}$$

a result that we shall need shortly. Also we define $0! = 1$. Then in order for the last boxed formula to hold true for $r = 0$, we must have $\binom{n}{0} = 1$. Finally, that formula implies that

$$\binom{n}{r} = \binom{n}{n-r}$$

For example, $\binom{6}{4} = \binom{6}{2}$ and $\binom{50}{45} = \binom{50}{5}$. This corresponds to the fact that Pascal's triangle is symmetric about its vertical median.

THE BINOMIAL FORMULA

Here is the result toward which we have been aiming.

$$(x + y)^n = \binom{n}{0}x^n y^0 + \binom{n}{1}x^{n-1}y^1 + \binom{n}{2}x^{n-2}y^2 + \cdots + \binom{n}{n-1}x^1 y^{n-1} + \binom{n}{n}x^0 y^n$$

$$= \sum_{k=0}^{n} \binom{n}{k}x^{n-k}y^k = \sum_{k=0}^{n} \binom{n}{k}x^k y^{n-k}$$

It is called the **binomial formula,** and it plays a significant role in many parts of mathematics including calculus.

As one example, we may write

$$(x + y)^6 = \binom{6}{0}x^6 + \binom{6}{1}x^5 y + \binom{6}{2}x^4 y^2 + \binom{6}{3}x^3 y^3 + \binom{6}{4}x^2 y^4 + \binom{6}{5}xy^5 + \binom{6}{6}y^6$$

$$= x^6 + 6x^5 y + 15x^4 y^2 + 20x^3 y^3 + 15x^2 y^4 + 6xy^5 + y^6$$

■ **Example 2.** Obtain the complete expansion of $(2a - b^2)^6$.

Solution. We simply think of $2a$ as x and $-b^2$ as y, apply the expansion above, and simplify.

$$(2a + (-b^2))^6 = (2a)^6 + 6(2a)^5(-b^2) + 15(2a)^4(-b^2)^2$$
$$+ 20(2a)^3(-b^2)^3 + 15(2a)^2(-b^2)^4$$
$$+ 6(2a)(-b^2)^5 + (-b^2)^6$$
$$= 64a^6 - 192a^5 b^2 + 240a^4 b^4 - 160a^3 b^6$$
$$+ 60a^2 b^8 - 12ab^{10} + b^{12} \qquad ■$$

■ **Example 3.** Find and simplify the term in the expansion of $(2s + t)^{20}$ that involves s^5.

Solution. Note that the exponents in each term of $(x + y)^n$ sum to n and that the exponent on y agrees with the lower number in the corresponding binomial coefficient. Thus, the required term is

$$\binom{20}{15}(2s)^5 t^{15} = (15{,}504)(32)s^5 t^{15} = 496{,}128s^5 t^{15}$$

Here, we made use of the result of Example 1. ■

■ **Example 4.** If \$100 is invested at 12% interest compounded monthly, it will accumulate to $\$100(1.01)^{12}$ at the end of 1 year. Use the binomial formula to approximate this amount.

Solution.

$$100(1.01)^{12} = 100(1 + 0.01)^{12}$$

$$= 100\left(1^{12} + \binom{12}{1}1^{11}(.01) + \binom{12}{2}1^{10}(.01)^2 + \binom{12}{3}1^9(.01)^3 + \ldots\right)$$

$$= 100(1 + 12(.01) + 66(.01)^2 + 220(.01)^3 + \ldots)$$

$$= 100(1 + 0.12 + 0.0066 + 0.00022 + \ldots)$$

$$= 100(1.12682\ldots) = \$112.68$$

The answer \$112.68 is accurate to the nearest penny; the last nine terms of the expansion do not add up to as much as a quarter of a penny. ∎

PROOF OF THE BINOMIAL FORMULA

It does not take much faith to believe that the binomial formula is valid for $(x + y)^{10}$, since the result is easily checked. Would it be valid for $(x + y)^{10000}$? A rigorous proof that establishes the result for any exponent n should erase any lingering doubts. This is exactly what mathematical induction is designed to do. Let P_n be the statement that the binomial formula holds for the integer n.

We note that

$$(x + y)^1 = \binom{1}{0}x^1y^0 + \binom{1}{1}x^0y^1 = x + y$$

so P_1 is true.

Next, suppose the P_k is true; that is, suppose that

$$(x + y)^k = \binom{k}{0}x^k + \binom{k}{1}x^{k-1}y + \binom{k}{2}x^{k-2}y^2 + \cdots + \binom{k}{k}y^k$$

We must show that this supposition implies that P_{k+1} is true, namely, that

$$(x + y)^{k+1} = \binom{k+1}{0}x^{k+1} + \binom{k+1}{1}x^ky + \binom{k+1}{2}x^{k-1}y^2 + \cdots + \binom{k+1}{k+1}y^{k+1}$$

To do this, we multiply both sides of the equality for $(x + y)^k$ by $x + y$. On the right side, we accomplish this by multiplying first by x, then by y, and adding the results as shown below. Then $(x + y)^{k+1}$ is equal to

$$\binom{k}{0}x^{k+1} + \boxed{\binom{k}{1}} x^ky + \boxed{\binom{k}{2}} x^{k-1}y^2 + \cdots + \boxed{\binom{k}{k}} xy^k$$

$$+ \boxed{\binom{k}{0}} x^ky + \boxed{\binom{k}{1}} x^{k-1}y^2 + \cdots + \boxed{\binom{k}{k-1}} xy^k + \binom{k}{k}y^{k+1}$$

The first and last coefficients have the right values, namely, 1. It remains to show that the coefficients in the rectangles have the required sum. For example, we want

$$\binom{k}{1} + \binom{k}{0} = \binom{k+1}{1} \qquad \binom{k}{2} + \binom{k}{1} = \binom{k+1}{2}$$

and, more generally, we want

$$\binom{k}{r} + \binom{k}{r-1} = \binom{k+1}{r}$$

If we can establish the boxed result, we will be done. Incidentally, this result corresponds to the fact in Pascal's triangle that we get an interior term by adding the adjacent neighbors above it.

Here is the required demonstration.

$$\binom{k}{r} + \binom{k}{r-1} = \frac{k!}{r!(k-r)!} + \frac{k!}{(r-1)!(k-r+1)!}$$

$$= \frac{k!}{(r-1)!(k-r)!}\left(\frac{1}{r} + \frac{1}{k-r+1}\right)$$

$$= \frac{k!}{(r-1)!(k-r)!}\left(\frac{k+1}{r(k-r+1)}\right)$$

$$= \frac{(k+1)!}{r!(k-r+1)!} = \binom{k+1}{r}$$

TEASER SOLUTION

```
          1
        1   1
      1   2   1
    1   3   3   1
  1   4   6   4   1
1   5  10  10   5   1
```

Figure 1

We have reproduced the first few rows of Pascal's triangle in Figure 1 (recall that the first row is labeled row 0). Note that the row sums are 1, 2, 4, 8, 16, 32, ... Our conjecture is that the sum of the numbers in row n of the triangle is 2^n. To prove this, simply substitute $x = 1$ and $y = 1$ in the binomial formula. The result is important enough to highlight.

$$\binom{n}{0} + \binom{n}{1} + \binom{n}{2} + \binom{n}{3} + \cdots + \binom{n}{n} = 2^n$$

PROBLEM SET 8.5

A. Skills and Techniques

Evaluate each of the following (by hand the easy way).

1. $\binom{11}{2}$

2. $\binom{12}{3}$

3. $\binom{11}{9}$

4. $\binom{12}{9}$

5. $\binom{50}{48}$

6. $\binom{60}{58}$

7. $\dbinom{100}{97}$

8. $\dbinom{100}{100}$

9. $\dbinom{60}{20} - \dbinom{60}{40}$

10. $\dbinom{9}{6} + \dbinom{9}{7}$

11. $\left(\dbinom{8}{6} + \dbinom{8}{7}\right)^2$

12. $\dbinom{5}{0} + \dbinom{5}{1} + \dbinom{5}{2} + \dbinom{5}{3} + \dbinom{5}{4} + \dbinom{5}{5}$

In Problems 13–20, expand and simplify.

13. $(x + y)^3$
14. $(x - y)^3$
15. $(x - 2y)^3$
16. $(3x + b)^3$
17. $(c^2 - 3d^3)^4$
18. $(xy - 2z^2)^4$
19. $(ab^2 - bc)^5$
20. $\left(ab + \dfrac{1}{a}\right)^5$

Write the first three terms of each expansion in Problems 21–24 in simplified form.

21. $(x + y)^{20}$
22. $(x + y)^{30}$
23. $\left(x + \dfrac{1}{x^5}\right)^{20}$
24. $\left(xy^2 + \dfrac{1}{y}\right)^{14}$

25. Find the term in the expansion of $(y^2 - z^3)^{10}$ that involves z^9.
26. Find the term in the expansion of $(3x - y^3)^{10}$ that involves y^{24}.
27. Find the term in the expansion of $(2a - b)^{12}$ that involves a^3.
28. Find the term in the expansion of $\left(x^2 - \dfrac{2}{x}\right)^5$ that involves x^4.

In Problems 29–32, use the first four terms of a binomial expansion to find an approximate value of the given expression.

29. $20(1.02)^8$
30. $100(1.002)^{20}$
31. $500(1.005)^{20}$
32. $200(1.04)^{10}$
33. Bacteria multiply in a certain medium so that by the end of k hours their number N is $N = 100(1.02)^k$. Approximate the number of bacteria after 20 hours.
34. Do Problem 33 assuming that $N = 1000(1.01)^k$.
35. Find the constant term in the expansion of $(3x^2 + 1/(3x))^{12}$.

36. Without using a calculator, show that $(1.0003)^{10} > 1.003004$.
37. Without using a calculator, show that $(1.01)^{50} > 1.5$.
38. Without using a calculator, find $(0.999)^{10}$ accurate to six decimal places.

B. Applications and Extensions

39. Expand and simplify.

    (a) $\dfrac{(x + h)^3 - x^3}{h}$ (b) $\dfrac{(x + h)^4 - x^4}{h}$

40. In each of the following, find the term in the expanded and simplified form that does not involve h (a procedure very important in calculus).

    (a) $\dfrac{(x + h)^n - x^n}{h}$

    (b) $\dfrac{(x + h)^{10} + 2(x + h)^4 - x^{10} - 2x^4}{h}$

41. In the expansion of the trinomial $(x + y + z)^n$, the coefficient of $x^r y^s z^t$ where $r + s + t = n$ is $n!/(r!\,s!\,t!)$.
    (a) Expand $(x + y + z)^3$.
    (b) Find the coefficient of the term $x^2 y^4 z$ in the expansion of $(2x + y + z)^7$.
42. Find the sum of all the coefficients in the expansion of the trinomial $(x + y + z)^n$.

43. Find a simple formula for $\displaystyle\sum_{k=0}^{n} \dbinom{n}{k} 2^k$. (*Hint:* Let $x = 1$ and $y = 2$ in the binomial formula.)
44. Find the sum of all the coefficients in $(4x^3 - x)^6$ after it is expanded and simplified. (*Hint:* This is a simple problem when looked at in the right way.)

45. Evaluate $\dbinom{30}{0} + \dbinom{30}{2} + \dbinom{30}{4} + \cdots + \dbinom{30}{30}$
    (*Hint:* Begin by substituting $x = 1$ and $y = -1$ in the binomial formula.)
46. Verify the formula
$$\dbinom{n}{0}^2 + \dbinom{n}{1}^2 + \dbinom{n}{2}^2 + \cdots + \dbinom{n}{n}^2 = \dbinom{2n}{n}$$
for $n = 10$, 15, and 20 by writing a program (or modifying an existing one) to calculate the left side.
47. Write programs (or modify existing ones) to calculate each of the following for any n and then apply them to calculate the value for $n = 20$ and 30.

    (a) $\displaystyle\sum_{k=1}^{n} 1/k!$

    (b) $\displaystyle\sum_{k=1}^{n} 4^k/k!$

    (c) $\displaystyle\sum_{k=1}^{n} k \dbinom{n}{k}$

    (d) $\displaystyle\sum_{k=1}^{n} 1/\dbinom{n}{k}$

48. Challenge. Let $P(x)$ be the nth-degree polynomial defined by

$$P(x) = 1 + x + \frac{x(x-1)}{2!} + \frac{x(x-1)(x-2)}{3!}$$
$$+ \cdots + \frac{x(x-1)(x-2)\cdots(x-n+1)}{n!}$$

Find a simple formula for each of the following.
(a) $P(k), k = 0, 1, 2, \ldots, n$
(b) $P(n + 1)$
(c) $P(n + 2)$

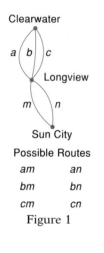

TEASER   The Senior Birdwatchers' Club, consisting of four women and two men, is about to hold its annual business meeting. In how many different ways can they line up for their group picture if the two men refuse to stand next to each other?

8.6   COUNTING ORDERED ARRANGEMENTS

Clearwater

a b c

Longview

m n

Sun City

Possible Routes

am	an
bm	bn
cm	cn

Figure 1

Counting is probably the first mathematical activity that most of us learned; we feel pretty confident of our ability to count accurately. Yet most of you are likely to have difficulty giving the correct answer to the Teaser above unless you have learned some counting principles that we are going to discuss in this section. The first of these principles is called the **multiplication principle.**

To lead up to this principle, consider the following problem. Suppose there are three roads a, b, and c from Clearwater to Longview and two roads m and n from Longview to Sun City. How many different routes can you choose from Clearwater to Sun City going through Longview? Figure 1 clarifies this situation. For each of the three choices from Clearwater to Longview, you have two choices from Longview to Sun City. Thus, you have $3 \cdot 2$ routes from Clearwater to Sun City. Here is the general principle.

MULTIPLICATION PRINCIPLE
Suppose that an event H can occur in h ways, and, after it has occurred, event K can occur in k ways. Then the number of ways in which both H **and** K can occur is hk.

This principle, stated for two events, extends in an obvious way to three or more events.

■ **Example 1.**   The Senior Birdwatchers' Club, mentioned in the opening Teaser, will elect a president, a vice-president, and a secretary. In how many different ways can they do this if the president is to be female and the vice-president male? Assume that no person can fill two positions and that the secretary can be of either sex.

Solution. The election can be considered to consist of three successive events.

P: Elect a female president
V: Elect a male vice-president
S: Elect a secretary of either sex

Event P can occur in 4 ways (there are 4 women). Event V can occur in 2 ways (there are 2 men). S can occur in 4 ways (after electing the president and vice-president, there are 4 people from which to elect a secretary). Thus, the entire election process can be accomplished in $4 \cdot 2 \cdot 4 = 32$ ways. ■

■ **Example 2.** Steve has 2 sport coats, 4 pairs of pants, 6 shirts, and 9 ties. How many different outfits can he wear consisting of one item of each kind?

Solution. Steve can wear

$$2 \cdot 4 \cdot 6 \cdot 9 = 432$$

different outfits. ■

PERMUTATIONS

Permutations
of *ART*

ART
ATR
RAT
RTA
TAR
TRA

Figure 2

To permute a set of objects is to rearrange them. Thus, a **permutation** of a set of objects is an arrangement of those objects in some order (Figure 2). Take the set of letters in the word *FACTOR* as an example. Imagine that each of these six letters is printed on a small card so that the letters can be arranged at will. Then we may form words such as *COTARF*, *TRAFOC*, and *FRACTO*, none of which is in a dictionary, but all of which are perfectly good words from our point of view. Let's call them code words. How many six-letter code words can be made from the letters of the word *FACTOR*; that is, how many permutations of six objects are there?

Think of this as the problem of filling six slots.

We may fill the first slot in 6 ways. Having done that, we may fill the second slot in 5 ways, the third in 4 ways, and so on. By the multiplication principle, we can fill all 6 slots in

$$6 \cdot 5 \cdot 4 \cdot 3 \cdot 2 \cdot 1 = 720$$

ways.

What if we want to make three-letter code words from the letters of the word *FACTOR*, words like *ACT*, *COF*, and *TAC*? How many such words can be made? This is the problem of filling 3 slots with 6 letters available. We can fill the first slot in 6 ways, then the second in 5 ways, and then the third in 4 ways. Therefore we can make $6 \cdot 5 \cdot 4 = 120$ three-letter code words from the word *FACTOR*.

■ **Example 3.** Suppose that the nine letters of the word *LOGARITHM* are written on nine cards.

(a) How many nine-letter code words can be made?

(b) How many four-letter code words can be made?

Solution.

(a) $9 \cdot 8 \cdot 7 \cdot 6 \cdot 5 \cdot 4 \cdot 3 \cdot 2 \cdot 1 = 362{,}880$

(b) $9 \cdot 8 \cdot 7 \cdot 6 = 3024$ ∎

CALCULATOR HINT

Many calculators allow automatic calculation of both $n!$ and $_nP_r$. For example, both symbols occur in the MATH menu on the TI-81 and TI-82. However, you will be unlikely to need this feature to make the relatively simple calculations required in this section.

Consider the corresponding general problem. Suppose that from n distinguishable objects, we select r of them and arrange them in a row. The resulting arrangement is called a **permutation of n things taken r at a time.** The number of such permutations is denoted by the symbol $_nP_r$. Thus

$$_6P_3 = 6 \cdot 5 \cdot 4 = 120$$
$$_6P_6 = 6 \cdot 5 \cdot 4 \cdot 3 \cdot 2 \cdot 1 = 720$$
$$_8P_2 = 8 \cdot 7 = 56$$

and in general

$$_nP_r = n(n-1)(n-2) \cdots (n-r+2)(n-r+1)$$

Note two things about the permutation symbol.

1. $_nP_n = n!$ (see Section 8.5 for a discussion of $n!$)

2. $_nP_r = \dfrac{n!}{(n-r)!}$

THE ADDITION PRINCIPLE IN COUNTING

Let's consider another problem for the Senior Birdwatchers' Club. In how many ways could they elect their three officers if the president is to be of one sex and the vice-president and secretary of the other? This means that the president should be female and the other two officers male, *or* the president should be male and the other two officers female. To answer a question like this, we need another principle.

ADDITION PRINCIPLE

Let H and K be disjoint events, that is, events that cannot happen simultaneously. If H can occur in h ways and K in k ways, then H **or** K can occur in $h + k$ ways.

This principle generalizes to three or more disjoint events (that is, to events no two of which can happen simultaneously).

Applying this principle to the question at hand, we define H and K as follows.

H: Elect a female president, male vice-president, and male secretary

K: Elect a male president, female vice-president, and female secretary

Clearly H and K are disjoint. From the multiplication principle,

H can occur in $4 \cdot 2 \cdot 1 = 8$ ways
K can occur in $2 \cdot 4 \cdot 3 = 24$ ways

Then by the addition principle, H or K can occur in $8 + 24 = 32$ ways.

■ **Example 4.** How many code words of any length can be made from the letters of *FACTOR*, assuming these letters are written on six cards (so that letters cannot be repeated)?

Solution. We immediately translate the question into asking about 6 disjoint events: make 6-letter words, or 5-letter words, or 4-letter words, or 3-letter words, or 2-letter words, or 1-letter words. We can do this in the following number of ways.

$$
\begin{aligned}
_6P_6 + {} &_6P_5 + {_6P_4} + {_6P_3} + {_6P_2} + {_6P_1} \\
&= 6 \cdot 5 \cdot 4 \cdot 3 \cdot 2 \cdot 1 + 6 \cdot 5 \cdot 4 \cdot 3 \cdot 2 + 6 \cdot 5 \cdot 4 \cdot 3 + 6 \cdot 5 \cdot 4 + 6 \cdot 5 + 6 \\
&= \quad\quad 720 \quad\quad + \quad\quad 720 \quad\quad + \quad 360 \quad + 120 \; + 30 \; + 6 \\
&= 1956
\end{aligned}
$$
■

Students sometimes find it hard to decide whether to multiply or to add in a counting problem. Notice that the words **and** and **or** are in boldface type in the statements of the multiplication principle and of the addition principle. They are the key words; **and** goes with multiplication; **or** goes with addition.

VARIATIONS ON THE ARRANGEMENT PROBLEM

Most counting problems rely on the two fundamental principles that we have enunciated. But many variations need to be considered.

■ **Example 5.** Suppose that the letters of the word *COMPLEX* are printed on seven cards. How many three-letter code words can be formed from these letters if
(a) the first and last letters must be consonants (that is, *C, M, P, L,* or *X*);
(b) all vowels used (if any) must occur in the right-hand portion of a word (that is, a vowel cannot be followed by a consonant)?

Solution.
(a) Let c denote consonant, v vowel, and a any letter. We must fill the three slots below.

We begin by filling the two restricted slots, which can be done in $5 \cdot 4 = 20$ ways. Then we fill the unrestricted slot using one of the 5 remaining letters. It can be done in 5 ways. There are $20 \cdot 5 = 100$ code words of the required type. The following diagram summarizes the procedure.

Fill first   Fill third   Fill second

5		5		4
c		*a*		*c*

$$5 \cdot 5 \cdot 4 = 100$$

(b) We want to count words of the form *cvv*, *ccv*, or *ccc*. Note the use of the addition principle (as well as the multiplication principle) in the following solution.

5	2	1	or	5	4	2	or	5	4	3
c	*v*	*v*		*c*	*c*	*v*		*c*	*c*	*c*

$$5 \cdot 2 \cdot 1 + 5 \cdot 4 \cdot 2 + 5 \cdot 4 \cdot 3 = 10 + 40 + 60 = 110 \quad \blacksquare$$

■ **Example 6.** Lucy has 3 identical red flags, 1 white flag, and 1 blue flag. How many different 5-flag signals could she display from the flagpole of her small boat?

Solution. If the 3 red flags were distinguishable, the answer would be $_5P_5 = 5! = 120$. Pretending they are distinguishable leads to counting an arrangement such as *RRBRW* six times, corresponding to the 3! ways of arranging the 3 red flags (Figure 3). For this reason, we must divide by 3!. Thus the number of signals Lucy can make is

$$\frac{5!}{3!} = \frac{5 \cdot 4 \cdot 3 \cdot 2 \cdot 1}{3 \cdot 2 \cdot 1} = 20 \quad \blacksquare$$

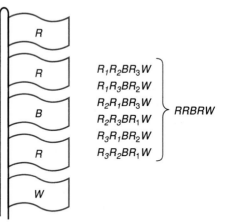

Figure 3

This result can be generalized. For example, given a set of *n* objects in which *j* are of one kind, *k* of a second kind, and *m* of a third kind, then the number of distinguishable permutations is

$$\frac{n!}{j! \, k! \, m!}$$

■ **Example 7.** Suppose that the 11 letters of *MISSISSIPPI* are written on cards. How many 11-letter code words can be formed?

Solution. If all letters were different, the answer would be 11! To take care of the 4 identical *I*s, the 4 identical *S*s, and the 2 identical *P*s, we must divide twice by 4! and by 2!. The answer is

$$\frac{11!}{4!\,4!\,2!} = 34{,}650 \qquad \blacksquare$$

TEASER SOLUTION

M	M	W	W	W	W
W	M	M	W	W	W
W	W	M	M	W	W
W	W	W	M	M	W
W	W	W	W	M	M

Figure 4

We want to count the number of arrangements of 4 women and 2 men that have the men separated. One way to proceed is to find the number of arrangements that have the two men together and subtract this number from the number of ways of arranging 6 people without restriction, that is, from 6! = 720. Figure 4 shows the 5 configurations that put the two men next to each other. The people in each of these configurations can be arranged in 2! · 4! = 48 ways. Thus, the total number of ways of arranging the 6 people with the 2 men together is 5 · 48 = 240. We conclude that the number of ways with the men separated is 720 − 240 = 480.

PROBLEM SET 8.6

A. Skills and Techniques

Evaluate each expression.

1. (a) $_{10}P_3$
   (b) $_6P_4$
2. (a) $_{20}P_2$
   (b) $_8P_3$
3. $_9P_5/_5P_5$
4. $_{15}P_{13}/_{13}P_{13}$

Problems 5–23 require use of the multiplication and addition principles.

5. In how many ways can a president and a secretary be chosen from a group of 6 people?
6. Suppose that a club consists of 3 women and 2 men. In how many ways can a president and a secretary be chosen if
   (a) the president is to be female and the secretary, male?
   (b) the president is to be male and the secretary, female?
   (c) the president and secretary are to be of opposite sex?
7. A box contains 12 cards numbered 1 through 12. Suppose one card is drawn from the box. Find the number of ways each of the following can occur.
   (a) The number drawn is even.
   (b) The number is greater than 9 or less than 3.

8. Suppose that two cards are drawn in succession from the box in Problem 7. Assume that the first card is not replaced before the second one is drawn. In how many ways can each of the following occur?
   (a) Both numbers are even.
   (b) The two numbers are both even or both odd.
   (c) The first number is greater than 9 and the second number is less than 3.
9. Do Problem 8 with the assumption that the first card is replaced before the second one is drawn.
10. In how many ways can a president, a vice-president, and a secretary be chosen from a group of 10 people?
11. How many 4-letter code words can be made from the letters of the word *EQUATION*? (Letters are not to be repeated.)
12. How many 3-letter code words can be made from the letters of the word *PROBLEM* if
    (a) letters cannot be repeated?
    (b) letters can be repeated?
13. Five roads connect Cheer City and Glumville. Starting at Cheer City, how many different ways can Smith drive to Glumville and return? That is, how many different round-trips can he make? How many different round-trips can he make if he wishes to return by a different road than he took to Glumville?
14. Filipe has 4 ties, 6 shirts, and 3 pairs of trousers. How many different outfits can he wear? Assume that he wears one of each kind of article.

15. Papa's Pizza Place offers 3 choices of salad, 20 kinds of pizza, and 4 different desserts. How many different 3-course meals can one order?

16. Maxikota license plate numbers consist of 3 letters followed by 3 digits (for example, AFF033). How many different plates could be issued? (You need not multiply out your answer.)

17. The letters of the word *CREAM* are printed on 5 cards. How many 3-, 4-, or 5-letter code words can be formed?

18. How many code words of all lengths can be made from the letters of *LOGARITHM*, repeated letters not allowed?

19. Frigid Treats sells 10 flavors of ice cream and makes cones in one- to three-dip sizes. How many different looking cones can it make?

20. Using the letters of the word *FACTOR* (without repetition), how many 4-letter code words can be formed
(a) starting with *R*?
(b) with vowels in the two middle positions?
(c) with only consonants?
(d) with vowels and consonants alternating?
(e) with all the vowels (if any) in the left-hand portion of a word (that is, a vowel cannot be preceded by a consonant)?

21. Using the letters of the word *EQUATION* (without repetition), how many 4-letter code words can be formed
(a) starting with *T* and ending with *N*?
(b) starting and ending with a consonant?
(c) with vowels only?
(d) with three consonants?
(e) with all the vowels (if any) in the right-hand portion of the word?

22. Three brothers and 3 sisters are lining up to be photographed. How many arrangements are there
(a) altogether?
(b) with brothers and sisters in alternating positions?
(c) with the 3 sisters standing together?

23. A baseball team is to be formed from a squad of 12 people. Two teams made up of the same 9 people are different if at least some of the people are assigned different positions. In how many ways can a team be formed if
(a) there are no restrictions?
(b) only 2 of the people can pitch and these 2 cannot play any other position?
(c) only 2 of the people can pitch but they can also play any other position?

24. How many different signals consisting of 8 flags can be made using 4 white flags, 3 red flags, and 1 blue flag?

25. How many different signals consisting of 7 flags can be made using 3 white, 2 red, and 2 blue flags?

26. How many different 5-letter code words can be made from the 5 letters of the word *MIAMI*?

27. How many different 11-letter code words can be made from the 11 letters of the word *ABRACADABRA*?

28. In how many different ways can a^4b^6 be written without using exponents? (*Hint:* One way is *aaaabbbbbb*.)

29. In how many different ways can a^3bc^6 be written without using exponents?

30. Consider the part of a city map shown in Figure 5. How many different shortest routes (no backtracking, no cutting across blocks) are there from *A* to *C*? Note that the route shown might be given the designation *EENENNNE*, with *E* denoting *East* and *N* denoting *North*.

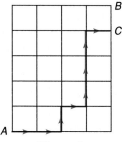

Figure 5

31. How many different shortest routes are there from *A* to *B* in Problem 30?

32. Simplify
(a) $\dfrac{(n + 1)! - n!}{n!}$
(b) $\dfrac{(n + 1)! + n!}{(n + 1)! - n!}$

B. Applications and Extensions

33. Obtain a nice formula for
$$\frac{1}{2!} + \frac{2}{3!} + \frac{3}{4!} + \cdots + \frac{n}{(n + 1)!}$$

Hint: Show first that
$$\frac{k}{(k + 1)!} = \frac{1}{k!} - \frac{1}{(k + 1)!}$$

34. Obtain a nice formula for $1 \cdot 1! + 2 \cdot 2! + 3 \cdot 3! + \cdots + n \cdot n!$ (*Hint:* $k \cdot k! = (k + 1)! - k!$.)

35. A telephone number has 10 digits consisting of an area code (three digits, first is not 0 or 1, second is 0 or 1), an exchange (three digits, first is not 0 or 1, second is not 0 or 1), and a line number (four digits, not all are zeros). How many such 10-digit numbers are there?

36. If Minnesota license plates have 3 digits followed by 3 letters or 3 letters followed by 3 digits or any list of 3 to 7 letters, how many such license plates are possible?

37. How many different integers are there between 0 and 60,000 that use only the digits 1, 2, 3, 4, or 5?

38. Consider making 6-digit numbers from the digits 1, 2, 3, 4, 5, and 6 without repetition.
(a) How many such numbers are there?
(b) Find the sum of these numbers.

39. In how many ways can 6 people be seated at a round table? (We consider two arrangements of people at a round table to be the same if everyone has the same people to the left and right in both arrangements.)

40. A husband and wife plan to invite 4 couples to dinner. The dinner table is rectangular. They decide on a seating arrangement in which the hostess will sit at the end nearest the window, the host at the opposite end, and 4 guests on each side. Furthermore, no man shall sit next to another man, nor shall he sit next to his own wife. In how many ways can this be done?

41. Suppose that n teams enter a tournament in which a team is eliminated as soon as it loses a game.

Since $n \geq 2$ is arbitrary, a number of byes may be needed. How many games must be scheduled to determine a winner? (*Hint:* There is a clumsy way to do this problem, but there is also a very elegant way.)

42. Challenge. Here is an old problem. Suppose we start with an ordered arrangement $(a_1, a_2, \ldots, a_n)$ of n objects. Let d_n be the number of derangements of this arrangement. By a **derangement,** we mean a permutation that leaves no object fixed. For example, the derangements of *ABC* are *BCA* and *CAB*. Show each of the following.
(a) $d_1 = 0$, $d_2 = 1$, and $d_n = (n - 1)d_{n-1} + (n - 1)d_{n-2}$. (*Hint:* To derange the n objects, we may either derange the first $n - 1$ objects and then exchange a_n with one of them, or we may exchange a_n with a_j and then derange the remaining $n - 2$ objects.)
(b) $d_n = n!\left(\dfrac{1}{2!} - \dfrac{1}{3!} + \dfrac{1}{4!} - \dfrac{1}{5!} + \cdots + (-1)^n\dfrac{1}{n!}\right)$
(c) Write a program (or modify an existing one) to calculate d_n for any n. Then calculate d_8, d_{10}, and d_{12}.

TEASER A simple chessboard with its 64 small squares is shown in the accompanying figure. How many rectangles (one is shown with bold boundaries) are determined by this board?

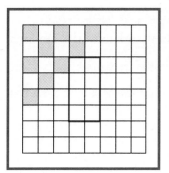

8.7 COUNTING UNORDERED COLLECTIONS

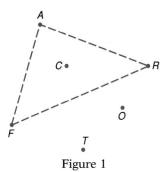

Figure 1

The question in the Teaser is simple once one discovers a clever way to rephrase it and applies the right tools. But that is the rub. How can the question be rephrased, and what are the right tools?

Let us begin with a simpler geometric counting problem. How many triangles are determined by the 6 points shown in Figure 1? Each choice of 3 of these points determines a triangle, but the order in which we choose these points does not matter. For example, *FAR, FRA, AFR, ARF, RAF,* and *RFA* all determine the same triangle, namely, the one shown by dashed lines. The question about triangles is very different from the question about 3-letter code words raised in Section 8.6; yet there is a connection.

We learned that we can make

$$_6P_3 = 6 \cdot 5 \cdot 4 = 120$$

3-letter code words from the letters of *FACTOR*. However, every triangle determined by the points in Figure 1 can be labeled with $3! = 6$ different code words. To find the number of triangles, we should therefore divide the number of code words by $3!$. We conclude that the number of triangles that can be drawn is

$$\frac{_6P_3}{3!} = \frac{6 \cdot 5 \cdot 4}{3 \cdot 2 \cdot 1} = 20$$

COMBINATIONS

An unordered collection of objects is called a **combination** of those objects. If we select r objects from a set of n distinguishable objects, the resulting subset is called a **combination of n things taken r at a time.** The number of such combinations is denoted by $_nC_r$. Thus, $_6C_3$ is the number of combinations of 6 things taken 3 at a time. We calculated this number in the triangle problem; its value is 20.

In general, if $1 \le r \le n$, the combination symbol $_nC_r$ is given by the following formula.

$$_nC_r = \frac{_nP_r}{r!} = \frac{n(n-1)(n-2)\cdots(n-r+2)(n-r+1)}{r(r-1)(r-2)\cdots 2 \cdot 1}$$

You will note that this is exactly the same formula that defines the binomial coefficient $\binom{n}{r}$. Thus, $_nC_r = \binom{n}{r}$, and we will use these two symbols interchangeably. The easy way to remember the formula in the box is to note that the denominator is the product of r integers from r down to 1 and that the numerator has the same number of factors starting with n and working down. The answer is always an integer, so the denominator must divide evenly into the numerator.

■ **Example 1.** Evaluate (a) $_{12}C_4$ (b) $_{10}C_8$ (c) $_{10}C_2$

Solution.

(a) $_{12}C_4 = \dfrac{\cancel{12} \cdot 11 \cdot \overset{5}{\cancel{10}} \cdot 9}{\cancel{4} \cdot \cancel{3} \cdot \cancel{2} \cdot 1} = 495$

(b) $_{10}C_8 = \dfrac{\overset{5}{\cancel{10}} \cdot 9 \cdot \cancel{8} \cdot \cancel{7} \cdot \cancel{6} \cdot \cancel{5} \cdot \cancel{4} \cdot \cancel{3}}{\cancel{8} \cdot \cancel{7} \cdot \cancel{6} \cdot \cancel{5} \cdot \cancel{4} \cdot \cancel{3} \cdot \cancel{2} \cdot 1} = 45$

(c) $_{10}C_2 = \dfrac{\overset{5}{\cancel{10}} \cdot 9}{\cancel{2} \cdot 1} = 45$ ■

Notice that $_{10}C_8 = {_{10}C_2}$. This result is not surprising, since in selecting a subset of 8 objects out of 10, we are also selecting 2 objects to leave behind.

The same reasoning says that

$$_nC_r = {}_nC_{n-r}$$

a fact that also follows from the formula

$$_nC_r = \frac{n!}{r!(n-r)!}$$

Since $0! = 1$, consistency requires that $_nC_0 = 1$.

■ **Example 2.** In how many ways could the majority leader select a group of 15 senators to serve on an ad hoc committee if 20 are eligible for appointment?

Solution. This is a combination problem; the answer is $_{20}C_{15}$. Here is how we calculate this number.

$$_{20}C_{15} = {}_{20}C_5 = \frac{20 \cdot 19 \cdot 18 \cdot 17 \cdot 16}{5 \cdot 4 \cdot 3 \cdot 2 \cdot 1} = 15{,}504 \qquad ■$$

COMBINATIONS VERSUS PERMUTATIONS

Whenever we consider the problem of counting the number of ways of selecting r objects from n objects, we are faced with a question. Is the notion of order significant? If the answer is yes, it is a permutation problem; if no, it is a combination problem.

Consider the Birdwatchers' Club of six people discussed in the previous section. Suppose the club wishes to select a president, vice-president, and secretary (with no restrictions). Is order significant? Yes. The selection can be done in

$$_6P_3 = 6 \cdot 5 \cdot 4 = 120$$

ways.

But suppose the Birdwatchers' Club decides simply to choose an executive committee of three members. Is order relevant? No. A committee consisting of Filipe, Celia, and Amanda is the same as a committee consisting of Celia, Amanda, and Filipe. A 3-member committee can be chosen from 6 people in

$$_6C_3 = \frac{6 \cdot 5 \cdot 4}{3 \cdot 2 \cdot 1} = 20$$

ways.

The words *arrangement, lineup,* and *signal* all suggest order. The words *set, committee, group,* and *collection* do not.

■ **Example 3.** From a set of 20 books, Meg will select 4 to read on her vacation. In how many ways can she make this selection?

Solution. Since order is irrelevant, she can do this in $_{20}C_4 = 4845$ ways. ■

■ **Example 4.** From a set of 20 books, a jury will select four to be awarded first, second, third, and fourth prizes for the quality of their illustrations. In how many ways can this be done?

Solution. Clearly order is important. The jury can make the selection in $_{20}P_4 = 116,280$ ways. ■

■ **Example 5.** From a set of 20 books, how many subsets (of all sizes) can be formed?

Solution. There are subsets of size 0, 1, 2, . . . , 20. Thus, the number of subsets is

$$_{20}C_0 + {}_{20}C_1 + {}_{20}C_2 + {}_{20}C_3 + \cdots + {}_{20}C_{20}$$

We have seen this problem before (Section 8.5) but in connection with the binomial coefficients. In fact, the indicated sum is just the sum of the numbers in row 20 of Pascal's triangle $\left(\text{remember that } {}_nC_r = \binom{n}{r}\right)$. Thus, the sum is 2^{20}. We conclude that a set of 20 elements has $2^{20} = 1,048,576$ subsets. ■

ANOTHER APPROACH

Think about Example 5 this way. To select a subset from the set of 20 books, we have to make a sequence of 20 decisions; that is, for each book we must decide whether to include it in the subset. Since each decision can be made in two ways, the sequence of decisions can be made in 2^{20} ways. The same reasoning shows that a set of n elements has 2^n subsets.

HARDER COMBINATION PROBLEMS

Sometimes we can break a counting task into parts, each of which is a combination problem. Then we put the parts together using the multiplication principle and the addition principle of Section 8.6.

■ **Example 6.** A committee of 4 is to be selected from a group of 3 seniors, 4 juniors, and 5 sophomores. In how many ways can this be done if
(a) there are no restrictions on the selection?
(b) the committee must have 2 sophomores, 1 junior, and 1 senior?
(c) the committee must have at least 3 sophomores?
(d) the committee must have at least 1 senior?

Solution.

(a) $_{12}C_4 = \dfrac{\cancel{12} \cdot 11 \cdot \overset{5}{\cancel{10}} \cdot 9}{\cancel{4} \cdot \cancel{3} \cdot \cancel{2} \cdot 1} = 495$

(b) Two sophomores can be chosen in $_5C_2$ ways, 1 junior in $_4C_1$ ways, and 1 senior in $_3C_1$ ways. By the multiplication principle of counting, the committee can be chosen in

$$_5C_2 \cdot {}_4C_1 \cdot {}_3C_1 = 10 \cdot 4 \cdot 3 = 120$$

ways. We used the multiplication principle because we chose 2 sophomores *and* 1 junior *and* 1 senior.

(c) At least 3 sophomores means 3 sophomores and 1 nonsophomore *or* 4 sophomores. The word *or* tells us to use the addition principle of counting. We get

$$_5C_3 \cdot {}_7C_1 + {}_5C_4 = 10 \cdot 7 + 5 = 75$$

(d) Let x be the number of selections with at least one senior and let y be the number of selections with no seniors. Then $x + y$ is the total number of selections, that is, $x + y = 495$ (see part (a)). We calculate y rather than x because it is easier.

$$y = {}_9C_4 = \frac{9 \cdot 8 \cdot 7 \cdot 6}{4 \cdot 3 \cdot 2 \cdot 1} = 126$$

$$x = 495 - 126 = 369 \qquad \blacksquare$$

CALCULATOR HINT

We suggest that students calculate small combination symbols by hand. Combination symbols with large subscripts are easily evaluated on most graphics calculators. On the TI-81 or TI-82, use the MATH menu and the **PRB** option. Using these, we find that

$${}_{52}C_{13} \approx 6.350135596 \times 10^{11}$$

■ **Example 7.** A standard deck consists of 52 cards. There are 4 suits (spades, clubs, hearts, diamonds), each with 13 cards (2, 3, 4, . . . , 10, jack, queen, king, ace). A bridge hand consists of 13 cards.
(a) How many different possible bridge hands are there?
(b) How many of them have exactly 3 aces?
(c) How many of them have no aces?
(d) How many of them have cards from just 3 suits?

Solution.
(a) The order of the cards in a hand is irrelevant; it is a combination problem. We can select 13 cards out of 52 in ${}_{52}C_{13}$ ways, a number so large we will not bother to calculate it.
(b) The three aces can be selected in ${}_4C_3$ ways, the 10 remaining cards in ${}_{48}C_{10}$ ways. The answer (using the multiplication principle) is ${}_4C_3 \cdot {}_{48}C_{10}$.
(c) From 48 nonaces, we select 13 cards; the answer is ${}_{48}C_{13}$.
(d) We think of this as no clubs, or no spades, or no hearts, or no diamonds and use the addition principle.

$$ {}_{39}C_{13} + {}_{39}C_{13} + {}_{39}C_{13} + {}_{39}C_{13} = 4 \cdot {}_{39}C_{13} \qquad \blacksquare$$

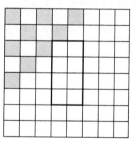

Figure 2

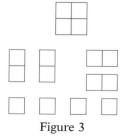

Figure 3

TEASER SOLUTION

We have reproduced a chessboard in Figure 2. Four line segments (the edges) determine a rectangle. We can choose the vertical edges in ${}_9C_2$ ways (we select 2 out of the 9 vertical lines) and, similarly, we can choose the horizontal edges in ${}_9C_2$ ways. Thus, the standard 8 by 8 chessboard determines

$${}_9C_2 \cdot {}_9C_2 = 36 \cdot 36 = 1296$$

rectangles.

What about an n by n chessboard? How many rectangles does it determine? Completely analogous reasoning shows that this chessboard determines

$${}_{n+1}C_2 \cdot {}_{n+1}C_2 = \left(\frac{n(n + 1)}{2} \right)^2$$

rectangles. For example, a 2 by 2 chessboard determines $3^2 = 9$ rectangles. They are shown in Figure 3.

See Problems 47 and 48 for more counting problems related to the chessboard.

PROBLEM SET 8.7

A. Skills and Techniques

Evaluate the symbols in Problems 1–4 by hand.

1. $_{20}C_3$
2. $_{50}C_{48}$
3. $_8C_4$
4. $_{14}C_4$
5. In how many ways can a committee of 3 be selected from a class of 8 students?
6. In how many ways can a committee of 5 be selected from a class of 8 students?
7. A political science professor must select 4 students from her class of 12 students for a field trip to the state legislature. In how many ways can she do it?
8. The professor of Problem 7 was asked to rank the top 4 students in her class of 12. In how many ways could that be done?
9. A police chief needs to assign officers from the 10 available to control traffic at junctions A, B, and C. In how many ways can he do it?
10. If 12 horses are entered in a race, in how many ways can the first 3 places (win, place, show) be taken?
11. A basket contains 30 apples. In how many ways can a person select
    (a) a sample of 4 apples?
    (b) a sample of 3 apples that excludes the 5 rotten ones?
12. From a group of 9 cards numbered 1, 2, 3, . . . , 9, how many
    (a) sets of 3 cards can be drawn?
    (b) 3-digit numbers can be formed?
13. How many games will be played in a 10-team league if each team plays every other team exactly twice?
14. Determine the maximum number of intersection points for a group of 15 lines.
15. How many 10-member subsets does a set of size 12 have?
16. How many subsets (of all sizes) does a set of size 5 have? The empty set is a subset.
17. From a class of 6 members, in how many ways can a committee of any size be selected (including a committee of one)?
18. From a penny, a nickel, a dime, a quarter, and a half dollar, how many different sums can be made?
19. An investment club has a membership of 4 women and 6 men. A research committee of 3 is to be formed. In how many ways can this be done if
    (a) there are to be 2 women and 1 man on the committee?
    (b) there is to be at least 1 woman on the committee?
    (c) all 3 are to be of the same sex?

20. A senate committee of 4 is to be formed from a group consisting of 5 Republicans and 6 Democrats. In how many ways can this be done if
    (a) there are to be 2 Republicans and 2 Democrats on the committee?
    (b) there are to be no Republicans on the committee?
    (c) there is to be at most one Republican on the committee?
21. Suppose that a bag contains 4 black and 7 white balls. In how many ways can a group of 3 balls be drawn from the bag consisting of
    (a) 1 black and 2 white balls?
    (b) balls of just one color?
    (c) at least 1 black ball?
    Note: Assume that the balls are distinguishable; for example, they may be numbered.
22. John is going on a vacation trip and wants to take 5 books with him from his personal library, which consists of 6 science books and 10 novels. In how many ways can he make his selection if he wants to take
    (a) 2 science books and 3 novels?
    (b) at least 1 science book?
    (c) 1 book of one kind and 4 books of the other kind?

Problems 23–28 deal with bridge hands.

23. How many of the possible hands have only red cards? (*Note:* Half of the cards are red.)
24. How many of the hands have only honor cards (aces, kings, queens, and jacks)?
25. How many of the hands have one card of each kind (1 ace, 1 king, 1 queen, and so on)?
26. How many of the hands have exactly 2 kings?
27. How many of the hands have 2 or more kings?
28. How many of the hands have exactly 2 aces and 2 kings?

Problems 29–32 deal with poker hands, which consist of 5 cards.

29. How many different poker hands are possible?
30. How many of them have exactly 2 hearts and 2 diamonds?
31. How many have 2 pairs of different kinds (for example, 2 aces and 2 fives)?
32. How many are 5-card straights (for example, 7, 8, 9, 10, jack in same suit)? An ace may count either as the highest or the lowest card.

B. Applications and Extensions

33. From 5 representatives of labor, 4 representatives of business, and 3 representatives of the general public, how many different mediation committees can be formed with 2 people from each of the three groups?

34. In how many ways can a group of 12 people be split into three nonoverlapping committees of size 5, 4, and 3, respectively?

35. A class of 12 people will select a president, a secretary, a treasurer, and a program committee of 3 with no overlapping of positions. In how many ways can this be done?

36. A committee of 4 is to be formed from a group of 4 freshmen, 3 sophomores, 2 juniors, and 6 seniors. In how many ways can this be done if
(a) each class must be represented?
(b) freshmen are excluded?
(c) the committee must have exactly two seniors?
(d) the committee must have at least one senior?

37. A test consists of 10 true-false items.
(a) How many different sets of answers are possible?
(b) How many of these have exactly 4 right answers?

38. An ice cream parlor has 10 different flavors. How many different double-dip cones can be made if
(a) the two dips must be of different flavors but the order of putting them on the cone does not matter?
(b) the two dips must be different and order does matter?
(c) the two dips need not be different but order does matter?
(d) the two dips need not be different and order does not matter?

39. Mary has a penny, a nickel, a dime, a quarter, a half dollar, and a silver dollar in her purse. How many different possible sums of money (consisting of at least one coin) could she give to her daughter Tosha?

40. In how many ways can 8 presents be split between John and Margaret if
(a) each is to get 4 presents?
(b) John is to get 5 presents and Margaret 3 presents?
(c) There are no restrictions on how the presents are split?

41. Let 10 fixed points on a circle be given. How many convex polygons can be formed that have vertices chosen from among these points?

42. Write a nice formula for

$$ {}_nC_0 + {}_nC_1 + {}_nC_2 + \cdots + {}_nC_n $$

and use it to calculate this sum for $n = 30$.

43. How many subsets of size 3 or more does a set of 15 objects have?

44. Show that

$$ {}_nC_0 \, {}_nC_n + {}_nC_1 \, {}_nC_{n-1} + {}_nC_2 \, {}_nC_{n-2} + \cdots + {}_nC_n \, {}_nC_0 = {}_{2n}C_n $$

by counting the number of ways of drawing n balls from an urn that has n red and n black balls.

45. Find a nice formula for

$$ \sum_{j=0}^{n} ({}_nC_j)^2. $$

(*Hint:* See Problem 44.)

46. Note that ${}_nC_{r-1} + {}_nC_r = {}_{n+1}C_r$ since the corresponding fact is true for binomial coefficients. Use this to obtain a nice formula for

$$ S = {}_{n+1}C_1 + {}_{n+2}C_2 + {}_{n+3}C_3 + \cdots + {}_{n+k}C_k $$

(*Hint:* Add ${}_{n+1}C_0$ to S on the left; then use the result above repeatedly to collect two terms on the left.)

47. Use the diagrams in Figure 4 to derive formulas for

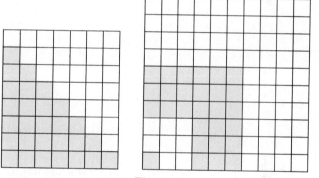

Figure 4

(a) $1 + 2 + 3 + \cdots + n$
(b) $1^3 + 2^3 + 3^3 + \cdots + n^3$

48. **Challenge.** The number of squares determined by the $n \times n$ chessboard is

$$ 1^2 + 2^2 + 3^2 + \cdots + n^2 $$

Verify this for $n = 1, 2,$ and 3. Then find a way to demonstrate the general result.

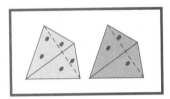

TEASER   A tetrahedral die has the 1, 2, 3, and 4 spots painted on its four sides. Two such dice are to be tossed. What is the probability that the hidden faces will show a total of 6 or more?

8.8   INTRODUCTION TO PROBABILITY

To introduce the basic notions of probability, teachers have traditionally used coins, cubical dice, and decks of cards (Figure 1); we will too. We assume that our coins, dice, and decks are fair. That is, we assume that a coin is as likely to fall showing heads as tails, that each of the six faces of a tossed die has the same chance of showing, and that each of the 52 cards in a well-shuffled deck has the same likelihood of being drawn.

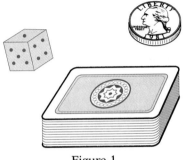

Figure 1

With this background, we are ready to introduce the main notion of this section, the probability of an event. Consider the experiment of tossing a standard die, which can result in six equally likely outcomes. Of these six outcomes, three show an even number of spots. We therefore say that the probability that an even number will show is 3/6.

Or consider the experiment of picking a card from a well-shuffled deck. Thirteen of the 52 cards are diamonds. Thus we say that the probability of getting a diamond is 13/52.

More generally, if an experiment can result in any one of n equally likely outcomes and if exactly m of them result in event E, then we say that the **probability** of E is m/n. We write this statement as

$$P(E) = \frac{m}{n}$$

■ **Example 1.**   One card is to be drawn at random from a deck with 99 cards, numbered 1, 2, 3, . . . , 99. Calculate the probability that
(a) the card will show an even number
(b) the card will show a number greater than 11
(c) the sum of the digits on the card will exceed 11

Solution.
(a) There are 49 even-numbered cards; thus the required probability is 49/99.
(b) There are 88 cards numbered above 11; thus the required probability is 88/99 = 8/9.
(c) The cards meeting the criterion are those with the numbers 39, 48, 49, 57, 58, 59, and so on. There are $1 + 2 + 3 + \cdots + 7 = 28$ of these cards. The required probability is 28/99. ∎

PROPERTIES OF PROBABILITY

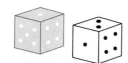

Figure 2

We use two dice, one colored and the other white, to illustrate the major properties of probabilities (Figure 2). Since one die can fall in 6 ways, two dice can fall in $6 \times 6 = 36$ ways. The 36 equally likely outcomes are shown in Figure 3. As you read the next several paragraphs, you will want to refer to this diagram.

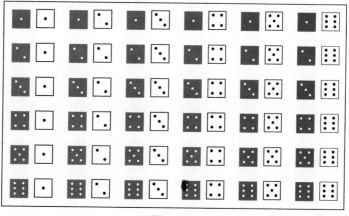

Figure 3

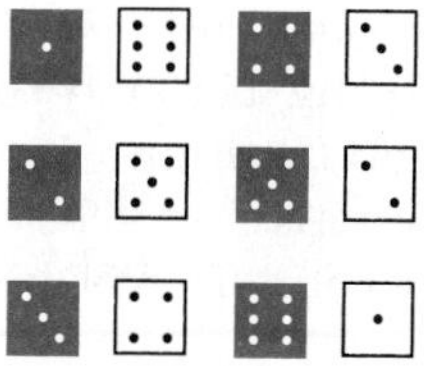

Most questions about a pair of dice have to do with the total number of spots showing after a toss. For example, what is the probability of getting a total of 7? Six of the 36 outcomes result in this event (Figure 4). Therefore, the required probability is 6/36 = 1/6.

Six ways to get 7.
P (getting 7) $= \frac{6}{36}$

Figure 4

■ **Example 2.** Let T denote the total number of spots showing after two fair dice are tossed. Calculate the following probabilities.
(a) $P(T = 11)$
(b) $P(T = 12)$
(c) $P(T > 7)$
(d) $P(T < 13)$
(e) $P(T = 13)$

Two ways to get 11.
P (getting 11) $= \frac{2}{36}$

Figure 5

Solution.

(a) $P(T = 11) = \dfrac{2}{36} = \dfrac{1}{18}$ (see Figure 5)

(b) $P(T = 12) = \dfrac{1}{36}$

(c) $P(T > 7) = \dfrac{15}{36} = \dfrac{5}{12}$

(d) $P(T < 13) = \dfrac{36}{36} = 1$

(e) $P(T = 13) = \dfrac{0}{36} = 0$ ∎

The last two events in Example 2 are worthy of comment. The event "getting a total less than 13" is certain to occur; we call it a **sure event.** The probability of a sure event is always 1. However, the event "getting a total of 13" cannot occur; it is called an **impossible event.** The probability of an impossible event is 0.

Next, consider the event "getting 7 or 11," which is important in the dice game called craps. From the diagram, we see that 8 of the 36 outcomes give a total of 7 or 11, so the probability of this event is 8/36. But note that we could have calculated this probability by adding the probability of getting 7 and the probability of getting 11.

$$P(T = 7 \; or \; T = 11) = P(T = 7) + P(T = 11)$$

$$\frac{8}{36} = \frac{6}{36} + \frac{2}{36}$$

It would appear that we have found a very useful property of probability. When an event is described by using the conjunction *or,* we may find its probability by adding the probabilities of the two parts of the conjunction. A different example, however, makes us take a second look. Note that

$$P(T \text{ is odd } or \; T > 7) \neq P(T \text{ is odd}) + P(T > 7)$$

$$\frac{27}{36} \neq \frac{18}{36} + \frac{15}{36}$$

Why is it that we can add probabilities in one case but not in the other? The reason is a simple one. The events "getting 7" and "getting 11" are *disjoint* (they cannot both happen). But the events "getting an odd total" and "getting over 7" overlap (a number such as 9 satisfies both conditions).

Considerations like these lead us to the main properties of probability.

FOUR PROPERTIES OF PROBABILITY

1. $P(\text{impossible event}) = 0$; $P(\text{sure event}) = 1$
2. $0 \leq P(A) \leq 1$ for any event A.
3. $P(A \text{ or } B) = P(A) + P(B)$, provided that A and B are disjoint (that is, cannot both happen at the same time).
4. $P(A) = 1 - P(\text{not } A)$

Property 4 deserves comment since it follows directly from properties 1 and 3. The events "A" and "not A" are certainly disjoint. Thus, from property 3,

$$P(A \text{ or not } A) = P(A) + P(\text{not } A)$$

The event "*A or* not *A*" must occur; it is a sure event. Hence from property 1,

$$1 = P(A) + P(\text{not } A)$$

which is equivalent to property 4.

■ **Example 3.** Calculate the probability of getting a total *T* less than 11, when tossing two dice.

Solution.

$$P(T < 11) = 1 - P(T \geq 11) = 1 - \frac{3}{36} = \frac{33}{36} = \frac{11}{12}$$

■

RELATION TO SET LANGUAGE

Most American students are introduced to the language of sets in the early grades. Even so, a brief review may be helpful. In everyday language, we talk of a bunch of grapes, a class of students, a herd of cattle, a flock of birds, or perhaps even a team of toads, a passel of possums, or a gaggle of geese. Why are there so many words to express the same idea? We do not know, but we do know that mathematicians prefer to use one word—**set.** The objects that make up a set are called **elements,** or **members,** of the set.

Sets can be put together in various ways (Figure 6). We have $A \cup B$ (read "*A* **union** *B*"), which consists of the elements in *A or B*. The set $A \cap B$ (read "*A* **intersection** *B*") is made up of the elements that are in both *A and B*. We have the notion of an **empty set** $\varnothing$ and of a **universe** *S* (the set of all elements under discussion). Finally, we have *A'* (read "*A* **complement**"), which is composed of all elements in the universe that are not in *A*.

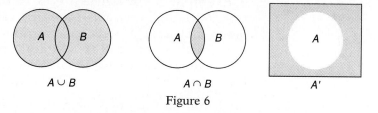

$$A \cup B \qquad\qquad A \cap B \qquad\qquad A'$$

Figure 6

The language of sets and the language of probability are very closely related, as the list below suggests.

Probability Language	**Set Language**
Outcome	Element
Event	Set
Or	Union
And	Intersection
Not	Complement
Impossible event	Empty set
Sure event	Universe

If we borrow the notation of set theory, we can state the laws of probability in a very succinct form.

$$
\begin{array}{ll}
\textbf{1.} & P(\varnothing) = 0; \quad P(S) = 1 \\
\textbf{2.} & 0 \le P(A) \le 1 \\
\textbf{3.} & P(A \cup B) = P(A) + P(B), \text{ provided that } A \cap B = \varnothing \\
\textbf{4.} & P(A) = 1 - P(A')
\end{array}
$$

GAMBLING AND CARD PROBLEMS

Although gamblers may not realize it, they illustrate the laws of probability.

■ **Example 4.** Gamblers often use the language of odds rather than of probability. For example, Las Vegas may report that the odds in favor of the Yankees winning the seventh game of a World Series are 3 to 2. Translate this statement into probability language.

Solution. To say that the odds are 3 to 2 in favor of the Yankees means that if the game were played many times, the Yankees would win 3 times to every 2 times their opponent won. Thus the Yankees would win 3/5 of the times, which means the probability of the Yankees winning is 3/5. In general, if the odds in favor of event E are m to n, then the probability of E is $m/(m + n)$. Conversely, if the probability of E is j/k, then the odds in favor of E are j to $k - j$. ■

■ **Example 5.** A standard deck consists of 52 cards. There are 4 suits (spades, clubs, hearts, and diamonds), each with 13 cards (2, 3, . . . , 10, jack, queen, king, ace). A poker hand consists of 5 cards. If a poker hand is to be dealt from a standard deck, what is the probability it will be
(a) a diamond flush (that is, all diamonds)?   (b) a flush?

Solution.
(a) Recall our study of combinations from Section 8.7. There are $_{52}C_5$ possible 5-card hands, of which $_{13}C_5$ consists of all diamonds. Thus

$$
P(\text{diamond flush}) = \frac{_{13}C_5}{_{52}C_5} = \frac{\dfrac{13 \cdot 12 \cdot 11 \cdot 10 \cdot 9}{5 \cdot 4 \cdot 3 \cdot 2 \cdot 1}}{\dfrac{52 \cdot 51 \cdot 50 \cdot 49 \cdot 48}{5 \cdot 4 \cdot 3 \cdot 2 \cdot 1}} \approx 0.000495
$$

(b) A flush is a diamond flush or a heart flush or a club flush or a spade flush. Thus

$$
\begin{aligned}
P(\text{flush}) &= P(\text{diamond flush}) + P(\text{heart flush}) \\
&\quad + P(\text{club flush}) + P(\text{spade flush}) \\
&\approx 0.0005 + 0.0005 + 0.0005 + 0.0005 \\
&= 0.002
\end{aligned}
$$
■

A Fifth Property of Probability

We have already used without comment one rather obvious generalization of property 3. If A, B, and C are three mutually disjoint events, then

$$P(A \cup B \cup C) = P(A) + P(B) + P(C)$$

Clearly this kind of result also holds for four or more mutually disjoint events.

We are interested in another type of generalization of property 3, namely, a result for the case where two events A and B overlap (can occur at the same time), as in Figure 7. We claim that for any events A and B, overlapping or not,

5. $P(A \cup B) = P(A) + P(B) - P(A \cap B)$

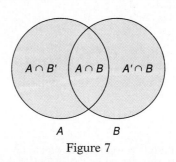

Figure 7

To see that this is true, note that the three events $A \cap B'$, $A' \cap B$, and $A \cap B$ are mutually disjoint and that together they make up $A \cup B$. Thus,

$$
\begin{aligned}
P(A \cup B) &= P(A \cap B') + P(A' \cap B) + P(A \cap B) \\
&= (P(A \cap B') + P(A \cap B)) + (P(A' \cap B) + P(A \cap B)) - P(A \cap B) \\
&= P(A) + P(B) - P(A \cap B)
\end{aligned}
$$

■ **Example 6.**   Determine the probability of getting an honor card (jack, queen, king, ace) or a diamond in drawing one card from a well-shuffled deck.

Solution. Let H denote the event "get an honor card" and D the event "get a diamond." The event of interest is $H \cup D$.

$$
\begin{aligned}
P(H \cup D) &= P(H) + P(D) - P(H \cap D) \\
&= \frac{16}{52} + \frac{13}{52} - \frac{4}{52} = \frac{25}{52}
\end{aligned}
$$
■

Teaser Solution

A pair of tetrahedral dice can fall in $4 \times 4 = 16$ equally likely ways. Let us label these ways by ordered pairs $(1, 1)$, $(1, 2)$, $(1, 3)$, . . . , $(4, 3)$, $(4, 4)$ with the two numbers in an ordered pair indicating the number of spots face down on the first and second die, respectively. Of these 16 outcomes, the pairs $(2, 4)$, $(3, 3)$, $(3, 4)$, $(4, 2)$, $(4, 3)$, and $(4, 4)$ exhibit a total of 6 or more. Thus the probability of this event is $6/16 = 3/8$.

A. Skills and Techniques

1. An ordinary die is tossed. What is the probability that the number of spots on the upper face will be
   (a) three
   (b) greater than 3
   (c) less than 3
   (d) an even number
   (e) an odd number

2. Nine balls, numbered 1, 2, ... , 9, are in a bag. If one is drawn at random, what is the probability that its number is
   (a) 9
   (b) greater than 5
   (c) less than 6
   (d) even
   (e) odd

3. A penny, a nickel, and a dime are tossed. List the 8 possible outcomes of this experiment. What is the probability of
   (a) 3 heads
   (b) exactly 2 heads
   (c) more than 1 head

4. A coin and a die are tossed. Suppose that one side of the coin has 1 on it and the other 2. List the 12 possible outcomes of this experiment, for example, (1, 1), (1, 2), (1, 3). What is the probability of
   (a) a total of 4
   (b) an even total
   (c) an odd total

5. Two ordinary dice are tossed. What is the probability of
   (a) a double (both showing the same number)
   (b) the number on one of the dice being twice that on the other
   (c) the numbers on the two dice differing by at least 2

6. Two regular octahedra (polyhedra having eight identical faces) have faces numbered 1, 2, ... , 8. Suppose that they are tossed and we record the outcomes by listing the numerals on the bottom faces.
   (a) How many outcomes are there?
   (b) What is the probability of a sum of 7?
   (c) What is the probability of a sum less than 7?

7. What is wrong with each of the following statements?
   (a) Since there are 50 states, the probability of being born in Wyoming is 1/50.
   (b) The probability that a person smokes is 0.45, and that he or she drinks, 0.54. Therefore the probability that he or she smokes or drinks is 0.54 + 0.45 = 0.99.
   (c) The probability that a certain candidate for president of the United States will win is 3/5, and that he or she will lose, 1/4.
   (d) Two football teams A and B are evenly matched; therefore, the probability that A will win is 1/2.

8. During the past 30 years, Professor Witquick has given only 100 As and 200 Bs in Math 13 to the 1200 students who registered for the class. On the basis of these data, what is the probability that a student who registers next year
   (a) will get an A or a B
   (b) will not get either an A or a B

9. A poll was taken of all administrators, faculty, and students at Podunk University on the question of coeducational dormitories, with the following results.

	Administrators	Faculty	Students	Total
For	4	16	100	120
Against	3	32	100	135
Unsure	3	2	40	45
Total	10	50	240	300

   On the basis of this poll, what is the probability that
   (a) a randomly chosen faculty member will favor coed dorms
   (b) a randomly chosen student will be against coed dorms
   (c) a person selected at random at Podunk University will favor coed dorms
   (d) a person selected at random at Podunk University will be a faculty member who is against coed dorms

10. The well-balanced spinner shown in Figure 8 is spun. What is the probability that the pointer will stop at
    (a) red
    (b) green
    (c) red or green
    (d) not green

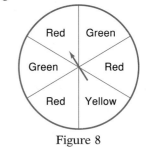

Figure 8

11. A numbered five-volume set of books is placed on a shelf at random. What is the probability they will be in the correct left-to-right order?

12. Four balls numbered 1, 2, 3, and 4 are placed in a bag, mixed, and drawn out, one at a time. What is the probability that they will be drawn in the order 1, 2, 3, 4?

13. Suppose that in a Vikings/Steelers Super Bowl, the odds in favor of the Vikings winning are given as 2 to 9. What is the probability that they will lose the Super Bowl game? (*Note:* Ties are not possible in Super Bowls.)

14. What is the probability that David will marry Jane if he says the odds in favor are 1 to 8?

15. What are the odds in favor of getting two heads when tossing two coins?

16. What are the odds in favor of getting 7 or 11 in tossing two dice?

17. From a standard deck, one card is drawn. What is the probability that it will be
    (a) red
    (b) a spade
    (c) an ace
    (*Note:* Two of the suits are red and two are black.)

18. Two cards are drawn from a standard deck (there are $_{52}C_2$ ways of doing it). What is the probability that both will be
    (a) red
    (b) of the same color
    (c) aces

19. Three cards are drawn from a standard deck. What is the probability that
    (a) all will be red
    (b) all will be diamonds
    (c) exactly one will be a queen
    (d) all will be queens

20. A poker hand is drawn from a standard deck. What is the probability of drawing at least one ace? (*Hint:* Look at the complementary event consisting of no aces.)

21. What is the probability of getting a poker hand consisting of all kings and queens?

22. What is the probability of getting a full-house poker hand (two cards of one kind, three of another)?

B. Applications and Extensions

23. If three dice are tossed together, what is the probability that the total obtained will be
    (a) 18
    (b) 16
    (c) greater than 4

24. A purse contains 3 nickels, 1 dime, 1 quarter, and 1 half dollar. If three coins are drawn from the purse at random, what is the probability that their value is
    (a) 15¢
    (b) 40¢
    (c) $1
    (d) more than 50¢

25. A single card is drawn from a standard deck. Find the probability of getting the following.
    (a) An honor card (*A, J, Q, K*)
    (b) A black card or the queen of hearts (hearts are red)
    (c) A red card or the queen of hearts

26. A coin is tossed four times. Find the probability of getting
    (a) no heads
    (b) at least 1 tail
    (c) exactly 3 heads

27. If four balls are drawn at random from a bag containing 5 red and 3 black balls, what is the probability of getting each of the following?
    (a) One black ball and 3 red balls
    (b) Two balls of each color
    (c) At least 1 black ball
    (d) At least 1 red ball

28. If 4 cards are drawn at random from a standard deck, what is the probability of getting each of the following?
    (a) One card from each suit
    (b) Three clubs and 1 diamond
    (c) Two kings and 2 queens

29. A box contains 15 cards numbered 1 through 15. Three cards are drawn at random from the box. Find the probability of each event.
    (a) All 3 numbers are even
    (b) At least 1 number is odd
    (c) The product of the 3 numbers is even

30. A die has been loaded so that the probabilities of getting 1, 2, 3, 4, 5, and 6 are 1/3, 1/4, 1/6, 1/12, 1/12, and 1/12, respectively. Assume that the usual properties of probability are still valid in this situation where the outcomes are not equally likely. Find the probability of rolling each of the following.
    (a) An even number
    (b) A number less than 5
    (c) An even number or a number less than 5

31. Three men and 4 women are to be seated in a row at random. Find the probability of each event.
    (a) The men and women alternate.
    (b) The women are together.
    (c) The end positions are occupied by men.

32. Six shoes are to be picked from a pile consisting of 10 identical left shoes and 7 corresponding identical right shoes. What is the probability of getting
    (a) three pairs
    (b) exactly two pairs

33. A single die is to be tossed four times in succession. Find the probability of each event.
    (a) The numbers 1, 2, 3, and 4 appear in that order.
    (b) The numbers 1, 2, 3, and 4 appear in any order.
    (c) At least one 6 appears.
    (d) The same number appears each time.

34. If A and B overlap, then the fifth property of probability says that

    $$P(A \cup B) = P(A) + P(B) - P(A \cap B)$$

    Determine the corresponding formula for

    $$P(A \cup B \cup C)$$

35. The letters of *MATHEMATICS* are written on 11 cards and arranged in a row at random. What is the probability that they spell *MATHEMATICS?*

36. Let S be the set of 25 ordered pairs that are the coordinates of the 25 points shown in Figure 9. If an ordered pair (x, y) is chosen at random from S, what is the probability that
    (a) $x + y = 4$
    (b) $x + y < 5$
    (c) neither x nor y is 5
    (d) $y > x$

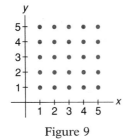

Figure 9

37. Consider the set of triangles that can be formed with vertices from the points $(1, 1)$, $(2, 1)$, $(3, 1)$, $(2, 3)$, $(3, 3)$, and $(4, 3)$. If a triangle is chosen at random from this set, what is the probability that it
    (a) is right-angled
    (b) has area 2
    (*Hint:* A picture will help.)

38. A careless secretary typed 4 letters and 4 envelopes and then inserted the letters at random in the envelopes. Find the probability of each of the following.

(a) No letter went into the correct envelope.
(b) At least 1 letter went into the correct envelope.
(c) Exactly 1 letter went into the correct envelope.
(d) At least 2 letters went into the correct envelopes.
(e) Exactly 3 letters went into the correct envelopes.

39. The careless secretary of Problem 38 typed n letters and n envelopes $(n > 2)$ and then inserted the letters at random. Find the probabilities of each of the following.
    (a) Exactly $n - 1$ letters went into the correct envelopes.
    (b) No letter went into the correct envelope (see Problem 42 of Section 8.6).
    (c) Exactly 1 letter went into the correct envelope. (*Note:* In calculus, it is shown that the answer to (b) converges to $1/e \approx 0.368$ as n increases and that this convergence is very rapid. In fact, 0.368 is a fine approximation for $n \geq 6$. Thus, the probability that 6 letters all go into the wrong envelopes is about the same as the probability that 600 letters all go into the wrong envelopes, a result most people find surprising.)

40. **Challenge.** Three sticks are chosen at random from 8 sticks of lengths $1, 2, \ldots, 8$. What is the probability that a triangle can be formed from them?

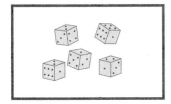

TEASER   In the game of Yahtzee, which involves tossing five dice, what is the probability of getting four of a kind in one toss?

8.9   INDEPENDENCE IN PROBABILITY PROBLEMS

A perfectly balanced coin has shown nine tails in a row. What is the probability that it will show heads on the tenth flip? Some people argue that a mystical law of averages makes the appearance of a head practically certain. It is as if the coin had a memory and a conscience; it must atone for falling on its face nine times in a row. Such thinking is pure nonsense. The probability of showing tails on the tenth flip is 1/2, just as it was on each of the previous nine flips. The outcome of any flip is independent of what happened on previous flips.

Here is a different question, not to be confused with the one just answered. If one plans to flip a coin 10 times, what is the probability of getting all tails?

To answer, we reason that there are 2 possibilities on the first flip, 2 on the second, and so forth. By the multiplication principle for counting (see Section 8.6), there are $2 \cdot 2 \cdot 2 \cdot 2 \cdot 2 \cdot 2 \cdot 2 \cdot 2 \cdot 2 \cdot 2$, or 1024, possible outcomes of this experiment, only one of which consists of all tails. The probability of 10 tails is 1/1024, a very unlikely event indeed.

So what do we mean by saying two events are independent? Informally, we say that event A is independent of event B if the occurrence of A does not influence the occurrence of B. Thus, having gotten tails on the first toss of a coin does not influence the likelihood of getting tails (or heads) on the second toss. On the other hand, having gotten a good grade in mathematics may very well affect the likelihood of getting a passing grade in physics. These latter two events are dependent.

DEPENDENCE VERSUS INDEPENDENCE

Perhaps we can make the distinction between dependence and independence clear by describing an experiment in which we again toss two dice—one colored and the other white. Let A, B, and C designate the following events.

Only 6 outcomes if we know that colored die shows 6

Figure 1

A: colored die shows 6
B: white die shows 5
C: total on the two dice is greater than 7

Consider first the relationship between B and A. It seems quite clear that the chance of B occurring is not affected by whether or not A has occurred. In fact, if we let $P(B \mid A)$ denote the probability of B given that A has occurred, then (see Figure 1)

$$P(B \mid A) = \frac{1}{6}$$

But this is equivalent to the answer we get if we calculate $P(B)$ without any knowledge of A. For then we look at all 36 outcomes for two dice and note that 6 of them have the white die showing 5; that is,

$$P(B) = \frac{6}{36} = \frac{1}{6}$$

We conclude that $P(B \mid A) = P(B)$, just as we expected.

The relation between C and A is very different; C's chances are greatly improved if A has occurred. From Figure 1, we see that

$$P(C \mid A) = \frac{5}{6}$$

But if we do not know whether or not A has occurred and calculate $P(C)$ by looking at the 36 outcomes for two dice, we find

$$P(C) = \frac{15}{36} = \frac{5}{12}$$

Clearly $P(C \mid A) \neq P(C)$.

In terms of the symbol $P(B \mid A)$, we can now state the **multiplication rule for probabilities,** namely,

$$P(A \cap B) = P(A)P(B \mid A) = P(B)P(A \mid B)$$

In words, the probability of both A and B occurring is equal to the probability that A will occur multiplied by the probability that B will occur given that A has occurred. And in the case $P(B \mid A) = P(B)$, this takes the particularly simple form

$$P(A \cap B) = P(A)\,P(B)$$

which we take to be the formal definition of what it means for two events A and B to be **independent.**

■ **Example 1.** Suppose that two cards are to be drawn (one after another) from a well-shuffled deck. Find in two different ways the probability that both will be spades.

Solution. Method 1: On the basis of previous knowledge from Section 8.8, we respond:

$$P(\text{two spades}) = \frac{{}_{13}C_2}{{}_{52}C_2} = \frac{\dfrac{13 \cdot 12}{2 \cdot 1}}{\dfrac{52 \cdot 51}{2 \cdot 1}} = \frac{13 \cdot 12}{52 \cdot 51}$$

Method 2: Consider the events

A: getting a spade on the first draw
B: getting a spade on the second draw

Our interest is in $P(A \cap B)$. According to the rule above, it it given by

$$P(A \cap B) = P(A)P(B \mid A) = \frac{13}{52} \cdot \frac{12}{51}$$

which naturally agrees with our earlier answer. ■

■ **Example 2.** Suppose that from a standard deck, we draw one card, place it back in the deck, reshuffle, and then draw a second card. What is the probability that both cards will be spades?

Solution. Let A and B have the meanings ascribed in Example 1. Now A and B are independent and

$$P(A \cap B) = P(A)P(B) = \frac{13}{52} \cdot \frac{13}{52} \qquad ■$$

■ **Example 3.** Recall two general laws of probability. The first appeared near the end of Section 8.8; the second came in the present section.

$$P(A \cup B) = P(A) + P(B) - P(A \cap B)$$

$$P(A \cap B) = P(A)P(B \mid A)$$

If $P(A) = 0.7$, $P(B) = 0.3$, and $P(B \mid A) = 0.4$, calculate
(a) $P(A \cap B)$ (b) $P(A \cup B)$ (c) $P(A \mid B)$

Solution.
(a) $P(A \cap B) = P(A)P(B \mid A) = (0.7)(0.4) = 0.28$
(b) $P(A \cup B) = P(A) + P(B) - P(A \cap B) = 0.7 + 0.3 - 0.28 = 0.72$
(c) $P(A \mid B) = P(A \cap B)/P(B) = 0.28/0.3 \approx 0.93$ ■

URNS AND BALLS

For reasons not entirely clear, teachers have always illustrated the central ideas of probability by talking about urns (vases) containing colored balls. Most of us have never seen an urn containing colored balls, but it won't hurt to use a little imagination.

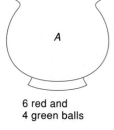

6 red and
4 green balls

■ **Example 4.** Consider two urns A and B, A containing 6 red balls and 4 green balls, and B containing 18 red balls and 2 green balls (Figure 2).
(a) If a ball is drawn from each urn, what is the probability that both will be red?
(b) If an urn is chosen at random and then a ball drawn, what is the probability that it will be red?

Solution.
(a) $P(\text{red from } A \textit{ and } \text{red from } B) = P(\text{red from } A) \cdot P(\text{red from } B)$

$$= \frac{6}{10} \cdot \frac{18}{20} = \frac{3}{5} \cdot \frac{9}{10} = \frac{27}{50}$$

18 red and
2 green balls

Figure 2

(b) First describe the desired event as

"choose A *and* draw red"   or   "choose B *and* draw red"

The events in quotation marks are disjoint, so their probabilities can be added. We obtain the answer

$$\frac{1}{2} \cdot \frac{6}{10} + \frac{1}{2} \cdot \frac{9}{10} = \frac{15}{20} = \frac{3}{4}$$

You should note the procedure we use. We describe the event using the words *and* and *or*. When we determine probabilities, *and* translates into *times*, and *or* into *plus*.   ■

Figure 3

■ **Example 5.** An urn contains 2 red, 3 green, and 5 black balls (Figure 3). Two balls are drawn at random. What is the probability that both are green
(a) if we replace the first ball before the second is drawn
(b) if we do not replace the first ball before the second is drawn

Solution.
(a) Here we have independence.

$$P(\text{green } and \text{ green}) = P(\text{green}) \cdot P(\text{green})$$
$$= \frac{3}{10} \cdot \frac{3}{10} = \frac{9}{100}$$

(b) Now we have dependence. The outcome of the first draw does affect what happens on the second draw.

$$P(\text{green } and \text{ green}) = P(\text{first ball green}) \cdot P(\text{second ball green} \mid \text{first ball green})$$
$$= \frac{3}{10} \cdot \frac{2}{9} = \frac{6}{90} = \frac{1}{15}$$ ∎

A HISTORICAL EXAMPLE

The two French mathematicians Fermat and Pascal are usually given credit for originating the theory of probability. This is how it happened: The famous gambler, Chevalier de Méré, was fond of a dice game in which he would bet that a 6 would appear at least once in four throws of a die. He won more often than he lost for, though he probably did not know it,

$$P(\text{at least one 6}) = 1 - P(\text{no 6s})$$
$$= 1 - \frac{5}{6} \cdot \frac{5}{6} \cdot \frac{5}{6} \cdot \frac{5}{6} \approx 1 - 0.48 = 0.52$$

Growing tired of this game, de Méré introduced a new one played with two dice. He then bet that at least one double 6 would appear in 24 throws of two dice. Somehow (perhaps he noted that 4/6 = 24/36) he thought he should do just as well as before. But he lost more often than he won. Mystified, he proposed it as a problem to Pascal, who in turn wrote to Fermat. Together they produced the following explanation.

$$P(\text{at least one double 6}) = 1 - P(\text{no double 6s})$$
$$= 1 - \left(\frac{35}{36}\right)^{24} \approx 1 - 0.51 = 0.49$$

From this humble, slightly disreputable origin grew the science of probability.

THE BINOMIAL DISTRIBUTION

In tossing a regular die, let us consider getting 1 or 2 a success (denoted by S) and getting anything else a failure (denoted by F). Then

$$P(S) = \frac{1}{3} \quad \text{and} \quad P(F) = \frac{2}{3}$$

Imagine tossing this die 8 times. What is the probability of getting exactly 3 successes?

Two sequences of outcomes that contain 3 successes are SSSFFFFF and SFSFSFFF. Note that

$$P(SSSFFFFF) = \frac{1}{3}\frac{1}{3}\frac{1}{3}\frac{2}{3}\frac{2}{3}\frac{2}{3}\frac{2}{3}\frac{2}{3} = \left(\frac{1}{3}\right)^3\left(\frac{2}{3}\right)^5$$

$$P(SFSFSFFF) = \frac{1}{3}\frac{2}{3}\frac{1}{3}\frac{2}{3}\frac{1}{3}\frac{2}{3}\frac{2}{3}\frac{2}{3} = \left(\frac{1}{3}\right)^3\left(\frac{2}{3}\right)^5$$

In fact, all sequences of outcomes that contain 3 successes have this same probability. How many such sequences are there? As many as the number of code words that can be formed from 3 Ss and 5 Fs. But in Section 8.6 we learned that this number is $8!/(3! \cdot 5!)$ or, equivalently, it is $\binom{8}{3}$. We conclude that the probability of exactly 3 successes is $\binom{8}{3}(1/3)^3(2/3)^5 \approx 0.273$.

We can generalize to any experiment with two outcomes S and F with $P(S) = p$ and $P(F) = q = 1 - p$. If we conduct n independent trials of this experiment, the probability of exactly r successes is given by

$$P(r \text{ successes in } n \text{ trials}) = \binom{n}{r}p^r q^{n-r}$$

Because the expression $\binom{n}{r}p^r q^{n-r}$ obtained above is the general term in the binomial expansion of $(p + q)^n$, the distribution of probabilities obtained from repeated trials of a two-outcome experiment is called the **binomial distribution.**

■ **Example 6.** Experience has shown that 95% of the microscopes made by a certain manufacturer pass the stringent performance test of the ABC laboratory on delivery but that 5% of them fail and have to be sent back. If the ABC laboratory buys a lot of 20 microscopes from this manufacturer, what is the probability of the event M that at least 18 of them will pass the test?

Solution. That at least 18 should pass the test means that 18 or 19 or 20 must pass. We conclude that

$$P(M) = P(18 \text{ pass}) + P(19 \text{ pass}) + P(20 \text{ pass})$$
$$= \binom{20}{18}(0.95)^{18}(0.05)^2 + \binom{20}{19}(0.95)^{19}(0.05) + \binom{20}{20}(0.95)^{20}$$
$$\approx 0.9245$$

■

TEASER SOLUTION

First let us agree that tossing five dice is equivalent to tossing one die five times. Also to get 4 of a kind will mean to get at least 4 ones or at least 4 twos, and so on (there are 6 possibilities). Since

$$P(\text{at least 4 ones}) = P(4 \text{ ones and 1 non-one}) + P(5 \text{ ones})$$
$$= \binom{5}{4}\left(\frac{1}{6}\right)^4\left(\frac{5}{6}\right) + \binom{5}{5}\left(\frac{1}{6}\right)^5$$

the probability of getting four of a kind is

$$6\left(5\left(\frac{1}{6}\right)^4\left(\frac{5}{6}\right) + \left(\frac{1}{6}\right)^5\right) \approx 0.02006$$

PROBLEM SET 8.9

A. Skills and Techniques

1. Toss a balanced die three times in succession. What is the probability of getting all 1s?
2. Toss a fair coin four times in succession. What is the probability of getting all heads?
3. Spin the two spinners pictured in Figure 4. What is the probability that

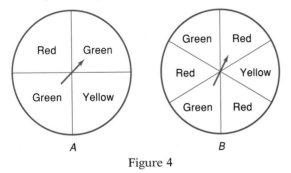

Figure 4

   (a) both will show red (that is, *A* shows red *and B* shows red)
   (b) neither will show red (that is, *A* shows not red *and B* shows not red)
   (c) spinner *A* will show red and spinner *B* not red
   (d) spinner *A* will show red and spinner *B* red or green
   (e) just one of the spinners will show green
4. The two boxes shown in Figure 5 are shaken thoroughly and a ball is drawn from each. What is the probability that

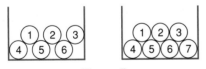

Figure 5

   (a) both will be 1s
   (b) exactly one of them will be a 2 (2 from first box and not 2 from second box, or not 2 from first box and 2 from second box)
   (c) both will be even
   (d) exactly one of them will be even
   (e) at least one of them will be even
5. In each case, indicate whether or not the two events seem independent to you. Explain.
   (a) Getting an *A* in math and getting an *A* in physics
   (b) Getting an *A* in math and winning a tennis match
   (c) Getting a new shirt for your birthday and stubbing your toe the next day
   (d) In tossing two dice, getting an odd total and getting a 5 on one of the dice

   (e) Being a woman and being a doctor
   (f) Walking under a ladder and having an accident the next day
6. In each case indicate whether or not the two events are necessarily disjoint.
   (a) Getting an *A* in math and getting an *A* in physics
   (b) Getting an *A* in Math 101 and getting a *B* in Math 101
   (c) In tossing two dice, getting an odd total and getting the same number on both dice
   (d) In tossing two dice, getting an odd total and getting a 5 on one of the dice
   (e) The sun shining on Tuesday and the weather being rainy on Tuesday
   (f) Not losing the college football game and not winning it
7. A machine produces bolts that are put in boxes. It is known that 1 box in 10 will have at least one defective bolt in it. Assuming that the boxes are independent of each other, what is the probability that a customer who ordered 3 boxes will get all good bolts?
8. Suppose that the probability of being hospitalized during a year is 0.152. Assuming that family members are hospitalized independently of each other, what is the probability that no one in a family of five will be hospitalized this year? Do you think the assumption of independence is reasonable?
9. Given that $P(A) = 0.8$, $P(B) = 0.5$, and $P(A \cap B) = 0.4$, find
   (a) $P(A \cup B)$
   (b) $P(B \mid A)$
   (c) $P(A \mid B)$
10. Given that $P(A) = 0.8$, $P(B) = 0.4$, and $P(B \mid A) = 0.3$, find
    (a) $P(A \cap B)$
    (b) $P(A \cup B)$
    (c) $P(A \mid B)$
11. In Sudsville, 30% of the people are Catholics, 55% of the people are Democrats, and 90% of the Catholics are Democrats. Find the probability that a person chosen at random from Sudsville is
    (a) both Catholic and Democrat
    (b) Catholic or Democrat
    (c) Catholic given that he or she is Democrat
12. At Podunk U, 1/4 of the applicants fail the entrance examination in mathematics, 1/5 fail the one in English, and 1/9 fail both mathematics and English. What is the probability that an applicant will
    (a) fail mathematics or English?
    (b) fail mathematics, given that he or she failed English?
    (c) fail English, given that he or she failed mathematics?
13. Consider two urns, one with 3 red balls and 7 white balls, the other with 6 red balls and 6 white balls.

If an urn is chosen at random and then a ball is drawn, what is the probability it will be red?

14. Suppose that 4% of males are colorblind, that 1% of females are colorblind, and that males and females each make up 50% of the population. If a person is chosen at random, what is the probability that this person will be colorblind?

15. Refer to the urns of Example 4. Draw a ball from urn A. If it is red, replace it and draw 2 balls from urn A. If it is green, draw a second ball from urn B. Calculate the probability of finishing with 2 balls of the same color.

16. In Figure 5, choose a box at random and draw a ball. If it is odd, draw a second ball from that box; if even, draw a second ball from the opposite box. Calculate the probability that the two balls have the same parity, that is, both are odd or both are even.

Problems 17–22 refer to the urn of Example 5.

17. If 2 balls are drawn, what is the probability that both are black assuming
    (a) replacement between draws
    (b) nonreplacement

18. If 2 balls are drawn without replacement between draws, what is the probability of getting
    (a) red on the first draw and green on the second
    (b) a red and a green ball in either order
    (c) 2 red balls
    (d) 2 nonred balls

19. If 3 balls are drawn with replacement between draws, what is the probability of getting
    (a) 3 green balls
    (b) 3 red balls
    (c) 3 balls of the same color
    (d) 3 balls of all different colors

20. Answer the questions of Problem 19 if there is no replacement between draws.

21. Two balls are drawn. What is the probability that the second one drawn is red if
    (a) the first is replaced before the second is drawn
    (b) the first is not replaced before the second is drawn

22. If 4 balls are drawn without replacement between draws, what is the probability of getting
    (a) 4 balls of the same color
    (b) at least two different colors
    (c) 2 balls of one color and 2 of another

23. A balanced coin is tossed six times. Calculate the probability of getting
    (a) no heads
    (b) exactly one head
    (c) exactly two heads
    (d) exactly three heads
    (e) more than three heads

24. A fair coin is tossed eight times. Calculate the probability of getting
    (a) no tails
    (b) exactly one tail

(c) exactly two tails
(d) exactly three tails
(e) at most three tails

25. Experiments indicate that, for an ordinary thumbtack, the probability of its falling head down is 1/3 and head up, 2/3 (Figure 6). What is the probability in 12 tosses of its falling

Prob. = $\frac{1}{3}$

Prob. = $\frac{2}{3}$

Figure 6

(a) head up exactly four times
(b) head up exactly six times

26. On a true-false test of 20 items, Homer estimates that the probability of his getting any one item right is 3/4. What is the probability of his getting
    (a) exactly 19 right
    (b) at least 19 right

27. A certain machine will work if each of five parts operates successfully. If the probability of failure during a day is 1/10 for each part, what is the probability that the machine will work for a whole day without a breakdown? If only four of the five parts must operate successfully, what would the answer be?

28. A small airline has learned that 10% of those who make reservations for a given flight will fail to use their reservations. There are 22 reservations for a plane that has 20 seats. What is the probability that all who show will get seats? (*Hint:* Look at the complementary event, namely, 21 or 22 using their reservations.)

B. Applications and Extensions

29. Alonda Lake, a candidate for governor, estimates that the probability of winning her party's nomination is 2/3 and that if she wins the nomination, the probability of her winning the election is 5/8. Find the probability that
    (a) she will be nominated by her party and then lose the election
    (b) she will win the election

30. In the semifinals of a tennis tournament, A will play B and C will play D. The winners will meet in the finals. The probability that A will beat B is 2/3, that C will beat D is 5/6, that A will beat C (if they play) is 1/4, and that A will beat D (if they play) is 4/5. Find the probability that A will win the tournament.

31. An urn contains 4 red, 5 white, and 7 black balls. Two balls are drawn in succession. What is the probability of drawing 2 white balls if the first ball
    (a) is replaced before the second is drawn
    (b) is not replaced before the second is drawn

32. Four balls are drawn from the urn of Problem 31 without replacement between draws. What is the probability that
    (a) all are white
    (b) at least one is red
    (c) the second ball drawn is black

33. An urn contains 3 red, 2 white, and 5 black balls. If a ball is drawn at random, replaced, and then a second ball drawn, what is the probability that
    (a) both are red
    (b) one is white and the other is black
    (c) the two balls have the same color
    (d) the balls have different colors

34. Consider the urns of Problems 31 and 33. If one of these urns is picked at random and two balls are drawn without replacement between draws, what is the probability that
    (a) both are red
    (b) neither is white
    (c) one is red and the other is white

35. In tossing a pair of dice repeatedly, what is the probability of getting a 7 before an 11? (*Hint:* Consider the infinite geometric sum whose first term is the probability of getting 7 on the first toss, whose second term is the probability of getting neither 7 nor 11 on the first toss followed by 7 on the second toss, and so on.)

36. A coin is tossed. If a head appears, you draw two cards from a standard deck and, if a tail appears, you draw one card. Find the probability that
    (a) no spade is drawn
    (b) exactly one spade is drawn
    (c) at least one spade is drawn

37. A coin, loaded so that the probability of heads is 0.6, is tossed 8 times. Find the probability of getting each of the following.
    (a) At least 1 head
    (b) At least 7 tails

38. If the coin of Problem 37 is tossed repeatedly, what is the probability of getting a string of two heads before getting a string of two tails? (*Hint: HH* or *THH* or *HTHH* or *THTHH* or . . .)

39. Arlene, Betty, and Candy take turns (in that order) tossing a fair coin until one of them wins by getting heads. Find the probabilities of winning for each of them.

40. In a class of 23 students what is the probability that at least two people have the same birthday? Make a guess first. Then make a computation based on the following assumptions: (1) There are 365 days in a year; (2) one day is as likely as another

for a birthday; (3) the 23 students were born (chose their birthdays) independently of each other. (*Suggestion:* Look at the complementary event. For example, if there were only 3 people in the class, the probability that no two have the same birthday is $1 \cdot 364/365 \cdot 363/365$.)

41. Write a program to calculate

$$A = \sum_{k=0}^{m} \binom{n}{k} p^k (1 - p)^{n-k}$$

for any m, n, and p with $m \le n$. Then use it to calculate A in each case.
    (a) $m = 15, n = 31, p = \dfrac{1}{2}$
    (b) $m = 15, n = 15, p = \dfrac{2}{3}$
    (c) $m = 6, n = 20, p = \dfrac{1}{3}$
    (d) $m = 30, n = 70, p = \dfrac{2}{3}$

42. Use the program of Problem 41 to calculate the probability of getting at least 41 heads in tossing a fair coin 100 times.

43. In the World Series of baseball, the first team to win four games wins the series. Let us suppose that in a Dodgers-Yankees series the probability of the Dodgers winning any game is p (and the Yankees $1 - p$).
    (a) Show that the polynomial $f(p)$ that represents the probability that the Dodgers will win the series is given by

    $$f(p) = 35p^4 - 84p^5 + 70p^6 - 20p^7$$

    (b) Calculate the probability that the Dodgers will win the series if $p = 0.52$.
    (c) Calculate the probability that the series will go to 7 games if $p = 0.5$.

44. Graph the polynomial of Problem 43 and use it to determine the value of p that will give the Dodgers a probability of winning the series of (a) 90% (b) 99%

45. The Longlife Bulb Company packs 20 light bulbs in each of its boxes. (a) Write a formula for the probability A that a box will contain at most 1 defective bulb in terms of p, the probability that an individual light bulb is defective. Then determine the maximum p so that (b) $A \ge 0.95$ (c) $A \ge 0.99$.

46. **Challenge.** A stick of length 1 is divided into three pieces by cutting it at random at two places. What is the probability that a triangle can be formed from the three pieces? (*Hint:* Think of the problem this way. Imagine that the stick has a scale from 0 to 1 on it. You are to pick two numbers x and y on this interval at which to make the cuts; that is, you are to pick a point (x, y) in the unit square.)

CHAPTER 8 REVIEW PROBLEM SET

In Problems 1–10, write True or False in the blank. If false, tell why.

_____ **1.** The sequence 1, 1, 1, 1, 1, . . . is the only sequence that is both arithmetic and geometric.

_____ **2.** A recursion formula for a sequence relates the value of a term directly to its subscript.

_____ **3.** If a sum of money is invested at compound interest, its values at the end of successive years will form a geometric sequence.

_____ **4.** If a substance is growing so that the amount S after t years is given by $S = 3t + 14$, then the amounts at the end of successive years will form an arithmetic sequence.

_____ **5.** The symbols $0.99\overline{9}$ and 1 represent the same number.

_____ **6.** If P_1, P_2, and P_3 are true and the truth of P_k implies the truth of P_{k+3}, then P_n is true for all positive integers n.

_____ **7.** There are $_{10}P_2$ possible ways to respond to a 10-item true-false quiz.

_____ **8.** If m and n are positive integers with $n \geq m$, then $n!/(m!(n - m)!)$ is a positive integer.

_____ **9.** If the seventh and eighth terms in $(x + y)^n$ have the same coefficient, then $n = 13$.

_____ **10.** In probability theory, the words *disjoint* and *independent* have basically the same meaning.

Problems 11–17 refer to the sequences below.

(a) 3, 6, 9, 12, 15, . . .

(b) $-1, \dfrac{1}{2}, -\dfrac{1}{4}, \dfrac{1}{8}, -\dfrac{1}{16}, \dfrac{1}{32}, \ldots$

(c) $1, \sqrt{2}, 2, 2\sqrt{2}, 4, \ldots$

(d) $\pi, 3\pi, 5\pi, 7\pi, 9\pi, \ldots$

(e) 1, 1, 1, 3, 5, 9, 17, . . .

11. Which of these sequences are geometric?

12. Which of these sequences are arithmetic?

13. Write recursion formulas for sequences (d) and (e).

14. Find the 51st term of sequence (d).

15. Find the sum of the first 100 terms of sequence (a).

16. Find the sum of the first 10 terms of sequence (c).

17. Find the sum of all the terms of sequence (b).

18. Joe College (after taking precalculus) wants to negotiate a new allowance procedure with his parents. During each month, he would like 1¢ on day 1, 2¢ on day 2, 4¢ on day 3, 8¢ on day 4, and so on. Determine what Joe's allowance during September will be if he is successful.

19. Write $2.22\overline{2}$ as a ratio of two integers.

20. Calculate each sum. You don't need a calculator.

(a) $\displaystyle\sum_{i=1}^{50} (3i - 2)$

(b) $\displaystyle\sum_{i=1}^{\infty} 4\left(\dfrac{1}{3}\right)^i$

21. Suppose that $a_1 = a_2 = 1$ and $a_n = a_{n-1} + 2a_{n-2}$ for $n \geq 3$. Find the value of a_7.

22. Pat's starting annual salary is $30,000 and she has been promised an annual raise. Which would give her a better income (and by how much) during her tenth year: a straight $1800 raise each year or a 5% raise each year?

23. Which of the plans in Problem 22 would give Pat the larger total income during the 10 years and by how much?

In Problems 24–29 find the indicated terms and sums by using a calculator program. Here S_n represents the sum of the first n terms of the sequence $\{a_n\}$; that is, $S_n = a_1 + a_2 + \cdots + a_n$.

24. S_n for $n = 10, 20,$ and 30 of the sequence $a_n = n^3 - 5n^2$

25. $\displaystyle\sum_{k=1}^{n} \dfrac{\sqrt{k(k + 1)}}{k}$ for $n = 10, 50, 100,$ and 500

26. $\displaystyle\sum_{k=1}^{n} \dfrac{k^3 + k^2}{5k^4}$ for $n = 10, 20, 40,$ and 80

27. b_n and S_n for $n = 5, 10,$ and 15 of the sequence $b_n = -2b_{n-1}$ with $b_1 = 1$

28. a_n and S_n for $n = 15, 25,$ and 50 of the sequence $a_n = \sqrt{a_{n-1}a_{n-2}}$ with $a_1 = 1$ and $a_2 = 2$

29. (a) $\dfrac{1}{n}\displaystyle\sum_{k=1}^{n} \sin k$ (b) $\dfrac{1}{n}\displaystyle\sum_{k=1}^{n} \sin^2 k$, for $n = 10, 100,$ and 1000 (c) Conjecture what happens as $n \to \infty$ in cases (a) and (b).

30. After 2 years with a company, Jim signs a contract stating that his annual raise will be equal to 10% of his salary of 2 years before. If he earned $16,000 in each of his first 2 years, what will his salary be in his tenth year? What are his total earnings after 10 years?

31. Express the following in terms of factorials.

(a) $_{13}P_{10}$

(b) $_{13}C_{10}$

32. There are 10 candidates for a position. In how many ways can you rank your first, second, and third choices?

33. A local ice cream store offers 20 different flavors. How many different double-dip cones can it serve if

(a) the flavors must be different? (*Note:* We consider vanilla on top of chocolate to be the same as chocolate on top of vanilla.)

(b) the flavors may be the same?

34. How many 7-letter code words can be made from

(a) *ALGEBRA*

(b) *SEESAWS*

35. Pizza Heaven offers two types of crust, two types of cheese, five different meat toppings, and three different vegetable toppings. How many different pizzas can one order that have one crust, one cheese, three meats, and two vegetables?

36. King Arthur and 11 of his knights (including Sir Lancelot) wish to sit at the Round Table. In how many ways can they do this if
  (a) there are no special seats
  (b) King Arthur must sit at the chief chair with Sir Lancelot next to him

37. A basketball team has 15 members: 6 guards, 4 centers, and 5 forwards. How many starting lineups (2 guards, 1 center, and 2 forwards) are possible?

38. A baseball team has 12 members. The coach first chooses people to fill the 9 field positions and then prescribes a batting order. In how many ways can he do this?

39. How many license plates are there that consist of 2, 3, or 4 letters followed by enough digits to make a 7-symbol string?

40. A combination lock has the digits 0, 1, 2, ..., 9 written around its circular face. How many 3-digit combinations are possible if no combination can have the same or consecutive digits in the final two positions (9-0 is considered consecutive.) Thus 6-5-7 and 5-5-4 are legal combinations, but 6-5-5 and 6-9-0 are not.

41. Prove using mathematical induction that a set with n elements has 2^n subsets.

42. Prove using mathematical induction that $n! > 2^n$ for $n \geq 4$.

43. Prove or disprove: $n^2 - n + 11$ is prime for every positive integer n.

44. If P_2 is true and the truth of P_k implies the truth of P_{k+2}, what can you conclude about the sequence P_n of statements?

45. Write out the expansion of $(x + 2y)^6$.

46. Calculate $\sum_{r=0}^{6} {_6}C_r 2^r$. (*Hint:* Relate it to Problem 45.)

47. Find and simplify the term in the expansion of $(2x - 3y^2)^8$ that involves y^6.

48. A city is laid out in a uniform block pattern with north-south and east-west streets. How many different paths (meaning shortest paths along streets) are there from a corner to another one 5 blocks north and 3 blocks east?

49. Find $(1.00001)^8$ accurate to 15 decimal places.

50. How many subsets with an odd number of elements does the set $\{A, B, C, \ldots, X, Y, Z\}$ have?

51. Suppose that you are to pick a number at random from the first 100 positive integers. What is the probability that you will pick
  (a) an odd number
  (b) a multiple of 5
  (c) a number whose digits sum to more than 16
  (d) a number with no 1s digits

52. What is the probability of getting a royal flush (ace, king, queen, jack, 10 in one suit) for a five-card poker hand?

53. There are 5 red, 6 green, and 7 black balls in an urn. Two balls are to be drawn.
  (a) What is the probability that both will be red?
  (b) What is the probability that the two balls will be the same color?

54. Answer the questions of Problem 53 if the first ball is replaced before the second ball is drawn.

55. If the probability that Arnold will marry is 0.8, that he will graduate from college is 0.6, and that he will do both is 0.5, what is the probability that he will do one or the other (or both)?

56. At Westcott College, both English 11 and History 13 are required courses. On their first attempt, 30% of the students fail English 11, 20% fail History 13, and 8% fail both. Find the probability that a student will
  (a) fail one or the other of these courses (or both)
  (b) fail English 11, given that he or she has already failed History 13

57. What proportion of families with 4 children would you expect to have 2 boys and 2 girls? Assume that boys and girls are equally likely.

58. In tossing a fair die, call getting 1 or 2 a success.
  (a) Write an expression for the probability of getting 30 or fewer successes in tossing a die 100 times.
  (b) Use a calculator program to calculate this probability.

ANSWERS TO SELECTED PROBLEMS

PROBLEM SET 1.1 (p. 7)

1. 0.875
3. $0.\overline{285714}$
5. $0.\overline{8}$
7. 1.0625
9. $0.\overline{4117647058823529}$
11. $\dfrac{3}{5}$
13. $\dfrac{2}{3}$
15. $\dfrac{41}{333}$
17. $\dfrac{122}{99}$
19. Irrational
21. Irrational
23. Rational
25. Irrational
27. Irrational

29. Rational (2s repeat)
31. -10
33. -15
35. 2.445990027
37. $1.086514262 \times 10^{-10}$
39. 21.76559238
41. 4.471438755
43. $\pi r^2 h \approx \pi(1.5)^2(26) \approx 184$ cubic inches
45. $2(2\pi r)w \approx 4\pi(10)(3) \approx 377$ square feet
47. Number of female children under 18 + number of their mothers + number of other females $\approx$ 33,000 + 30,000 + 60,000 $\approx$ 120,000
49. 1, 0.3, 0.35. *Conjecture:* Such decimals can also be written as terminating decimals.
51. Suppose $z = x + y$ is rational. Then $y = z - x$ is rational, which is a contradiction.
53. $1 + 2 + 4 + 8 + \cdots + 2^{n-1} \approx 2^n$. $\$2^{30} \approx \$1.07$ billion.

PROBLEM SET 1.2 (p. 13)

1. $3.141592 < 3.14159\overline{2} < \pi < 3.1416 < \dfrac{22}{7}$
3. $2222 < 222^2 < 22^{22} < 2^{222} < 2^{2^{22}}$
5. Let $x = \dfrac{m}{n}$ and $y = \dfrac{p}{q}$.
   Then $\dfrac{x + y}{2} = \dfrac{m/n + p/q}{2} = \dfrac{mq + np}{2nq}$
7. Suppose $z = \sqrt{2}/n$ is rational. Then $\sqrt{2} = nz$ is rational, a contradiction.

9. There is no such number.
11. 5.5
13. $\dfrac{101}{20} = 5.05$
15. $\dfrac{2}{3}x + \dfrac{1}{3}y < \dfrac{8}{15}x + \dfrac{7}{15}y < \dfrac{1}{2}x + \dfrac{1}{2}y < 0.49x + 0.51y$
17. $\dfrac{145}{25} = 5.8$

19. 3.77 feet from left end
21. 3.275
23. 2.22
25. 3.5
27. $0.11
29. 2.12
31. $378.84 < A < 381.64$
33. $\bar{x} = 12.9, s = 1.50$
35. $\bar{x} = 12.9, s = 1.50$
37. Multiply both sides by bd.

39. $a \le b \Rightarrow a^2 \le ab \le b^2$ ($\Rightarrow$ is the symbol for *implies*)

41. $\left(a - \dfrac{1}{a}\right)^2 \ge 0 \Rightarrow a^2 - 2 + \dfrac{1}{a^2} \ge 0 \Rightarrow a^2 + 2 + \dfrac{1}{a^2} \ge 4$

$\Rightarrow \left(a + \dfrac{1}{a}\right)^2 \ge 4 \Rightarrow a + \dfrac{1}{a} \ge 2$

43. Let $w_1 + w_2 + w_3 = 1$ and $w_1 > 0, w_2 > 0, w_3 > 0$. Then $x = w_1 x + w_2 x + w_3 x \le w_1 x + w_2 y + w_3 z \le w_1 z + w_2 z + w_3 z = z$

PROBLEM SET 1.3 (p. 20)

1. $\sqrt{8} \approx 2.8284$
3. $\sqrt{102} \approx 10.0995$
5. 9.6421
7. y-axis symmetry

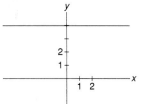

9. None

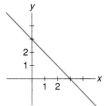

11. y-axis symmetry

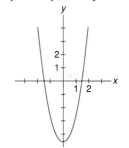

13. y-axis symmetry

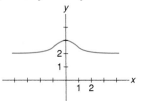

15. None

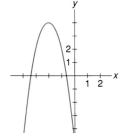

17. Origin symmetry
19. Origin symmetry
21. y-axis symmetry
23. All three symmetries
25. None
27. -4.1
29. $-3, 5$
31. 2.5
33. $-2, 3$
35. $-4.1926, 1.1926$
37. $\dfrac{3 \pm \sqrt{21}}{6}$ or $-0.2638, 1.2638$
39. $\dfrac{532}{15} \approx 35.4667$
41. $-1.8, 3.8$
43. 1.4
45. $-1.5, 0.8$
47. $(-2, -1)$
49. $(-\sqrt{5}, 0)$ and $(\sqrt{5}, 0)$
51. $\left(-\dfrac{3}{2}, \dfrac{27}{4}\right)$ and $(2, -2)$
53. $(-3, 1)$ and $\left(\dfrac{1}{2}, 1\right)$
55. $\sqrt{45} \approx 6.7082$
57. $\sqrt{245} \approx 15.6525$
59. $(3, 0)$; 6
61. $(1, 2)$; $\sqrt{11}$
63. 224 feet
65. 2.27 hours after noon, or at 2:16 P.M.
67. 6 miles per hour
69. $2\sqrt{53} + 6\pi \approx 33.41$
71. 98
73. 64.6 pounds

PROBLEM SET 1.4 (p. 27)

1.

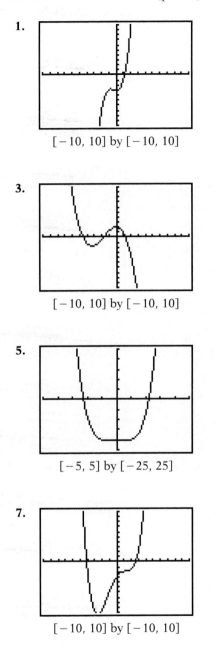

$[-10, 10]$ by $[-10, 10]$

3.

$[-10, 10]$ by $[-10, 10]$

5.

$[-5, 5]$ by $[-25, 25]$

7.

$[-10, 10]$ by $[-10, 10]$

9.

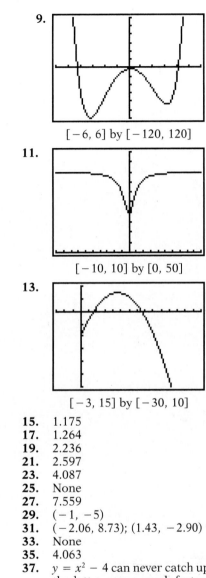

$[-6, 6]$ by $[-120, 120]$

11.

$[-10, 10]$ by $[0, 50]$

13.

$[-3, 15]$ by $[-30, 10]$

15. 1.175
17. 1.264
19. 2.236
21. 2.597
23. 4.087
25. None
27. 7.559
29. $(-1, -5)$
31. $(-2.06, 8.73); (1.43, -2.90)$
33. None
35. 4.063
37. $y = x^2 - 4$ can never catch up with $y = x^6$, because the latter grows much faster.
39. 2.192 by 22.808
41. 2.058 inches or 5.448 inches
43. 9.695 centimeters

PROBLEM SET 1.5 (p. 33)

1. $x > \dfrac{8}{5}$

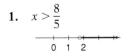

3. $x \geq \dfrac{7}{2}$

5. $x \le -\dfrac{10}{3}$

$$\begin{array}{ccccc} \hline & & & & \\ -5 & -4 & -3 & -2 & -1 \end{array}$$

7. $-12 \le x < 4.5$

$$\begin{array}{cccccc} \hline -12 & -8 & -4 & 0 & 4 & 8 \end{array}$$

9. $-5 < x < -\dfrac{11}{4}$

$$\begin{array}{cccccc} \hline -5 & -4 & -3 & -2 & -1 & 0 \end{array}$$

11. $-2 < x < -\dfrac{3}{4}$

$$\begin{array}{ccccc} \hline -3 & -2 & -1 & 0 & 1 \end{array}$$

13. $x < 3$ or $x > 5$
15. $-6 \le x \le 1$
17. $x < -3$ or $x > 0$
19. $-1 \le x \le 0$ or $x > 2$
21. $-3 < x < 1$ or $x > 4$
23. $-2.732 < x < 0.732$
25. $x > -2.627$
27. $-0.799 \le x \le 0$

29. $0 \le x \le 0.124$ or $x \ge 1.386$
31. $x < -3.06$
33. $x < -1.35$ or $0 < x < 1.35$
35. $x < 1.37$ or $x > 9.94$
37. $2.48 < x < 3.00$
39. $-3 \le x \le 3$
41. $-4 < x < 1$
43. $x < \dfrac{5}{2}$ or $x > \dfrac{13}{2}$
45. $x \le -1$ or $x \ge \dfrac{11}{3}$
47. $x \le -6$ or $x \ge -0.8$
49. $x < -3.73$ or $x > 4.24$
51. 2.56 centimeters $\le x \le$ 2.61 centimeters
53. $40.35 \le V \le 41.78$
55. $\$14,000 < x < \$41,000$
57. $1.08 < t < 5.80$; $1.08 < t < 1.88$ or $5.00 < t < 5.80$
59. $0 \le x < 1250$
61. (a) $4 - \sqrt{7} < x < 4 + \sqrt{7}$
(b) $6 - \sqrt{32} < y < 6 + \sqrt{32}$

(c) $x < 0$ or $x > \dfrac{32}{5}$

63. Total amount is $1/(1 + a) + 1 + a$, which is greater than 2 by Problem 62.

PROBLEM SET 1.6 (p. 41)

1. $-\dfrac{1}{9}$

3. 0

5. 1

7. 4

9. $y = -x - 1$

11. $y = \dfrac{2}{3}x - \dfrac{23}{3}$

13. $y = -\dfrac{1}{3}x - 2$

15. $y = -\dfrac{2}{3}x - 4$

17. $y = \dfrac{2\sqrt{2} - 1}{\sqrt{2} + 1}x - \dfrac{3\sqrt{2}}{\sqrt{2} + 1}$

19. $y = -\dfrac{25}{8}x + 5$

21. $y = -x - 1$

23. $\dfrac{18}{5}$

25. $\dfrac{5}{13}$

27. $\dfrac{7}{\sqrt{10}}$

29. $\left(3, \dfrac{10}{3}\right)$

31. $(x - 2)^2 + (y - 6)^2 = 25$

33. $S_1 = 1200 + 40x$; $S_2 = 1000 + 50x$; $x > 20$

35. $\$95,000$

37. (a) $k = 14$
(b) $k = 0$
(c) Impossible
(d) $k = -6$
(e) $k = 4$
(f) $k = \dfrac{4}{3}$

39. $y + 2 = \dfrac{3}{10}\left(x - \dfrac{7}{2}\right)$

41. The midpoints of both diagonals have coordinates $\left(\dfrac{a + b}{2}, \dfrac{c}{2}\right)$.

43. $(-13, 29)$
45. $a = 2$, $b = \sqrt{22}$
47. 6014.9 meters

PROBLEM SET 1.7 (p. 48)

1. Linear, $r = 0.98$

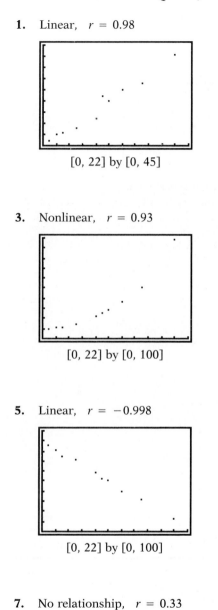

[0, 22] by [0, 45]

3. Nonlinear, $r = 0.93$

[0, 22] by [0, 100]

5. Linear, $r = -0.998$

[0, 22] by [0, 100]

7. No relationship, $r = 0.33$

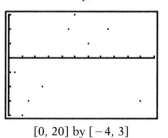

[0, 20] by [-4, 3]

9. Nonlinear, $r = 0.44$

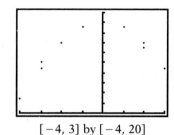

[-4, 3] by [-4, 20]

11. $S_1 = 9.5$, $S_2 = 8.3$

[0, 6] by [0, 6]

13. $S_1 = 8$, $S_2 = 8$

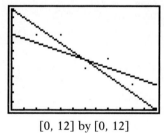

[0, 12] by [0, 12]

15. $y = 1.65 + 0.77x$
17. $y = 10.9 - 0.75x$
19. (a) $y = 13,590 - 945x$
    (b) \$8865; \$2250
21. (a) $y = 7180 - 540x$
    (b) \$4480; \$700
23. (a) $y = 14,690 - 575x$
    (b) \$11,815; \$7790
25. 0.9
27. 0.9
29. 0.9
31. -0.9

33. $y = 0.6 + 2.26x; r = 0.95$

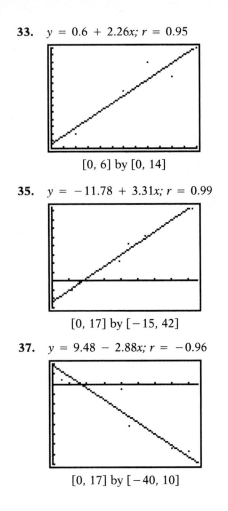

[0, 6] by [0, 14]

35. $y = -11.78 + 3.31x; r = 0.99$

[0, 17] by [−15, 42]

37. $y = 9.48 - 2.88x; r = -0.96$

[0, 17] by [−40, 10]

39. $V = -0.099 + 135.77I, R \approx 136$ ohms
41. Ball 1: (a) 6.54 cm/s; −7.7 cm   (b) 188.5 cm
Ball 2: (a) 2.3 cm/s; 55.5 cm   (b) 124.5 cm
Ball 3: (a) 19.98 cm/s; 0.1 cm   (b) 599.5 cm
43. (b) $r = -0.99$
(c) $E = 429.8 - 0.2157Y$
(d) 2.8 million

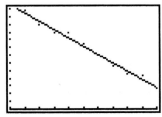

[1930, 1980] by [0, 13]

CHAPTER 1 REVIEW PROBLEM SET (p. 50)

1. F
2. T
3. F
4. T
5. T
6. F
7. F
8. F
9. T
10. T
11. A ratio p/q of integers p and q with $q \neq 0$.
12. Suppose $x = 4 - \sqrt{3}$ is rational. Then $\sqrt{3} = 4 - x$ is rational, contrary to what is given.
13. $-\dfrac{13}{8}$
14. (a) 11473.93777
(b) 2.096446942
(c) 1678620.795
(d) −0.1310254962

15. Cut out a 2-inch cube and count the number p of seeds in it. The number of seeds in the watermelon is approximately $Vp/8$ where V is the volume of the seeded (red) part of the melon measured in cubic inches. V could be estimated by assuming that the melon is a sphere of radius, say, 5 inches.
16. $54 \times 7 \times 8 \approx 3000$ cubic feet
17. $0.\overline{45}, 0.\overline{384615}$
18. $\dfrac{52}{111}, \dfrac{357}{110}$
19. $\dfrac{13}{4} > \dfrac{16}{5}$
20. $\sqrt{2} < 1.44 < 1.\overline{4} < \dfrac{29}{20} < \dfrac{130}{89}$
21. $3a < 2a + b$. Divide by 3.
22. Either a or b is 0.
23. 2.381
24. Fred: 2.889; Wilma: 2.944

25. (a) 2.667
(b) 4.571
26. (a) 13
(b) 7.654
27. (a) Origin symmetry

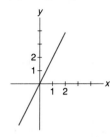

(b)

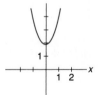

(c) y-axis symmetry

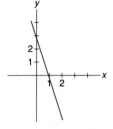

(d)

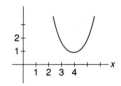

(e) Origin symmetry

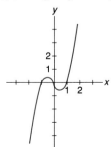

(f)

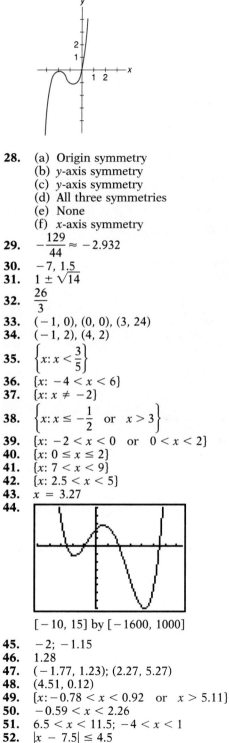

28. (a) Origin symmetry
(b) y-axis symmetry
(c) y-axis symmetry
(d) All three symmetries
(e) None
(f) x-axis symmetry
29. $-\dfrac{129}{44} \approx -2.932$
30. $-7, 1.5$
31. $1 \pm \sqrt{14}$
32. $\dfrac{26}{3}$
33. $(-1, 0), (0, 0), (3, 24)$
34. $(-1, 2), (4, 2)$
35. $\left\{x : x < \dfrac{3}{5}\right\}$
36. $\{x : -4 < x < 6\}$
37. $\{x : x \neq -2\}$
38. $\left\{x : x \le -\dfrac{1}{2} \quad \text{or} \quad x > 3\right\}$
39. $\{x : -2 < x < 0 \quad \text{or} \quad 0 < x < 2\}$
40. $\{x : 0 \le x \le 2\}$
41. $\{x : 7 < x < 9\}$
42. $\{x : 2.5 < x < 5\}$
43. $x = 3.27$
44.

$[-10, 15]$ by $[-1600, 1000]$

45. $-2; -1.15$
46. 1.28
47. $(-1.77, 1.23); (2.27, 5.27)$
48. $(4.51, 0.12)$
49. $\{x : -0.78 < x < 0.92 \quad \text{or} \quad x > 5.11\}$
50. $-0.59 < x < 2.26$
51. $6.5 < x < 11.5; -4 < x < 1$
52. $|x - 7.5| \le 4.5$
53. $A = 10x - x^2$

54. $V = x(12 - 2x)^2$
55. 9.82 feet
56. 4.2 miles per hour
57. $x - 4 = 0$
58. $4x + 3y - 17 = 0$
59. $5x + 7y = 0$
60. $3x - 2y + 12 = 0$
61. $x - 2y - 2 = 0$
62. $3x - 4y - 25 = 0$
63. $\sqrt{20} + \sqrt{72} + \sqrt{116} \approx 23.73$
64. $AB: 2x - y - 7 = 0$
$BC: x + y - 8 = 0$
$AC: 5x + 2y - 13 = 0$
65. $D = (4, 1)$, $E = (1, 4)$, DE has slope -1, $\overline{DE} = \sqrt{18} = \frac{1}{2}\overline{BC}$

66. $(x - 4)^2 + (y - 1)^2 = 5$
67. $\left(3, \dfrac{10}{3}\right)$
68. $\left(\dfrac{13}{2}, 4\right)$
69. Apple: $h = 6.893 + 0.113t; r = 0.967$
Elm: $h = 4.129 + 0.224t; r = 0.995$
Oak: $h = 13.970 + 0.031t; r = 0.947$
70. Elm; 0.224 meters per year
71. In 1998
72. 8.6 meters
73. In 2004 when $h \approx 9.7$ meters
74. $V = 50,000 - 4850t$

PROBLEM SET 2.1 (p. 58)

1. Rg: $\{-6, 0, 3, 6, 12, 18\}$
3. Rg: $\{-1, 0, 3, 15\}$
5. Rg: $\left\{\dfrac{1}{13}, \dfrac{1}{5}, 1, 2\right\}$
7. Rg: $\{0, 6, 20\}$
9. (a) 0
(b) $\sqrt{2} \approx 1.414$
(c) 9
(d) $\sqrt{\sqrt{2}} \approx 1.189$
(e) 4
11. (a) -1
(b) 0.6
(c) 3
(d) -0.471
(e) 1.039
13. (a) 0
(b) 4
(c) 3
(d) 48
(e) 3
15. $\{x: x \neq 0\}$
17. $\{x: x < -1$ or $x > 1\}$
19. Dm: $\mathbb{N}$, Rg: $\{0, 2, 7\}$
21. Dm: $\mathbb{R}$, Rg: $\{y: y \geq -2\}$
23. (a) 2
(b) 3, 5
(c) $3 < x < 5$
25. (a) -0.2
(b) 2
(c) $x \neq 2$
27. (a) $-4, 2, 3.5$
(b) $-4 < x < 2$ or $x > 3.5$
(c) $-2.5, 3$
(d) $-2.5 \leq x \leq 3$
(e) $-3, 2.5, 3.8$
(f) $-3 < x < 2.5$ or $x > 3.8$

29. Dm: $\mathbb{R}$, Rg: $\mathbb{R}$, $\{x: x < -\sqrt{2}$ or $0 < x < \sqrt{2}\}$

$[-10, 10]$ by $[-10, 10]$

31. Dm: $\{x: x \neq 0\}$, Rg: $\{y: y \neq 5\}$, $\{x: 0 < x < 0.4\}$

$[-10, 10]$ by $[-10, 10]$

33. Dm: $\mathbb{R}$, Rg: $\mathbb{R}$, $\{x: x < 20.82\}$

$[-10, 25]$ by $[-250, 100]$

35. Odd

37. Neither
39. Even
41.

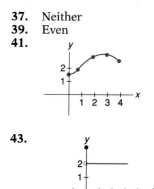

43.

45.

47. $f(x) = \frac{1}{3}x^2 + 6$

49. $f(x) = 14x^2$

51. $d = f(x) = \sqrt[3]{\frac{6x}{\pi}}$

53. $f(x) = \frac{\sqrt{3}}{36}x^2$

55. $f(x) = (x - 2)(x - 3)$

57. $d(t) = \sqrt{900t^2 + 1296t + 1296},$
$$d(6) \approx 203.65 \text{ miles}$$

59. $f(19.5) = 8, f(28) = 9,$ Dm: $\mathbb{R}$, Rg: $\mathbb{N} \cup \{0\}$

61. (a) -38

(b) $\dfrac{2a}{a - 1}$

(c) $\dfrac{2}{1 - 2a}$

(d) $\dfrac{-4}{(x + h - 2)(x - 2)}$

63. $f(f(x)) = x, f(f(3.4567)) = 3.4567$

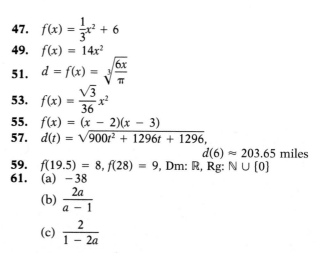

PROBLEM SET 2.2 (p. 64)

1. $(-2, 4)$

3. $(-6, -11.5)$

5. $(0.944, -1.467)$

7. (a) $C(x) = 3.00 + 0.40x$
(b) $C(12) = \$7.80$
(c) Less than 17.5 miles

9. (a) $V(t) = 85,000 + 6000t$
(b) $\$6000$ per year

11. (a) $V(t) = 98,000 - 14,000t$
(b) $\$42,000$
(c) $2.714 < t < 4.857$
(d) $-\$14,000$ per year

13. (a) $R(x) = 3.2x + 3$
(b) 10.3 inches
(c) $3.2\ R$ per inch

15. (a) $I(x) = 22,000 + 0.009x$
(b) $\$27,400$
(c) $\$2,444,444.44$

17. (a) $C(x) = 50 + 0.3x$
(c) $\$30$

19. (a) $R(x) = 0.8x$
(b)

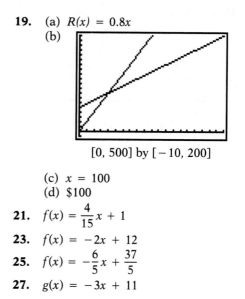

$[0, 500]$ by $[-10, 200]$

(c) $x = 100$
(d) $\$100$

21. $f(x) = \frac{4}{15}x + 1$

23. $f(x) = -2x + 12$

25. $f(x) = -\frac{6}{5}x + \frac{37}{5}$

27. $g(x) = -3x + 11$

29. (a) $C = \dfrac{5}{9}(F - 32)$

(b) $22.22°C$
(c) $F = -40°$

(d) $\dfrac{5}{9}$ degree C per degree F

31. (a) $20,000
(b) $65,090
(c) $20,000 + 90x$ dollars

33. 40,000, $600
35. $s(t) = 500 - 119t$, -119 mph

37. $f(x) = -\dfrac{3}{5}x + 7$ or $f(x) = -\dfrac{112}{105}x + \dfrac{28}{3}$

39. Let $f(x) = mx + q$ and $g(x) = rx + s$. Then $h(x) = f(g(x)) = f(rx + s) = m(rx + s) + q = (mr)x + (ms + q)$

41. Let $f(x) = mx + q$. Since f is odd, $m(-x) + q = -(mx + q)$. This implies that $q = -q$ and so $q = 0$.

43. Let $f(x) = mx + q$. Then

$$f\left(\frac{1}{3}x_1 + \frac{2}{3}x_2\right) = m\left(\frac{1}{3}x_1 + \frac{2}{3}x_2\right) + q$$

$$= \frac{1}{3}mx_1 + \frac{2}{3}mx_2 + \frac{1}{3}q + \frac{2}{3}q = \frac{1}{3}f(x_1) + \frac{2}{3}f(x_2)$$

PROBLEM SET 2.3 (p. 72)

1. $\sqrt{90} \approx 9.49$
3. 3.02
5.

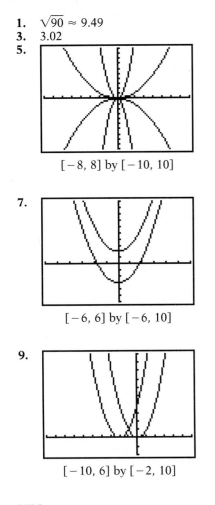

$[-8, 8]$ by $[-10, 10]$

7.

$[-6, 6]$ by $[-6, 10]$

9.

$[-10, 6]$ by $[-2, 10]$

11. Shifted right 3 units and down 3 units.
13. Flipped across x-axis, shifted right 3 units and up 2 units.
15. One
17. Two
19. One
21. Zero
23. $(-2, -23)$, opens up
25. $(2.5, 4.4375)$, opens down

27. $\left(\dfrac{1}{8}, -\dfrac{1}{16}\right)$, opens up

29. (a) $t = 5.5$ seconds
(b) 484 feet
(c) $t = 11.64$ seconds
(d) $5 < t \leq 6$

31. $v_0 = 8\sqrt{140} \approx 94.66$ feet per second

33. $y = \dfrac{1}{500}x^2 + 10$

35. 87.87 feet
37. 6
39. 0
41. 24
43. $f(x) = 4x^2$
45. $f(x) = 4(x - 2)^2$
47. $f(x) = 2x^2 - 3$
49. $f(x) = x^2 - 5x + 4$
51. $(5, -22.5)$
53. $c = 19$
55. Maximum height is quadrupled and the time to get there is doubled.
57. One meter per second
59. (a) $P = -10x^2 + 420x - 3600$
(b) $21
(c) $20 profit per dollar of price

PROBLEM SET 2.4 (p. 78)

1. (a) 6
   (b) −2
   (c) $1 \le x \le 3$
3. (a) 1.5
   (b) −1
   (c) $0 \le x \le 1, 4 \le x \le 5$
5. (3, 11)
7. (0, 4)
9. (2.1, −1)
11. (−0.62, 0.35), (1.62, −1.52)
13. (25, 8487.5), (0, 50)
15. (−1.38, 1.39), (2.49, −0.25)
17. (2.92, −219.30)
19. (−2.64, 4.87)
21. $-1.67 \le x \le 1$
23. $x \le 0, x \ge 5.43$
25. $-2.00 \le x \le 0, 1.00 \le x \le 3.00$
27. None
29. $-1.2 \le x \le 1.2$
31. $x \ge -2$

33. Everywhere
35. $f(x) = 1800x - 2x^2, 0 \le x \le 900$; 450 feet by 900 feet, with the long side parallel to the river
37. $f(x) = (150 + 2x^2)\left(\dfrac{50}{x}\right), x > 0$; 8.66 miles per hour
39. (1.59, 1.19)
41. 450
43. $x = 1.73, y = 6$
45. No maximum; minimum when $x = 3, y = 2$
47. 2.52 by 2.52 by 2.52
49. $700
51. $f(\alpha_1 x_1 + \alpha_2 x_2 + \alpha_3 x_3)$

$$= f\left((\alpha_1 + \alpha_2)\frac{\alpha_1 x_1 + \alpha_2 x_2}{\alpha_1 + \alpha_2} + \alpha_3 x_3\right)$$

$$\le (\alpha_1 + \alpha_2)f\left(\frac{\alpha_1 x_1 + \alpha_2 x_2}{\alpha_1 + \alpha_2}\right) + \alpha_3 f(x_3)$$

$$\le (\alpha_1 + \alpha_2)\left(\frac{\alpha_1}{\alpha_1 + \alpha_2}f(x_1) + \frac{\alpha_2}{\alpha_1 + \alpha_2}f(x_2)\right)$$

$$+ \alpha_3 f(x_3) = \alpha_1 f(x_1) + \alpha_2 f(x_2) + \alpha_3 f(x_3)$$

PROBLEM SET 2.5 (p. 84)

1. Left arm down, right arm up
3. Left arm up, right arm down
5. Both arms down
7.

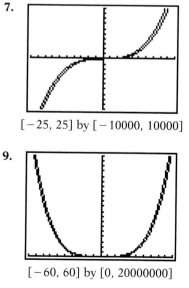

$[-25, 25]$ by $[-10000, 10000]$

9.

$[-60, 60]$ by $[0, 20000000]$

11.

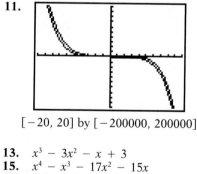

$[-20, 20]$ by $[-200000, 200000]$

13. $x^3 - 3x^2 - x + 3$
15. $x^4 - x^3 - 17x^2 - 15x$
17. $2x^5 - 22x^4 + 82x^3 - 122x^2 + 60x$
19. $6x^3 - 21x^2 + 6x + 9$
21. $x^4 + 2x^3 - 2x - 1$
23. $(x + 2)^3(x - 3)^2$
25. $(x - 1)(x - 2)(x - 5)$
27. $(x - 3)(x - 7)(x + 1)(x + 4)$
29. Crosses at −3, 2, and 4
31. Crosses at −2
33. Crosses at 0 and 1/2, touches at −2
35. Crosses and is tangent at −7/2, crosses at 1/3 and 0.
37. Touches at 0, crosses at −2 and 3.

39. $p(x) = (x + 2)(x + 1)(x)(x - 2)(x - 3)$
41. $p(x) = (x + 2)^2(x - 1)^2(x - 3)$
43. $p(x) = (x + 1)^3(x - 2)(x - 3)$
45. $p(x) = 2x^3 - 10x^2 + 4x + 16$
47. $k = 2$
49. Two
51. Five
53. Too many zeros
55. Too many turning points
57. No straight segments on the graph of a polynomial of degree greater than 1.

59. 13
61. 9 meters per second
63. $p\left(\dfrac{1}{c}\right) = \dfrac{1}{c^4} + \dfrac{3}{c^3} - \dfrac{5}{c^2} + \dfrac{3}{c} + 1$

$$= \dfrac{1 + 3c - 5c^2 + 3c^3 + c^4}{c^4} = \dfrac{p(c)}{c^4} = 0$$

If the coefficients are symmetric about a midpoint and $p(c) = 0$, then $p\left(\dfrac{1}{c}\right) = 0$.

PROBLEM SET 2.6 (p. 92)

1. $y = -8$
3. $y = 14$
5. $y = \dfrac{1}{3}$
7. $y = 0$
9. $y = \dfrac{81}{4}$
11. None
13. $x = -3, y = 1$
15. $x = -3, y = 0$
17. $x = 3, x = -1, y = 2$
19. $x = -2, x = 6, y = -3$
21. $x = -5, x = 4, y = \dfrac{9}{2}$
23. $x = -0.1, x = 0.1, y = 1$
25. $y = 2x$
27. $x = 2, y = -x + 3$
29. $x = 0, y = \dfrac{3}{2}x$
31. $x = -2, y = 2x - 8$
33. $y = 3x - 2$
35.

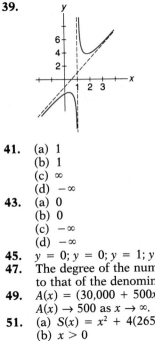

37.

39.

41. (a) 1
   (b) 1
   (c) ∞
   (d) $-\infty$
43. (a) 0
   (b) 0
   (c) $-\infty$
   (d) $-\infty$
45. $y = 0; y = 0; y = 1; y = x;$ none
47. The degree of the numerator is less than or equal to that of the denominator.
49. $A(x) = (30,000 + 500x)/x.$ $A(x) \to \infty$ as $x \to 0^+$ and $A(x) \to 500$ as $x \to \infty.$
51. (a) $S(x) = x^2 + 4(265/x)$
   (b) $x > 0$
   (c) 8.09 by 8.09 by 4.05
53. Dm: $\{x: x \neq 5\}$; Rg: $\{y: y \leq 1 \text{ or } y \geq 9\}$
55. $f(x) = 2x + 3 + \dfrac{2}{x - 3}$
57. $f(x) = -x + 2 + \dfrac{10}{(x + 2)(x - 1)}$

PROBLEM SET 2.7 (p. 99)

1. (a) 6
(b) -3
(c) 0
(d) 4

3. (a) $\frac{2}{3}$ (b) -2 (c) -2 (d) 0

5. (a) -7
(b) 0
(c) 2
(d) $-\frac{4}{3}$

7. $(f \circ g)(-5) = 12; (g \circ f)(-5) = 11$
$(f \circ g)(x) = 2(1 - x); (g \circ f)(x) = 1 - 2x$
Dm $(f \circ g)$: $\mathbb{R}$; Dm $(g \circ f)$: $\mathbb{R}$

9. $(f \circ g)(0) = -\frac{2}{3}; (g \circ f)(0) = -2$

$(f \circ g)(x) = \frac{3x - 2}{-3x + 3}; (g \circ f)(x) = \frac{5x - 2}{1 - x}$
Dm $(f \circ g)$: $\{x: x \neq 1\}$; Dm $(g \circ f)$: $\{x: x \neq 1\}$

11. (a) $\frac{1}{2} \cdot 2 = 1$

(b) $\frac{3}{2} - 1 = \frac{1}{2}$

(c) $f(0) = 2$
(d) $g(1) = 1$
(e) $f(0) = 2$

(f) $f(3) - 3 \cdot 2 = -\frac{11}{2}$

13. (a) If $g(x) = x^2 + 3$ and $f(x) = \sqrt{x}$, then $F = f \circ g$.
(b) If $h(x) = x^2, g(x) = x + 3$, and $f(x) = \sqrt{x}$, then $F = f \circ g \circ h$.

15. (a) If $g(x) = x - 4$ and $f(x) = x^2 + 3$, then $F = f \circ g$.
(b) If $h(x) = x - 4, g(x) = x^2$, and $f(x) = x + 3$, then $F = f \circ g \circ h$.

17. (a) If $g(x) = x + 1$ and $f(x) = 1/x^2$, then $F = f \circ g$.
(b) If $h(x) = x + 1, g(x) = x^2, f(x) = 1/x$, then $F = f \circ g \circ h$.

19.
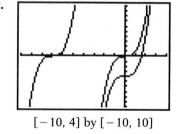
$[-10, 4]$ by $[-10, 10]$

21.
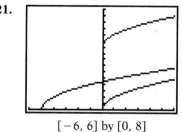
$[-6, 6]$ by $[0, 8]$

23. g is stretched vertically; h is shrunk vertically.
25. g is stretched vertically and is flipped across the x-axis; h is stretched horizontally.
27. g is shifted right; h is shrunk horizontally.
29. (a) 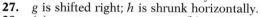 (b)
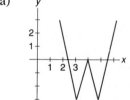

31. (a) (b)
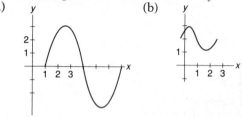

33.
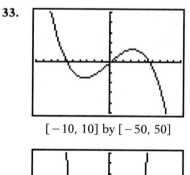
$[-10, 10]$ by $[-50, 50]$

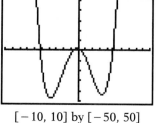
$[-10, 10]$ by $[-50, 50]$

35.

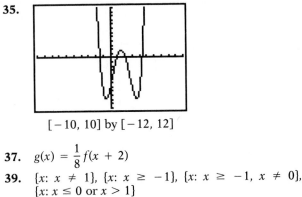

[−10, 10] by [−12, 12]

37. $g(x) = \frac{1}{8}f(x + 2)$

39. $\{x: x \neq 1\}, \{x: x \geq -1\}, \{x: x \geq -1, x \neq 0\},$
$\{x: x \leq 0 \text{ or } x > 1\}$

41. $f(x) = -3x - 2$

43. (a) $2x + h + 2$
  (b) $\dfrac{-1}{x(x + h)}$

45. (a) Odd
  (b) Odd
  (c) Even
  (d) Odd
  (e) Neither
  (f) Odd
  (g) Even
  (h) Even

47. (a) $d(t) = \sqrt{28.25t^2 - 32t + 16}$
  (b) After 4 hours and 18 minutes, so at 4:18 P.M.

49. Conjecture: $f(f(f(x))) = x$
  Proof: Straightforward algebra

PROBLEM SET 2.8 (p. 108)

1. $g(y) = \frac{1}{2}(y - 7)$

3. $g(y) = 3y + 4$

5. $g(y) = (y - 1)^2$

7. $g(y) = y^2 - 1$

9. $g(y) = \sqrt[3]{y} + 5$

11. $g(y) = \sqrt[3]{9y} + 3$

13. No

15. Yes

17. Yes

19. No

21. Yes

23. No

25. No

27. $f^{-1}(x) = -\frac{1}{2}(x - 2)$

29. $f^{-1}(x) = \dfrac{-3}{x - 1}$

31. $f^{-1}(x) = \dfrac{3x + 2}{x - 1}$

33. No inverse

35. $f^{-1}(x) = (x - 3)^2 + 3$

37. $f^{-1}(x) = \frac{1}{4}x - \frac{7}{16}$

[−10, 10] by [−7, 7]

39.

[−10, 10] by [−7, 7]

41.

[−10, 10] by [−7, 7]

43. $x \geq 0; f^{-1}(x) = \sqrt{5 - x}$

45. $x \geq 3; f^{-1}(x) = 3 + \sqrt{x - 5}$

47. $x \geq 1; f^{-1}(x) = 1 + \sqrt{\dfrac{10 - 3x}{x}}$

49. $f^{-1}(0) = -\frac{1}{2}, f^{-1}(-2) = -2, f^{-1}(3) = 3$

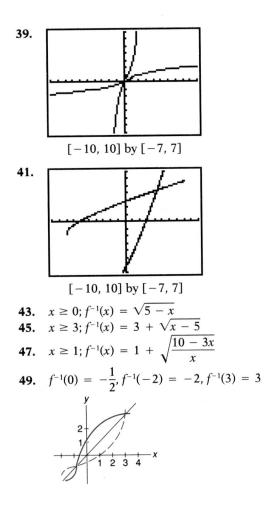

51. $f^{-1}(0) = -2, f^{-1}(-2) = 4, f^{-1}(3) = 1$

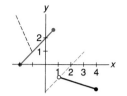

53. $f^{-1}(x) = \sqrt[3]{\dfrac{2 - 3x}{x - 1}}$

55. $a = 2$

57. The reflection of the graph over the line $y = x$ must be itself. Examples: $f(x) = x$ and $f(x) = -x$.

59. $a = -d$

PROBLEM SET 2.9 (p. 115)

1.

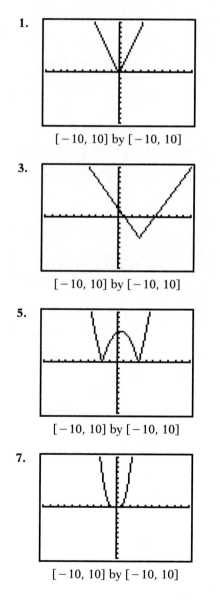

$[-10, 10]$ by $[-10, 10]$

3.

$[-10, 10]$ by $[-10, 10]$

5.

$[-10, 10]$ by $[-10, 10]$

7.

$[-10, 10]$ by $[-10, 10]$

9.

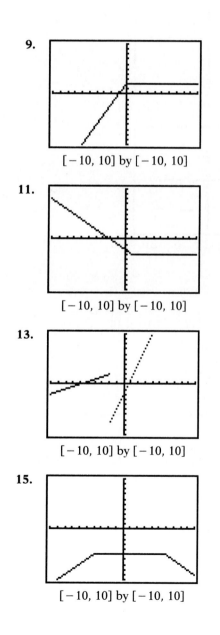

$[-10, 10]$ by $[-10, 10]$

11.

$[-10, 10]$ by $[-10, 10]$

13.

$[-10, 10]$ by $[-10, 10]$

15.

$[-10, 10]$ by $[-10, 10]$

17.

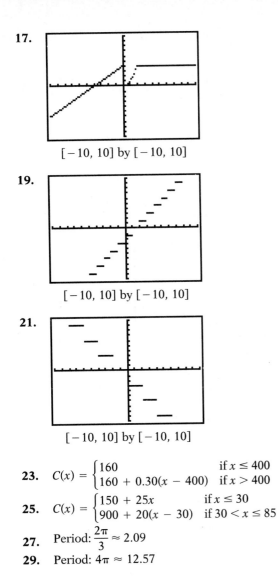

$[-10, 10]$ by $[-10, 10]$

19.

$[-10, 10]$ by $[-10, 10]$

21.

$[-10, 10]$ by $[-10, 10]$

23. $C(x) = \begin{cases} 160 & \text{if } x \le 400 \\ 160 + 0.30(x - 400) & \text{if } x > 400 \end{cases}$

25. $C(x) = \begin{cases} 150 + 25x & \text{if } x \le 30 \\ 900 + 20(x - 30) & \text{if } 30 < x \le 85 \end{cases}$

27. Period: $\dfrac{2\pi}{3} \approx 2.09$

29. Period: $4\pi \approx 12.57$

31. Period: 0.5
33. Not periodic
35. Period: 0.2
37. Period: 1
39. $C(x) = 25 + 25\,[[x]]$ cents; $0 < x < 6$
41. (a) $f(x) = |x + 2| + 1$
  (b) $f(x) = -2|x - 1| + 2$
43. (a) 3
  (b) 1
  (c) 2
  (d) $f(x) = \left[\left[\dfrac{1}{2}(x + 3)\right]\right]$
45. $A(x) = \begin{cases} x & \text{if } 0 \le x \le 1000 \\ 1000 + 0.2(x - 1000) & \text{if } 1000 < x \le 6000 \\ 2000 & \text{if } x > 6000 \end{cases}$

  Maximum: \$2000
47. $T(x) = \begin{cases} 0 & \text{if } 0 \le x \le 10{,}000 \\ 0.06(x - 10{,}000) & \text{if } 10{,}000 < x \le 20{,}000 \\ 600 + 0.09(x - 20{,}000) & \text{if } x > 20{,}000 \end{cases}$

49. (a) 0
  (b) 0.5
  (c) 2
51. (a) Rg: $\{y: -2 \le y < -1 \text{ or } 0 \le y < 1 \text{ or } 2 \le y < 3\}$
  (b) Yes
  (c) -0.5
53. $a \le -2$ or $a \ge 0$
55. (a)   (b) Period: 5

$[-10, 10]$ by $[-1, 1]$

CHAPTER 2 REVIEW PROBLEM SET (p. 118)

1. F
2. F
3. F
4. T
5. T
6. F
7. F
8. T
9. T
10. F

11. (a) $\dfrac{10}{9}$
  (b) 3
  (c) 96
  (d) Undefined
  (e) 2
  (f) -8
12. $\{x: x \ge 0, x \ne 1\}$
13. 1.5, 0
14. -2.9

15. $\{x: x \leq -2.3 \quad \text{or} \quad -1.5 \leq x \leq 2.3\}$

16. $\{x: -2 \leq x \leq 0.3\}$

17. $\{x: -1.4 < x < 2.5\}$

18. -2.6

19. -1.75

20. $-2.3, -1.5, 2.3$

21. (a) Odd
(b) Even
(c) Odd
(d) Neither
(e) Even
(f) Odd

22. Shifted 2 units left and 5 units up

23. (a) $(-1, 4)$, down
(b) $(-2, -4)$, up
(c) $(2, -7)$, down
(d) $(2, 6)$, up

24 -3.06

25. $(1, -5), (-1, 13)$

26. $(-1.50, 12.37), -1.50 \leq x \leq 0$

27. $(3.75, -9.20), 0.5 < x < 3$

28. 5.065

29. 24

30. $(x + 2)(x - 7)(x + 3)$

31. $(x - 1)(x - 2)(x - 3)^3$

32. (a) $1(2), -2(1), 3(1)$
(b) $0(3), -3(2), 3(2)$
(c) $0(1), -4(2)$
(d) $-1(3), 1(3)$

33. (a)   (b)

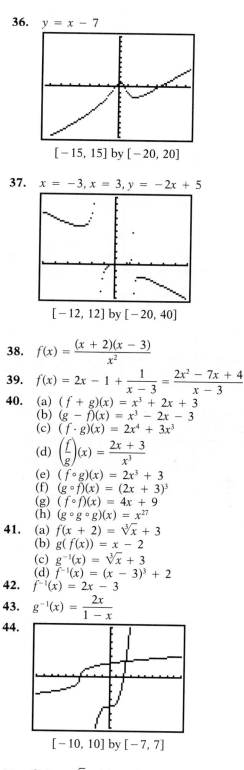

34. $x = -5, y = 2$

$[-12, 6]$ by $[-10, 10]$

35. $y = 3$

$[-10, 10]$ by $[-5, 5]$

36. $y = x - 7$

$[-15, 15]$ by $[-20, 20]$

37. $x = -3, x = 3, y = -2x + 5$

$[-12, 12]$ by $[-20, 40]$

38. $f(x) = \dfrac{(x + 2)(x - 3)}{x^2}$

39. $f(x) = 2x - 1 + \dfrac{1}{x - 3} = \dfrac{2x^2 - 7x + 4}{x - 3}$

40. (a) $(f + g)(x) = x^3 + 2x + 3$
(b) $(g - f)(x) = x^3 - 2x - 3$
(c) $(f \cdot g)(x) = 2x^4 + 3x^3$
(d) $\left(\dfrac{f}{g}\right)(x) = \dfrac{2x + 3}{x^3}$
(e) $(f \circ g)(x) = 2x^3 + 3$
(f) $(g \circ f)(x) = (2x + 3)^3$
(g) $(f \circ f)(x) = 4x + 9$
(h) $(g \circ g \circ g)(x) = x^{27}$

41. (a) $f(x + 2) = \sqrt[3]{x} + 3$
(b) $g(f(x)) = x - 2$
(c) $g^{-1}(x) = \sqrt[3]{x} + 3$
(d) $f^{-1}(x) = (x - 3)^3 + 2$

42. $f^{-1}(x) = 2x - 3$

43. $g^{-1}(x) = \dfrac{2x}{1 - x}$

44.

$[-10, 10]$ by $[-7, 7]$

45. $f(x) = \sqrt{x}, g(x) = x^3 - 7$. Other answers are possible.

46. $\dfrac{2x + 1}{x + 2} = 2 \Rightarrow 2x + 1 = 2x + 4 \Rightarrow 1 = 4$, a contradiction.

47. $\{x: -3 \le x \le 3, x \ne -2\}$

48. $\left\{x: x \ge \dfrac{5}{2}\right\}, f^{-1}(x) = \dfrac{1}{2}(\sqrt{x} + 5)$

49. $(f \cdot g)(-x) = f(-x)g(-x) = f(x)(-g(x)) = -f(x)g(x) = -(f \cdot g)(x);$ $(f \circ g)(-x) = f(g(-x)) = f(-g(x)) = f(g(x)) = (f \circ g)(x)$

50.

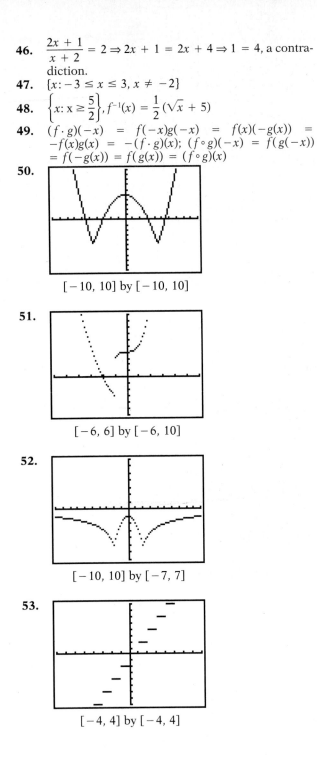

$[-10, 10]$ by $[-10, 10]$

51.

$[-6, 6]$ by $[-6, 10]$

52.

$[-10, 10]$ by $[-7, 7]$

53.

$[-4, 4]$ by $[-4, 4]$

54.

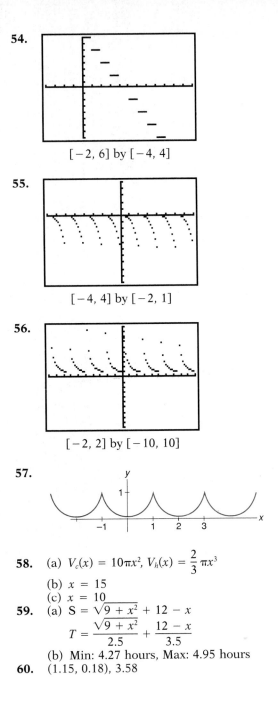

$[-2, 6]$ by $[-4, 4]$

55.

$[-4, 4]$ by $[-2, 1]$

56.

$[-2, 2]$ by $[-10, 10]$

57.

58. (a) $V_c(x) = 10\pi x^2$, $V_h(x) = \dfrac{2}{3}\pi x^3$

(b) $x = 15$

(c) $x = 10$

59. (a) $S = \sqrt{9 + x^2} + 12 - x$

$T = \dfrac{\sqrt{9 + x^2}}{2.5} + \dfrac{12 - x}{3.5}$

(b) Min: 4.27 hours, Max: 4.95 hours

60. $(1.15, 0.18)$, 3.58

PROBLEM SET 3.1 (p. 126)

1. $-\dfrac{1}{16}$

3. $\dfrac{1}{16}$

5. 1

7. 4

9. $\dfrac{a^8 - 1}{a^4}$

11. $\dfrac{1}{b^{27}}$

13. b^{12}

15. $\dfrac{9a^4c^2}{b^4}$

17. a^{11}

19. $\dfrac{8y^{18}}{27x^9}$

21. $\dfrac{1}{x^4}$

23. $x^2 + 2 + \dfrac{1}{x^2}$

25. $x^{2/3}$

27. $x^{2/3}$

29. $\dfrac{1}{y^{1/12}} = y^{-1/12}$

31. $-(x + y)^{3/2}$

33. $x^{3/2}y^{4/3}$

35. $(a - b)^2$

37. $a - 2a^{1/2}b^{1/2} + b$

39. 5

41. π

43. $\dfrac{a^4}{b^{11}}$

45. $\dfrac{x^{1/3}}{4}$

47. $\dfrac{16}{y^6}$

49. $\dfrac{3y^3 - 1}{y}$

51. $a^3 + 2a^{3/2}b^{3/2} + b^3$

53. $x^{9/2}$

55. $\dfrac{-3}{(x - 3)^{1/2}}$

57. $\dfrac{2}{(x^2 + 2)^{1/3}}$

59. 13.45231

61. 0.43544

63. 1.76256

65. 1.76256

67. 14.39693

69. $\dfrac{3^\pi}{\pi^3} < \pi^2 - 2^\pi < 2^{0.1} < (\sqrt{2})^{1/\pi}$

71. $0.5\pi^{\sqrt{2}} < (\sqrt{2})^{\sqrt{7.5}} < (\sqrt{3})^{\sqrt{3}} < 2^{\sqrt[9]{2\pi}}$

73. $x = -\dfrac{1}{2}$

75. $x = -\dfrac{3}{4}$

77. Conjecture: $\sqrt{2 + \sqrt{3}} = \dfrac{1}{2}(\sqrt{2} + \sqrt{6})$

Proof: Square both and simplify.

79. $n^{1/n} \to 1$ as $n \to \infty$

81. (a) $1 = 1^1$
(b) $\sqrt{2} = 2^{1/2}$
(c) $1 = (\sqrt{2})^0$

PROBLEM SET 3.2 (p. 132)

1. 3.62686

3. 0.68153

5.

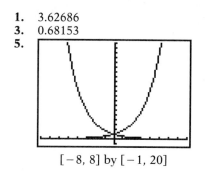

$[-8, 8]$ by $[-1, 20]$

7.

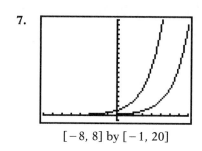

$[-8, 8]$ by $[-1, 20]$

9. Second function is shifted 2 units right and 2 units up.
11. Second function eventually grows faster.
13. (a) B
(b) C
(c) D
(d) A
15. $2 \cdot 2^5 < 4^5 < e^{10}$
17. $2^{2.3} < e^{2.3} < 2.3^6$
19. $50^6 < 3 \cdot 2^{50} < e^{50}$
21. 148.41316
23. 266.96284
25. $k = 2.485$
27. $k = 0.788$
29. (1.00, 1.47)
31. (6.00, 10.75)
33. $|x| < 3x^2 < x^3 < 1.1^x$ for large x.
35. Zero

37. One
39. Zero
41. Four
43. (0.693, 8.00)
45. (1.765, 17.12) and (5, 3125)
47. (1.631, 11.55) and (5.938, 7381)
49. $-0.693, 1.099, 2.303$
51. The slope of the graph of e^x at x is e^x.
53.

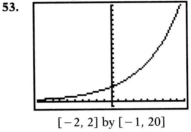

$[-2, 2]$ by $[-1, 20]$

PROBLEM SET 3.3 (p. 138)

1. 3
3. 2
5. 4
7. -5
9. $\dfrac{1}{3}$
11. -1
13. -3
15. 24
17. $2 \log x + \log y$
19. $\log (x + 1) - \log y$
21. $\log 7 + 3 \log x + 4 \log y - \log (y + 2)$
23. $\dfrac{1}{3} (\log 3 + 2 \log y + \log 4 + 4 \log x)$
25. $\dfrac{1}{2} (\log (x + y) - \log 3 - 2 \log y)$
27. $x + \dfrac{1}{2} \ln (x^2 + 1)$
29. $\log (xy^3)$
31. $\log \left(\dfrac{\sqrt{x + 2y}}{y^2} \right)$
33. $\log \left(\dfrac{y^4 (7 - x)^{3/2}}{\sqrt{3x + y}} \right)$
35. 4.75489
37. 2.07760
39. 3.58496

41. 0.92318
43. -0.07438
45. ± 0.73987
47. -2.51294
49. $t = \dfrac{\ln A_0 - \ln A}{k}$
51. $n = \dfrac{\ln (Ai + R) - \ln R}{\ln (1 + i)}$
53. $x = \dfrac{1}{2} e^y$
55. $x = \sqrt[3]{\dfrac{A}{B}}$
57. Dm: $\left\{ x: x > \dfrac{10}{3} \right\}$, Rg: $\mathbb{R}$, $f^{-1}(x) = \dfrac{1}{3}(e^x + 10)$
59. Dm: $\{x: x > -2\}$, Rg: $\mathbb{R}$, $f^{-1}(x) = \sqrt[3]{e^x - 8}$
61. $x = \dfrac{2 + 2e}{2e - 1}$
63. $x = -2, 3$
65. $x = e^{\pm \sqrt{\ln 10}}$
67. $x = e^e$
69. 1.310
71. 1.376
73. Use properties of exponents.
75. Let $x = t, a = s$, and $b = t$ in the change of base formula.

PROBLEM SET 3.4 (p. 145)

1. Exponential decay
3. Exponential decay
5. Neither
7. Exponential growth
9. Exponential growth
11. Exponential decay
13. $N = 50(1.03)^t$
15. $V = 15,000(0.8)^t$
17. $A = 50\left(\dfrac{1}{2}\right)^{t/10}$
19. 15,405,000
21. (a) $P = 6000\left(1 + \dfrac{r}{100}\right)^t$
    (b) After about 26 months
23. 26.9 million
25. 17.67 years
27. 31.55 years
29. 1.768 grams
31. 116.60 days

33. 889 years
35. (a) $N_0 = 100$
    (b) $L = 400$
    (c) 379
37. (a) $N = \dfrac{800}{1 + 39e^{-0.40707t}}$
    (c) 618
39. 10 times stronger
41. (a) 50 and 70
    (b) 10 and 7.25
    (c) 1.18 and 17.83
43. (a) $t > 90.28$
    (b) $t > 19$
45. 38,000 years
47. 2.3 years after stocking
49. (a) $P(20) = 8, P(50) = 15$
    (b) 1.198, 1.174, 1.132, 1.084
    (c) $\dfrac{P(x)}{f(x)} \to 1$ as $x \to \infty$

PROBLEM SET 3.5 (p. 152)

1. $518.00, $559.35
3. $7550.00, $8157.34
5. $1900.00, $2367.36
7. $2000.00, $2593.74
9. (a) $1790.85
   (b) $1814.02
   (c) $1819.40
   (d) $1822.03
   (e) $1822.12
11. (a) $1819.40
    (b) $2009.66
    (c) $2219.64
    (d) $2451.36
    (e) $2707.04
13. 10.3813%
15. 9.9645%

17. 9.4174%
19. 8.2% compounded quarterly
21. 12.3% compounded quarterly
23. 8.2% compounded quarterly
25. (a) 8.69 years
    (b) 100.56 months $\approx$ 8.38 years
    (c) 3049 days $\approx$ 8.35 years
27. 5.9463%
29. 8.6654%
31. 8.43 years, 15.56 years
33. 7.0924%
35. 6.7768%
37. 132.77 years
39. 9.4210%
41. $P > $12,440.38

PROBLEM SET 3.6 (p. 159)

1. (a)

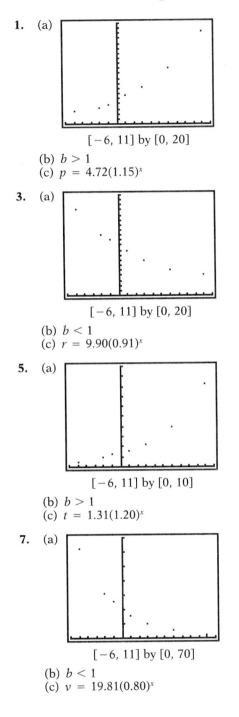

$[-6, 11]$ by $[0, 20]$

   (b) $b > 1$
   (c) $p = 4.72(1.15)^x$

3. (a)

$[-6, 11]$ by $[0, 20]$

   (b) $b < 1$
   (c) $r = 9.90(0.91)^x$

5. (a)

$[-6, 11]$ by $[0, 10]$

   (b) $b > 1$
   (c) $t = 1.31(1.20)^x$

7. (a)

$[-6, 11]$ by $[0, 70]$

   (b) $b < 1$
   (c) $v = 19.81(0.80)^x$

9. $f = -0.15 + 1.08 \ln a$

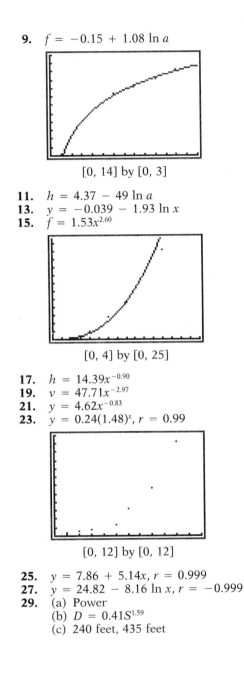

$[0, 14]$ by $[0, 3]$

11. $h = 4.37 - 49 \ln a$
13. $y = -0.039 - 1.93 \ln x$
15. $f = 1.53x^{2.60}$

$[0, 4]$ by $[0, 25]$

17. $h = 14.39x^{-0.90}$
19. $v = 47.71x^{-2.97}$
21. $y = 4.62x^{-0.83}$
23. $y = 0.24(1.48)^x, r = 0.99$

$[0, 12]$ by $[0, 12]$

25. $y = 7.86 + 5.14x, r = 0.999$
27. $y = 24.82 - 8.16 \ln x, r = -0.999$
29. (a) Power
    (b) $D = 0.41S^{1.59}$
    (c) 240 feet, 435 feet

31. (a) $P(y) = 13.46(1.0166)^y$
(b)

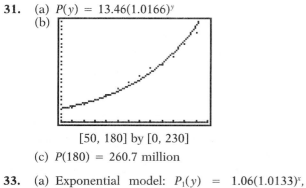

[50, 180] by [0, 230]

(c) $P(180) = 260.7$ million

33. (a) Exponential model: $P_1(y) = 1.06(1.0133)^x$, $r = 0.982$
Power model: $P_2(y) = 0.002x^{1.65}$, $r = 0.967$
(b) $P_1(180) = 11.43$, $P_2(180) = 10.53$

35. $L(t) = 0.37 + 2.12 \ln t$; $L(10) = 5.3$; $L(25) = 7.2$

37.

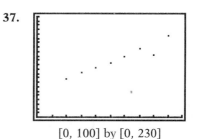

[0, 100] by [0, 230]

With outlier: $s = 63.45 + 1.23t$, $r = 0.960$, $s = 200$ in 2011
Without outlier: $s = 58.08 + 1.39t$, $r = 0.997$, $s = 200$ in 2003

39. (a) $C = 972.8(0.5275)^x$
(b) About 1.08 hours

CHAPTER 3 REVIEW PROBLEM SET (p. 162)

1. T
2. F
3. F
4. T
5. T
6. T
7. T
8. F
9. T
10. F
11. x^2
12. a^2b
13. $\dfrac{y^2}{xz^3}$
14. $6x^{2/3}$
15. $27a^{1/4}$
16. $\dfrac{2^{5/4}y^{3/4}}{x^2}$
17. Their graphs are reflections across the y-axis.
18. 5^x
19. x^3, $\dfrac{x^5 - 10x^2}{x + 100}$, e^x, 3^x
20. 1.53, 10.85
21. 10.22
22. $(-1.00, 3.74)$
23. (a) 5
(b) 0
(c) -2
(d) π
(e) $\dfrac{5}{3}$
(f) 12

24. (a) 64
(b) $\dfrac{1}{125}$
(c) 9
(d) π
(e) 2
(f) 16
25. $\log_4\left(\dfrac{(x^2 + 1)^3}{(x + 2)^2}\right)$
26. (a) ± 3
(b) 2
27. It is shifted 1 unit right and 2 units up.
28. $\dfrac{3}{2}$
29. 1.42483
30. 0.36342
31. 0.34657
32. 4.38
33. They are reflections in the line $y = x$.
34. $\ln (x + 5)$, $\log_2 x$, $\ln (x^2 - 5)$, $\dfrac{1}{2}x$
35. (a) 12.60 years
(b) 41.87 years
36. $y = 40e^{-0.06931t}$
37. 1161 years
38. 5.615 hours
39. (a) 93.78 days
(b) 50
40. (a) 5.761 years
(b) 5.529 years

41. (a) $1640
(b) $1884.54
(c) $1896.35

42. First Bank offers an effective rate of 8.5619%, but Second Bank offers the even better rate 8.5692%.

43. 21.73 years

44. 6.96104%

45. (a) $y = 5.05(1.0143)^t$
(b)

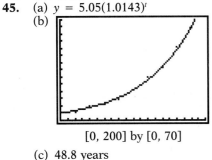

[0, 200] by [0, 70]

(c) 48.8 years
(d) 2010

46. (a) $y = 3.958(0.99847)^x$
(b)

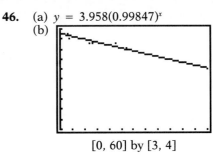

[0, 60] by [3, 4]

(c) 3:38.0
(d) 2034

47. (a) $y = 0.9211(1.1014)^t$
(b) $115.23
(c) Not to be trusted that far into the future, especially since the data base covers only 40 years.

PROBLEM SET 4.1 (p. 170)

1. 0.7351512728
3. 0.9540731638
5. 1.500627967
7. −0.4003190029
9. 0.9119557904
11. −0.4368582524
13. 1.401247955
15. 1.624408533
17. 59.933°
19. 53.386°
21. 76.252°
23. 16.968
25. 41.336
27. 34.273°
29. 66.605

31. $\beta = 47.7°$, $a \approx 23.6$, $b \approx 26.0$
33. $\alpha = 33.75°$, $a \approx 50.75$, $b \approx 75.95$
35. $\beta = 50.55°$, $b \approx 146.2$, $c \approx 189.3$
37. $\alpha \approx 40.85°$, $\beta \approx 49.15°$, $c \approx 16.31$
39. $\alpha \approx 52.9°$, $\beta \approx 37.1°$, $b \approx 30.4$
41. 19.41 feet
43. 6.82 feet
45. 725 feet
47. 37.94
49. 448 meters
51. 41.8°
53. (a) 76.75 miles
(b) 12.03°
55. $P = 24$; $A = 24\sqrt{3} \approx 41.57$
57. $\sqrt{3}$

PROBLEM SET 4.2 (p. 178)

1. $\frac{2}{3}\pi \approx 2.0944$

3. $\frac{7}{6}\pi \approx 3.6652$

5. $3\pi \approx 9.4248$

7. $\frac{34}{180}\pi \approx 0.5934$

9. $54.5\left(\frac{\pi}{180}\right) \approx 0.9512$

11. 240°

13. −120°
15. 495°
17. 171.89°
19. 0.8333
21. 2.705
23. 23.4, 105.3
25. 37.699, 169.646
27. 15.446 inches
29. 21.991, 1.047
31. IV

33. II
35. III
37. II
39. 104.720 inches per second
41. 2513.274 inches per minute
43. 5.712 miles per hour
45. 672.270
47. 2799 miles
49. IV, (0.86, −0.51)
51. 328.557, 2586.031

53. (a) 17.659
(b) 90.321
55. 21.991 square inches
57. (a) 753.98 inches per minute
(b) 15 revolutions per minute
59. 1036.73 miles per hour
61. 329.87 meters
63. 1.1519
65. 7460 miles

PROBLEM SET 4.3 (p. 185)

1. 0.23684
3. 0.99122
5. 0.90595
7. 0.25126
9. 0.20279
11. −0.94380
13. 0.21857
15. $-\dfrac{1}{2}$
17. $\dfrac{1}{2}\sqrt{3}$
19. $-\dfrac{1}{2}\sqrt{2}$
21. 1
23. 0
25. $\dfrac{1}{2}$
27. Plus
29. Minus
31. Plus
33. cos 1
35. sin 3
37. sin 1
39. sin 3
41. $\dfrac{12}{13}$
43. $\dfrac{1}{5}\sqrt{21}$
45. $-\dfrac{3}{4}$
47. $-\dfrac{4}{5}$
49. 1

51. (a) (0, 1)
(b) (−1, 0)
(c) (0.64726, 0.76227)
(d) (−0.92568, −0.37830)
53. 1.98289
55. (a) $\dfrac{1}{4}\pi, \dfrac{5}{4}\pi$

(b) $\dfrac{1}{6}\pi < t < \dfrac{1}{3}\pi$ or $\dfrac{2}{3}\pi < t < \dfrac{5}{6}\pi$

(c) $0 \le t \le \dfrac{1}{3}\pi$ or $\dfrac{2}{3}\pi \le t \le \dfrac{4}{3}\pi$ or $\dfrac{5}{3}\pi \le t \le 2\pi$

(d) $0 \le t < \dfrac{1}{4}\pi$ or $\dfrac{3}{4}\pi < t < \dfrac{5}{4}\pi$ or $\dfrac{7}{4}\pi < t \le 2\pi$
57. $\sin(\pi + t) = -\sin t;\ \cos(\pi + t) = -\cos t$
59. (a) $\dfrac{2}{3}\sqrt{2}$

(b) $\dfrac{1}{3}$

(c) $\dfrac{1}{3}$

(d) $\dfrac{1}{3}$

(e) $-\dfrac{1}{3}$

(f) $\dfrac{2}{3}\sqrt{2}$

(g) $-\dfrac{1}{3}$

(h) $-\dfrac{2}{3}\sqrt{2}$
61. −1

PROBLEM SET 4.4 (p. 192)

1. $\pi, 2, 0$

3. $6\pi, \dfrac{3}{2}, 0$

5. $\dfrac{2}{3}\pi, 1.5, 0$

7. $\dfrac{1}{2}\pi, 1, -\dfrac{1}{8}\pi$

9. $\pi, 2, \dfrac{1}{6}\pi$

11. $\dfrac{2}{3}\pi, 2, 0$

13. $y = 2 \sin 2x$
15. $y = 1 + \sin x$
17. $y = -\sin x$
19. $2\pi, 3.06$

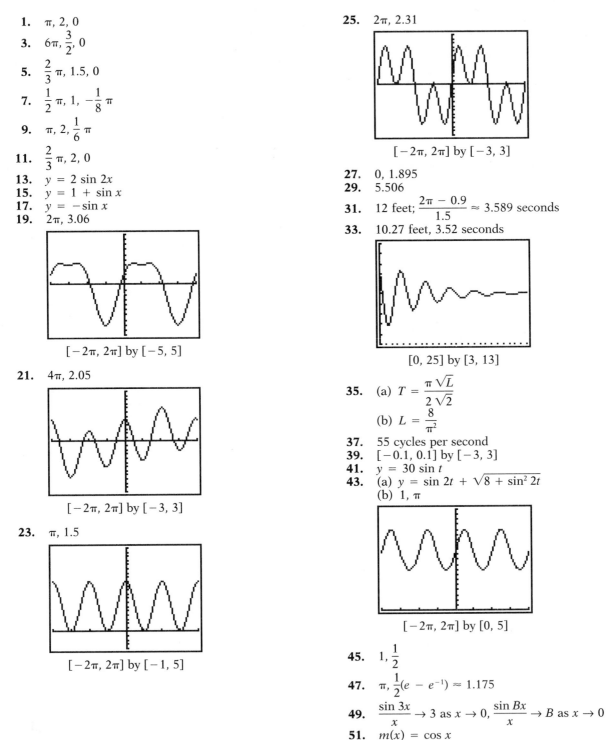

$[-2\pi, 2\pi]$ by $[-5, 5]$

21. $4\pi, 2.05$

$[-2\pi, 2\pi]$ by $[-3, 3]$

23. $\pi, 1.5$

$[-2\pi, 2\pi]$ by $[-1, 5]$

25. $2\pi, 2.31$

$[-2\pi, 2\pi]$ by $[-3, 3]$

27. $0, 1.895$
29. 5.506

31. 12 feet; $\dfrac{2\pi - 0.9}{1.5} \approx 3.589$ seconds

33. 10.27 feet, 3.52 seconds

$[0, 25]$ by $[3, 13]$

35. (a) $T = \dfrac{\pi \sqrt{L}}{2\sqrt{2}}$

(b) $L = \dfrac{8}{\pi^2}$

37. 55 cycles per second
39. $[-0.1, 0.1]$ by $[-3, 3]$
41. $y = 30 \sin t$
43. (a) $y = \sin 2t + \sqrt{8 + \sin^2 2t}$

(b) $1, \pi$

$[-2\pi, 2\pi]$ by $[0, 5]$

45. $1, \dfrac{1}{2}$

47. $\pi, \dfrac{1}{2}(e - e^{-1}) \approx 1.175$

49. $\dfrac{\sin 3x}{x} \to 3$ as $x \to 0$, $\dfrac{\sin Bx}{x} \to B$ as $x \to 0$

51. $m(x) = \cos x$

PROBLEM SET 4.5 (p. 200)

1. 0.95897
3. −19.10732
5. −15.89424
7. −53.78778
9. 27.98034
11. 3.79329
13. 3.18778
15. $\cos x = -\dfrac{4}{5}$, $\tan x = \dfrac{3}{4}$, $\cot x = \dfrac{4}{3}$, $\sec x = -\dfrac{5}{4}$,
    $\csc x = -\dfrac{5}{3}$
17. $\tan x = \dfrac{12}{5}$, $\sec x = \dfrac{13}{5}$, $\cos x = \dfrac{5}{13}$, $\sin x = \dfrac{12}{13}$,
    $\csc x = \dfrac{13}{12}$
19. $ab(1 - a^2)$
21. $\dfrac{a^2 + b^2}{a^2 b^2}$
23. $\dfrac{1 - a^2}{a^2}$
25. 1
27. 0
29. −1
31. $\mathbb{R}$, $\pi/2$

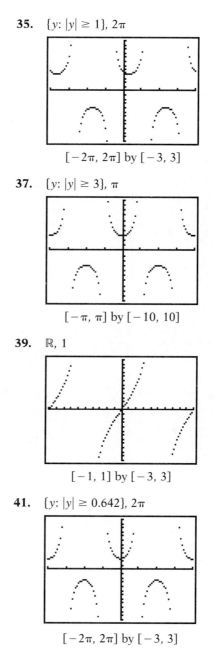

$[-\pi, \pi]$ by $[-3, 3]$

33. $\{y: |y| \geq 1\}$, 4π

$[-2\pi, 2\pi]$ by $[-3, 3]$

35. $\{y: |y| \geq 1\}$, 2π

$[-2\pi, 2\pi]$ by $[-3, 3]$

37. $\{y: |y| \geq 3\}$, π

$[-\pi, \pi]$ by $[-10, 10]$

39. $\mathbb{R}$, 1

$[-1, 1]$ by $[-3, 3]$

41. $\{y: |y| \geq 0.642\}$, 2π

$[-2\pi, 2\pi]$ by $[-3, 3]$

43. $\{y: y \geq -2\}$, π

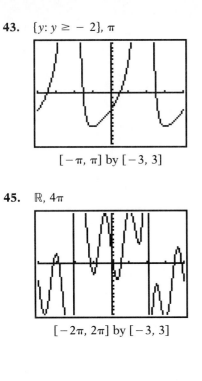

$[-\pi, \pi]$ by $[-3, 3]$

45. $\mathbb{R}$, 4π

$[-2\pi, 2\pi]$ by $[-3, 3]$

47. (a) 1

(b) $\sin \theta - \dfrac{1}{\cos \theta}$

(c) $1 + 2 \sin \theta \cos \theta$

(d) $\dfrac{1}{\sin \theta}$

(e) $\cos \theta + \sin \theta$

(f) $\dfrac{-(1 + \sin^2\theta)}{\cos^2\theta}$

49. $\dfrac{91}{60}$

51. (a) $\tan (t + \pi) = \tan t$
(b) $\cot (t + \pi) = \cot t$
(c) $\sec (t + \pi) = -\sec t$
(d) $\csc (t + \pi) = -\csc t$

53. 1.047

55. 0.789

57. Infinitely many

59. $y = -\dfrac{1}{3}\sqrt{3}(x - 5)$

61. (a) $y = -1.376x$
(b) $y - 2.5 = 0.315(x + 4.330)$

63. $L = 4 \csc \theta + 3 \sec \theta$; 9.866

PROBLEM SET 4.6 (p. 208)

1. $\dfrac{1}{3}\pi$

3. $\dfrac{2}{3}\pi$

5. $-\dfrac{1}{2}\pi$

7. π

9. Undefined

11. $-\dfrac{1}{4}\pi$

13. $\dfrac{1}{3}\pi$

15. 0.3

17. 0.30

19. 0

21. 0.23681

23. -0.23681

25. 1.27187

27. 2.06116

29. Undefined

31. 1.26263

33. 0.97319

35. -0.41602

37. Undefined

39. 0.90095

41. $\{x: -2 \leq x \leq 2\}$, $\left\{y: -\dfrac{1}{2}\pi \leq y \leq \dfrac{1}{2}\pi\right\}$

43. $\mathbb{R}$, $\left\{y: -\dfrac{1}{2}\pi < y < \dfrac{1}{2}\pi\right\}$

45. $\{x: -1 \leq x \leq 1\}$, $\{y: -1 \leq y \leq 1\}$

47. $\theta = \sin^{-1}\left(\dfrac{2}{x}\right)$

49. $\theta = \tan^{-1}\left[\dfrac{1}{6}(2 + x)\right] - \tan^{-1}\left(\dfrac{1}{6}x\right)$

51. $\{x: -1 \leq x \leq 1\}$, $\left\{\dfrac{1}{2}\pi\right\}$

53. $\left\{x: x \neq \left(n + \dfrac{1}{2}\right)\pi, n \text{ an integer}\right\}$, $\mathbb{R}$

55. $\left\{x: -\dfrac{1}{3} \leq x \leq \dfrac{1}{3}\right\}$, $\{y: \pi \leq y \leq 3\pi\}$

57. $\{x: -1 \leq x \leq 1\}$, $\{y: -3.206 \leq y \leq 3.206\}$

59. 0.944

61. (1.264, 1.574)

63. $\ll x \gg = \dfrac{1}{2\pi} \cos^{-1}(\cos 2\pi x)$

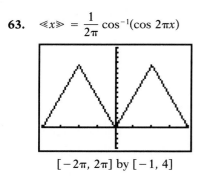

$[-2\pi, 2\pi]$ by $[-1, 4]$

65. $\sin^{-1} x = \tan^{-1}\left(\dfrac{x}{\sqrt{1-x^2}}\right)$

67. About 49 feet from A

CHAPTER 4 REVIEW PROBLEM SET (p. 210)

1. T

2. T

3. F

4. F

5. F

6. T

7. F

8. T

9. F

10. T

11. $b = 12,\ \alpha \approx 36.87°,\ \beta \approx 53.13°$

12. $\beta \approx 17.6°,\ a \approx 93.3,\ c \approx 97.9$

13. -1

14. $\dfrac{1}{3}\sqrt{3}$

15. 2

16. -1

17. $\dfrac{1}{2}\sqrt{2}$

18. -1

19. -1

20. -1

21. $\dfrac{2}{3}\pi,\ \dfrac{4}{3}\pi$

22. $\dfrac{3}{4}\pi,\ \dfrac{7}{4}\pi$

23. $\dfrac{1}{4}\pi,\ \dfrac{3}{4}\pi$

24. $\dfrac{1}{3}\pi,\ \dfrac{5}{3}\pi$

25. $\dfrac{-3}{\sqrt{40}}$

26. $-60°,\ 60°$

27. (a) $\dfrac{3}{5}$

(b) $-\dfrac{3}{5}$

(c) $-\dfrac{3}{5}$

(d) $\dfrac{4}{5}$

28. (a) $-\sqrt{1 - \sin^2 \theta}$

(b) $\dfrac{1}{\sin \theta}$

(c) $\dfrac{-\sin \theta}{\sqrt{1 - \sin^2 \theta}}$

29. (a) $-\dfrac{12}{5}$

(b) $\dfrac{13}{12}$

30. (a) $0 < t < \pi$

(b) $0 < t < \dfrac{1}{2}\pi,\ \pi < t < \dfrac{3}{2}\pi$

31. $\{y : -5 \le y \le 1\}$

32. $\tan(-t) = \dfrac{\sin(-t)}{\cos(-t)} = -\dfrac{\sin t}{\cos t} = -\tan t$

33. $a = 2,\ b = 2(4 - \sqrt{3})$

34. (a) $6,\ -2$

(b) $\dfrac{7}{6}\pi,\ \dfrac{11}{6}\pi$

35. (a) $\cos t$

(b) $-\cos t$

(c) $1 - \cos^2 t$

(d) $-\cos t$

36. $\dfrac{25}{16} + \dfrac{27}{125} \approx 1.7785$

37. $A = \dfrac{1}{2}\, ab = \dfrac{1}{2}\, a(a \cot \alpha) = \dfrac{1}{2}\, a^2 \cot \alpha$

38. (a) 1.72894
(b) 0.31133

39. Per: π, Amp: 3, PS: 0

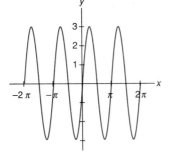

40. Per: 4π, Amp: 2.5, PS: $-\dfrac{1}{4}\,\pi$

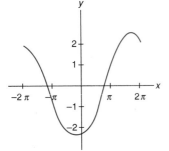

41. Per: $\dfrac{1}{2}\,\pi$, Amp: 1, PS: $-\dfrac{1}{8}\,\pi$

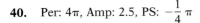

42. Per: π, Amp: 5, PS: $\dfrac{1}{4}\,\pi$

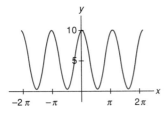

43. Per: 2π, Amp: 2.52

44. (a) Dm: $\{x: x \neq n\pi, n \text{ an integer}\}$, Rg: $\mathbb{R}$, Per: 1
(b) Dm: $\left\{x: x \neq \dfrac{1}{2}\, m\pi, m \text{ odd}\right\}$, Rg: $\mathbb{R}$, Per: π
(c) Dm: $\mathbb{R}$, Rg: $\{y: 1.5 \leq y \leq 3\}$, Per: π
(d) Dm: $\{x: x \neq 2n\pi, n \text{ an integer}\}$,
Rg: $\{y: |y| \geq 0.1\}$, Per: 4π

45. (a) 1.27, 2.01
(b) 0, 1.70, π

46. Max: 2.81, Min: -2.18

47. Dm: $\{x: x \neq 0\}$, Rg: $\{y: 0 < y < 3\}$

48. 0.39

49. (a) $\dfrac{1}{4}\,\pi$
(b) $\dfrac{2}{3}\,\pi$
(c) $-\dfrac{1}{4}\,\pi$
(d) $\dfrac{1}{4}\,\pi$

50. (a) 2.5
(b) $\dfrac{1}{4}\,\pi$
(c) $\dfrac{5}{3}$
(d) -5

51. $\dfrac{1}{4}\,\pi, \dfrac{5}{4}\,\pi$

52. (a) 4.36 feet per second
(b) 523.6 feet

53. 387.01

54. 13.11 feet

55. 0.76 feet

56. $x = 1.107$, $y = 2.344$

57. (a) $50\pi \approx 157.08$ meters
(b) 156.43 meters

58. 8

59. 73.475 centimeters

60. 1,459,380 square feet

61. Dm: $\left\{x: -\dfrac{1}{4}\,\pi \leq x \leq \dfrac{1}{4}\,\pi\right\}$; $f^{-1}(x) = 0.5 \sin^{-1} x$

62. $\cot^2 u = \csc^2 u - 1 = \dfrac{1 - \sin^2 u}{\sin^2 u}$
Thus, $\cot^2(\sin^{-1} x) = \dfrac{1 - x^2}{x^2}$
and
$\cot (\sin^{-1} x) = \pm\dfrac{\sqrt{1 - x^2}}{x}$
Finally, note that the two sides have the same sign.

63. 0.4460

64. $x = 3 \tan (\theta + 11°) - 3 \tan \theta$, 73.29°

PROBLEM SET 5.1 (p. 218)

1. $-x^2 - 7x$
3. $t^3 - 5t^2 + 2t - 1$
5. $9x^2 - 24xy + 16y^2$
7. $4x^2 - 16y^4$
9. $9a^4 + 6a^2bc + b^2c^2$
11. $3u^2 - 4uv - 4v^2$
13. $6x^4 - 2y^2z^2$
15. $8s^3 - 12s^2t + 6st^2 - t^3$
17. $x^5y - 4xy^3$
19. $x^2 + 2xy + y^2 - 16$
21. $3a^2bc(b^2 - 2ac)$
23. $x^2(xy - z)(xy + z)$
25. $(y - 6)(y + 2)$
27. $(3x + 4)^2$
29. $(2a - 3b)(4a^2 + 6ab + 9b^2)$
31. $2x(2u + v)(4u^2 - 2uv + v^2)$
33. $(ab - c)^3$
35. $\dfrac{1}{x - 5}$
37. $x^2 + x - 6$
39. $x^2 + xy + y^2$
41. $\dfrac{x}{x - 2}$

43. -4
45. $\dfrac{6y^2 + 9y + 2}{9y^2 - 1}$
47. (a) $2\sqrt{2}\,xy - 4y^2$
    (b) $8x^3 - y^3$
49. (a) $(x + 2y + 3)^2$
    (b) $(m - n + 4)(m - n + 1)$
51. $\dfrac{x^2 - ax + a^2}{x^2 + ax + a^2}$
53. $\dfrac{x - 5}{3x + 6}$
55. $-x$
57. $-\dfrac{nm + m^2}{n}$
59. $(2m)^2 + (m^2 - 1)^2 = 4m^2 + m^4 - 2m^2 + 1 = m^4 + 2m^2 + 1 = (m^2 + 1)^2$
61. $\dfrac{50}{99}$
63. $-\dfrac{5}{2}, \dfrac{487}{2}$

PROBLEM SET 5.2 (p. 224)

1. $\cos x$
3. $\sin^2 t$
5. $\cot u$
7. $\cos^2 u$
9. $2 \cot u$
11. $\dfrac{1 + \sin^2 x}{\cos x}$
13. 0
15. $\cos x = -\sqrt{1 - \sin^2 x}, \cot x = \dfrac{-\sqrt{1 - \sin^2 x}}{\sin x}$
    $\csc x = \dfrac{1}{\sin x}, \tan x = \dfrac{-\sin x}{\sqrt{1 - \sin^2 x}}$
    $\sec x = \dfrac{-1}{\sqrt{1 - \sin^2 x}}$
17. $\cos x = \dfrac{1}{\sec x}, \cot x = \dfrac{-1}{\sqrt{\sec^2 x - 1}}$
    $\csc x = \dfrac{-\sec x}{\sqrt{\sec^2 x - 1}}, \tan x = -\sqrt{\sec^2 x - 1}$
    $\sin x = -\dfrac{\sqrt{\sec^2 x - 1}}{\sec x}$
19. $\dfrac{\sec^2 x - 1}{\sec^2 x} = \dfrac{1/\cos^2 x - 1}{1/\cos^2 x} = 1 - \cos^2 x = \sin^2 x$

21. $\dfrac{\sec t - 1}{\tan t} = \dfrac{(\sec t - 1)(\sec t + 1)}{(\tan t)(\sec t + 1)}$
    $= \dfrac{\sec^2 t - 1}{(\tan t)(\sec t + 1)}$
    $= \dfrac{\tan^2 t}{(\tan t)(\sec t + 1)} = \dfrac{\tan t}{\sec t + 1}$
23. $\dfrac{\tan^2 u}{\sec u + 1} = \dfrac{\sec^2 u - 1}{\sec u + 1} = \sec u - 1$
    $= \dfrac{1}{\cos u} - 1 = \dfrac{1 - \cos u}{\cos u}$
25. $\dfrac{\sin t + \cos t}{\tan^2 t - 1} = \dfrac{\sin t + \cos t}{(\sin^2 t - \cos^2 t)/\cos^2 t}$
    $= \dfrac{\cos^2 t}{\sin t - \cos t}$
27. $(1 + \tan^2 t)(\cos t + \sin t) = (\sec^2 t)(\cos t + \sin t)$
    $= (\sec t)(1 + \sec t \sin t) = (\sec t)(1 + \tan t)$
29. $2\sec^2 y - 1 = \dfrac{2}{\cos^2 y} - 1 = \dfrac{2 - \cos^2 y}{\cos^2 y} =$
    $\dfrac{1 + \sin^2 y}{\cos^2 y}$

31. $\dfrac{\sin z}{\sin z + \tan z} = \dfrac{\sin z}{\sin z + \sin z/\cos z}$

$\qquad = \dfrac{\sin z \cos z}{(\sin z)(\cos z + 1)} = \dfrac{\cos z}{1 + \cos z}$

33. $f(x) = \sin^2 x + 2 + \csc^2 x + \cos^2 x + 2$

$\qquad + \sec^2 x - \tan^2 x - 2 - \cot^2 x$

$\qquad = 2 + \sin^2 x + \cos^2 x + \csc^2 x$

$\qquad - \cot^2 x + \sec^2 x - \tan^2 x$

$\qquad = 2 + 1 + 1 + 1 = 5$

35. $f(x) = \dfrac{1 + \cos x + 1 - \cos x}{1 - \cos^2 x}$

$\qquad = \dfrac{2}{\sin^2 x} = 2 \csc^2 x \geq 2$

37. $\sin x$

39. $(\csc t + \cot t)^2 = \left(\dfrac{1 + \cos t}{\sin t}\right)^2$

$\qquad = \dfrac{(1 + \cos t)^2}{1 - \cos^2 t} = \dfrac{1 + \cos t}{1 - \cos t}$

41. $\dfrac{1 + \tan x}{1 - \tan x} = \dfrac{(\cos x + \sin x)/\cos x}{(\cos x - \sin x)/\cos x}$

$\qquad = \dfrac{\cos x + \sin x}{\cos x - \sin x}$

43. $\dfrac{2 \tan u}{1 - \tan^2 u} + \dfrac{1}{\cos^2 u - \sin^2 u}$

$\qquad = \dfrac{2 \sin u/\cos u}{(\cos^2 u - \sin^2 u)/\cos^2 u}$

$\qquad + \dfrac{1}{\cos^2 u - \sin^2 u} = \dfrac{2 \sin u \cos u + 1}{\cos^2 u - \sin^2 u}$

$\qquad = \dfrac{(\cos u + \sin u)^2}{(\cos u - \sin u)(\cos u + \sin u)}$

$\qquad = \dfrac{\cos u + \sin u}{\cos u - \sin u}$

45. $\dfrac{\cos^3 t + \sin^3 t}{\cos t + \sin t}$

$\qquad = \dfrac{(\cos t + \sin t)(\cos^2 t - \cos t \sin t + \sin^2 t)}{\cos t + \sin t}$

$\qquad = 1 - \sin t \cos t$

47. $\left(\dfrac{1 - \cos \theta}{\sin \theta}\right)^2 = \dfrac{(1 - \cos \theta)^2}{1 - \cos^2 \theta} = \dfrac{1 - \cos \theta}{1 + \cos \theta}$

49. $(\csc t - \cot t)^4 (\csc t + \cot t)^4$

$\qquad = ((\csc t - \cot t)(\csc t + \cot t))^4$

$\qquad = (\csc^2 t - \cot^2 t)^4 = 1^4 = 1$

51. $\sin^6 u + \cos^6 u = (\sin^2 u + \cos^2 u)(\sin^4 u - \sin^2 u$

$\cos^2 u + \cos^4 u) = (\sin^4 u + 2 \sin^2 u \cos^2 u + \cos^4 u)$

$- 3 \sin^2 u \cos^2 u = (\sin^2 u + \cos^2 u)^2 - 3 \sin^2 u$

$\cos^2 u = 1 - 3 \sin^2 u \cos^2 u$

53. $\cot 3x = \dfrac{1}{\tan 3x}$

$\qquad = \dfrac{1 - 3 \tan^2 x}{3 \tan x - \tan^3 x} = \dfrac{1 - 3/\cot^2 x}{3/\cot x - 1/\cot^3 x}$

$\qquad = \dfrac{\cot^3 x - 3 \cot x}{3 \cot^2 x - 1}$

$\qquad = \dfrac{3 \cot x - \cot^3 x}{1 - 3 \cot^2 x}$

PROBLEM SET 5.3 (p. 231)

1. (a) $\dfrac{1}{2}(\sqrt{2} + 1)$

(b) $\dfrac{1}{4}(\sqrt{6} + \sqrt{2})$

3. (a) $-\dfrac{1}{2}(\sqrt{2} + \sqrt{3})$

(b) $\dfrac{1}{4}(-\sqrt{6} + \sqrt{2})$

5. $\dfrac{1}{4}(\sqrt{6} + \sqrt{2})$

7. $\dfrac{1}{4}(\sqrt{6} - \sqrt{2})$

9. $\dfrac{1}{2}$

11. $\dfrac{1}{2}\sqrt{2}$

13. $\dfrac{1}{2}\sqrt{3}$

15. $\cos 1$

17. $\sin \alpha$

19. $\cos t$

21. $\sqrt{3} \sin u$

23. $\sin 5\theta$

25. (a) $\dfrac{7}{25}$

(b) 0

27. (a) $-\dfrac{56}{33}$

(b) $-\dfrac{16}{33}$

29. $40.60°$

31. $55.49°$

33. 128.31°

35. $\tan(s + \pi) = \dfrac{\tan s + \tan \pi}{1 - \tan s \tan \pi} = \tan s$

37. $\cot(u + v) = \dfrac{\cos(u + v)}{\sin(u + v)}$

$= \dfrac{(\cos u \cos v - \sin u \sin v)/(\sin u \sin v)}{(\sin u \cos v + \cos u \sin v)/(\sin u \sin v)}$

$= \dfrac{\cot u \cot v - 1}{\cot u + \cot v}$

39. $\dfrac{\cos 2t}{\sin t} + \dfrac{\sin 2t}{\cos t} = \dfrac{\cos 2t \cos t + \sin 2t \sin t}{\sin t \cos t}$

$= \dfrac{\cos t}{\sin t \cos t} = \csc t$

41. $\dfrac{\sin(x + h) - \sin x}{h}$

$= \dfrac{\sin x \cos h + \cos x \sin h - \sin x}{h}$

$= \sin x \dfrac{\cos h - 1}{h} + \cos x \dfrac{\sin h}{h}$

43. $\sin(x + y)\sin(x - y) = (\sin x \cos y + \cos x \sin y)(\sin x \cos y - \cos x \sin y)$
$= \sin^2 x \cos^2 y - \cos^2 x \sin^2 y$
$= \sin^2 x (\cos^2 y + \sin^2 y) - \sin^2 y(\sin^2 x + \cos^2 x)$
$= \sin^2 x - \sin^2 y$

45. $f(t) = \left(\sin t \cos \dfrac{2}{3}\pi - \cos t \sin \dfrac{2}{3}\pi\right)^2$

$+ \sin^2 t + \left(\sin t \cos \dfrac{2}{3}\pi + \cos t \sin \dfrac{2}{3}\pi\right)^2$

$= 2 \sin^2 t \left(-\dfrac{1}{2}\right)^2 + 2 \cos^2 t \left(\dfrac{1}{2}\sqrt{3}\right)^2 + \sin^2 t$

$= \dfrac{3}{2}\sin^2 t + \dfrac{3}{2}\cos^2 t = \dfrac{3}{2}$

47. $\alpha + \beta = \dfrac{1}{4}\pi$

49. $\alpha + \beta = \pi - \gamma$
$\sin(\alpha + \beta) = \sin(\pi - \gamma) = -\cos \pi \sin \gamma = \sin \gamma$
$\cos(\alpha + \beta) = \cos(\pi - \gamma) = \cos \pi \cos \gamma = -\cos \gamma$

51. $\gamma = \pi - (\alpha + \beta)$ and so $\cos \gamma = -\cos(\alpha + \beta) = -\cos \alpha \cos \beta + \sin \alpha \sin \beta$. Use this to replace $\cos \gamma$; then simplify.

PROBLEM SET 5.4 (p. 238)

1. $\dfrac{120}{169}$

3. $\dfrac{120}{119}$

5. $\dfrac{7}{10}\sqrt{2}$

7. $\dfrac{1}{5}$

9. $\dfrac{-23256}{105625}$

11. $-\dfrac{24}{25}$

13. $-\dfrac{120}{119}$

15. $\dfrac{3}{10}\sqrt{10}$

17. 3

19. $\dfrac{8}{81}\sqrt{5}$

21. $-\dfrac{9}{35}$

23. $\dfrac{1}{2}\sqrt{2 - \sqrt{3}}$

25. $\dfrac{\sqrt{2}}{2 + \sqrt{2}} = \sqrt{2} - 1$

27. $\dfrac{1}{2}\sqrt{2 + \sqrt{2}}$

29. $2 \sin 8t$

31. $\cos^2 x$

33. $-\cos 8t$

35. $-\tan 2t$

37. $4 \cos^3 \theta - 3 \cos \theta$

39. $(\sin t + \cos t)^2 = \sin^2 t + 2 \sin t \cos t + \cos^2 t = 1 + 2 \sin t \cos t$

41. $\dfrac{1}{8}(1 - \cos 4t) = \dfrac{1}{8}(2 \sin^2 2t)$

$= \dfrac{1}{4}(2 \sin t \cos t)^2 = \sin^2 t \cos^2 t$

43. $\tan 2u = \tan(u + u) = \dfrac{\tan u + \tan u}{1 - \tan u \tan u}$

$= \dfrac{2 \tan u}{1 - \tan^2 u}$

45. $\dfrac{2 \tan x}{1 + \tan^2 x} = \dfrac{2 \sin x/\cos x}{\sec^2 x}$

$= 2 \sin x \cos x = \sin 2x$

47. $\cos 4\theta = 2 \cos^2 2\theta - 1 = 2(2 \cos^2 \theta - 1)^2 - 1$
$= 8 \cos^4 \theta - 8 \cos^2 \theta + 1$

49. $10 \sin\left(2t + \dfrac{1}{3}\pi\right)$

51. $4 \sin\left(t + \dfrac{5}{6}\pi\right)$

53. $\cos^4 z - \sin^4 z = (\cos^2 z - \sin^2 z)(\cos^2 z + \sin^2 z)$
$= \cos 2z$

55. $1 + \dfrac{1 - \cos 8t}{1 + \cos 8t} = 1 + \dfrac{2 \sin^2 4t}{2 \cos^2 4t}$

$\qquad = 1 + \tan^2 4t = \sec^2 4t$

57. $\tan \dfrac{1}{2} \theta - \sin \theta = \dfrac{\sin \theta}{1 + \cos \theta} - \sin \theta$

$\qquad = \left(\sin \theta \right) \left(\dfrac{1}{1 + \cos \theta} - 1 \right)$

$\qquad = \left(\sin \theta \right) \left(\dfrac{- \cos \theta}{1 + \cos \theta} \right)$

$\qquad = \left(- \sin \theta \right) \left(\dfrac{1/\sec \theta}{1 + 1/\sec \theta} \right)$

$\qquad = - \dfrac{\sin \theta}{1 + \sec \theta}$

59. $(3 \cos t - \sin t)(\cos t + 3 \sin t)$

$\qquad = 3 \cos^2 t + 8 \sin t \cos t - 3 \sin^2 t$

$\qquad = 3 \cos 2t + 4 \sin 2t$

61. $2(\cos 3x \cos x + \sin 3x \sin x)^2 = 2(\cos 2x)^2$

$\qquad = 1 + 2 \cos^2 2x - 1 = 1 + \cos 4x$

63. $\tan 3t = \tan (2t + t) = \dfrac{\tan 2t + \tan t}{1 - \tan 2t \tan t}$

Now replace $\tan 2t$ by $2 \tan t/(1 - \tan^2 t)$ and simplify.

65. $\dfrac{3}{4} + \dfrac{1}{4} \cos 4u = \dfrac{3}{4} + \dfrac{1}{4}(2 \cos^2 2u - 1)$

$\qquad = \dfrac{1}{2} + \dfrac{1}{2}(2 \cos^2 u - 1)^2$

$\qquad = \dfrac{1}{2} + \dfrac{1}{2}(4 \cos^4 u - 4 \cos^2 u + 1)$

$\qquad = \cos^4 u + (\cos^4 u - 2 \cos^2 u + 1)$

$\qquad = \cos^4 u + (\cos^2 u - 1)^2$

$\qquad = \cos^4 u + \sin^4 u$

67. $\cos^2 x + \cos^2 2x + \cos^2 3x$

$\qquad = \dfrac{1}{2}(1 + \cos 2x + 1 + \cos 4x) + \cos^2 3x$

$\qquad = 1 + \cos 3x \cos x + \cos^2 3x$

$\qquad = 1 + \cos 3x (2 \cos 2x \cos x)$

$\qquad = 1 + 2 \cos x \cos 2x \cos 3x$

69. Use double-angle identities.

71. $2 \cos x - 4 \sin x \sin 2x = 2 \cos 3x$

73. $32 \cos^6 t - 48 \cos^4 t + 18 \cos^2 t - 1 = \cos 6t$

PROBLEM SET 5.5 (p. 244)

1. $- \sin 28°$

3. $\cos 64°$

5. $\tan 34°$

7. $- \cot 62°$

9. $\sin (\pi - 3) = \sin (0.14159)$

11. $\tan (3.33 - \pi) = \tan (0.18841)$

13. $\cos (7.2832 - 2\pi) = \cos (1.00001)$

15. $0, \pi, 2\pi$

17. $\dfrac{1}{3} \pi, \dfrac{5}{3} \pi$

19. $3.378, 6.046$

21. $2.172, 5.314$

23. None

25. $\dfrac{1}{4} \pi, \dfrac{3}{4} \pi, \dfrac{5}{4} \pi, \dfrac{7}{4} \pi$

27. $\dfrac{1}{3} \pi, \dfrac{5}{3} \pi$

29. $0, \pi, \dfrac{7}{6} \pi, \dfrac{11}{6} \pi, 2\pi$

31. $1.107, \dfrac{3}{4} \pi, 4.249, \dfrac{7}{4} \pi$

33. $2.897, 6.038, 2.820, 5.961$

35. $\dfrac{1}{2} \pi, \dfrac{3}{2} \pi$

37. $\dfrac{1}{6} \pi, \dfrac{5}{6} \pi$

39. $\dfrac{1}{18} \pi, \dfrac{5}{18} \pi, \dfrac{13}{18} \pi, \dfrac{17}{18} \pi, \dfrac{25}{18} \pi, \dfrac{29}{18} \pi$

41. $1.231, 5.052$

43. $1.052, 2.089$

45. $\{x: 0.588 \leq x \leq 3.730\}$

47. $\{x: 0.203 \leq x \leq 1.763\}$

49. $0, \dfrac{2}{3} \pi, 2\pi$

51. $0, \dfrac{1}{6} \pi, \dfrac{5}{6} \pi, \pi, \dfrac{7}{6} \pi, \dfrac{11}{6} \pi, 2\pi$

53. $\dfrac{1}{2} \pi, \dfrac{7}{12} \pi, \dfrac{11}{12} \pi, \dfrac{3}{2} \pi, \dfrac{19}{12} \pi, \dfrac{23}{12} \pi$

55. No solution

57. π

59. (a) 15 inches

(b) $\tan \theta = \dfrac{2}{3}$

(c) $\theta \approx 0.588$ radians

61. (a) $T(213) = 15.49$

(b) 122 days

63. $\left\{ x: \dfrac{1}{6} \pi \leq x \leq \dfrac{1}{2} \pi \quad \text{or} \quad \dfrac{5}{6} \pi \leq x \leq \dfrac{3}{2} \pi \right\}$

65. $\dfrac{1}{6} \pi, \dfrac{1}{3} \pi, \dfrac{2}{3} \pi, \dfrac{5}{6} \pi$

PROBLEM SET 5.6 (p. 252)

1. $\gamma = 55.81°$, $b \approx 20.66$, $c \approx 17.31$
3. $\beta = 53.52°$, $a \approx 56.89$, $c = 56.89$
5. $\beta \approx 42.00°$, $\gamma \approx 23.22°$, $c \approx 20.22$
7. $\beta \approx 17.734°$, $\gamma \approx 131.999°$, $c \approx 12.472$
9. $\beta_1 \approx 56.516°$, $\gamma_1 \approx 93.217°$, $c_1 \approx 10.126$
   $\beta_2 \approx 123.484°$, $\gamma_2 \approx 26.249°$, $c_2 \approx 4.486$
11. No triangle
13. 198.8
15. 267.1
17. 10.4
19. 76.6 meters
21. 44.7°
23. 78.4°
25. 695 square feet
27. 1769 feet
29. 255 feet
31. One configuration: $A = 15.99$
    A less obvious configuration: $A = 2.05$
33. 12:42.36, 7.5395
35. Max A: 40.5

37. $6r^2 \sin \phi \, (\cos \phi + \sqrt{3} \sin \phi)$

PROBLEM SET 5.7 (p. 259)

1. $a \approx 13.32$, $\beta \approx 74.09°$, $\gamma \approx 45.68°$
3. $c \approx 33.09$, $\alpha \approx 19.77°$, $\beta \approx 39.78°$
5. $a \approx 61.61$, $\beta \approx 0.81°$, $\gamma \approx 0.58°$
7. $\gamma \approx 70.31°$, $\beta \approx 59.73°$, $\alpha \approx 49.96°$
9. No solution
11. 17.1
13. 8.51
15. 98.8 meters
17. 24 miles
19. 106.1°
21. 41.7°
23. 42.6 miles
25. 30.12, 20.08, 20.08
27. 19.21
29. (c) $r = \dfrac{2}{3}\sqrt{6} \approx 1.633$

31. $a_1^2 = \left(\dfrac{1}{2}a\right)^2 + b^2 - ab \cos \gamma$

   $= \dfrac{1}{4}a^2 + b^2 - ab \, \dfrac{(a^2 + b^2 - c^2)}{2ab}$

   $= -\dfrac{1}{4}a^2 + \dfrac{1}{2}b^2 + \dfrac{1}{2}c^2$

   Similarly,

   $b_1^2 = -\dfrac{1}{4}b^2 + \dfrac{1}{2}a^2 + \dfrac{1}{2}c^2$

   $c_1^2 = -\dfrac{1}{4}c^2 + \dfrac{1}{2}a^2 + \dfrac{1}{2}b^2$

33. 12:31.46
35. 3.8502

PROBLEM SET 5.8 (p. 266)

1.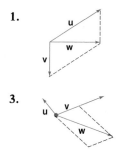

3.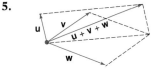

5.

7. 18.48 miles
9. 243.7 kilometers, S43.9°W
11. N2.68°E, 479 miles per hour
13. 15.87, S7.5°W

15. (a) $\overrightarrow{AD} - \overrightarrow{AB}$

  (b) $\dfrac{1}{2}\overrightarrow{AD} + \dfrac{1}{2}\overrightarrow{AB}$

  (c) $\overrightarrow{AB} - \dfrac{1}{2}\overrightarrow{AD}$

  (d) $\overrightarrow{AD} - \dfrac{1}{2}\overrightarrow{AB}$

17. $4\mathbf{i} - 24\mathbf{j}$, -33, $-\dfrac{33}{65}$

19. $3\mathbf{i} + \mathbf{j}$, 10, $\dfrac{2}{\sqrt{5}}$

21. $101.4°$

23. $5\mathbf{i} + 2\mathbf{j}$, $4\mathbf{i} - 3\mathbf{j}$, 14

25. $-4\mathbf{i} - 5\mathbf{j}$, $-6\mathbf{i} + 5\mathbf{j}$, -1

27. $-5\mathbf{i} + 5\sqrt{3}\,\mathbf{j}$

29. $\dfrac{4}{3}$

31. $\dfrac{3}{5}\mathbf{i} - \dfrac{4}{5}\mathbf{j}$

33. 100 units

35. $325\sqrt{2} \approx 460$ dyne-centimeters

37. N10.33°E

39. 69.09, 49.62

41. 0

43. $\dfrac{1}{2}\sqrt{7} \approx 1.32$ miles per hour

47. Amp: 60, Per: 3

49. $|\mathbf{u} \cdot \mathbf{v}| = \|\mathbf{u}\|\,\|\mathbf{v}\|\,|\cos\theta| \le \|\mathbf{u}\|\,\|\mathbf{v}\|$
  with equality when $\mathbf{u}$ and $\mathbf{v}$ are parallel.

51. $0 = (\mathbf{u} + \mathbf{v}) \cdot (\mathbf{u} - \mathbf{v}) = \mathbf{u} \cdot \mathbf{u} - \mathbf{v} \cdot \mathbf{v} = \|\mathbf{u}\|^2 - \|\mathbf{v}\|^2$
  $\Rightarrow \|\mathbf{u}\| = \|\mathbf{v}\|$

53. $(1368, 515)$

55. $\dfrac{1}{4}(-1 + \sqrt{5})$

CHAPTER 5 REVIEW PROBLEM SET (p. 269)

1. T
2. F
3. F
4. T
5. T
6. F
7. T
8. F
9. T
10. T
11. (a) $5ab^2c^2(c - 2a)(c + 2a)$
  (b) $x^2(x^2 - 2y)^2$
  (c) $(\tan u - 6)(\tan u + 1)$
  (d) $(\sin t + 2\cos t)(\sin^2 t - 2\sin t\cos t + 4\cos^2 t)$

12. (a) $\dfrac{x + 6}{x + 3}$

  (b) $\dfrac{17x}{2x - 14}$

  (c) $\sec^2 t$

  (d) $\dfrac{\cos x - \sin x}{\cos x + \sin x}$

13. (a) $\dfrac{(x - y)(4x - 1)}{x + y}$

  (b) -1

14. (a) $-1 + 3\cos^2 t$
  (b) $1 - \cos t$
  (c) $(2\cos^2 t)(1 - \cos^2 t)(1 + \cos t)$

15. $\cot\theta\cos\theta = \dfrac{\cos^2\theta}{\sin\theta} = \dfrac{1 - \sin^2\theta}{\sin\theta} = \csc\theta - \sin\theta$

16. $\sec t - \cos t = \dfrac{1}{\cos t} - \cos t = \dfrac{1 - \cos^2 t}{\cos t}$

  $= \sin t\,\dfrac{\sin t}{\cos t} = \sin t\tan t$

17. $\left(\cos\dfrac{1}{2}t + \sin\dfrac{1}{2}t\right)^2$

  $= \cos^2\dfrac{1}{2}t + \sin^2\dfrac{1}{2}t + 2\sin\dfrac{1}{2}t\cos\dfrac{1}{2}t = 1 + \sin t$

18. $\sec^4\theta - \sec^2\theta = (\sec^2\theta)(\sec^2\theta - 1)$
  $= (1 + \tan^2\theta)\tan^2\theta$
  $= \tan^2\theta + \tan^4\theta$

19. $\tan u + \cot u = \tan u + \dfrac{1}{\tan u} = \dfrac{\tan^2 u + 1}{\tan u}$

  $= \dfrac{\sec^2 u\cos u}{\sin u} = \sec u\csc u$

20. $\dfrac{1 - \cos x}{\sin x} = \dfrac{(1 - \cos x)(1 + \cos x)}{(\sin x)(1 + \cos x)}$

  $= \dfrac{\sin^2 x}{(\sin x)(1 + \cos x)} = \dfrac{\sin x}{1 + \cos x}$

21. $\cos 120° = -\dfrac{1}{2}$

22. $\sin\dfrac{1}{2}\pi = 1$

23. $-\cos 225° = \dfrac{1}{2}\sqrt{2}$

24. $\tan 45° = 1$

25. $\dfrac{2\sin^2\theta}{1 - 2\sin^2\theta}$

26. $1 - 8\sin^2 t + 8\sin^4 t$

27. (a) $-\dfrac{5}{13}$ (b) $\dfrac{120}{169}$ (c) $-\dfrac{119}{169}$ (d) $-\dfrac{3}{2}$

28. (a) $-\dfrac{1}{3}\sqrt{5}$ (b) $-\dfrac{3}{5}$ (c) $-\dfrac{1}{15}(6 + 4\sqrt{5})$

29. $\cos\left(u + \dfrac{1}{3}\pi\right)\cos(\pi - u)$

$\qquad - \sin\left(u + \dfrac{1}{3}\pi\right)\sin(\pi - u) = \cos\dfrac{4}{3}\pi = -\dfrac{1}{2}$

30. $\dfrac{\cos 5t}{\sin t} - \dfrac{\sin 5t}{\cos t} = \dfrac{\cos t \cos 5t - \sin t \sin 5t}{(1/2)(2\sin t \cos t)}$

$\qquad = \dfrac{2\cos 6t}{\sin 2t}$

31. $\csc 2t + \cot 2t = \dfrac{1}{\sin 2t} + \dfrac{\cos 2t}{\sin 2t}$

$\qquad = \dfrac{2\cos^2 t}{2\sin t \cos t} = \cot t$

32. $\sin 3\theta = \sin(\theta + 2\theta)$

$\qquad = \sin\theta\cos 2\theta + \cos\theta\sin 2\theta$

$\qquad = (\sin\theta)(1 - 2\sin^2\theta) + 2\cos^2\theta\sin\theta$

$\qquad = \sin\theta - 2\sin^3\theta + 2(1 - \sin^2\theta)\sin\theta$

$\qquad = 3\sin\theta - 4\sin^3\theta$

33. $\dfrac{\sin(\alpha - \beta)}{\cos\alpha\cos\beta} = \dfrac{\sin\alpha\cos\beta - \cos\alpha\sin\beta}{\cos\alpha\cos\beta}$

$\qquad = \tan\alpha - \tan\beta$

34. $\dfrac{1 - \tan^2(u/2)}{1 + \tan^2(u/2)} = \dfrac{1 - \sin^2(u/2)/\cos^2(u/2)}{1 + \sin^2(u/2)/\cos^2(u/2)}$

$\qquad = \dfrac{\cos^2(u/2) - \sin^2(u/2)}{\cos^2(u/2) + \sin^2(u/2)} = \cos u$

35. (a) 0.96 (b) -0.02

36. (a) $0.4 + 0.3\sqrt{3}$ (b) -3

37. $\dfrac{5}{6}\pi, \dfrac{11}{6}\pi$

38. $\dfrac{1}{3}\pi, \dfrac{1}{2}\pi, \dfrac{3}{2}\pi, \dfrac{5}{3}\pi$

39. $\dfrac{3}{2}\pi$

40. $0, \dfrac{1}{4}\pi, \dfrac{3}{4}\pi, \pi, \dfrac{5}{4}\pi, \dfrac{7}{4}\pi, 2\pi$

41. $\dfrac{1}{12}\pi, \dfrac{5}{12}\pi, \dfrac{13}{12}\pi, \dfrac{17}{12}\pi$

42. $\dfrac{1}{6}\pi, \dfrac{1}{3}\pi, \dfrac{2}{3}\pi, \dfrac{5}{6}\pi, \dfrac{7}{6}\pi, \dfrac{4}{3}\pi, \dfrac{5}{3}\pi, \dfrac{11}{6}\pi$

43. $\dfrac{1}{3}\pi, \dfrac{5}{3}\pi$

44. $\dfrac{1}{3}\pi, \dfrac{2}{3}\pi, \dfrac{3}{4}\pi, \dfrac{4}{3}\pi, \dfrac{5}{3}\pi, \dfrac{7}{4}\pi$

45. $4\tan t \cos^2 t \csc(2t + \pi)$

$\qquad = 4\dfrac{\sin t \cos^2 t}{\cos t \sin(2t + \pi)}$

$\qquad = \dfrac{-4\sin t \cos t}{\sin 2t} = \dfrac{-4\sin t \cos t}{2\sin t \cos t} = -2$

46. 4.00, 5.43

47. $n = 3$

48. $\tan^{-1}\left(\dfrac{x}{\sqrt{1 - x^2}}\right) = \sin^{-1}x$

49. $\gamma = 105°, a \approx 5.1764, b \approx 7.3205$

50. $\alpha \approx 28.96°, \beta \approx 46.56°, \gamma \approx 104.48°$

51. $\alpha \approx 26.03°, \gamma \approx 11.97°, c \approx 31.67$

52. $\alpha \approx 32.5°, \beta \approx 109.9°, c \approx 13.2$

53. 71.8

54. $12\sqrt{5} \approx 26.83$

55. (a) 273.2 yards (b) 136.6 yards

56. $\beta_1 \approx 58.99°, \gamma_1 \approx 81.01°, c_1 \approx 2.30$
$\quad\ \beta_2 \approx 121.01°, \gamma_2 \approx 18.99°, c_2 \approx 0.76$

57. 2.0267

58. $-(3\sqrt{3} + 8)\mathbf{i} + 15\mathbf{j}$

59. (a) 13 (b) 25 (c) 36
$\quad\ $ (d) 83.64° (e) $\dfrac{24}{25}\mathbf{i} + \dfrac{7}{25}\mathbf{j}$

60. (a) $\dfrac{36}{25} = 1.44$

$\quad\ $ (b) $1.3824\mathbf{i} + 0.4032\mathbf{j}$

61. $\mathbf{w} = 72\mathbf{i} + 186\mathbf{j}$

62. (a) 797.8 foot-pounds
$\quad\ $ (b) 768 foot-pounds

63. (a) 2π (b) 1 (c) π

64. $2(t + \sin t), t = 0.511$

PROBLEM SET 6.1 (p. 277)

1. $(2, -1)$

3. $(-2, 4)$

5. $(1, -2)$

7. $(0, 0, -2)$

9. $(1, 4, -1)$

11. $(2, 1, 4)$

13. $(0, 0, 0)$

15. $(5, 6, 0, -1)$

17. $(15z - 110, 4z - 32, z)$

19. $(2y - 3z - 2, y, z)$

21. $(-z, 2z, z)$

23. Inconsistent

25. $\left(-z + \dfrac{2}{5}, z + \dfrac{16}{5}, z\right)$

27. $\left(\dfrac{1}{2}, -\dfrac{1}{3}\right)$

29. (e^2, e^4)

31. $a = \dfrac{3}{2}, b = 6$

33. 285

35. $y = 2x^3 - 3x + 4$

37. $x^2 + y^2 - 4x - 3y = 0$

39. $(10, 0), (2, 4)$

41. $(5, -7), (6, 0)$
43. $(-1, 2), (1, 2)$
45. $\left(\dfrac{2}{5}\sqrt{5}, \dfrac{4}{5}\sqrt{5}\right)$
47. 24 by 7 in meters

49. $(4.98, 0.23)$
51. $(-2.39, 4.46), (9.06, 0.65)$
53. $(-3.38, 4.38), (4.38, -3.38)$
55. 19.26

PROBLEM SET 6.2 (p. 284)

1. $\begin{bmatrix} 2 & -1 & 4 \\ 1 & -3 & -2 \end{bmatrix}$

3. $\begin{bmatrix} 1 & -2 & 1 & 3 \\ 2 & 1 & 0 & 5 \\ 1 & 1 & 3 & -4 \end{bmatrix}$

5. $\begin{bmatrix} 2 & -3 & -4 \\ 3 & 1 & -2 \end{bmatrix}$

7. $\begin{bmatrix} 1 & 0 & 0 & 5 \\ 1 & 2 & -1 & 4 \\ 3 & -1 & -5 & -13 \end{bmatrix}$

9. Unique solution
11. No solution
13. Unique solution
15. Infinitely many solutions
17. No solution
19. No solution
21. $(1, 2)$

23. $\left(\dfrac{2}{3}y + \dfrac{1}{3}, y\right)$
25. $(1, 4, -1)$
27. $\left(\dfrac{16}{3}z + \dfrac{32}{3}, -\dfrac{7}{3}z - \dfrac{2}{3}, z\right)$
29. $(3, 0, 0)$
31. $(4.40, 1.23, -0.96)$
33. $(0, 2, 1)$
35. $\left(\dfrac{21}{2}z - 48, -5z + 26, z\right)$
37. $a = -4, b = 8, c = 0$
39. $(x + 5)^2 + (y + 4)^2 = 25, (-5, -4), 5$
41. $\alpha = 80°, \beta = 30°, \gamma = 110°, \delta = 50°$
43. $a = -3, b = 2, c = 5$
45. A: 10, B: 40, C: 50
47. No solution

PROBLEM SET 6.3 (p. 291)

1. $\begin{bmatrix} 8 & 4 \\ 1 & 10 \end{bmatrix}, \begin{bmatrix} -4 & -6 \\ 5 & 4 \end{bmatrix}, \begin{bmatrix} 6 & -3 \\ 9 & 21 \end{bmatrix}$

3. $\begin{bmatrix} 5 & 4 & 4 \\ 8 & 3 & -6 \end{bmatrix}, \begin{bmatrix} 1 & -8 & 6 \\ 0 & -3 & 0 \end{bmatrix}, \begin{bmatrix} 9 & -6 & 15 \\ 12 & 0 & -9 \end{bmatrix}$

5. $\begin{bmatrix} 14 & 7 \\ 4 & 36 \end{bmatrix}, \begin{bmatrix} 27 & 29 \\ 5 & 23 \end{bmatrix}$

7. $\begin{bmatrix} -3 & -4 & 2 \\ 8 & 22 & -13 \\ -2 & 0 & 9 \end{bmatrix}, \begin{bmatrix} 0 & 5 & -17 \\ 13 & 10 & 3 \\ 1 & -3 & 18 \end{bmatrix}$

9. **AB** not possible; $\mathbf{BA} = \begin{bmatrix} 7 & 2 & -7 & 6 \\ 15 & 2 & -11 & 16 \end{bmatrix}$

11. $\mathbf{AB} = \begin{bmatrix} 2 \\ 16 \\ -2 \end{bmatrix}$; **BA** not possible

13. $\mathbf{AB} = \mathbf{BA} = \begin{bmatrix} 0 & 0 \\ 0 & 0 \end{bmatrix}$

15. $\begin{bmatrix} -4 & 7 & 9 \\ -5 & -5 & 8 \end{bmatrix}$

17. $\mathbf{A}(\mathbf{B} + \mathbf{C}) = \mathbf{AB} + \mathbf{AC}$
$= \begin{bmatrix} -7 & -3 \\ 39 & 34 \end{bmatrix}$; the distributive property

19. 93.5917

21. $\begin{bmatrix} -22 & 22 & 48 \\ 36 & -31 & -18 \\ 43 & -34 & -13 \end{bmatrix}$

23. $\begin{bmatrix} 88 & -120 & -192 \\ -984 & 1120 & 1488 \\ -1620 & 1884 & 2440 \end{bmatrix}$

25. $\begin{bmatrix} 6847 & -6510 & -6801 \\ -11895 & 11905 & 13854 \\ -15954 & 16038 & 18994 \end{bmatrix}$

27. (a) **AB** and (g) $\mathbf{A}^2$
29. Let **A** be $m \times n$ and **B** be $p \times q$. Since **AB** makes sense, $n = p$. Since **BA** makes sense, $q = m$. Thus, **AB** is $q \times q$ and **BA** is $p \times p$.
31. It consists of zeros.
33. It is a Heisenberg matrix.

35. $\mathbf{B}^n = \begin{bmatrix} 1 & 0 & 3n \\ 0 & 1 & 0 \\ 0 & 0 & 1 \end{bmatrix}$

37. $\begin{bmatrix} a & 0 & 0 \\ 0 & b & 0 \\ 0 & 0 & c \end{bmatrix}^n = \begin{bmatrix} a^n & 0 & 0 \\ 0 & b^n & 0 \\ 0 & 0 & c^n \end{bmatrix}$

39. (a) $\begin{bmatrix} 1 & 0 & 0 & 1 \\ 1 & 0 & 0 & 1 \\ 0 & 1 & 1 & 0 \\ 0 & 0 & 1 & 0 \end{bmatrix}$

   (b) The ij element of $\mathbf{U}^2$ gives the number of ways of getting a message from i to j via one other person.

   (c) No

   (d) Same as (b) with two intermediate people

   (e) Yes

41. (a) $\mathbf{A}^n = \dfrac{1}{14}\begin{bmatrix} 5 & 9 \\ 5 & 9 \end{bmatrix}$

   $= \begin{bmatrix} 0.3571428571 & 0.6428571429 \\ 0.3571428571 & 0.6428571429 \end{bmatrix}$

   (b) $\mathbf{A}^n = \dfrac{1}{225}\begin{bmatrix} 81 & 87 & 57 \\ 81 & 87 & 57 \\ 81 & 87 & 57 \end{bmatrix}$

   (c) $\mathbf{A}^n = \dfrac{1}{999}\begin{bmatrix} 190 & 209 & 346 & 254 \\ 190 & 209 & 346 & 254 \\ 190 & 209 & 346 & 254 \\ 190 & 209 & 346 & 254 \end{bmatrix}$

PROBLEM SET 6.4 (p. 299)

1. Yes, $\begin{bmatrix} -1 & -3 \\ 1 & 2 \end{bmatrix}$

3. No

5. Yes, $\begin{bmatrix} -\dfrac{2}{3} & \dfrac{1}{2} \\ \dfrac{1}{3} & 0 \end{bmatrix}$

7. Yes, $\begin{bmatrix} -4 & 3 & -4 \\ \dfrac{1}{2} & -\dfrac{1}{2} & 1 \\ 6 & -4 & 6 \end{bmatrix}$

9. No

11. Yes, $\begin{bmatrix} -\dfrac{4}{7} & \dfrac{2}{7} & \dfrac{3}{7} \\ \dfrac{6}{7} & -\dfrac{3}{7} & -\dfrac{1}{7} \\ \dfrac{5}{7} & \dfrac{1}{7} & -\dfrac{2}{7} \end{bmatrix}$

13. Yes, $\begin{bmatrix} -\dfrac{1}{9} & \dfrac{1}{9} & \dfrac{8}{9} \\ \dfrac{10}{9} & -\dfrac{1}{9} & -\dfrac{26}{9} \\ \dfrac{1}{9} & -\dfrac{1}{9} & \dfrac{1}{9} \end{bmatrix}$

15. No

17. Yes, $\begin{bmatrix} -\dfrac{1}{2} & \dfrac{1}{2} & \dfrac{1}{2} & \dfrac{1}{2} \\ \dfrac{1}{2} & 0 & 0 & -\dfrac{1}{2} \\ \dfrac{1}{2} & 0 & -\dfrac{1}{2} & 0 \\ \dfrac{1}{2} & -\dfrac{1}{2} & 0 & 0 \end{bmatrix}$

19. $\left(\dfrac{5}{7}, \dfrac{10}{7}, -\dfrac{1}{7}\right)$

21. $\left(\dfrac{47}{9}, -\dfrac{128}{9}, \dfrac{7}{9}\right)$

23. $\left(\dfrac{1}{6}, \dfrac{1}{6}, \dfrac{5}{6}\right)$

25. $(2.27272, -4.09091, -0.18182, 0.54545)$

27. $\begin{bmatrix} 1.920 & 3.594 \\ -0.449 & -1.906 \\ -0.080 & -0.528 \\ -0.215 & -1.016 \\ -0.182 & -0.631 \end{bmatrix}$

29. $\begin{bmatrix} \dfrac{1}{2} & 0 & 0 \\ 0 & \dfrac{1}{3} & 0 \\ 0 & 0 & -\dfrac{1}{4} \end{bmatrix}$

31. $(\mathbf{AB})(\mathbf{B}^{-1}\mathbf{A}^{-1}) = \mathbf{A}(\mathbf{BB}^{-1})\mathbf{A}^{-1} = \mathbf{AA}^{-1} = \mathbf{I}$
$(\mathbf{B}^{-1}\mathbf{A}^{-1})(\mathbf{AB}) = \mathbf{B}^{-1}(\mathbf{A}^{-1}\mathbf{A})\mathbf{B} = \mathbf{B}^{-1}\mathbf{B} = \mathbf{I}$

33. If $\mathbf{AB} = \mathbf{0}$, then $\mathbf{A}^{-1}\mathbf{AB} = \mathbf{A}^{-1}\mathbf{0}$; that is, $\mathbf{B} = \mathbf{0}$.

35. $\mathbf{A}^n = \begin{bmatrix} 1 & 0 & 0 \\ n & 1 & 0 \\ n & 0 & 1 \end{bmatrix}$

37. (b) $\mathbf{A}^{-1} = \mathbf{A}^2$, $\mathbf{B}^{-1} = \mathbf{B}^3$
(c) If $\mathbf{C}$ is an $n \times n$ matrix that has all zero entries except for 1s in the first diagonal above the main diagonal and a 1 in the lower left corner, then $\mathbf{C}^{-1} = \mathbf{C}^{n-1}$.

39. (a) $\begin{bmatrix} 5 \\ -1 \\ 0 \\ 0 \end{bmatrix}$ (b) $\begin{bmatrix} 8 \\ -17 \\ 14 \\ -4 \end{bmatrix}$ (c) $\begin{bmatrix} 32 \\ -79 \\ 68 \\ -20 \end{bmatrix}$

41. (a) $\begin{bmatrix} 4 & -6 \\ -6 & 12 \end{bmatrix}$, $\begin{bmatrix} 9 & -36 & 30 \\ -36 & 192 & -180 \\ 30 & -180 & 180 \end{bmatrix}$
(b) $\begin{bmatrix} 1 \\ 1 \\ 1 \end{bmatrix}$

PROBLEM SET 6.5 (p. 307)

1. -8
3. 8
5. -50
7. 0
9. -159
11. 6
13. 0
15. 1
17. 72
19. 4
21. 39
23. -960
25. 5.075
27. -72
29. 57
31. $11{,}732$
33. $-0.26, 0.58, 6.68$
35. $-2.36, 0.47$
37. (a) 12 (b) -12 (c) 36 (d) 12
39. $\frac{3}{4}$, 2
41. 0
43. (a) -32 (b) $-\frac{1}{2}$ (c) -2 (d) -216
45. -14
47. (a) $1, -1$ (b) $-3, 2$
49. The determinant consists of a sum of products of the entries.
51. Each determinant has the value 1.
53. $\mathbf{A}_2^{-1} = \begin{bmatrix} 2 & -1 \\ -1 & 1 \end{bmatrix}$ $\mathbf{A}_3^{-1} = \begin{bmatrix} 3 & -3 & 1 \\ -3 & 5 & -2 \\ 1 & -2 & 1 \end{bmatrix}$

$\mathbf{A}_4^{-1} = \begin{bmatrix} 4 & -6 & 4 & -1 \\ -6 & 14 & -11 & 3 \\ 4 & -11 & 10 & -3 \\ -1 & 3 & -3 & 1 \end{bmatrix}$

$\mathbf{A}_5^{-1} = \begin{bmatrix} 5 & -10 & 10 & -5 & 1 \\ -10 & 30 & -35 & 19 & -4 \\ 10 & -35 & 46 & -27 & 6 \\ -5 & 19 & -27 & 17 & -4 \\ 1 & -4 & 6 & -4 & 1 \end{bmatrix}$

55. (a) $D = 93$, $\mathbf{B} = \begin{bmatrix} -3 & -57 & 39 \\ 12 & -51 & 30 \\ 7 & 9 & 2 \end{bmatrix}$, $\mathbf{A}^{-1} = \frac{1}{93}\mathbf{B}$

(b) $D = -183$,
$\mathbf{B} = \begin{bmatrix} -44 & -3 & 39 & 32 \\ -229 & -78 & 282 & 283 \\ -178 & -87 & 216 & 196 \\ 128 & 42 & -180 & -143 \end{bmatrix}$,

$\mathbf{A}^{-1} = -\frac{1}{183}\mathbf{B}$

57. (a) $\begin{bmatrix} 25 & -300 & 1050 & -1400 & 630 \\ -300 & 4800 & -18900 & 26880 & -12600 \\ 1050 & -18900 & 79380 & -117600 & 56700 \\ -1400 & 26880 & -117600 & 179200 & -88200 \\ 630 & -12600 & 56700 & -88200 & 44100 \end{bmatrix}$

(b) $\begin{bmatrix} 5 \\ -120 \\ 630 \\ -1120 \\ 630 \end{bmatrix}$ (c) $\begin{bmatrix} 15.5 \\ -309 \\ 1423.8 \\ -2296 \\ 1197 \end{bmatrix}$

Small changes in $\mathbf{C}$ produce large changes in the solution. If the numbers in $\mathbf{C}$ are measurements, the solution is meaningless.

PROBLEM SET 6.6 (p. 316)

1.

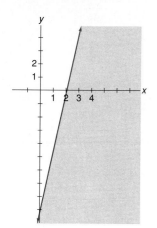

3.

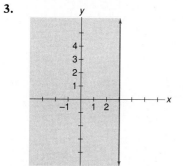

5.

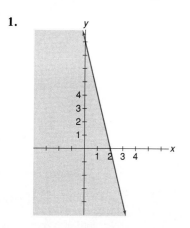

7.

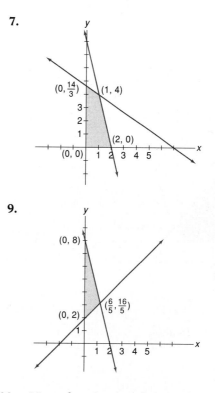

9.

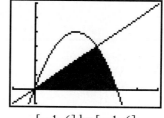

11. Max value: 6, min value: 0

13. Max value: $-\dfrac{4}{5}$, min value: -8

15. Min value of 14 at $(2, 2)$

17. Min value of 4 at $\left(\dfrac{3}{2},\ 1\right)$

19. Max value of 4 at $(2, 0)$

21. Max value of 9 at $(3, 3)$

$[-1, 6]$ by $[-1, 6]$

23. Max value of 80 + 3 ln 8 at (8, ln 8)

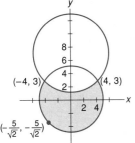

[−1, 10] by [−1, 3]

25. Max value of 29 at (−1, 10)

[−2, 4] by [−5, 12]

27. Max value of $\dfrac{11}{2}$ at $\left(\dfrac{9}{4}, \dfrac{13}{4}\right)$; min value of 0 at (0, 0)

29. Max value of 14 at (4, 3)

Min value of $-10\sqrt{2}$ at $\left(\dfrac{-5}{\sqrt{2}}, \dfrac{-5}{\sqrt{2}}\right)$

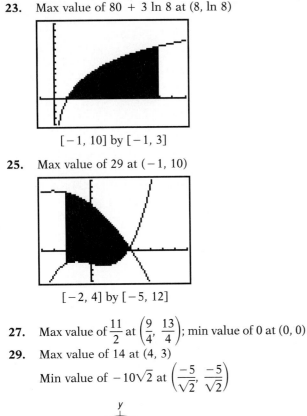

31. 266
33. Two camper units and 6 house trailers
35. Ten pounds of type A and 5 pounds of type B
37.

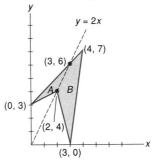

On A, $|y - 2x| + y + x = 2y - x$, which achieves its maximum value of 9 at (3, 6)
On B, $|y - 2x| + y + x = 3x$, which achieves its maximum value of 12 at (4, 7)
Therefore, the maximum value on the entire polygon is 12.
39. Max value of about 160.5 at (8.7, 5.3)

CHAPTER 6 REVIEW PROBLEM SET (p. 318)

1. T
2. F
3. T
4. F
5. T
6. T
7. T
8. F
9. F
10. F
11. (3, −1)

12. No solution
13. $\left(\dfrac{1}{2}, \dfrac{1}{3}\right)$
14. (−2, −3), (−2, 3), (2, −3), (2, 3)
15. (−4, 3, −2)
16. No solution
17. (2, −1, 3)
18. $\left(\dfrac{5}{7}z + 3, \dfrac{4}{7}z - 1, z\right)$
19. No solution

20. $(0, \log_3 4)$

21. $\begin{bmatrix} 4 & -1 & 10 \\ 8 & -1 & 13 \end{bmatrix}$

22. $\begin{bmatrix} -2 & 29 \\ 9 & 14 \end{bmatrix}$

23. $\begin{bmatrix} -16 \\ 11 \\ -27 \end{bmatrix}$

24. $\begin{bmatrix} 2 & 9 & -16 \\ 18 & -3 & 10 \\ 10 & -15 & 30 \end{bmatrix}$

25. $\begin{bmatrix} 5b & 2b \\ -3b & 4b \end{bmatrix}$

26. $\begin{bmatrix} 0 & 0 \\ 0 & 0 \end{bmatrix}$

27. $\begin{bmatrix} 5 & 3 \\ 2 & 4 \end{bmatrix}$

28. $\begin{bmatrix} 1 & -4 & 9 \\ 0 & 4 & 3 \\ 1 & -4 & 12 \end{bmatrix}$

29. $\begin{bmatrix} 5 & 1 & -4 \\ \frac{1}{2} & \frac{1}{2} & -\frac{1}{2} \\ -1 & 0 & 1 \end{bmatrix}$

30. $(6, -1, -2)$

31. -2

32. 0

33. 44

34. -10

35. 0

36. 56

37. 2

38. 38

39. $\begin{bmatrix} -2 & -4 & -6 & -8 \\ -1 & -3 & -8 & -13 \\ 0 & -5 & -13 & -24 \\ 1 & -10 & -24 & -44 \end{bmatrix}$

40. $\begin{bmatrix} 943 & 1742 & 2819 & 4253 \\ 2163 & 3994 & 6463 & 9749 \\ 3383 & 6246 & 10107 & 15245 \\ 4603 & 8498 & 13751 & 20741 \end{bmatrix}$

41. 0

42. 1

43. $\begin{bmatrix} -1 & 0 & 2 & -1 \\ 3 & -4 & 3 & -1 \\ 0 & 4 & -4 & 1 \\ -1 & -1 & 1 & 0 \end{bmatrix}$

44. Does not exist

45. $(-12, 10, 7, -6)$

46. $(-40, 18, -1, 3)$

47. $(1, 7, -14)$

48. $(0.6, 0.4, -0.2, -0.5)$

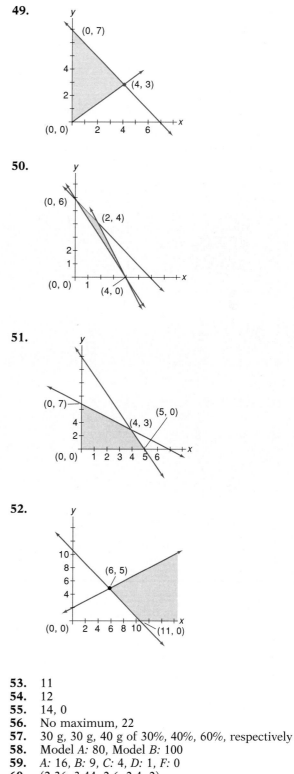

49.

50.

51.

52.

53. 11

54. 12

55. 14, 0

56. No maximum, 22

57. 30 g, 30 g, 40 g of 30%, 40%, 60%, respectively

58. Model *A:* 80, Model *B:* 100

59. *A:* 16, *B:* 9, *C:* 4, *D:* 1, *F:* 0

60. $(2.36, 3.44, 2.6, 2.4, 2)$

PROBLEM SET 7.1 (p. 327)

1. $y^2 = -24x$
3. $x^2 = 20y$
5. $x^2 = -16y$
7. $x^2 = \frac{1}{2}y$
9. $y^2 = \frac{2}{3}x$
11. $\left(0, \frac{1}{2}\right); y = -\frac{1}{2}$
13. $\left(-\frac{1}{8}, 0\right); x = \frac{1}{8}$
15. $\left(0, -\frac{9}{8}\right); y = \frac{9}{8}$
17. $8(x - 4) = (y + 3)^2$
19. $-6\left(y + \frac{1}{2}\right) = (x - 2)^2$
21. $2(y - 1) = (x + 1)^2$
23. $(-2, 3), (-1, 3)$
25. $(-2, 3), \left(-2, \frac{5}{2}\right)$

27. $(-4, 4), (-3, 4)$
29. $(-1, 6), \left(-1, \frac{23}{4}\right)$
31. $\left(-\frac{1}{2}, 1\right), \left(-\frac{1}{2}, \frac{15}{16}\right)$
33. $\left(\frac{1}{2}, 4\right), (1, 4)$
35. $(0.22, 0.44), (2.28, 4.56)$
37. $2\sqrt{2}$
39. $(0, 4)$
41. 28.8 feet
43. 2.5 feet
45. 50 centimeters
47. $\frac{9}{4}$
49. $80\sqrt{5} \approx 178.9$ meters
51. $5\sqrt{10} \approx 15.81$
53. 123.22
55. $8\sqrt{3}\, p$

PROBLEM SET 7.2 (p. 335)

1. $(0, \pm10), (0, \pm8)$
3. $(\pm4, 0), (\pm\sqrt{7}, 0)$
5. $(\pm6, 0), (\pm4, 0)$
7. $(0, \pm4), \left(0, \pm\frac{8}{3}\sqrt{2}\right)$
9. $(0, \pm k), \left(0, \pm\sqrt{3}\frac{k}{2}\right)$
11. $\frac{x^2}{25} + \frac{y^2}{9} = 1$
13. $\frac{x^2}{16} + \frac{y^2}{52} = 1$
15. $\frac{x^2}{49} + \frac{y^2}{4} = 1$
17. $\frac{x^2}{81} + \frac{y^2}{9} = 1$
19. $\frac{x^2}{36} + \frac{y^2}{32} = 1$
21. $\frac{(x - 3.5)^2}{12.25} + \frac{y^2}{4} = 1$

23. $\frac{(x - 2)^2}{31.25} + \frac{(y + 2)^2}{56.25} = 1$
25. $(-3, 4), (-3, 4 \pm 6)$
27. $(-1, 3), (-1, 3 \pm 5\sqrt{2})$
29. $(-5, -1), (-5 \pm 6, -1)$
31. 2.75 feet
33. 155,000,000 kilometers
35. 17.545
37. $(\pm1.653, 2.732)$
39. 20,000 miles, 18,330 miles
41. $\pi\sqrt{77} \approx 27.567$
43. (a) 647.85 square feet, (b) 737.60 square feet
45. $\frac{x^2}{49} + \frac{y^2}{81} = 1$
47. $(-4.60, 5.73), (7.53, -2.36)$
49. $(-5.07, 5.42), (1.30, 6.91)$
51. $\frac{1}{20}$

PROBLEM SET 7.3 (p. 342)

1.

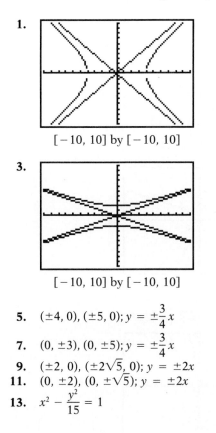

$[-10, 10]$ by $[-10, 10]$

3.

$[-10, 10]$ by $[-10, 10]$

5. $(\pm 4, 0), (\pm 5, 0); y = \pm \dfrac{3}{4}x$

7. $(0, \pm 3), (0, \pm 5); y = \pm \dfrac{3}{4}x$

9. $(\pm 2, 0), (\pm 2\sqrt{5}, 0); y = \pm 2x$

11. $(0, \pm 2), (0, \pm\sqrt{5}); y = \pm 2x$

13. $x^2 - \dfrac{y^2}{15} = 1$

15. $\dfrac{x^2}{25} - \dfrac{y^2}{100} = 1$

17. $\dfrac{y^2}{9} - \dfrac{4x^2}{9} = 1$

19. $\dfrac{x^2}{9} - \dfrac{y^2}{27} = 1$

21. $\dfrac{(y-4)^2}{16} - \dfrac{(x-1)^2}{9} = 1$

23. $(x-6)^2 - \dfrac{(y-4)^2}{8} = 1$

25. $(-2 \pm 5, 3), (-2 \pm 3, 3)$

27. $(-3, \pm 2\sqrt{5}), (-3, \pm 2)$

29. Vertical hyperbola; $(1, -3); 2\sqrt{7}$

31. Vertical ellipse; $(1, 3); 2\sqrt{11}$

33. Horizontal hyperbola; $(-1, 3); 2\sqrt{2}$

35. $\dfrac{32}{3}$

37. $\dfrac{(y-5/2)^2}{4} - \dfrac{4x^2}{9} = 1$

39. $\sqrt{106}$

41. $(2\sqrt{3}, \sqrt{3}), (-2\sqrt{3}, -\sqrt{3})$

43. $\dfrac{x^2}{9} - \dfrac{y^2}{81} = 1$

45. 18,600,000 miles

47. $5\sqrt{3}$

49. $\dfrac{y^2}{(1100)^2} - \dfrac{x^2}{3(1100)^2} = 1$

PROBLEM SET 7.4 (p. 350)

1. Circle
3. The point $(-6, 1)$
5. Horizontal hyperbola
7. Parabola opening down
9. Circle
11. Vertical ellipse
13. The point $(2, 4)$
15. Intersecting lines
17. Parabola turning down
19. $4u^2 + v^2 = 16$
21. $u^2 + 2uv + v^2 - 8u + 8v = 0$
23. $u^2 + 3v^2 = 8$
25. $-2u^2 + 3v^2 = 6$

27. $-u^2 + \dfrac{v^2}{4} = 1$

29. $\dfrac{-u^2}{36} + \dfrac{v^2}{4} = 1$

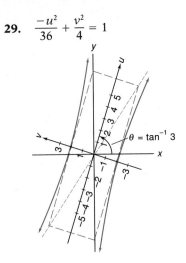

31. $v^2 = 6u$

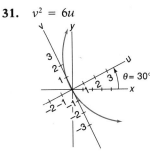

33. $\dfrac{u^2}{2} + \dfrac{v^2}{8} = 1$

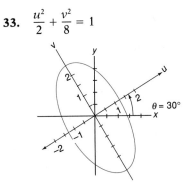

35. $v^2 + 4v + 3 = 0$

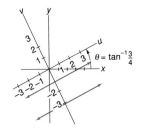

37. $y = -x \pm \sqrt{10 - 2x^2}$; ellipse

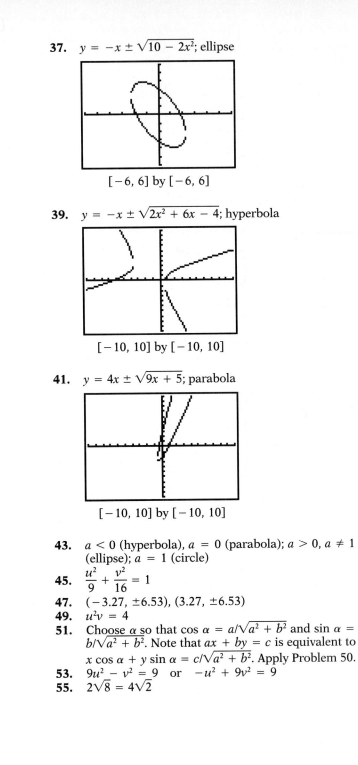

$[-6, 6]$ by $[-6, 6]$

39. $y = -x \pm \sqrt{2x^2 + 6x - 4}$; hyperbola

$[-10, 10]$ by $[-10, 10]$

41. $y = 4x \pm \sqrt{9x + 5}$; parabola

$[-10, 10]$ by $[-10, 10]$

43. $a < 0$ (hyperbola), $a = 0$ (parabola); $a > 0$, $a \neq 1$ (ellipse); $a = 1$ (circle)

45. $\dfrac{u^2}{9} + \dfrac{v^2}{16} = 1$

47. $(-3.27, \pm6.53)$, $(3.27, \pm6.53)$

49. $u^2v = 4$

51. Choose α so that $\cos \alpha = a/\sqrt{a^2 + b^2}$ and $\sin \alpha = b/\sqrt{a^2 + b^2}$. Note that $ax + by = c$ is equivalent to $x \cos \alpha + y \sin \alpha = c/\sqrt{a^2 + b^2}$. Apply Problem 50.

53. $9u^2 - v^2 = 9$ or $-u^2 + 9v^2 = 9$

55. $2\sqrt{8} = 4\sqrt{2}$

PROBLEM SET 7.5 (p. 358)

1. Line segment

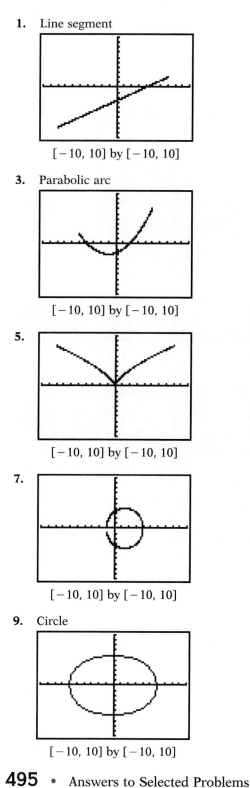

$[-10, 10]$ by $[-10, 10]$

3. Parabolic arc

$[-10, 10]$ by $[-10, 10]$

5.

$[-10, 10]$ by $[-10, 10]$

7.

$[-10, 10]$ by $[-10, 10]$

9. Circle

$[-10, 10]$ by $[-10, 10]$

11. Ellipse

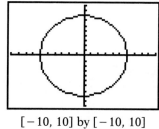

$[-10, 10]$ by $[-10, 10]$

13. Semi-ellipse

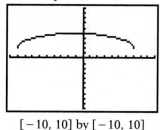

$[-10, 10]$ by $[-10, 10]$

15. Hyperbola

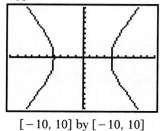

$[-10, 10]$ by $[-10, 10]$

17.

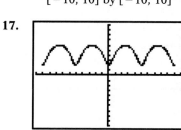

$[-55, 55]$ by $[-10, 10]$

19. $y = \dfrac{2}{3}x + \dfrac{11}{3}$; line

21. $y = \dfrac{1}{4}x^2 - \dfrac{1}{2}x + 2$; parabola

23. $\dfrac{x^2}{4} + \dfrac{y^2}{16} = 1$; ellipse

25. $\dfrac{x^2}{4} - \dfrac{y^2}{9} = 1$; hyperbola

27. $x = -1 + 8t,\ y = 2 + 4t$

29. $x = 1 + t, y = 2 + 3t$

31. $x = t, y = t^2 - 3t + 1$

33. $x = 5 \cos t, y = 3 \sin t$

35. $x = 2 + 4 \cos t, y = 3 + 4 \sin t$

37. $x = \sec t, y = 2 \tan t$

39. $x = 15(t - \sin t), y = 15(1 - \cos t)$

41. $x = \dfrac{5}{2}\sqrt{2} \cos t - 2\sqrt{2} \sin t,$

$y = \dfrac{5}{2}\sqrt{2} \cos t + 2\sqrt{2} \sin t$

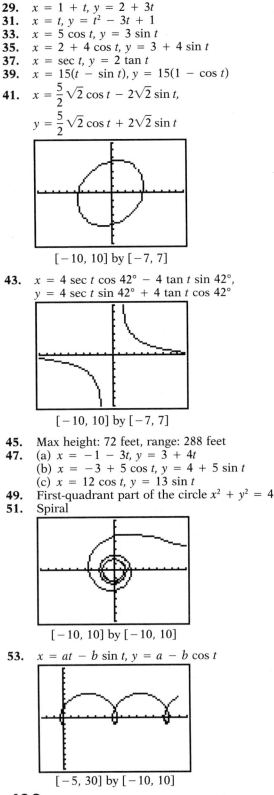

$[-10, 10]$ by $[-7, 7]$

43. $x = 4 \sec t \cos 42° - 4 \tan t \sin 42°,$
$y = 4 \sec t \sin 42° + 4 \tan t \cos 42°$

$[-10, 10]$ by $[-7, 7]$

45. Max height: 72 feet, range: 288 feet

47. (a) $x = -1 - 3t, y = 3 + 4t$
(b) $x = -3 + 5 \cos t, y = 4 + 5 \sin t$
(c) $x = 12 \cos t, y = 13 \sin t$

49. First-quadrant part of the circle $x^2 + y^2 = 4$

51. Spiral

$[-10, 10]$ by $[-10, 10]$

53. $x = at - b \sin t, y = a - b \cos t$

$[-5, 30]$ by $[-10, 10]$

55. (a)

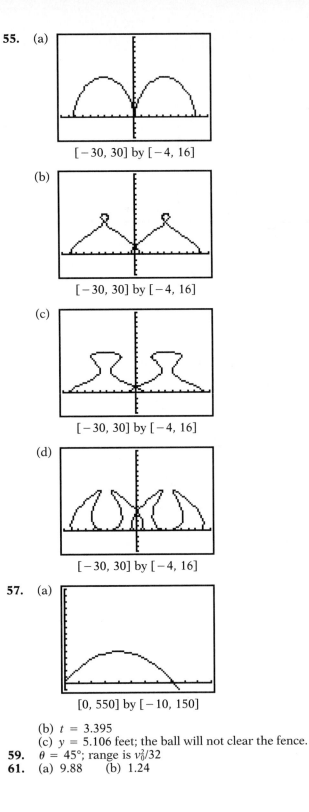

$[-30, 30]$ by $[-4, 16]$

(b)

$[-30, 30]$ by $[-4, 16]$

(c)

$[-30, 30]$ by $[-4, 16]$

(d)

$[-30, 30]$ by $[-4, 16]$

57. (a)

$[0, 550]$ by $[-10, 150]$

(b) $t = 3.395$
(c) $y = 5.106$ feet; the ball will not clear the fence.

59. $\theta = 45°$; range is $v_0^2/32$

61. (a) 9.88   (b) 1.24

PROBLEM SET 7.6 (p. 366)

1, 3, 5, 7.

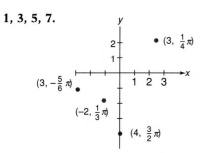

9.

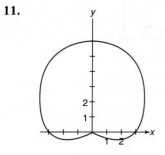

11.

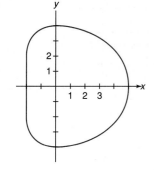

13.

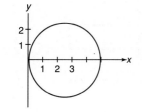

15.

17.

19.

$[-10, 10]$ by $[-10, 10]$

21.

$[-10, 10]$ by $[-7, 7]$

23.

$[-10, 10]$ by $[-7, 7]$

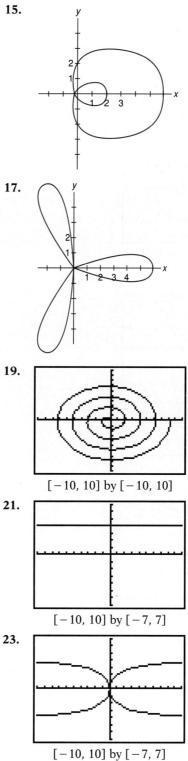

25.

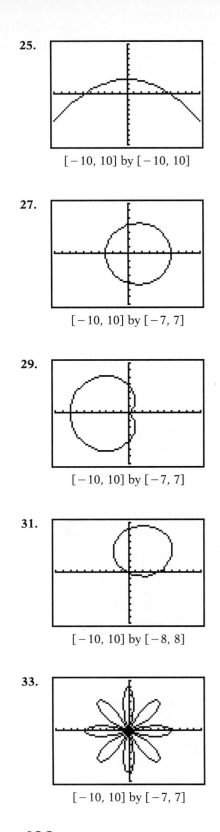

$[-10, 10]$ by $[-10, 10]$

27.

$[-10, 10]$ by $[-7, 7]$

29.

$[-10, 10]$ by $[-7, 7]$

31.

$[-10, 10]$ by $[-8, 8]$

33.

$[-10, 10]$ by $[-7, 7]$

35. (a) $(2\sqrt{2}, 2\sqrt{2})$
(b) $(3\sqrt{3}, 3)$
(c) $(-3, 0)$
(d) $(0, -2)$
(e) $(4\sqrt{3}, -4)$
(f) $(-1, -\sqrt{3})$
(g) $(-1.87, -0.70)$
(h) $(0.85, 3.91)$
(i) $(2.67, 2.98)$

37. $r = 5$

39. $r = 2 \cos \theta$

41. $r = 2 \sin \theta$

43. $x^2 + y^2 = 64$

45. $x^2 + y^2 = 3x$

47. $(x^2 + y^2)^{3/2} = x^2 - y^2$

49. $(x^2 + y^2)^2 = x^2$

51. $2x^2 + y^2 = 3$

53. (a) $\left(2, \dfrac{1}{3}\pi\right), \left(2, \dfrac{5}{3}\pi\right)$

(b) $\left(3, \dfrac{1}{3}\pi\right), (0, \pi)$

57. $A = \dfrac{1}{2}(b^2 - a^2)(\beta - \alpha)$

59.

$[-10, 10]$ by $[-12, 12]$

61. $b = 0.5$, ellipse

$[-12, 12]$ by $[-12, 12]$

$b = 0.8$, ellipse

$[-12, 12]$ by $[-12, 12]$

$b = 1.0$, parabola

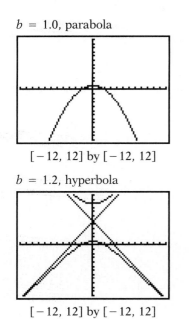

$[-12, 12]$ by $[-12, 12]$

$b = 1.2$, hyperbola

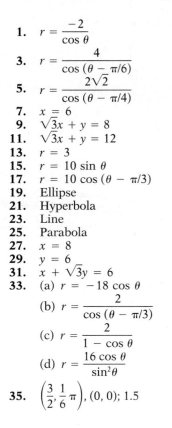

$[-12, 12]$ by $[-12, 12]$

$b = 1.5$, hyperbola

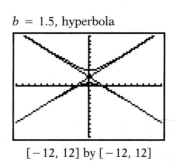

$[-12, 12]$ by $[-12, 12]$

63. The graph has n petals if n is odd, $2n$ petals if n is even.

65. $M = 4\pi$

PROBLEM SET 7.7 (p. 374)

1. $r = \dfrac{-2}{\cos \theta}$

3. $r = \dfrac{4}{\cos (\theta - \pi/6)}$

5. $r = \dfrac{2\sqrt{2}}{\cos (\theta - \pi/4)}$

7. $x = 6$

9. $\sqrt{3}x + y = 8$

11. $\sqrt{3}x + y = 12$

13. $r = 3$

15. $r = 10 \sin \theta$

17. $r = 10 \cos (\theta - \pi/3)$

19. Ellipse

21. Hyperbola

23. Line

25. Parabola

27. $x = 8$

29. $y = 6$

31. $x + \sqrt{3}y = 6$

33. (a) $r = -18 \cos \theta$

    (b) $r = \dfrac{2}{\cos (\theta - \pi/3)}$

    (c) $r = \dfrac{2}{1 - \cos \theta}$

    (d) $r = \dfrac{16 \cos \theta}{\sin^2\theta}$

35. $\left(\dfrac{3}{2}, \dfrac{1}{6}\pi\right)$, $(0, 0)$; 1.5

37. Major diameter: $\dfrac{2ed}{1 - e^2}$

    minor diameter: $\dfrac{2ed\sqrt{1 - e^2}}{1 - e^2}$

39. $e = 0.5$

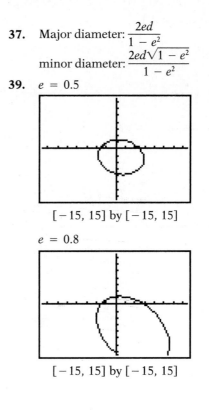

$[-15, 15]$ by $[-15, 15]$

$e = 0.8$

$[-15, 15]$ by $[-15, 15]$

$e = 1.0$

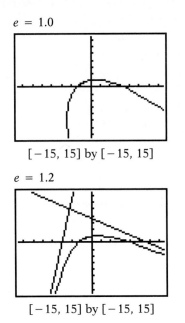

[−15, 15] by [−15, 15]

$e = 1.2$

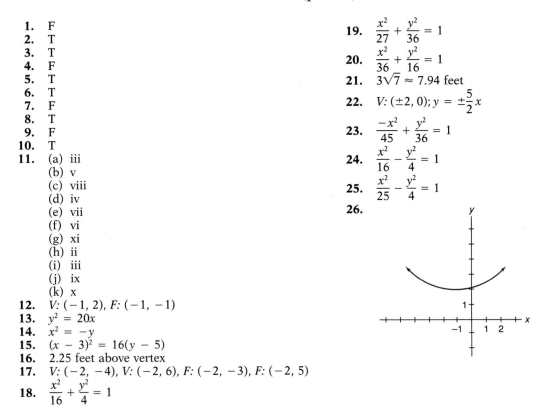

[−15, 15] by [−15, 15]

$e = 1.5$

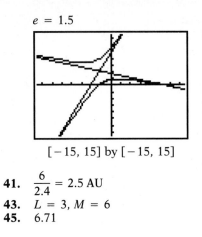

[−15, 15] by [−15, 15]

41. $\dfrac{6}{2.4} = 2.5$ AU

43. $L = 3, M = 6$

45. 6.71

CHAPTER 7 REVIEW PROBLEM SET (p. 376)

1. F
2. T
3. T
4. F
5. T
6. T
7. F
8. T
9. F
10. T
11. (a) iii
   (b) v
   (c) viii
   (d) iv
   (e) vii
   (f) vi
   (g) xi
   (h) ii
   (i) iii
   (j) ix
   (k) x
12. $V: (-1, 2), F: (-1, -1)$
13. $y^2 = 20x$
14. $x^2 = -y$
15. $(x - 3)^2 = 16(y - 5)$
16. 2.25 feet above vertex
17. $V: (-2, -4), V: (-2, 6), F: (-2, -3), F: (-2, 5)$
18. $\dfrac{x^2}{16} + \dfrac{y^2}{4} = 1$

19. $\dfrac{x^2}{27} + \dfrac{y^2}{36} = 1$

20. $\dfrac{x^2}{36} + \dfrac{y^2}{16} = 1$

21. $3\sqrt{7} \approx 7.94$ feet

22. $V: (\pm 2, 0); y = \pm\dfrac{5}{2}x$

23. $\dfrac{-x^2}{45} + \dfrac{y^2}{36} = 1$

24. $\dfrac{x^2}{16} - \dfrac{y^2}{4} = 1$

25. $\dfrac{x^2}{25} - \dfrac{y^2}{4} = 1$

26.

27.

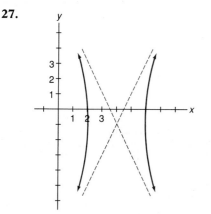

28.

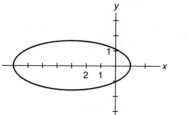

29. $(x - 7)^2 + (y + 1)^2 = 25$; circle

30. $\dfrac{(x - 5)^2}{4} + \dfrac{(y + 2)^2}{9} = 1$; vertical ellipse

31. $9(x - 3)^2 - 4(y - 2)^2 = 0$; intersecting lines

32. $\dfrac{u^2}{36} + \dfrac{v^2}{4} = 1$; ellipse

33. $u^2 - \dfrac{v^2}{12} = 1$; $2\sqrt{13}$

34. $(-0.53, 2.44)$, $(1.13, 7.19)$

35. $(-1.99, 5.94)$, $(3.16, 6.85)$

36. 8.09

37. Line; $3x + 4y + 7 = 0$

38. Ellipse; $\dfrac{x^2}{16} + \dfrac{y^2}{9} = 1$

39. Parabola; $y = 3x^2 + x - 4$

40. Upper semi-ellipse; $\dfrac{x^2}{81} + \dfrac{y^2}{9} = 1$

41. Left branch of a hyperbola; $\dfrac{x^2}{25} - y^2 = 1$

42. Circle; $x^2 + (y - 2)^2 = 4$

43. Circle

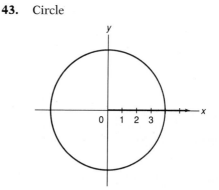

44. Vertical line

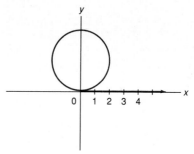

45. Circle

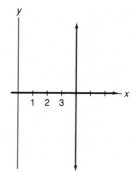

46. Three-leaved rose

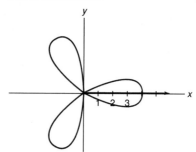

47. Parabola

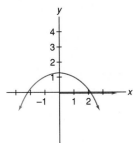

48. Cardioid

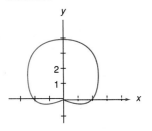

49. $r = 2 \sin \theta - 4 \cos \theta$
50. $x^3 - y^3 = 4xy$
51. (a) vii
    (b) iv
    (c) i
    (d) ii
    (e) vi
    (f) v
    (g) iii
    (h) iv
52. (a) $\dfrac{1}{2}$ (b) 1
53. 9.63
54. 0.94

PROBLEM SET 8.1 (p. 385)

1. 35
3. 121
5. 14
7. -32
9. 81
11. $a_n = n^3, a_{10} = 1000$
13. $a_n = \dfrac{n}{n+1}, a_{10} = \dfrac{10}{11}$
15. $a_n = 1 + (n-1)3 = 3n - 2, a_{10} = 28$
17. $a_n = -5n + 17, a_{10} = -33$
19. $a_n = n^2 - 1, a_{10} = 99$
21. 1920
23. 1365
25. -660
27. 10,100

29. 6633
31. 4095
33. 382.5 centimeters
35. 3.8, 4.6, 5.4, 6.2
37. 11, 2772
39. 4, 2, 5
41. 105
43. 44,850
45. 2440
47. $1950
49. $6930
51. 2475
53. $5400\pi \approx 16,965$ inches
55. $p_n = \dfrac{3}{2}n^2 - \dfrac{1}{2}n$

PROBLEM SET 8.2 (p. 392)

1. 5120
3. $\dfrac{128}{81}$
5. 1.48595

7. 80.97564
9. 3,486,784,400
11. 767,520,036.5
13. 4394.70705

15. 9
17. 500
19. 20,200
21. $\dfrac{23}{99}$
23. $\dfrac{32}{99}$
25. $\dfrac{41}{333}$
27. $\dfrac{173}{55}$
29. 102,400
31. \$2,147,483,647
33. \$215.89; 11.91 years
35. \$1564.55
37. 30 feet

39. (a) Geometric
(b) Neither
(c) Arithmetic
(d) Geometric
(e) Neither
(f) Geometric
41. \$4 billion including the original \$1 billion
43. $\dfrac{3\sqrt{3}}{14} \approx 0.37115$
45. $\dfrac{4}{5}$; no
47. (a) $S = \dfrac{r}{(1-r)^2}$ (b) $S = \dfrac{3}{4}$
49. $\dfrac{1000}{9} \approx 111.11$ yards
51. 125 miles

PROBLEM SET 8.3 (p. 400)

1. $a_n = a_{n-1} + a_{n-2}$ with $a_1 = 0$ and $a_2 = 2$
3. $a_n = a_{n-1} + n - 1$ with $a_1 = 0$
5. $a_n = n^3 - 1$
7. $a_n = a_{n-1} + a_{n-2} + a_{n-3}$ with $a_1 = a_2 = a_3 = 1$
9. $\dfrac{5}{7}$
11. 240
13. 1
15. 32,768
17. 62
19. $-25,050$
21. 1
23. -1
25. 393,750; 950,000
27. 7; 84.875
29. 0.99990001; 0.999999

31. 524,288; 6.33825×10^{29}
33. 2.71828; 2.71828
35. 524,289; 6.33825×10^{29}
37. 3.666671753; 3.666666667
39. 110; 85,850
41. 94.15146; 644.72862
43. 89.36892; 694.47720
45. 63.34496; 63.44900
47. 15046.53681
49. 30225.9919
51. 920
53. 570
55. $M = 6$
57. 3819.06 feet; N60.62°E
59. (a) $\displaystyle\sum_{k=1}^{n} k(k+9)$ (b) 13,640

PROBLEM SET 8.4 (p. 408)

Note: In the text, several proofs by mathematical induction are given in complete detail. To save space, we show only the key step here, namely, that P_{k+1} is true if P_k is true.

1. $(1 + 2 + \cdots + k) + (k + 1) = k(k + 1)/2 + k + 1 = (k(k + 1) + 2(k + 1))/2 = (k + 1)(k + 2)/2$
3. $(3 + 7 + \cdots + (4k - 1)) + (4k + 3) = k(2k + 1) + (4k + 3) = 2k^2 + 5k + 3 = (k + 1)(2k + 3)$
5. $(1 \cdot 2 + 2 \cdot 3 + \cdots + k(k + 1)) + (k + 1)(k + 2)$
$= \dfrac{1}{3} k(k + 1)(k + 2) + (k + 1)(k + 2)$
$= \dfrac{1}{3}(k + 1)(k + 2)(k + 3)$

7. $(2 + 2^2 + \cdots + 2^k) + 2^{k+1} = 2(2^k - 1) + 2^{k+1}$
$= 2^{k+1} - 2 + 2^{k+1} = 2(2^{k+1} - 1)$
9. P_n is true for $n \geq 8$.
11. P_1 is true.
13. P_n is true for odd positive n.
15. P_n is true for every positive integer n.
17. P_n is true when n is a nonnegative integral power of 4.

19. $x^{2k+2} - y^{2k+2} = x^{2k}(x^2 - y^2) + (x^{2k} - y^{2k})y^2$. Now $(x + y)$ is a factor of both $x^2 - y^2$ and $x^{2k} - y^{2k}$, the latter by assumption.

21. $(k + 1)^2 - (k + 1) = k^2 + 2k + 1 - k - 1 = (k^2 - k) + 2k$. Now 2 divides $k^2 - k$ by assumption and clearly divides $2k$.

23. $N = 4$. If $k + 5 < 2^k$, then
$$k + 6 < 2^k + 1 < 2^k + 2^k = 2^{k+1}.$$

25. $N = 1$. Since $k + 1 < 10k$,
$$\log (k + 1) < 1 + \log k < 1 + k.$$

27. $N = 1$. Multiply both sides of $(1 + x)^k \geq 1 + kx$ by $(1 + x)$: $(1 + x)^{k+1} \geq (1 + x)(1 + kx) = 1 + (k + 1)x + kx^2 > 1 + (k + 1)x$.

29. (a) $(1 + 2 + 3 + \cdots + k) + (k + 1) = \frac{1}{2}k(k + 1) +$

$k + 1 = (k + 1)\left(\frac{1}{2}k + 1\right) = \frac{1}{2}(k + 1)(k + 2)$

(c) $(1^3 + 2^3 + \cdots + k^3) + (k + 1)^3 = \frac{1}{4}k^2(k + 1)^2 +$

$(k + 1)^3 = ((k + 1)^2/4)(k^2 + 4(k + 1))$

$= \frac{1}{4}(k + 1)^2(k + 2)^2$

(d) $(1^4 + 2^4 + \cdots + k^4) + (k + 1)^4 = \frac{1}{30}k(k +$

$1)(6k^3 + 9k^2 + k - 1) + (k + 1)^4 = ((k + 1)/30)(6k^4 + 9k^3 + k^2 - k + 30k^3 + 90k^2 + 90k + 30) = ((k + 1)/30)(6k^4 + 39k^3 + 91k^2 +$

$89k + 30) = \frac{1}{30}(k + 1)(k + 2)(6k^3 + 27k^2 +$

$37k + 15) = \frac{1}{30}(k + 1)(k + 2)(6(k + 1)^3 +$

$9(k + 1)^2 + (k + 1) - 1)$

31. (a) 15,250 (b) 220 (c) 4355 (d) $2n(n + 1)^2$

33.
$$\left(1 - \frac{1}{4}\right)\left(1 - \frac{1}{9}\right) \cdots \left(1 - \frac{1}{k^2}\right)\left(1 - \frac{1}{(k + 1)^2}\right)$$
$$= ((k + 1)/2k)(1 - 1/(k + 1)^2)$$
$$= ((k + 1)/2k)((k^2 + 2k)/(k + 1)^2)$$
$$= (k + 2)/(2(k + 1))$$

35. Let $S_k = 1/(k + 1) + 1/(k + 2) + \cdots + 1/(2k)$ and assume that $S_k > 3/5$. Then $S_{k+1} = 1/(k + 2) + 1/(k + 3) + \cdots + 1/(2k) + 1/(2k + 1) + 1/(2k + 2) = S_k + 1/(2k + 1) + 1/(2k + 2) - 1/(k + 1) > S_k + 2/(2k + 2) - 1/(k + 1) = S_k > 3/5$.

37. The statement is true when $n = 3$ since it asserts that the angles of a triangle have a sum of $180°$. Now any $(k + 1)$-sided convex polygon can be dissected into a k-sided polygon and a triangle. Its angles add up to $(k - 2)180° + 180° = (k - 1)180°$.

39. (i) $R_1 = 2 = (1^2 + 1 + 2)/2$.
(ii) Assume $R_k = (k^2 + k + 2)/2$. Then $R_{k+1} =$
$$R_k + k + 1 = \frac{1}{2}(k^2 + k + 2) + k + 1 = \frac{1}{2}(k^2 +$$
$$3k + 4) = \frac{1}{2}((k + 1)^2 + (k + 1) + 2).$$

41. $f_1^2 + f_2^2 + \cdots + f_n^2 + f_{n+1}^2 = f_n f_{n+1} + f_{n+1}^2 = f_{n+1}(f_n + f_{n+1}) = f_{n+1}f_{n+2}$.

43. Assume the equality holds for a_k and a_{k+1}. Then $a_{k+2} = (a_k + a_{k+1})/2$
$$= \frac{2}{3}\left[\left(1 - \left(-\frac{1}{2}\right)^k\right) + 1 - \left(-\frac{1}{2}\right)^{k+1}\right)\frac{1}{2}\right]$$
$$= \frac{2}{3}\left[1 - \frac{1}{2}\left(-\frac{1}{2}\right)^k - \frac{1}{2}\left(-\frac{1}{2}\right)^{k+1}\right]$$
$$= \frac{2}{3}\left[1 - \left(-\frac{1}{2}\right)^{k+2}\right].$$
Check also that the result is true for $n = 0$ and $n = 1$.

PROBLEM SET 8.5 (p. 414)

1. 55
3. 55
5. 1225
7. 161,700
9. 0
11. 1296
13. $x^3 + 3x^2y + 3xy^2 + y^3$
15. $x^3 - 6x^2y + 12xy^2 - 8y^3$
17. $c^8 - 12c^6d^3 + 54c^4d^6 - 108c^2d^9 + 81d^{12}$
19. $a^5b^{10} - 5a^4b^9c + 10a^3b^8c^2 - 10a^2b^7c^3 + 5ab^6c^4 - b^5c^5$
21. $x^{20} + 20x^{19}y + 190x^{18}y^2$
23. $x^{20} + 20x^{14} + 190x^8$
25. $-120y^{14}z^9$
27. $-1760a^3b^9$
29. 23.433
31. 552.446

33. 149
35. $\dfrac{55}{9}$
37. $1.01^{50} > 1 + 50(0.01) = 1.5$
39. (a) $3x^2 + 3xh + h^2$
    (b) $4x^3 + 6x^2h + 4xh^2 + h^3$
41. (a) $x^3 + y^3 + z^3 + 3x^2y + 3x^2z + 3y^2z + 3xy^2 + 3xz^2 + 3yz^2 + 6xyz$
    (b) 420
43. 3^n
45. 2^{29}
47. (a) 1.718281828; 1.718281828
    (b) 53.59814993; 53.59815003
    (c) 10,485,760; $1.610612736 \times 10^{10}$
    (d) 1.112933094; 1.071848988

PROBLEM SET 8.6 (p. 421)

1. (a) 720   (b) 360
3. 126
5. 30
7. (a) 6   (b) 5
9. (a) 36   (b) 72   (c) 6
11. 1680
13. 25; 20
15. 240
17. 300
19. 1110
21. (a) 30 (b) 180 (c) 120 (d) 120 (e) 450
23. (a) $12 \cdot 11 \cdot 10 \cdot 9 \cdot 8 \cdot 7 \cdot 6 \cdot 5 \cdot 4$
    (b) $2 \cdot 10 \cdot 9 \cdot 8 \cdot 7 \cdot 6 \cdot 5 \cdot 4 \cdot 3$
    (c) $2 \cdot 11 \cdot 10 \cdot 9 \cdot 8 \cdot 7 \cdot 6 \cdot 5 \cdot 4$

25. 210
27. 83,160
29. 840
31. 126
33. $1 - \dfrac{1}{(n + 1)!}$
35. $8 \cdot 2 \cdot 10 \cdot 8 \cdot 8 \cdot 10(10^4 - 1)$
37. 3905
39. $5! = 120$
41. The number of games equals the number of losers, which is $n - 1$.

PROBLEM SET 8.7 (p. 428)

1. 1140
3. 70
5. 56
7. 495
9. 720
11. (a) $_{30}C_4 = 27,405$ (b) $_{25}C_3 = 2300$
13. $2 \cdot _{10}C_2 = 90$
15. $_{12}C_{10} = _{12}C_2 = 66$
17. 63
19. (a) 36   (b) 100   (c) 24
21. (a) 84   (b) 39   (c) 130
23. $_{26}C_{13} = 10,400,600$
25. $4^{13} = 67,108,864$
27. $_4C_2 \cdot _{48}C_{11} + _4C_3 \cdot _{48}C_{10} + _4C_4 \cdot _{48}C_9 \approx$
                        $1.634111724 \times 10^{11}$

29. $_{52}C_5 = 2,598,960$
31. $_{13}C_2 \cdot _4C_2 \cdot _4C_2 \cdot 44 = 123,552$
33. $_5C_2 \cdot _4C_2 \cdot _3C_2 = 180$
35. $12 \cdot 11 \cdot 10 \cdot _9C_3 = 110,880$
37. (a) 1024   (b) 210
39. 63
41. $_{10}C_3 + _{10}C_4 + \cdots + _{10}C_{10} =$
              $2^{10} - _{10}C_2 - _{10}C_1 - _{10}C_0 = 968$
43. $2^{15} - _{15}C_2 - _{15}C_1 - _{15}C_0 = 32,647$
45. $_{2n}C_n$
47. (a) $\dfrac{n(n + 1)}{2}$ (b) $\left(\dfrac{n(n + 1)}{2}\right)^2$

PROBLEM SET 8.8 (p. 436)

1. (a) $\dfrac{1}{6}$ (b) $\dfrac{1}{2}$ (c) $\dfrac{1}{3}$ (d) $\dfrac{1}{2}$ (e) $\dfrac{1}{2}$
3. (a) $\dfrac{1}{8}$ (b) $\dfrac{3}{8}$ (c) $\dfrac{1}{2}$
5. (a) $\dfrac{1}{6}$ (b) $\dfrac{1}{6}$ (c) $\dfrac{5}{9}$
7. (a) States have different size populations.
    (b) The two events are not disjoint.
    (c) The two events are complementary; their probabilities should sum to 1.
    (d) Ties are possible in football.

9. (a) $\dfrac{8}{25}$ (b) $\dfrac{5}{12}$ (c) $\dfrac{2}{5}$ (d) $\dfrac{8}{75}$
11. $\dfrac{1}{120}$
13. $\dfrac{9}{11}$
15. 1 to 3
17. (a) $\dfrac{1}{2}$ (b) $\dfrac{1}{4}$ (c) $\dfrac{1}{13}$

19. (a) $_{26}C_3/_{52}C_3 \approx 0.118$
(b) $_{13}C_3/_{52}C_3 \approx 0.013$
(c) $_4C_1 \cdot _{48}C_2/_{52}C_3 \approx 0.204$
(d) $_4C_3/_{52}C_3 \approx 0.00002$

21. $_8C_5/_{52}C_5 \approx 0.0002$

23. (a) $\dfrac{1}{216}$ (b) $\dfrac{1}{36}$ (c) $\dfrac{53}{54}$

25. (a) $\dfrac{4}{13}$ (b) $\dfrac{27}{52}$ (c) $\dfrac{1}{2}$

27. (a) $\dfrac{3}{7}$ (b) $\dfrac{3}{7}$ (c) $\dfrac{13}{14}$ (d) 1

29. (a) $\dfrac{1}{13}$ (b) $\dfrac{12}{13}$ (c) $\dfrac{57}{65}$

31. (a) $\dfrac{1}{35}$ (b) $\dfrac{4}{35}$ (c) $\dfrac{1}{7}$

33. (a) $\dfrac{1}{1296}$ (b) $\dfrac{1}{54}$ (c) $\dfrac{671}{1296}$ (d) $\dfrac{1}{216}$

35. $\dfrac{8}{11!} \approx 0.0000002$

37. (a) $\dfrac{4}{9}$ (b) $\dfrac{1}{3}$

39. (a) 0
(b) $\dfrac{1}{2!} - \dfrac{1}{3!} + \dfrac{1}{4!} + \cdots + (-1)^n \dfrac{1}{n!}$
(c) $\dfrac{1}{2!} - \dfrac{1}{3!} + \dfrac{1}{4!} + \cdots + (-1)^{n-1} \dfrac{1}{(n-1)!}$

PROBLEM SET 8.9 (p. 444)

1. $\dfrac{1}{216}$

3. (a) $\dfrac{1}{8}$ (b) $\dfrac{3}{8}$ (c) $\dfrac{1}{8}$ (d) $\dfrac{5}{24}$ (e) $\dfrac{1}{2}$

5. (a) No. Most would agree that doing well in physics is heavily dependent on skill in mathematics.
(b) Yes. Some might disagree, but we think these two events are quite unrelated.
(c) Yes. New shirts and stubbed toes have nothing to do with each other.
(d) No. Check the mathematical condition for independence.
(e) No. At the present time, it is more likely that a doctor is a man.
(f) Yes. Let old superstitions die.

7. $\left(\dfrac{9}{10}\right)^3 = 0.729$

9. (a) 0.9 (b) 0.5 (c) 0.8

11. (a) 0.27 (b) 0.58 (c) $\dfrac{0.27}{0.55} \approx 0.49$

13. $\dfrac{2}{5}$

15. $\dfrac{8}{25}$

17. (a) $\dfrac{1}{4}$ (b) $\dfrac{2}{9}$

19. (a) $\left(\dfrac{3}{10}\right)^3 = 0.027$
(b) $\left(\dfrac{2}{10}\right)^3 = 0.008$
(c) $0.027 + 0.008 + 0.125 = 0.16$
(d) $6\left(\dfrac{2}{10}\right)\left(\dfrac{3}{10}\right)\left(\dfrac{5}{10}\right) = 0.18$

21. (a) $\dfrac{1}{5}$ (b) $\dfrac{1}{5}$

23. (a) $\dfrac{1}{64}$ (b) $\dfrac{6}{64}$ (c) $\dfrac{15}{64}$ (d) $\dfrac{20}{64}$ (e) $\dfrac{22}{64}$

25. (a) $_{12}C_4 \left(\dfrac{2}{3}\right)^4 \left(\dfrac{1}{3}\right)^8 \approx 0.0149$
(b) $_{12}C_6 \left(\dfrac{2}{3}\right)^6 \left(\dfrac{1}{3}\right)^6 \approx 0.1113$

27. 0.590, 0.919

29. (a) $\dfrac{1}{4}$ (b) $\dfrac{5}{12}$

31. (a) $\dfrac{25}{256}$ (b) $\dfrac{1}{12}$

33. (a) $\dfrac{9}{100}$ (b) $\dfrac{1}{5}$ (c) $\dfrac{19}{50}$ (d) $\dfrac{31}{50}$

35. $\dfrac{3}{4}$

37. (a) $1 - (0.4)^8 \approx 0.9993$
(b) $8(0.6)(0.4)^7 + (0.4)^8 \approx 0.0085$

39. Arlene: $\dfrac{4}{7}$, Betty: $\dfrac{2}{7}$, Candy: $\dfrac{1}{7}$

41. (a) 0.5 (b) 1 (c) 0.479 (d) 0.000037

43. (a) $f(p) = p^4 + \binom{4}{1} p^3(1-p)p$
$+ \binom{5}{2} p^3(1-p)^2 p + \binom{6}{3} p^3(1-p)^3 p$
Now expand and simplify.
(b) 0.5437 (c) $\dfrac{5}{16} = 0.3125$

45. (a) $(1-p)^{20} + 20p(1-p)^{19}$
(b) 0.018 (c) 0.0076

506 • Answers to Selected Problems

1. F
2. F
3. T
4. T
5. T
6. T
7. F
8. T
9. T
10. F
11. (b), (c)
12. (a), (d)
13. $d_n = d_{n-1} + 2\pi$; $e_n = e_{n-1} + e_{n-2} + e_{n-3}$
14. 101π
15. 15,150
16. $31(\sqrt{2} + 1)$
17. $-\dfrac{2}{3}$
18. $(2^{30} - 1)\cent = \$10,737,418.23$
19. $\dfrac{20}{9}$
20. (a) 3725　(b) 2
21. 43
22. The 5% yearly raise gives her \$339.85 more in the 10th year.
23. The \$1800 yearly raise gives her \$3663.22 more over 10 years.
24. 1100, 29,750, 168,950
25. 11.3199, 52.0958, 102.4387, 503.2404
26. 0.8957, 1.0388, 1.1798, 1.3196
27. $b_5 = 16$, $b_{10} = -512$, $b_{15} = 16,384$
    $S_5 = 11$, $S_{10} = -341$, $S_{15} = 10,923$
28. $a_{15} = 1.587356281$, $a_{25} = 1.587401008$,
    $a_{50} = 1.587401052$
    $S_{15} = 23.5277$, $S_{25} = 39.4017$, $S_{50} = 79.0868$
29. (a) 0.1411, -0.001272, 0.0008140
    (b) 0.5001, 0.5027, 0.5002
    (c) $\to 0$, $\to 0.5$
30. \$32,488; \$227,129.60
31. (a) $\dfrac{13!}{3!}$　(b) $\dfrac{13!}{10!\,3!}$
32. 720
33. (a) 190　(b) 210

34. (a) 2520　(b) 420
35. 120
36. (a) $11! = 39,916,800$
    (b) $1 \cdot 2 \cdot 10! = 7,257,600$
37. $_6C_2 \cdot 4 \cdot {_5}C_2 = 600$
38. $_{12}P_9 \cdot 9! \approx 2.897 \times 10^{13}$
39. $26^2 10^5 + 26^3 10^4 + 26^4 10^3 = 700,336,000$
40. 800
41. The statement is true for a set with 1 element. Assuming that a set with k elements has 2^k subsets, we consider a set A with $k + 1$ elements. Think of A as $\{x\} \cup B$, where B is a set with k elements. Subsets of A are either subsets of B (of which there are 2^k) or subsets containing x together with a subset of B (of which there are 2^k). Thus, A has $2^k + 2^k = 2^{k+1}$ subsets.
42. $4! > 2^4$. Assume $k! > 2^k$ for some $k \geq 4$. Then $(k + 1)! = (k + 1)k! > (k + 1)2^k > 2 \cdot 2^k = 2^{k+1}$
43. False; $11^2 - 11 + 11$ is not prime.
44. P_n is true for even positive n.
45. $x^6 + 12x^5y + 60x^4y^2 + 160x^3y^3 + 240x^2y^4 + 192xy^5 + 64y^6$
46. $3^6 = 729$
47. $-48,384x^5y^6$
48. 56
49. 1.000080002800056
50. $2^{25} = 33,554,432$
51. (a) $\dfrac{1}{2}$　(b) $\dfrac{1}{5}$　(c) $\dfrac{3}{100}$　(d) $\dfrac{4}{5}$
52. $\dfrac{4}{_{52}C_5}$
53. (a) $\dfrac{5}{18} \cdot \dfrac{4}{17} \approx 0.06536$　(b) $\dfrac{92}{306} \approx 0.3007$
54. (a) $\left(\dfrac{5}{18}\right)^2 \approx 0.07716$　(b) $\dfrac{110}{324} \approx 0.3395$
55. 0.9
56. (a) 0.42　(b) 0.4
57. $\dfrac{3}{8}$
58. (a) $\displaystyle\sum_{k=0}^{30} \binom{100}{k}\left(\dfrac{1}{3}\right)^k\left(\dfrac{2}{3}\right)^{100-k}$　(b) 0.2766

INDEX